Agricultural Technology: Emerging Trends

Agricultural Technology: Emerging Trends

Editor: Caroline Walters

STATES
ACADEMIC PRESS
www.statesacademicpress.com

States Academic Press,
109 South 5th Street,
Brooklyn, NY 11249, USA

Visit us on the World Wide Web at:
www.statesacademicpress.com

ISBN: 978-1-63989-034-7 (Hardback)

Cataloging-in-Publication Data

Agricultural technology : emerging trends / edited by Caroline Walters.
p. cm.
Includes bibliographical references and index.
ISBN 978-1-63989-034-7
1. Agricultural innovations. 2. Agricultural industries--Appropriate technology.
3. Agriculture--Technology transfer. 4. Agriculture. I. Walters, Caroline.
S494.5.I5 A37 2022
338.1--dc23

Table of Contents

Preface

The practice and science of cultivating plants along with livestock is known as agriculture. Innovative practices and technological developments in the field of agriculture have resulted in increased crop yields and a better quality of crop. Agricultural technology refers to the technology and techniques which are used to control the growth and harvesting of agricultural products. Some of these techniques are tilling, soil fertilization, green manuring, composting, liming, etc. The machinery used in agriculture also falls in this category. It includes equipment such as bale wrapper, chaff cutter, manure spreader, drum chopper, mower, swayer, etc. Agricultural technology aims to design advanced agricultural machinery and structures to improve the sustainability and efficacy of agricultural practices. This book strives to provide a fair idea about this discipline and to help develop a better understanding of the latest advances within this field. The various advancements in agricultural technology are glanced at and their applications as well as ramifications are looked at in detail. This book is a vital tool for all researching and studying agricultural science.

Various studies have approached the subject by analyzing it with a single perspective, but the present book provides diverse methodologies and techniques to address this field. This book contains theories and applications needed for understanding the subject from different perspectives. The aim is to keep the readers informed about the progresses in the field; therefore, the contributions were carefully examined to compile novel researches by specialists from across the globe.

Indeed, the job of the editor is the most crucial and challenging in compiling all chapters into a single book. In the end, I would extend my sincere thanks to the chapter authors for their profound work. I am also thankful for the support provided by my family and colleagues during the compilation of this book.

Editor

1

Plant growth promoting bacteria and humic substances: crop promotion and mechanisms of action

Fábio Lopes Olivares[1], Jader Galba Busato[2], Alessandra M. de Paula[2], Lívia da Silva Lima[1], Natália Oliveira Aguiar[1] and Luciano Pasqualoto Canellas[1*]

Abstract

High-external input agriculture is one of the most disruptive human activities, which have been justified by the current economic paradigm due to high productivity and the need to feed a growing population. However, we are dangerously close to the edge of the planet resources and both hunger and food insecurity has increased. Limiting the use of non-renewable chemical fertilizers and pesticides, changing water management, enhancing diversity and considering the often-neglected social dimension of agriculture are the bases to other chemical and biological technologies to agriculture. Biological inputs can stimulate the substitution of chemical inputs without questioning the current fundaments or can be adopted as a turning point to intensify the harsh processes of transition to more environmental friendly agriculture. The debate is open and our contribution is to develop the scientific basis for biological inputs that, unlike soluble fertilizers and pesticides, depend on a number of factors for its success in promoting crop yield. In this review, we showed the results obtained with the combined use of diazotrophic endophytic bacteria and humic substances in diverse crops (sugarcane, maize, tomato, common beans and pineapple), presenting the main morphological and physiological changes induced by biological technology. A snapshot of the state of the art of the use of plant growth promoting bacteria together with humic substances was provided, showing their potential especially when plants are subjected to moderate to severe abiotic stress. The number of studies reporting the combined use of plant growth promoting bacteria and humic substances is surprisingly low. There is an open avenue for research and encouraging debate is the goal. To overcome the conventional agriculture, maintaining productivity levels is more than scientific challenge, is a humanitarian duty. The biological inputs can help in this purpose.

Keywords: Bacterial inoculants, Humic acids, Abiotic stress, Biostimulant, Sustainable agriculture

Introduction

Worldwide pesticide pollution, pest adaptation and resistance, loss of soil fertility and organic carbon, soil erosion, decreasing biodiversity, desertification, starving people in poor nations, increasing food prices [1] make conventional agriculture one of most disruptive human activities. But current economic paradigm kicks the can down the road. The debate about how to overcome the negative consequences of this model of agriculture is open, but even economic arguments can be questioned by anyone. The trade balance of the peripheral economies is strongly dependent of the agricultural commodities and the production costs have an important role. Most part of the fertilizers and agrochemicals used in Brazil and others development countries are imported and quoted in dollars, reaching to the farmers with a high price. For example, urea has been sold at a price higher than US$ 400 per ton, which is very expensive for the smallholder farmers. Simple like this: sovereignty and food security are dependent of the industrial fertilizers trade. Furthermore, the influence of fertilizers in the

*Correspondence: lucianocanellas@gmail.com
[1] Núcleo de Desenvolvimento de Insumos Biológicos para a Agricultura (NUDIBA), Universidade Estadual do Norte Fluminense Darcy Ribeiro (UENF), Av. Alberto Lamego, 2000, Campos dos Goytacazes, Rio de Janeiro 28013-602, Brazil
Full list of author information is available at the end of the article

production costs is not small and reaches about 30% [2]. In Brazil, the bigger agricultural country, the price of fertilizers increased by 37% in the 2015–2016 seasons growing compared to the previous agricultural year. Look this exemplary case considering the maize: in the last 5 years, the fertilizers increased near threefold while the price paid to the producer varied only 66% according to the Brazilian official harvest survey [2]. For us, in the tropical zone, it sounds very strange to hear that the application of chemical fertilizers is an inexpensive practice [3]. In addition, the main part of Brazil (and tropical countries) is covered by high weathered soils with low cation exchange capacity, cations nutrients removal by leaching process and consequently high acidity and aluminum toxicity [4]. In these conditions, you need a large amount of correctives of acidity and fertilizers which, in turn, has limited efficiency in these soil conditions. Thereby, economic reasons are moving sectors of traditional agriculture closer to those who do agriculture with less negative impact on the environment for socio-environmental reasons. Thereafter, increasing attention is being paid to the downside of high-input agricultural systems and much research is aimed at developing alternative ways to produce food, wood and plant for energy production in a sustainable and environmentally sound way [5].

Biological techniques are promising approach to enhance the efficiency of nutrient use, reducing the environmental and financial costs of agricultural production. The case of rhizobia/soybean in Brazil is an emblematic example of the biological inputs impact on economy and environment services. Over the years, the development of this technology allowed to abolish almost completely the need for nitrogen fertilizers application in the soybean crop. However, the main crops used to human food and bioenergy production are non-leguminous plants (rice, wheat, maize, sugarcane), i.e., non-nodulating species of plant growth promoting bacteria (PGPB) that colonize surface and interior of the plant tissues. In this case, epiphytic and endophytic PGPB can play a relevant role in the nutrient efficiency use through biofertilization and biostimulation mechanisms [6]. Plant growth-promoting bacteria (PGPB) are a diverse group of bacteria capable of promoting growth and yield of many crops as a result of several effects on host including wide variety of mechanisms such biological nitrogen fixation, phosphate solubilization, alleviation of abiotic stress, siderophore production, rhizosphere engineering, production of 1-aminocyclopropane-1-carboxylate deaminase (ACC), quorum sensing (QS) signal interference and inhibition of biofilm formation, phytohormone production, exhibiting antifungal activity, induction of systemic resistance, promoting beneficial plant–microbe symbioses and interference with pathogen toxin production [7]. The potential benefits of PGPB to promote plant growth is enormous and the global market for bio-inoculants is growing at an estimated rate of ~ 10% per annum [8].

The number of PGPB used to bioinoculant formulation is relatively low considering the natural biodiversity. In a critical review about formulations used for producing inoculants for plants from 1998 to 2014, it was found [9] that around of 30 genera and 60 species already have been used including: *Azospirillum* genus (*A. brasilense, A. lipoferum, A. amazonense, A. rugosum* and *A. zeae*); *Azotobacter* genus (*A. chroococcum, A. vinelandii* and *Azotobacter* sp.); *Bacillus* genus (*B. licheniformis, B. megaterium, B. mycoides, B. pumilus, B. polymyxa, B. subtilis* and *Bacillus* sp.); *Burkholderia* genus (*B. cepacia, B. pyrrocinia, B. phytofirmans, B. tropica* and *B. silvatlantica); Chlorella* genus (*C. vulgaris* and *C. sorokiniana*); *Herbaspirillum* genus (*H. seropedicae; H. rubrisubalbicans* and *H. frisingense*); *Pseudomonas* genus (*P. aeruginosa, P. corrugate, P. fluorescens, P. jessenii, P. putida, P. striata, P. alcaligenes, P. aureofaciens, P. chlororaphis, P. synxantha* and *Pseudomonas* spp.); *Pantoea* genus (*P. agglomerans, P. dispersa* and *Pantoea* sp.); *Streptomycetes* genus (*S. griseoluteus, S. griseoviridis* and *Streptomycetes* sp.); *Coniochaeta ligniaria; Beijerinckia* sp.; *Brevibacillus brevis; Enterobacter cloacae; Gluconacetobacter diazotrophicus, Klebsiella pneumoniae, Lysobacter gummosus, Methylobacterium oryzae, Micrococcus* sp., *Mycobacterium phlei, Paenibacillus polymyxa, Pichia guilermondii, Raoultella terrigena, Rhodobacter capsulatus, Rhodococcus rhodochrous, Serratia plymuthica, Sinorhizobium meliloti, Sphingobacterium canadense, Stenotrophomonas maltophilia, Thiobacillus* sp., and *Acinetobacter calcoaceticus.*

Nitrogen-fixing bacteria of the species *Gluconacetobacter diazotrophicus, Herbaspirillum seropedicae, Herbaspirillum rubrisubalbicans and Burkholderia tropica* have been isolated from genotypes of grass species in high numbers [10] and have been as a multispecies inoculant recommended for sugarcane [11]. All of these bacteria live inside of plant host, i.e., are diazotrophic endophytic bacteria (DEB). The feasibility of the inoculation using DEB in sugarcane grown in soils with low to medium levels of fertility was previously demonstrated [12]. It is important to emphasize here that in soil with high natural fertility or high rates application of nitrogen fertilizers, the results related to the increase of crop production may be levelled out or minimized by the large nutrient availability.

The adoption of this technology is far below the potential shown by scientific research. The scientific efforts are mainly concentrated on isolation and selection of plant-beneficial microorganisms and their further application in soil–plant system in controlled conditions [13] and little attention are dedicated to formulation procedures, delivering application and effectiveness under field conditions. Several vehicles for PGPB delivery to crops are used including ethylcellulose, modified starch, carboxymethylcellulose, corn starch, chitosan, alginate, polymeric inoculants, EB™ (mainly clay and wood particles), vermiculite, perlite, clay pellets, talc, activated carbon filters, torrefied grass fibers, turf, local soils, kaolin, clay minerals, loess soil, clay soils, poultry manure, and banana waste, wastewater sludge, lignin corn straw, wheat or oat bran, fibers from brewer's spent barley grain, grape bagasse, cork compost, peat, sugar, coir dust/coco peat, compost from different raw materials and vermicompost/earthworm compost, lignite, charcoal, gum, residues of *Azolla* and sawdust. All these vehicles were listed by Bashan et al. (2014) in reference [9] and, curiously, humic substances are not directly mentioned.

Humic substances (HS) are the major component of soil, water and sediment organic matter and chemically can be depicted as a collection of diverse, relatively low molecular mass components of organic molecules forming dynamic associations stabilized by hydrophobic interactions and hydrogen bonds on the nanometer scale [14]. The procedure consisting in the sequential fractionation of humic domains into either aqueous or organo soluble components and final unextractable humic residue, followed by molecular characterization suggested that humic suprastructures in soil are arranged in multimolecular layer [15]. Furthermore, it was possible to conclude that HS are spatially arranged in descending order of polarity, meaning that highly polar supramolecular subunits shield less polar subunits against the free soil solution and form layers of descending polarity', i.e., free soil solution [16]. This organization can protect bacteria cells and labile organic molecules against degradation in the hydrophobic domain, as elegantly demonstrated [17, 18]. The presence of organic acids, like those produced by root exudation, can modify the supramolecular arrangement, releasing bioactive molecules which can affect plant growth. The enhancement of organic acids exudation by maize due to HS treatment were independently demonstrated [19, 20] and the cross-talk involving plant roots and humic matter was recently approached [21]. The physiological effects of HS is widely documented [22, 23] and summarized as a result of enhance nutrient use efficiency, aiding assimilation of both macro and micronutrients and promoting plant growth by induction of carbon, nitrogen, and secondary metabolism [21].

HS are relatively recalcitrant to bacteria degradation and affect rooting including induction of lateral root formation and root hair initiation. The anatomical and physiological changes may favor the fitness of the mutualistic interaction by increasing rhizosphere population and chemotaxis, bacteria attachment and survival on plant surface as well as endophytic colonization (Fig. 1).

For this and other physiological modifications, we indicate the HS as a candidate to suitable vehicle for PGPB. The aim of this work is to produce an overview on the papers published in peer-review journals at Scopus and Web of Science data base using PGPB in combination with HS as key-word. In the first part, we showed the main results obtained in the Nudiba (UENF), Brazil, followed by the description of some mechanisms that underline the success of the combination of DEB and HS. The main results are summarized in the graphical abstract. After, the state of the art of this biotechnology is presented. We intend to demonstrate at the end of this review that co-inoculation can be an excellent, but very little explored tool to aid crop production, especially under stress conditions (nutritional, saline or toxicity of some element).

DEB and HS

Effects on sugarcane

The first report using the approach considering the application of cell suspension of the DEB together with soluble HS at low concentration was in a sugarcane seed pieces [24]. This crop, in the north of Rio de Janeiro State, Brazil, has the endemic incidence of *Leifsonia xyli* subsp. *xyli*, the causal agent of ratoon stunning disease, the main sugarcane crop disease [25]. The unique treatment, besides the introduction of sugarcane disease-resistance varieties, is the heat treatment which consists to subjecting seed pieces to a temperature of 50.5 °C for 2 h. As this procedure is not selective, there is a substantial drawback in the entire population of beneficial and harmful microbial after heat treatment, offering an opportunity to introduce selected microorganisms. The seed pieces after heat treatment were immersed for 12 h into a container filled with DEB (10^8 cells mL^{-1} of *Herbaspirillum seropedicae* strain HRC54) and humic acids (HA) (20 mg C $L^{-1)}$ isolated from vermicompost, transplanted to a pot containing 3 kg of substrate grown for 45 days. The main results were the root improvement (from 60 to 118%, in length, and from 33 to 233%, in surface area) by treatments compared to control, with more pronounced effect in plants under heat treatment. Likewise, the positive effect of the combined treatment (*H. seropedicae* and HA in suspension) on shoot and root biomass was significant compared to the control with heat-treated cuts. For heat-treated cuts, bacteria inoculation combined with HA increased the size of *H. seropedicae* population on roots. The practical limitations of this type of inoculation (after heat treatment) in Brazil are obvious considering its large scale production that reach almost nine millions of ha cultivated with sugarcane [2]. Hence, it was necessary to evaluate others forms of DEB-HA application and it was demonstrated a clear advantage of foliar application in respect to furrow [26]. The biostimulant manufactured with K^+-humate isolated from vermicompost and *H. seropedicae* strain HRC54, *H. rubrisubalbicans* strain HCC103 and *G. diazotrophicus* strain PAL5 had shown better performance when was applied at 60 days after sugarcane emergence, enhancing 37% of the stem yield when compared to control. The field assay using strip plot design with a large parcel area confirms the promotion of sugarcane yields by biostimulant during two consecutive ratoons increasing 19 and 18% of yield, representing 11 and 13 tons ha more than the control. In both experiments were not observed changes in the polarized sugar content and consequently the stems production is linearly related with sugar yields. A noticeable increase was observed in a year with severe drought stress occurrence showing the positive effect of the combined use on mitigation of abiotic stress damage. The field experimentation at commercial sugarcane plantation was accompanied by greenhouse experimentation where sugarcane was submitted to controlled drought stress and its recovery was evaluated [27]. The effect of biostimulant using the mixture of different DEB strains and HA suspension on shoot and root biomass was also observed in pot assay. The antioxidant enzymes superoxide dismutase, catalase and ascorbate peroxidase activities remained higher after rehydration only in plants treated with HA. Plants treated with only HA or DEB together with HA exhibited increased transpiration, stomatal conductance and net photosynthesis than the plants treated with DEB. The DEB-treated plants exhibited drought resistance that resembled 'delayed stress onset' mechanism, which is a term for preserving water in the plants tissues. Water preservation in plants treated with DEB was corroborated by higher relative water content than control plants at the end of the drought period. It was also observed an osmotic adjustment induce by the biostimulant treatment and indicated by a new pattern of metabolic response after rehydration, including generally enhanced carbohydrates and proteins and specific changes induced by HA-enhancing aromatic compounds, whereas DEB exhibited enhanced fatty acids and other aliphatic H species. The authors concluded that HA assist with drought stress recovery by inducing antioxidant enzyme activity, whereas DEB induced preservation of leaf water potential and relative water content by closing stomata efficiently, resulting in plant water preservation.

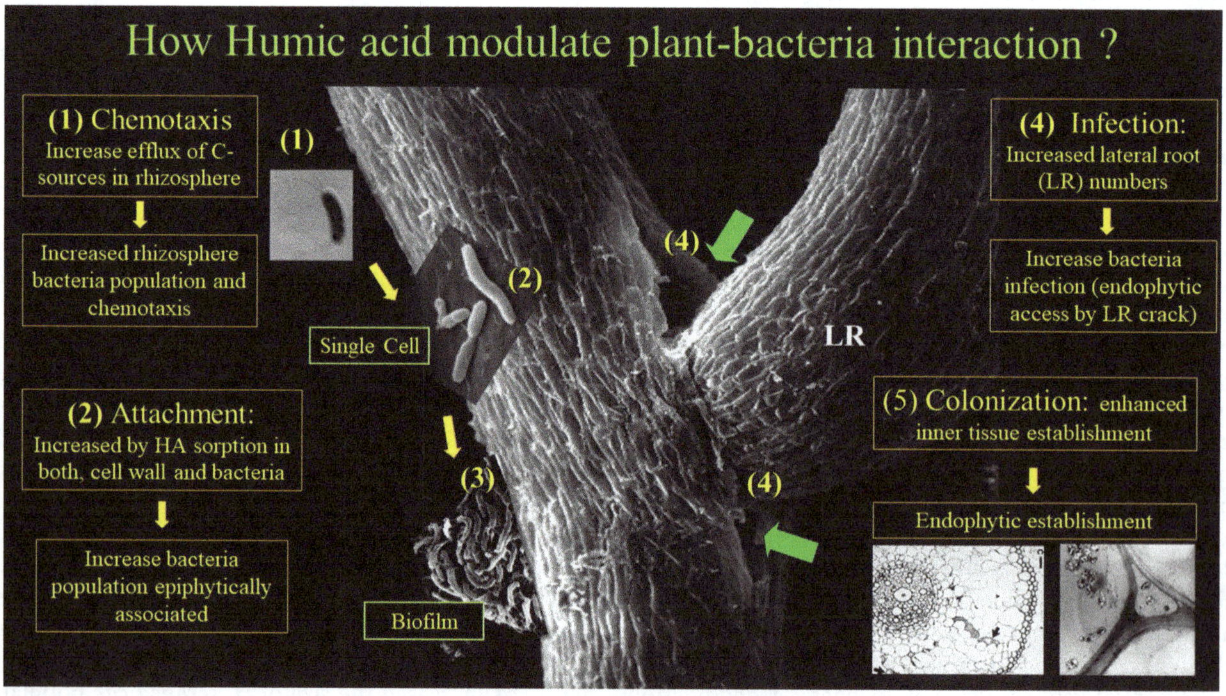

Fig. 1 Schematic representation of predominant events that characterize the endophytic interaction between diazotrophic bacteria and plant host: (1) chemotaxis from soil to rhizosphere; (2) attachment on plant root surface; (3) development of aggregates or biofilms; (4) lateral root cracks as main infection point and (5) apoplastic spread and endophytic colonization. Followed by how the application of humic acid combined with endophytic diazotrophic bacteria could modulate different steps of the biological interaction that results in increased epiphytic and endophytic population associated to the host plant

Effects on common beans

The same qualitative results were obtained with common beans, that is greater recovery from water stress in plants co-inoculated with *Rhizobium tropici* strains 'BR322', 'BR520', and 'BR534 and *H. seropedicae* strain HRC 54 in the presence of HA [28]. Two independent experiments were conducted in a greenhouse using *Phaseolus vulgaris* cv. Grafite and cv. Bonus, which originated in Brazil and Mozambique, respectively. The soil humidity was maintained at field capacity in pots, and the water suppression was induced at the pre-flowering stage. After 12 days, the water was restored, and the beans were evaluated. In the experiment with 'Grafite' beans, the relative water content of the foliar disk was significantly higher in the co-inoculated treatment, as were the numbers and masses of nodules. The phenylalanine ammonia lyase activity was induced by drought, and its activity was higher in co-inoculated plant leaves. The recuperation of the plants after water stress was clearly improved by the co-inoculation with rhizobia and *H. seropedicae* in the presence of HA-like substances.

Effects on maize

Coating maize seed with DEB and HA was previously reported [29]. The germination was not affected by seed coating that promoted root growth enhancing 44% the root length in respect to control. The number of diazotrophic bacteria cells was also increased from 2.4 to 6.7 $\log^{-1}$ cell per g of fresh root tissues (*H. seropedicae* strain Z67 = BR 11175). In that work, it was possible to observe that the treatment of seed coating only with HA also induce the increased native population of diazotrophs in the roots. This observation was corroborated when DEB and HA was used in maize seedlings in suspension (2×10^9 cell of *H. seropedicae* strain Z67 per mL plus 20 mg C of HA) [30]. The number of viable bacterial cells was higher in root tissues when the inoculation was in the presence of soluble humic substances. The physiological changes induced by co-inoculation was evaluated including enhancement of plasma membrane H^+-ATPase activity, alteration of carbohydrate and nitrogen metabolism, and greater net photosynthesis. In the field experiment, the treatments were applied by foliar spray at v6 developmental stage and when DEB and HS were singly applied grain yield increased about 20% in respect to control plants while applied together the increase of yield reach 65%. The soil used in this field experimentation was classified as Ultisol that showed low natural fertility (CEC < 5 $\text{cmol}_c\ \text{kg}^{-1}$) and were used low level of fertilizers (50 kg of N ha^{-1}). The level of maize yield is considered very low (2800 kg ha^{-1}) and typical from subsistence agriculture founded in the region where the experiment was conducted (Macaé, Rio de Janeiro State, Brazil). However, the enhance promoted by one foliar application of biostimulant (4480 kg ha^{-1}) is enough to promote the food security for smallholders. We repeat the field experimentation in other place for two consecutive growth season using an Inceptisol also located in North of Rio de Janeiro State with medium level of natural fertility and varying the urea amount used in nitrogen fertilization at coverture (0–200 kg N ha^{-1}) [31]. The maize yield increased independently of urea doses in parcels with application of a combination of humic substances isolated from vermicompost at 50 mg L^{-1} and *H. seropedicae* strain HRC 54 (5×10^8 cells mL^{-1}) at equivalent rate of 450 L ha^{-1} and applied at v6 stage. In the second year of experimentation, the increase of grain yield promoted by co-inoculation was also significant, but only until 75 kg N ha^{-1}. At higher N concentrations no effect of inoculation was observed in a typical and known effect [5]. The main difference between this two growth seasons was that the first year was marked by a severe drought stress while in the second the rain season had normal precipitation. The co-inoculation effect in maize yield was accentuated in the drought stress event. It was also reported that when more lately the foliar spray application of DEB humic acids was done best results were achieved [31]. The original idea of seed covering or foliar application at plant seedlings (7–15 days old) has been superseded by later application. It was also observed that maize parcels inoculated with DEB and HS showed higher root systems left in the soil after harvest in a significant contribution to soil organic matter maintenance. This important environmental service almost is not counted and should be considered as a carbon sequestration strategy.

Effects on tomatoes

Other approach with significant results was when DEB and HS were used in a substrate to seedlings growth knowing as biofortification strategy. Tomato is one of the main horticulture plants in Brazil and we applied the DEB and HS in the commercial substrate or in the substrate manufactured with vermicompost naturally enriched with PGPB and humic matter with high biological activity [32]. The biofortification of the substrate boost the initial development of tomato seedlings and when transplanting to experimental field the gains obtained in the initial phase of plant development were maintained. The foliar spray application of DEB and HS promoted additional effect on fruit production and disease-resistance resulting in greater yield [32]. It was observed the enhancement of the N_2-fixing population associated with rhizosphere soil, root and leaf tissues in the combined use of bacteria and humates by foliar spray.

Effects on pineapple

The initial performance of tissue-cultured pineapple variety *Vitória* in response to application of HA isolated from vermicompost and PGPB (*Burkholderia* sp. UENF 114111 and *Burkholderia silvatlantica* strain UENF 117111) during greenhouse acclimatization was studied by Baldotto and colleagues [33]. The period of pineapple acclimatization is critical due to the low growth rate of roots and shoots. In this phase, structural and physiological adjustments of plantlets to ex vitro conditions are crucial for the success in the subsequent phase, i.e., field transplantation. The basal leaf axils were treated with HA while roots were immersed in bacterial medium, resulting in improved shoot growth (147%). The isolated application of HA increased root growth by 50%, bacteria by 81% and the combined treatment by 105%. Pineapple growth was influenced by inoculation with both *Burkholderia*, and further improved in combination with humic acids, resulting in higher shoot and root biomass as well as nutrient contents (N 132%, P 131%, K 80%) than in uninoculated plantlets. The stability and increased consistency of the host plant response to bacterization in the presence of HS indicate a promising biotechnological tool to improve growth and adaptation of pineapple plantlets to the ex-vitro environment.

We showed some examples of successful use of biotechnology process using DEB and HS as inoculant applied directly on different plants. Now, we intended to depict the main mechanisms responsible for the synergist effect observed in so different plants crops as sugarcane, maize, tomato, and pineapple with different growth and developmental phases and environmental conditions.

Mechanisms responsible for the success of DEB and humic substances

Morphological adaptations

Our previous laboratory and field experiments commented above were strictly based on well recognized endophytic bacteria as part of the biostimulant formulation combined with HS. This is important issue not always considered. According to the professor Günter Neumann [34], from the University of Hohenheim (Stuttgart, Germany), about 70% of field experiments using microorganisms as bio-effectors did not show significant results. The use of endophytic bacteria drives the interaction with the host plant forcing the bacteria to compete for infection sites. The successful entry into the host plant by endophytes is made through root tips, root cracks at the point of emergence of lateral roots, injured sites on the root epidermis, stomata apertures, and damaged trichomes. The most prominent morphological modification in plants induced by both HS and DEB includes the promotion of lateral roots emergence (Fig. 2A). The morphological changes of the root system triggered by HS comprise increases of lateral root formation sites, root hair density and length [21], as well as overall surface area available for bacteria attachment (Fig. 2B). We could expect an increase in the population size plant-associated coupled with humate since the number of openings due to the emergence of lateral roots increases (Fig. 2C). In addition, the extreme heterogeneity of the root and leave surface after the contact with soluble humic acids facilitates the anchoring of microorganisms (Fig. 2D) and for overall picture (see Fig. 1).

Controlling plant root architecture is a fundamental part of plant development enabling to respond environmental conditions changes. Variations in the surface area of plant root systems are brought about largely by variations in root branching [35]. Underpinning each stage of lateral roots development is the hormone auxin [36]. It is known that HS can simulating the behavior of plant hormones and the presence of indole acetic acid has been demonstrated using different approaches. The auxin-like effect of HS was confirmed by a molecular approach [37]. The auxin-like compounds produced by PGPB is also well known and often reported as a positive Salkowski reaction and HPLC [38].

Other important morphological adaptation in roots induced by HS and scarcely reported is on the border cells formation and release from tips [39]. Border cells constitute the first living boundary in plant-soil ecosystems and play an important role in environmental sensing and regulate the interaction of living microorganisms of the rhizosphere [39]. HA can modulate the border cell number and shape and the association of border cells with mucilage, which serves to protect these cells and enhance viability. In addition, micrographs showing the packaging of border cells were obtained at all HA concentrations used and are indicative of the spontaneous self-organization of humic particles in suspension. We demonstrate that HA at the root tip surface can increase *H. seropedicae* attachment, aggregation and root tip colonization, thus changing the natural behavior of the bacterial-plant interaction [38]. It is possible that adsorbed HA can change the density and increase the number of hydrophobic domains, or modulate border cells secretion pattern through HA induction or neutralization of the biosynthesis of chemoattractant/antimicrobial compounds which results in increased bacterial cell adhesion to root tips. According to Olivares and Canellas [38] observations, exogenous HA increased the bacterial population surrounding border cells.

The induction of lateral roots, the increase of surface roughness of leaves and roots and the protect packing by HS particles as important morphological changes that contributes to the PGPB population increase promoted by HS.

Physiological changes

Although plant physiology is a very broad field of study, the first thing that comes to mind is related to the plant functioning including processes like photosynthesis, respiration, nutrient uptake, hormone control of development, stress adaption, plant water relationship among others. We started by exudation process since the infection process of *H. seropedicae* begins bacteria stimulation in rhizosphere coupled with their attraction to the roots. The host provides carbon sources for the bacteria. It was previously observed an increase of organic acids exudates into the rhizosphere of maize plants treated with HA [18, 19], which could be used as carbon substrate to support the bacteria population growth associated with the root system. Furthermore, one of the main mechanism of DEB to solubilize phosphorus is the efficient production of organic acids mainly oxalic, citric, succinic and ascorbic acids [40]. The significant efflux of organic anions in roots treated with humic acids was observed [41] Organic acids are one the main source of carbon to PGPB and the rhizodeposition of labile compounds induced by HA in the inoculation medium can attract the bacteria. Interestingly, the supramolecular structure disassembling by organic acid that was well reported, release bioactive compound that modulates root architecture and increase epiphytic and endophytic bacteria colonization. Hypothetically, such virtuous cycle has been established over the coevolution process involving organic matter-microorganism and plants and deciphering different aspects of this cross-talk represent a powerful tool to change agriculture paradigm.

Moreover, significant changes in root exudation profiles were induced by *H. seropedicae* and HA, including the increase on fatty acid exudation from seedlings treated only with HA and on nitrogenous compounds and terpenes from seedlings treated only with *H. seropedicae* [42]. Seedlings treated with *H. seropedicae* in combination with HA enhance the amount and diversity of nitrogenous compounds exuded, most of which had heterocyclic structures. These changes in the rhizosphere chemical nature can affect the bacteria attraction and favored the host colonization process. The chemical changes induced by HS and DEB in the rhizosphere have an important role in the cross-talk relationship among environment and root growth. The observed increase in the root surface induced by both DEB and HA can be linked with stimulation of plasma membrane H^+-ATPase according the acid growth mechanisms induced by auxins and auxin-like compounds [43] present in the humic suprastructure or synthesized by DEB. H^+-ATPase activation could also be mechanistically linked to the root hairs proliferation since it was demonstrated that cell growth depends on extracellular acidification in maize roots [44]. The DEB and HA-induced stimulation of the H^+-ATPase activity [29] suggest that the modifications are not restricted to root architecture, but are also extended to major biochemical components. Indeed, the driving force for most nutrient uptake is the electrochemical gradient across the plasma membrane, which is primarily generated by the H^+-ATPase. Previous report had shown that, in some circumstances, when bacteria gain entrance into the plant tissue, it could occur by cell wall-hydrolyzing enzymes that support the process of bacteria invasion and dissemination in the host [45]. These enzymes had increased activity under low pH, which is compatible with H^+-ATPase induction by HS, which reinforce our hypothesis that HS can make the delivery process of the DEB more efficient to the host plant.

The main function of the plasma membrane H^+ ATPase is to generate a proton electrochemical gradient, thereby providing the driving force for the uptake and efflux of ions and metabolites across the plasma membrane [46]. Thus, it is not surprise that HA from vermicompost and *Burkholderia* spp., separately and even more when combined, promoted the growth of the aerial part, root system and increased the contents of N, P, K, Ca and Mg in pineapple [32]. The enhancement of nutrient uptake is one of more significant effect of the co-inoculation in plants, especially nitrogen and phosphorus. The activities of enzymes linked to nitrogen assimilation in maize seedlings were promoted by the *H. seropedicae* and HA application [29]. The treatment combining *H. seropedicae* and different concentration of HS induced nitrate reductase (NR) activity in leaves between 120 and 230% in relation to the control. Nitrogen is the major limiting factor in plant growth and productivity, therefore it is of great importance to study the enzyme activity related to its assimilation. The NR enzyme, together with nitrite reductase, catalyze the two-step reduction of nitrate (NO_3^-) to ammonium (NH_4^+), which is rapidly incorporated into organic compounds through the activity of the enzyme glutamine synthase, which is also stimulated by co-inoculation. The carbohydrate metabolism was also changed by co-inoculation especially the decrease in the concentration of reducing sugar in leaves and interpreted as a metabolic intensification to nitrogen assimilation. Olivares and colleagues [31] also observed the modification on secondary metabolism in tomato co-inoculated with DEB and HA, including the enhance on the phenylalanine ammonia lyase (PAL) activity. The first evidence that HA can modify PAL activity in maize was provided by Schiavon et al. [47] and Ertani et al. [48]. This enzyme catalyzes the first committed step in the biosynthesis of phenolics by converting phenylalanine to trans-cinnamic acid and tyrosine to p-coumaric acid. Phenolic compounds are used by plants in a several mechanisms of defense

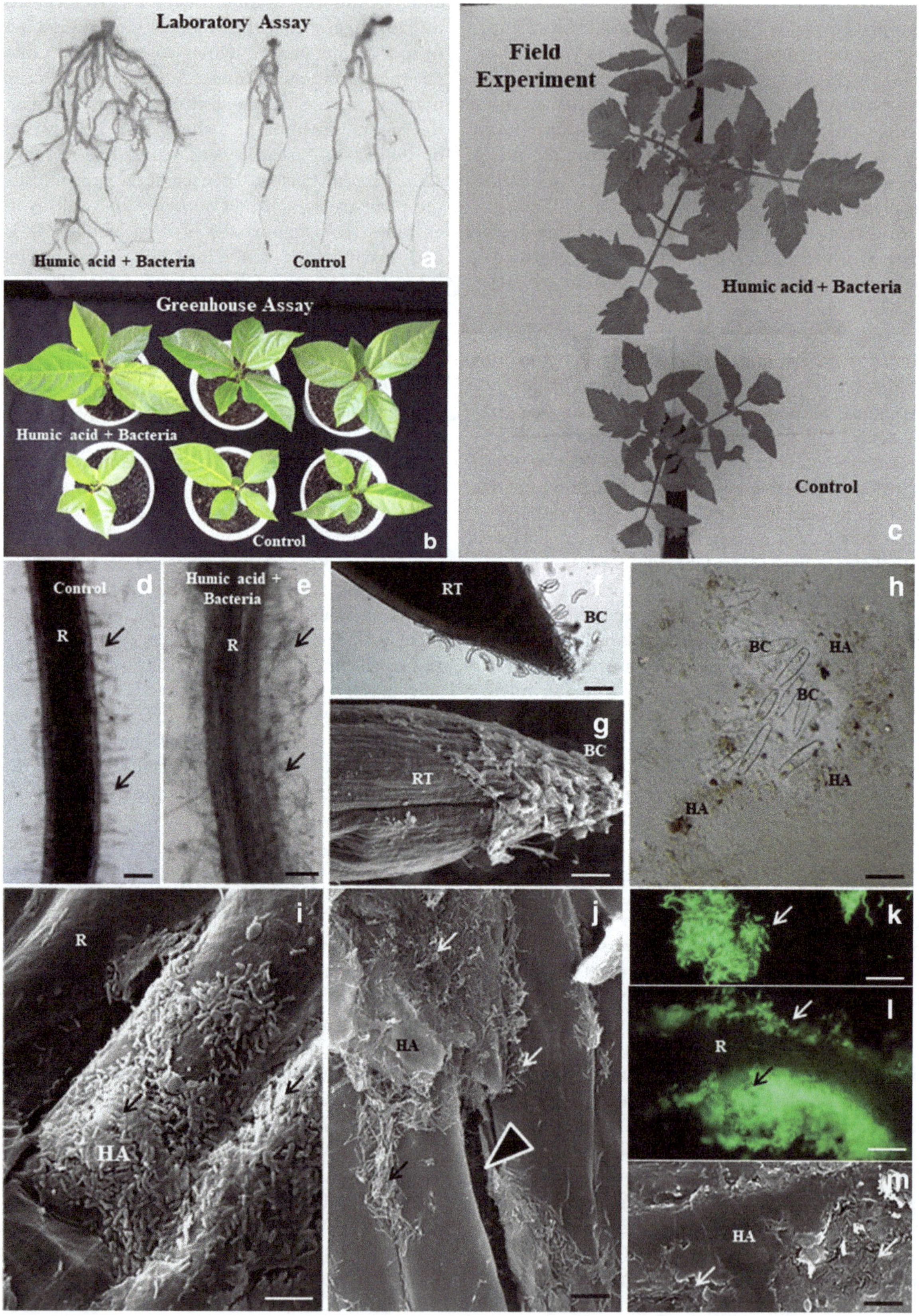
Laboratory Assay
Humic acid + Bacteria
Control
a
Greenhouse Assay
Humic acid + Bacteria
Control
b
Field Experiment
Humic acid + Bacteria
Control
c
Control
d
R
Humic acid + Bacteria
e
R
RT
BC
f
g
BC
RT
h
BC
HA
BC
HA
HA
i
R
HA
j
HA
k
l
R
m
HA

(See figure on previous page.)

Fig. 2 Basic and applied aspects related to the biological input designed with combined use of diazotrophic endophytic bacteria (DEB) and humic acid (HA). **A–C** Showing positives effect of the technology on root branching, root biomass and shoot biomass from lab to field experiments. **D**, **E** Light microscopy (LM) of maize root segments (R) comparing treated plants (DEB + HA) against mock plants. Note clear increase in density and length of root hairs (arrows). **F** LM from maize root tip (RT) showing border cells (BC) released trigged by HA. **G** Scanning electron microscopy (SEM) from the maize root tip (RT) showing root cap eroded by released border cells (BC). **H** LM from released border cells (BC) encapsulated by self-organized humic particles in suspension (HA). **I** SEM from maize root (R) surface colonized by *Herbaspirillum seropedicae* strain HRC54 aggregates in association with humic acid particles. **J** SEM from maize root surface close to the region of a lateral root emerged. Note crack developed in the epidermal layer (arrow head) and *H. seropedicae* strain HRC54 attached to the surface close to the entrance point (white arrows). **K** Epifluorescent microscopy (EM) of *H. seropedicae* strain RAM 10 linked with a green fluorescent protein marker (arrow) forming a cluster around to a humic acid particle. **L** EM of a maize root segment colonized by a very active *H. seropedicae* strain RAM 10 biofilm (arrows). **M** SEM of humic acid (HA) plates associated with *H. seropedicae* strain HRC54 attached to the sugarcane leaf surface after foliar spray. Bar equal 100, 100, 200, 200, 50, 5, 5, 10, 150 and 20 μM, respectively, for **D–M**

including osmoprotection, UV scavenging and activation of defense pathway against biotic harmful agents.

Finally, it is possible return and extend the conclusion of Nardi and colleagues [21] saying that plant co-inoculation with DEB and HS, besides induction of lateral root formation and root hair initiation, enhance nutrient use efficiency, aiding assimilation and promoting plant growth by the induction of carbon, nitrogen, and secondary metabolism.

PGPB + HS: why is it so little explored?

We search for another results including DEB and HA and we did not find a vast literature. The main bacteria used in association with HS are free living microorganisms that need to compete at rhizosphere level for survival. Soil being a heterogeneous, unpredictable environment, the inoculated bacteria must to overcome complex biotic and abiotic impairing factors to establish a niche for survival amongst the competitors and predators [49]. Very few studies have been considering the effect of HS as a carrier that promotes PGPB fitness in the soil-rhizosphere-plant continuum system. Table 1 summarizes the twelve reports founded in the database from two giants of scientific publishers (Web of Sciences from Thompson Reuters and Scopus from Elsevier). This is very little considering the potential of biological inputs and inoculation technology in the modern agriculture either in the processes of chemical inputs substitution, or in the adaptation to new technologies of production or even of ecological intensification. One possibility for explain this low number may lie in our fragmented way of doing research. Despite humus is a direct product of microbiological activity, the soil chemistry and soil microbiology have established faint scientific connections around the world. Increase in the sustainable intensification of agriculture is important developments towards new ways of training future scientist [50, 51]. Inter-disciplinarily and boundary-crossing in terminology and concepts are needed. The use of these products in commercial crops also seems limited. It is necessary also point to the fact that the answers waited results from biological mechanisms, with an action also dependent on the conditions of the growing environment. However, some agricultural areas are present in degradation conditions so deep that it is necessary, first and foremost, to recover them using reconstruction minimum levels of fertility and soil organic matter content prior start the use of biostimulants. One of the first questions that researchers who employ biological products hear from farmers is: is this product efficient? As, in general, the farmer demands for immediate results, this may cause frustration. Although important, this is not the focus of this work and we return to Table 1.

The combined used of PGPB and HS was used to (i) promote crop production, (ii) mitigation of plant stress effect and (iii) microorganism protection. When applied directly to plants, the crop promotion was attributed mainly due enhance of nutrient efficiency use [52–58]. Schoebitz and colleagues [53] reported the use of consortium of very diverse microorganisms combined (or not) with K-humate. The growth of blueberry and N and K uptake increased and the authors find a very interesting and significant change on the bacterial structure community at rhizosphere when HA was used. Previously, this kind of change on bacteria diversity at maize rhizosphere induced by HA presence [51] opening a wide avenue to study changes in the microbiota and the interaction of these microbiota changes with crop productivity. Two papers found no significant differences on yield by PGPB and HS treatments [55, 56]. Working with five different N concentration at field, it was found increase on wheat leaves, but not in grain yield [59] and when PGPB and HA was applied directly to soil it was observed significant increase on soil P availability, but without increase of soybean yield [60].

Two independent papers using very diverse type of plants found qualitative similar results when PGPB and HA were used under saline conditions [61, 62]. In the first report, the mitigation of salinity effects was

Table 1 Study references for plant growth promoting bacteria used in combination with humic substances

References	Treatments	Main results
Valero et al. [58]	Humic acid biologically extracted from low-rank coal inoculated with a mixed microbial inoculum of coal solubilizing bacteria (*Bacillus mycoides*, *Acinetobacter baumannii*, and *Microbacterium* sp.)	Edaphic substrate improvements (cation exchange capacity and microbial activity) and increase in the dry weight and the total contents of N and P in the maize seedlings
Bacilio et al. [61]	Humic acids and *Pseudomonas stutzeri* were inoculated in chili pepper	Mitigation of salinity effects
Esringü et al. [62]	Hungarian vetch; three PGPB strains—*Agrobacterium rubi* strain A1, *Bacillus subtilis* strain OSU-142 and *Bacillus megaterium* strain M3; commercial K^+-humate	Mitigation of salinity effects
Schoebitz et al. [53]	Blueberry seedlings: microbial consortium (*Bacillus subtilis*, *B. licheniformis*, *B. megaterium*, *B. polymyxa*, *B. macerans*, *Pseudomonas fluorescens*, *P. putida*, *Nocardia corallina*, *Saccharomyces cerevisiae* and *Trichoderma viride*). 15 g of humic substances from Leonardite (powder form) was added per seedling	Enhance on plant biomass;—enhance on nutrient uptake (N and K);—changes on the bacteria community at rhizosphere
Ahmad et al. [57]	*Acinetobacter pittii* and humic acids	Enhance on crop yield and nutrient efficiency use
Verma et al. [52]	Cabbage: *Pseudomonas fluorescens* and humic acid	Nutrient efficiency use; crop yield; plant quality
Rodrigues et al. [59]	Wheat: *Azospirillum brasilense* and humic acids	Without results on grain yield
Befrozfar et al. [55]	*Pseudomonas putida*, *Azospirillum lipoferum*, *Azotobacter chroococcum* and humic acids	Growth promotion
Winarso et al. [60]	Like humic acids isolated from straw compost and *P. putida* 27.4B. application directly to the soil and not to the plant	Enhance of P availability in soil; without effect on soybean growth
Magdi et al. [54]	Bioinoculant with N fixer (*Azospirillum* and *Azotobacter*) a P-solubilizer (*B. megatherium*) bacteria with humic acids applied to cowpea by drip irrigation	Nutrient efficiency use
Rekha et al. [56]	*Pseudomonas putida* (estirpe CC-FR2-4) *Bacillus subtilis* (strain CC-pg104); humic acids	Growth promotion; changes in enzymatic activity
Young et al. [63]	Bacillus subtilis CCpg104 was attempted with alginate by enriching the bead microenvironment with humic acid	Preservation of cell viability; P solubilization and enhance of root and shoot growth of lettuce plants under gnotobiotic (sterile) condition

attributed due improvements in K^+/Na^+ and Ca^{2+}/Na^+ ratios by chili peppers under inoculation with *Pseudomonas stutzeri* or HA treatment [61]. Changes in the ionic balance is the first and fast plant adaptation against stress including due promoted by high Na^+ concentration. The increase on H^+-ATPase activity induced by HS is well reported and necessary to energize K^+ and Ca^{2+} transporters. The other paper where the effect of salinity was observed in Hungarian Vetch and the mitigation was attributed to promotion of plant growth parameters [62]. Finally, we would like to do mention to the work of Young et al. published in 2006 [63] where the authors proposed the encapsulation of PGPB *Bacillus subtilis* CCpg104 in alginate enriched with HA. The inoculation of encapsulated PGPB-HA enhanced the growth of lettuce under gnotobiotic (sterile) condition in respect to control (no-inoculated) but the differences were not significant compared with non-encapsulated inoculation. However, the time of cell viability was larger in the encapsulated PGPB offer promising alternatives to increase the shelf-life of the inoculant and increase the protection against the deleterious effects of soil environment after delivering.

Conclusion

Overall agriculture's finger print is already immense. Struik and co-authors summarizing wrote [64, 65]:

> *"The societal costs of current ways of using technology for producing food also are substantial. From an agronomic and environmental perspective these include depletion and spillage of resources such as water, degradation of agro-ecosystems and natural ecosystems, decline in ecosystem services, loss of biodiversity, emission of greenhouse gases and toxic waste, post-harvest loss, among others, all contributing to agriculture's ecological footprint".*

We agree with assumption that solutions for food security must be holistic and must address issues such as food accessibility that filling the lack of attention to justice [64]. Furthermore, the role of biological inputs in this debate also should be questioned, including the concept that this type of technology is primitive and backward. Here, we reported the results and expose the main action mechanism of biostimulant manufactured with diazotrophic endophytic bacteria mixed with HS used to enhance crop yield. The main benefit observed is the possibility of reducing the level of nutrients maintaining the level of production, in addition to mitigate the effects of different types of stress. The numbers of scientific reports considering the terms "plant growth promoting bacteria + humic substances" is very low considering the huge potential of this kind of biological inputs.

Authors' contributions

LPC and FLO wrote the first draft of the manuscript and took care of the revisions of the review; JGB, AMP, LSL and NOA contributed to the section PGPB + humic substances: why is it so little explored? All authors read and approved the final manuscript.

Author details

[1] Núcleo de Desenvolvimento de Insumos Biológicos para a Agricultura (NUDIBA), Universidade Estadual do Norte Fluminense Darcy Ribeiro (UENF), Av. Alberto Lamego, 2000, Campos dos Goytacazes, Rio de Janeiro 28013-602, Brazil. [2] Faculdade de Agronomia e Medicina Veterinária, Universidade de Brasília, Campus Universitário Darcy Ribeiro, Caixa Postal 4508, CEP 70910-970 Brasília, DF, Brazil.

Funding

The work in NUDIBA laboratory (UENF, Brazil) was supported by Conselho Nacional de Desenvolvimento Científico e Tecnológico (CNPq), Fundação de Amparo à Pesquisa do Estado do Rio de Janeiro (FAPERJ), Instituto Nacional de Ciência e Tecnologia (INCT) para a Fixação Biológica de Nitrogênio and International Foundation of Science (IFS).

References

1. Lichfouse E, Navarrete M, Debaeke P, Souchre V, Alberola C. Agronomy for sustainable agriculture: a review. Agron Sustain Agric. 2009;29:1–6.
2. Conab (Companhia Nacional de Abastecimento) web site with methodology and results of agriculture production costs. http://www.conab.gov.br/conteudos.php?a=1546&t. Accessed Aug 2017.
3. Halpern M, Bar-Tal A, Ofek M, Minz D, Muller T, Yermiyahu U. The use of biostimulants for enhancing nutrient uptake. In: Sparks DL, editor. Adv Agron; 2015. p. 141–74.
4. Camargo MN, Jacomine PKT, Carvalho APJ, Olmos IL. The Brazilian classification of Latosols. In: International soil classification workshop: classification, characterization and utilization of oxisols, EMBRAPA, SMSS, AID, UPR, Rio de Janeiro, Brazil. 1988. p. 190–202.
5. Thonar C, Lekfeldt JDS, Cozzolino V, Kundel D, Kulhánek M, Mosimann C, Neumann G, Piccolo A, Symanczik S, Walder F. Potential of three microbial bio-effectors to promote maize growth and nutrient acquisition from alternative phosphorous fertilizers in contrasting soils. Chem Biol Technol Agric. 2017;4:7.
6. Adesemoye AO, Torbert HA, Kloepper JW. Plant growth-promoting rhizobacteria allow reduced application rates of chemical fertilizers. Microb Ecol. 2009;58:921–9.
7. Bhattacharyya PN, Jha DK. Plant growth promoting rhizobacteria (PGPR): emergence in agriculture. World J Microbiol Biotechnol. 2012;28:1327–50.
8. Owen D, Williams AP, Griffith GW, Withers PJA. Use of commercial bioinoculants to increase agricultural production through improved phosphorus acquisition. Appl Soil Ecol. 2015;86:41–54.
9. Bashan Y, de Bashan LE, Prabhu SR, Hernandez JP. Advances in plant growth promoting bacterial inoculant technology: formulations and practical perspectives (1998–2013). Plant Soil. 2014;378:1–33.
10. Döbereiner J, Baldani VLD, Reis VM. Endophytic occurrence of diazotrophic bacteria in non-leguminous crops. In: Vanderleyden J, Zamaroczy M, del Gallo FM, editors. Azospirillum VI and related microorganisms. Berlin: Springer; 1995. p. 3–14.
11. Embrapa. https://www.embrapa.br/en/busca-de-produtos-processos-e-servicos/-/produto-servico/550/estirpes-de-bacterias-fixadoras-de-nitrogenio-para-inoculacao-em-cana-de-acucar.
12. Oliveira ALM, Canuto EL, Urquiaga S, Reis VM, Baldani JI. Yield of micropropagated sugarcane varieties in different soil types following inoculation with diazotrophic bacteria. Plant Soil. 2006;284:23–32.
13. Vassilev N, Vassileva M, Lopez A, Martos V, Reyes A, Maksimovic I, Löbermann B, Malusà E. Unexploited potential of some biotechnological techniques for biofertilizer production and formulation. Appl Microbiol Biotechnol. 2015;99:4983–96.
14. Piccolo A. The supramolecular structure of humic substances: a novel understanding of humus chemistry and implications in soil science. Adv Agron. 2002;75:57–134.

15. Drosos M, Nebbioso A, Mazzei P, Vinci G, Spaccini R, Piccolo A. Molecular zoom into soil Humeome by a direct sequential chemical fractionation of soil. Sci Total Environ. 2017;586:807–16.
16. Fischer T. Humic supramolecular structures have polar surfaces and unpolar cores in native soil. Chemosphere. 2017;183:437–43.
17. Spaccini R, Piccolo A, Haberhauer G, Gerzabek MH. Transformation of organic matter from maize residues into labile and humic fractions of three European soils as revealed by 13C distribution and CPMAS-NMR spectra. Eur J Soil Sci. 2000;51:583–94.
18. Spaccini R, Piccolo A, Conte P, Haberhauer G, Gerzabek MH. Increased soil organic carbon sequestration through hydrophobic protection by humic substances. Soil Biol Biochem. 2002;34:1839–51.
19. Canellas LP, Teixeira Junior LRL, Dobbss LB, Silva CA, Medici LO, Zandonadi DB, Façanha AR. Humic acids cross interactions with root and organicacids. Ann Appl Biol. 2008;153:157–66.
20. Puglisi E, Fragoulis G, Del Re AA, Spaccini R, Piccolo A, Gigliotti G, Said-Pullicino D, Trevisan M. Carbon deposition in soil rhizosphere following amendments with compost and its soluble fractions, as evaluated by combined soil-plant rhizobox and reporter gene systems. Chemosphere. 2008;73:1292–9.
21. Nardi S, Ertani A, Ornella F. Soil root crosstalking: the role of humic substances. J Plant Nutr Soil Sci. 2017;180:5–13.
22. Nardi S, Carletti P, Pizzeghello D, Muscolo A. Biological activities of humic substances. In: Senesi N, Xing B, Huang PM, editors. Biophysico-chemical processes involving natural nonliving organic matter in environmental systems. Hoboken: Wiley; 2009. p. 305–39.
23. Canellas LP, Olivares FL. Physiological responses to humic substances as plant growth promoter. Chem Biol Technol Agric. 2014;1:3.
24. Marques Júnior RB, Canellas LP, Silva LG, Olivares FL. Rooting of micro seed pieces by combined use of humic substances and endophytic diazotrophic bacteria in sugar cane. R Bras Ci Solo. 2008;32:1121–8.
25. Carneiro JB Jr, Silveira SFS, Souza Filho GA, Olivares FL, Giglioti EA. Especificidade de anti-soro policlonal à *Leifsonia xyli* subsp. *Xyli*. Fitopatol Bras. 2004;29:614–9.
26. Silva SF, Olivares FL, Canellas LP. The biostimulant manufactured using diazotrophic endophytic bacteria and humates is effective to increase sugarcane yield. Chem Biol Technol Agric. 2017;4:24.
27. Aguiar NO, Medici LO, Olivares FL, Dobbss LB, TorresNetto A, Silva SF, Novotny EH, Canellas LP. Metabolic profile and antioxidant responses during drought stress recovery in sugarcane treated with humic acids and endophytic diazotrophic bacteria. Ann Appl Biol. 2016;168:203–13.
28. da Piedade Melo A, Olivares FL, Médici LO, Torres-Neto A, Dobbss LB, Canellas LP. Mixed rhizobia and *Herbaspirillum seropedicae* inoculations with humic acid-like substances improve water-stress recovery in common beans. Chem Biol Technol Agric. 2017;4:6.
29. Conceição PM, Vieira HD, Canellas LP, Marques-Júnior RB, Olivares FL. Recobrimento de sementes de milho com ácidos húmicos e bactérias diazotróficas endofíticas. Pesq Agropec Bras. 2008;43:545–8.
30. Canellas LP, Martínez-Balmori D, Médici LO, Aguiar NO, Campostrini E, Rosa RC, Façanha A, Olivares FL. A combination of humic substances and *Herbaspirillum seropedicae* inoculation enhances the growth of maize (*Zea mays* L.). Plant Soil. 2013;366:119–32.
31. Canellas LP, Silva SF, Olk D, Olivares FL. Foliar application of *Herbaspirillum seropedicae* and humic acid increase maize yields. J Food Agric Environ. 2015;13:146–53.
32. Olivares FL, Aguiar NO, Rosa RCC, Canellas LP. Substrate biofortification in combination with foliar sprays of plant growth promoting bacteria and humic substances boosts production of organic tomatoes. Sci Hortic. 2015;183:100–8.
33. Baldotto LBE, Baldotto MA, Canellas LP, Bressan-Smith R, Olivares FL. Growth promotion of pineapple 'Vitória' by humic acids and *Burkholderia* spp. during acclimatization. R Bras Ci Solo. 2010;34:1593–600.
34. Neumann G. Perspectives and limitations for the use of BIOEFFECTORS in crop production. Oral Lecture at Summer School "Organic matter and Biological Agriculture" 3–6 July 2017 Lecture hall 8, Department of Agricultural Sciences (DIA), Università degli Studi di Napoli FEDERICO II Via Università 100, 80055-Portici. http://www.carbonfarm.eu.
35. Nibau C, Gibbs DJ, Coates JC. Branching out in new directions: the control of root architecture by lateral root formation. New Phytol. 2008;179:595–614.
36. Casimiro I, Beeckman T, Graham N, Bhalerao R, Zhang H, Casero P, Sandberg G, Bennett MJ. Dissecting *Arabidopsis* lateral root development. Trends Plant Sci. 2003;8:165–71.
37. Trevisan S, Botton A, Vaccaro S, Vezzaroa A, Quaggiotti S, Nardia S. Humic substances affect *Arabidopsis* physiology by altering the expression of genes involved in primary metabolism, growth and development. Environ Exp Bot. 2011;74:45–55.
38. Frankenberger WT Jr, Arshad M. Phytohormones in soils: microbial production and function. New York: Marcel Dekker, Inc; 1995.
39. Canellas LP, Olivares FL. Production of border cells and colonization of maize root tips by *Herbaspirillum seropedicae* are modulated by humic acid. Plant Soil. 2017;417:403–13.
40. Driouch A, Cannesan MA, Dardelle F, Durand C, Plancot B, Brnard S, Follet-Gueye ML, Vicré-Gibouin M. Unit is strength: the power of border cells and border-like cells in relation with plant defense. In: Vivanco JM, Baluska F, editors. Secretions and exudates in biological systems signaling and communication in plants. Berlin: Spring; 2012. p. 91–108.
41. Busato JG, Lima LS, Aguiar NO, Canellas LP, Olivares FL. Changes in labile phosphorus forms during maturation of vermicompost enriched with phosphorus-solubilizing and diazotrophic bacteria. Bioresour Technol. 2012;110:390–5.
42. Ramos AC, Olivares FL, Silva LS, Aguiar NO, Canellas LP. Humic matter elicits proton and calcium fluxes and signaling dependent on Ca^{2+}-dependent protein kinase (CDPK) at early stages of lateral plant root development. Chem Biol Technol Agric. 2015;1:1–12.
43. Lima LS, Olivares FL, Oliveira RR, Vega MRG, Aguiar NO, Canellas LP. Root exudate profiling of maize seedlings inoculated with *Herbaspirillum seropedicae* and humic acids. Chem Biol Technol Agric. 2014;1:23.
44. Rayle DL, Cleland RE. The acid growth theory of auxin-induced cell elongation is alive and well. Plant Physiol. 1992;99:1271–4.
45. Peters WS, Felle HH. The correlation of profiles of surface pH and elongation growth in maize roots. Plant Physiol. 1999;121:905–12.
46. James EK, Olivares FL. Infection and colonization of sugarcane and other graminaceous plants by endophytic diazotrophs. Crit Rev Plant Sci. 1998;17:77–119.
47. Schiavon M, Pizzeghello D, Muscolo A, Vaccaro S, Francioso O, Nardi S. High molecular size humic substances enhance phenylpropanoid metabolism in maize (*Zea mays* L.). J Chem Ecol. 2010;36:662–9.
48. Ertani A, Schiavon M, Altissimo A, Franceschi A, Nardi S. Phenol-containing organic substances stimulate phenylpropanoid metabolism in Zea mays. J Plant Nutr Soil Sci. 2011;174:496–503.
49. Lucy M, Reed E, Glick BR. Applications of free living plant growth promoting rhizobacteria. Antonie Leeuwenhoek. 2004;86:1–25.
50. Struik PC, Kuyper TW. Sustainable intensification in agriculture: the richer shade of green: a review. Agron Sustain Dev. 2017;37:39.
51. Puglisi E, Pascazio S, Suciu N, Cattani I, Fait G, Spaccini R, Crecchio C, Piccolo A, Trevisan M. Rhizosphere microbial diversity as influenced by humic substance amendments and chemical composition of rhizodeposits. J Geochem Explor. 2013;129:82–94.
52. Verma R, Maurya BR, Meena VS. Integrated effect of bio-organics with chemical fertilizer on growth, yield and quality of cabbage (*Brassica oleracea* var capitata). Indian J Agric Sci. 2014;84:914–8.
53. Schoebitz M, López MD, Serrí H, Martínez O, Zagal E. Combined application of microbial consortium and humic substances to improve the growth performance of blueberry seedlings. J Soil Sci Plant Nutr. 2016;16:1010–23.
54. Magdi TA, Selim EM, El-Ghamry AM. Integrated effect of bio and mineral fertilizers and humic substances on growth, yield and nutrients contents of fertigated cowpea (*Vigna unguiculata* L.) grown on sandy soils. J Agron. 2011;10:34–9.
55. Befrozfar MR, Habibi D, Asgharzadeh A, Sadeghi-Shoae M, Tookalloo MR. Vermicompost, plant growth promoting bacteria and humic acid can affect the growth and essence of basil (*Ocimum basilicum* L.). Ann Biol Res. 2013;4:8–12.

56. Rekha PD, Lai WA, Arun AB, Young CC. Effect of free and encapsulated *Pseudomonas putida* CC-FR2-4 and *Bacillus subtilis* CC-pg104 on plant growth under gnotobiotic conditions. Bioresour Technol. 2007;98:447–51.
57. Ahmad S, Daur I, Al-Solaimani SG, Mahmood S, Bakhashwain AA, Madkour MH, Yasir M. Effect of rhizobacteria inoculation and humic acid application on canola (*Brassica Napus* L.) Crop. Pak J Bot. 2016;48:2109–20.
58. Valero N, Melgarejo LM, Ramírez R. Effect of low-rank coal inoculated with coal solubilizing bacteria on edaphic materials used in post-coal-mining land reclamation: a greenhouse trial. Chem Biol Technol Agric. 2016;3(1):1–10.
59. Rodrigues LFSO, Guimarães VF, Silva MB, Pinto Junior AS, Klein J, Costa ACPR. Características agronômicas do trigo em função de *Azospirillum brasilense*, ácidos húmicos e nitrogênio em casa de vegetação. Rev Bras Eng Agríc Amb. 2014;18:31–7.
60. Winarso S, Didik SD, Handayanto E. Effects of humic compounds and phosphate solubilizing bacteria on phosphorus availability in an acid soil. J Ecol Nat Environ. 2011;3:232–40.
61. Bacilio M, Moreno M, Bashan Y. Mitigation of negative effects of progressive soil salinity gradients by application of humic acids and inoculation with *Pseudomonas stutzeri* in a salt-tolerant and a salt-susceptible pepper. Appl Soil Ecol. 2016;107:394–404.
62. Esringü A, Kaynar D, Turan M, Ercisli S. Ameliorative effect of humic acid and plant growth-promoting rhizobacteria (PGPR) on Hungarian vetch plants under salinity stress. Commun Soil Sci Plant Anal. 2016;47:602–18.
63. Young CC, Rekha PD, Lai WA, Arun AB. Encapsulation of plant growth-promoting bacteria in alginate beads enriched with humic acid. Biotechnol Bioeng. 2006;95:73–86.
64. Struik PC, Kuyper TW, Brussaard L, Leeuwis C. Deconstructing and unpacking scientific controversies in intensification and sustainability: why the tensions in concepts and values? Curr Opin Environ Sustain. 2014;8:80–8.
65. Loos J, Abson DJ, Chappell MJ, Hanspach J, Mikulcak F, Tichit M, Fischer J. Putting meaning back into "sustainable intensification". Front Ecol Environ. 2014;2:356–61.

2

Effect of ethephon application on the cellular maturity of *Olea europaea* L. and on the extractability of phenolic compounds in virgin olive oil

Youssef Moustakime*, Zakaria Hazzoumi and Khalid Amrani Joutei

Abstract

Background: The effect of the ethephon application at fruit set and veraison on cell maturity of *Olea europaea* L. olives and on the extractability of phenolic compounds (PC) in virgin olive oil has been studied.

Methods: The ethephon was sprayed on olive trees of the Moroccan Picholine at the fruit set stage and on other olive trees at the veraison stage at a concentration of 100 mg/L. The effect of these treatments was evaluated by the fruit yield and the determination of the period of olives growth and ripening. The extractability of the olive oil and the diffusion of the PCs in the fat as well as the embrittlement of the parietal structures are also monitored.

Results and conclusions: A chemical thinning is entrained out by ethephon when applied at fruit set (16%) or veraison (12%). It also results in precocious ripening of fruits treated at fruit set (about 2 weeks) and a precocity of the harvest similar to that induced in olives treated at veraison (estimated at 6 weeks). At this stage, a significant increase in the fat accumulation (more than 26 g/100 g in the olives treated compared to 7.7 g/100 g olive pulp in the control) and quantities of extracted oils (87 and 83%, respectively, in the olives treated at fruit set and those treated at veraison compared to 70% in the control) as well as a significant improvement in the extractability of the diffusible PCs in these oils at maturity (540 and 590 mg/kg, respectively, in the treated olives at fruit set and those treated at veraison compared to 216 mg/kg in the control oil).

Keywords: *Olea europaea* L., Ethephon, Olive oil, Phenolic compounds, Ripening, Chemical thinning

Background

The ethylene (C_2H_4) physiology and biochemistry and the use of this phytohormone in agricultural practices have been deeply studied for many decades [1]. The demonstration of the ethylene biosynthetic pathway by Adams and Yang in 1971 and the recent application of molecular biology to unravel the complexities of the biosynthesis and ethylene action have stimulated much research in this field [2–5]. Endogenous or exogenous ethylene leads to numerous responses such as stimulation of pineapple flowering, accelerated maturation of climacteric fruits, abortion of flowers, and development of red spots in lettuce [6].

Moreover, numerous tests have shown the ability of ethylene-releasing chemicals to cause the precocious abscission of the olive trees' fruits [7]. Recent studies on thinning olives using ethephon (ethylene-releasing product) have shown that it can induce precocious pigmentation of drupes when pulverized just after fruit set [8]. Ethylene-releasing chemicals cause the precocious ripening of climacteric fruits, both attached and detached, but only if they are treated during the pre-climacteric period [9]. It also causes precocious ripening of non-climacteric fruits (i.e., cherries and grapes) when applied only a few days before normal ripening [10–13].

*Correspondence: youssef.moustakime@usmba.ac.ma
Laboratory of Bioactive Molecules: Structure and Function, Faculty of Science and Technology Fez, B.P. 2202-Road of Imouzzer, Fez, Morocco

In addition, several studies show that biosynthesis and the accumulation of phenolic compounds are dependent on hormonal factors: generally, plant growth stimulators (IAA, cytokinins, etc.) have a negative effect on these processes, whereas growth inhibitors (ABA, ethylene, etc.) favor them.

This work aims to study olives ripening and the extractability of phenolic compounds in virgin olive oil by carrying out a treatment with ethephon. The effects of this treatment were assessed by determining changes in physical (weight, diameter, humidity) and biochemical (anthocyanin content, oil accumulation, percentage of oil low, and its content of phenolic compounds) parameters during the ripening of olives. The embrittlement of the parietal structures of fruit estimated by changes in levels of protopectins and the endogenous enzyme activities during the ripening of olives was also determined.

Methods

Plant material

This study was carried out on non-irrigated olive trees of the Moroccan Picholine located in the botanical garden of the Faculty of Science and Technology of Fez-Morocco. These olive trees are of the same age (25 years), the same exposure to the sun, and the same type of soil. For these olive trees, fruit setting stage was estimated at the third week of June 2014.

Treatment of olive trees

Two treatments with ethephon at 100 mg/L were carried out by spraying before sunrise during the development and the growth of olives. The trees of this experiment were divided into three lots of 12 trees each: the first was treated with ethephon at fruit set; the second was treated with ethephon at the veraison; and the third and the third untreated as a control. Periodic samples were taken from the veraison to the total maturity of the olives to follow the evolution of the parameters studied.

Moisture determination of olives

The moisture content of olive drupes was determined gravimetrically according to the UNE standard Spanish method [14].

Determination of total oil content

The extraction of the total oil content is carried out using hexane in a Soxhlet extractor for 4 h from the dryer used for the determination of humidity content material [15]. After that, the traces of solvent are removed by evaporation. The determination of the oil content is determined by the weight of dry matter (% dry weight) and the weight of wet matter (% wet weight).

Extraction and determination of phenolic compounds (PC)

The protocol used for extracting phenolic compounds from olives is that of Brenes et al. [16] with some modifications. Olive pulp (10 g) was mixed with 30 mL of a solution of methanol:water 80/20 (v/v). The mixture was centrifuged for 5 min at 3500 rpm and then filtered. This extraction was repeated three times. The extracts are collected after evaporation of the organic solvent. Phenolic compounds are extracted with ethyl acetate (5 × 20 mL). The extraction of oil polyphenols is performed with methanol according to the method of Vázquez-Roncero and his collaborators [17].

The assay of total polyphenols is based on the reduction of phosphomolybdic acid of Folin–Ciocalteu agent by polyphenols in alkaline medium [18].

The assay of anthocyanins is based on the method used by Ribereau-Gayon and Stonestreet [19] implementing discoloration of flavylium cations by sulfur dioxide.

Isolation and determination of total polysaccharides

The isolation of pectins is achieved from 50 g olive pulp according to the procedure of Saulnier and Thibault [20] which implements the precipitation of insoluble material in alcohol (MIA) by several washings with ethanol 95°. From this MIA, successive extractions with water, sodium oxalate, hydrochloric acid, and soda are done.

The assay principle of the pectic substances is based on the colorimetric determination of the galacturonic acid content of the pectic chains hydrolyzed in hot acidic medium of different fractions isolated in the presence of 3-hydroxydiphenyl [21].

Extraction and measurement of endogenous enzyme activities

One gram of olive pulp was homogenized for 30 s in 25 mL potassium phosphate buffer (0.05 M, pH 6.6) containing 0.2 g of Triton X-100 using a Polytron homogenizer. 25 mg polyvinylpyrrolidone (PVPP) was added and the suspension was centrifuged at 4 °C for 15 min at 13,000 rpm. The supernatant is filtered through glass wool and used as a source of crude enzyme [22]. The various activities are determined as follows:

Polygalacturonase (PG) is assayed by the method of Somogyi [23].

The pectinesterase enzymes (PE) are assayed by the method of Baron [24].

Statistical analyses

An analysis of variance was performed for each parameter studied. The multiple comparison test averages Tukey post hoc is used to test for significant differences between treatments (at 5%). Univariate analysis was used to test for significant differences in treatment and their interaction for a single parameter. All statistical analyses were performed with IBM SPSS statistics, version 19. The results of every experiment are obtained from triplicates.

Results and discussion

Determination of the period of growth, maturation, and agronomic indexes during the ripening of the olives treated with ethephon

Influence of ethephon on the period of growth and maturation of the olives

Table 1 shows the influence of the application of ethephon on Moroccan Picholine olive trees to fruit set and veraison. It is noted that the fruits of the olive trees treated at the time of setting reached veraison 2 weeks earlier and their total maturity in a month and a half compared to those of the controls, respectively. The fruits of olives treated with ethephon at the veraison was matured earlier and reached full maturity also in a month and a half before the olives of the control, as the case of the treatment at fruit set.

Influence of ethephon on the number of fruits per branch

The exogenous application of ethephon at fruit set or at veraison on olive trees leads to a significant decrease in the number of fruits on the branches (Table 2). This reduction can reach 16% when the treatment is applied at fruit set and 12% when the treatment is applied at veraison.

Effect of ethephon on oil extractability and its polyphenol content during the olives ripening

During the olives ripening, the fat content increases in a regular manner from ripening until total maturity (Table 3). This increase follows the increase in weight and diameter and the decrease in olive moisture. Furthermore, fluctuations in the total polyphenols were observed during the ripening of olives, with their contents being increased and decreased at veraison and during the ripening, respectively, and then increased again at maturity.

The treatment of olive trees by ethephon at fruit set (Table 4) allows an accumulation of fat in olive significantly higher than that observed in the control olives and this is from veraison. This increase is concomitant with the increase in the weight and diameter of the olives so that at the total maturity of the fruit these two parameters register values higher than those of the control. The water content of the olives decreases more importantly than the control olives and records lower values from the fruit veraison. On the other hand, the phenolic compounds content in olives are quite high at veraison compared to those of the control. These levels then decrease during maturation before increasing slightly to the total maturity of the fruits, thus reaching a low concentration compared to the control.

The treatment of olive trees with ethephon veraison (Table 5) leads to an acceleration of fruit fat accumulation similar to that observed in olive fruits treated at fruit set (Table 4). At full maturity of the fruit, the accumulation of the fat reaches values significantly higher than those of the control and similar to those of the fruit set treatment. There is also an increase in the weight and diameter of the olives so that, at maturity, they have higher values than in the case of the control and the fruit set treatment. Moreover, the total polyphenol content in fruits increased at veraison and then decreased during maturation achieving a minimum recorded at total maturity. At this stage, the total polyphenol contents are less than those observed in fruits treated at fruit set and in controls.

Effect of ethephon on anthocyanin content

The accumulation of anthocyanins (Fig. 1) begins at veraison and their grades gradually increase to reach a maximum at total maturity. In fruits treated at fruit set, the appearance of anthocyanins begins 2 weeks earlier than in the controls and their contents increase regularly and rapidly to reach their maxima at total maturity with significantly higher levels than in the controls. When the treatment is applied to veraison, the anthocyanin synthesis is stimulated in an increased and accelerated manner

Table 1 Influence of ethephon on the period of growth and ripening of the olives

Treatment	Period of growth and ripening						
	June	July	August	September	October	November	December
Control							
Ethephon (N)							
Ethephon (V)							

Ethephon (N): treatment at fruit set; ethephon (V): treatment at veraison; (▪▪▪): growth period; (▬): ripening period

Table 2 Influence of ethephon applied at fruit set and at veraison on the yield of olive fruit

Treatment	Control	Ethephon (N)	Ethephon (V)
Number of fruits per branch	94.7 ± 7.4[a]	79.7 ± 2.5[b]	82.7 ± 2.1[b]
% of gain compared to control	-	− 15.6 ± 5.2[b]	− 12.2 ± 8.5[b]

[a, b] Values followed by different letters are significantly different ($P = 0.05$)

Ethephon (N): treatment at fruit set; ethephon (V): treatment at veraison

so that their contents reach very high values 40 days before those of the control. The rapid and increased synthesis observed in fruit treated at fruit set or veraison leads to a very important precocity in the olives ripening.

Influence of ethephon on the extraction of olive oil during the olives ripening

The olives accumulate the fat in their cells during maturation. This accumulation begins before veraison and reaches its maximum at the fruit's total maturity (Table 6). The treatment of olive trees at fruit setting results in a very rapid and significant accumulation of the fat in the olives in parallel with the early maturation and this from veraison. These treated olives have accumulated at maturity a 28% higher fat content compared to the control and 245% when the same sampling date is considered. Furthermore, the treatment of olives at veraison also stimulates the synthesis of the fat, thus recording values that are significantly higher than those of the control and almost similar to those observed when the treatment is applied to fruit set (28% higher at maturity and 246% if the same collection date is considered).

For all treatments, the quantities of extracted oils are very low at veraison, then increase during maturation to reach their maximum values at total maturity (Table 6). The treatment of olive trees at fruit set or veraison results in a rapid and significant increase in the quantities of extracted oils. These extracted quantities are, respectively, 34 and 29% higher than those of the mature control and 324 and 307% higher if the same sampling date is taken into account. This rapid increase is the result of the early maturity induced by ethephon estimated at about 6 weeks.

Under these conditions, the extractability of the oil estimated by the percentage of flow is maximum at maturity and may even exceed that of the control. It should be noted that at the maturity of fruits treated at setting and veraison, the flow rates reached almost 87 and 83%, respectively, and only 70% in the controls.

Effect of ethephon on the total polyphenol contents of extracted oils and their diffusion percentages

The diffusion of phenolic compounds in the oil from the pulp is very small and represents only a very small amount of the totality of these molecules present in the olives (Table 7). Nevertheless, this rate decreases during maturation and goes from 1.37% to veraison 0.3% at maturity.

Table 3 Agronomic indices of control olives during ripening

Parameter	Samples (control)						
	S1 (21/09)	S2 (03/10)	S3 (14/10)	S4 (25/10)	S5 (07/11)	S6 (19/11)	S7 (05/12)
Weight of 100 olives (g)	394.1 ± 7.6	401.4 ± 5.4	433.8 ± 7.1	443.4 ± 3.5	445.5 ± 2.2	451.12 ± 2.33	455.9 ± 3.4
Diameters (mm)	17.25 ± 1.32	17.25 ± 1.04	17.68 ± 1.3	17.7 ± 2.12	17.9 ± 1.9	17.93 ± 1.78	18 ± 2.7
Moisture content (%)	64.4 ± 0.13	63.6 ± 0.2	61.9 ± 0.16	58.7 ± 0.15	57 ± 0.11	56.42 ± 0.46	56 ± 0.14
Oil content of olives (g/100 g O.P)	2.56 ± 0.35	3.64 ± 0.19	4.51 ± 0.33	7.72 ± 0.39	14.31 ± 0.43	18.03 ± 0.43	20.8 ± 0.51
T.P content of olives (mg/g O.P)	58.8 ± 1.95	69.1 ± 2.1	50.7 ± 2.42	52.4 ± 1.51	54.2 ± 1.25	66.9 ± 1.7	69.9 ± 1.4

T.P total polyphenols, *O.P* olive pulp

Table 4 Influence of ethephon at fruit set on agronomic indices of olives during ripening

Parameter	Samples ethephon (N)				
	S1 (09/09)	S2 (21/09)	S3 (03/10)	S4 (14/10)	S5 (25/10)
Weight of 100 olives (g)	386 ± 9.2	391.5 ± 7.7	457.2 ± 5.7	481.8 ± 4.9	492.8 ± 3.7
Diameters (mm)	17 ± 1.2	17.1 ± 1.5	18.4 ± 0.9	19.1 ± 1.2	18.8 ± 1.3
Moisture content (%)	58.9 ± 0.8	56.4 ± 0.1	54.3 ± 0.2	53.2 ± 0.2	48.5 ± 0.2
Oil content of olives (g/100 g O.P)	4.3 ± 0.2	4.9 ± 0.5	8.6 ± 0.3	16.6 ± 0.5	26.6 ± 0.5
T.P content of olives (mg/g O.P)	75.2 ± 1.9	69.6 ± 1	58.8 ± 2.3	53.9 ± 1.5	59.1 ± 6.5

Ethephon (N): treatment at fruit set; *T.P* total polyphenols, *O.P* olive pulp

Table 5 Influence of ethephon at veraison on agronomic indices of olives during ripening

Parameter	Samples ethephon (V)			
	S1 (21/09)	S2 (03/10)	S3 (14/10)	S4 (25/10)
Weight of 100 olives (g)	392.4 ± 4.9	455.3 ± 4.7	487.5 ± 5.4	514.9 ± 4.4
Diameters (mm)	17.7 ± 0.8	19.1 ± 1.7	19.5 ± 2.1	19.2 ± 1.5
Moisture content (%)	63.5 ± 0.1	61.7 ± 0.2	59.5 ± 0.1	51.8 ± 0.1
Oil content of olives (g/100 g O.P)	2.6 ± 0.4	7.6 ± 0.2	16.9 ± 1.2	26.7 ± 0.2
T.P content of olives (mg/g O.P)	52.7 ± 1.8	71.9 ± 1	56.7 ± 2.4	52.9 ± 1

Ethephon (V): treatment at fruit set; *T.P* total polyphenols, *O.P* olive pulp

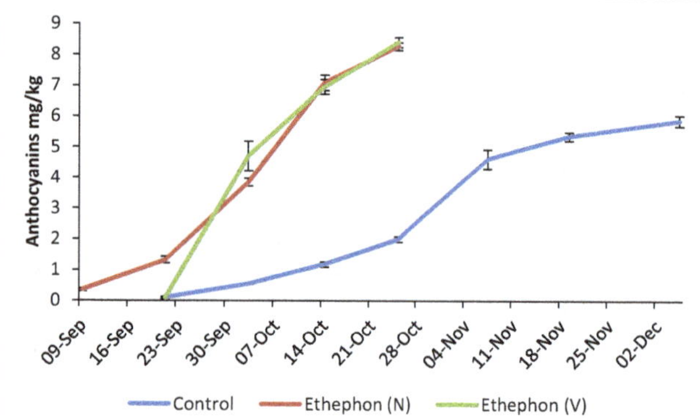

Fig. 1 Influence of ethephon on the anthocyanin contents during the olives ripening: ethephon (N) treatment at fruit set and ethephon (V) treatment at veraison

The olives treated with ethephon at fruit set show a significant increase in total polyphenol contents. On the other hand, this treatment allows to enrich the extracted oils with total polyphenols: at veraison, the oil contents in these compounds increase significantly with a gain of 65% compared to the control; these contents decrease during ripening so that at the olives that complete maturity, the extracted oils have contents of polyphenols significantly higher compared to the same stage of the control with an estimated gain of 150%.

In the case of the treatment of olives at veraison, the phenolic content increases significantly in olives a few days after the application of the treatment and then begins to decline regularly during maturation until reaching a minimum recorded at full maturity. This treatment also results in a significant increase in the diffusion of the total polyphenols in the oils at veraison, and then this diffusion gradually decreases during maturation, with minimum values at maturity which nevertheless remain very high compared with those of the control. The gain of polyphenol oils at full maturity is about 172% in comparison with the control at the same stage. This enrichment of the olive oil in total polyphenols is more important than in the case of the fruit setting treatment.

Effect of ethephon applied at fruit set or veraison on cellular embrittlement during the olives ripening

Influence of ethephon on the degradation of pectic substances constituting the cell wall during the olives ripening

At veraison, the constitutive pectins of the cell walls of the olives are mainly constituted by insoluble protopectins. Soluble pectins extracted by water and Na oxalate represent a small fraction compared to total pectins: the protopectins represent about 90% of the total pectin (Table 8). During the ripening of the drupe, the protopectin contents decrease regularly and very importantly to represent only about 25% of their initial contents at maturity. At the same time, there is a regular increase in the soluble pectin content, so that the difference between these two pectin fractions becomes very low at maturity and this is whatever the treatment used.

The application of ethephon, at fruit set or veraison, leads to a very rapid and intense degradation of protopectins (more than 80%) from veraison. After 45 days, the difference between the protopectin levels and the soluble pectin content disappears. Under these conditions, protopectins reached low levels precociously compared to the control and even a lower level than that recorded at the total maturity of the control.

Evolution of endogenous enzymatic activities of olives during maturation

It can be observed from Fig. 2 that the enzymatic activities of PE and PG generally increase during ripening: PE activities increase gradually to a maximum recorded in mid-maturation and then fall to a level lower than that observed at the veraison. PG activities decrease during veraison and then slowly increase during maturation to reach maximum at full maturity. The increase in these activities is accompanied by a decrease in protopectin levels and an increase in soluble pectin levels (Table 8).

The treatment of olive trees by ethephon at fruit setting or at veraison results in a drop in PE activities from veraison. This decrease is observed in the control olives toward the end of ripening. Similarly, PG activities increase rapidly at the beginning of ripening, as in the case of control olives at the end of maturation.

Discussion

Ethephon (2-chloroethylphosphonic acid) is a systemic plant growth regulator belonging to the phosphonate family. It is readily absorbed by the plant and releases

Table 6 Influence of ethephon on the total fat in the olives, the quantities of oil extracted, and their flow percentages during the olives ripening

Samples	Total fat (g/100 g olive pulp)			Quantities of oil extracted (g/100 g olive pulp)			% of low		
	Control	Ethephon (N)	Ethephon (V)	Control	Ethephon (N)	Ethephon (V)	Control	Ethephon (N)	Ethephon (V)
09-September	–	4.3 ± 0.2	–	–	1.3 ± 0.2	–	–	30.4 ± 3.2	–
21-September	2.56 ± 0.35[a]	4.9 ± 0.5[b]	2.6 ± 0.4[a]	0.65 ± 0.05[a]	1.72 ± 0.2[b]	0.6 ± 0.07[a]	25.8 ± 4.2[a]	35.3 ± 0.7[b]	23.2 ± 1.3[a]
03-October	3.64 ± 0.2[a]	8.6 ± 0.3[b]	7.6 ± 0.2[b]	1.28 ± 0.03[a]	4.72 ± 0.15[b]	4.22 ± 0.14[b]	35.2 ± 1.2[a]	55.1 ± 0.6[b]	55.8 ± 0.7[b]
14-October	4.51 ± 0.33[a]	16.6 ± 0.5[b]	16.9 ± 1.2[b]	2.55 ± 0.3[a]	14.15 ± 0.1[b]	13.42 ± 0.86[b]	56.5 ± 3.3[a]	85.4 ± 3.7[b]	79.4 ± 3.9[c]
25-October	7.72 ± 0.4[a]	26.6 ± 0.5[b]	26.7 ± 0.2[b]	5.43 ± 0.25[a]	23.06 ± 0.22[b]	22.1 ± 0.28[b]	70.4 ± 0.5[a]	86.8 ± 0.9[b]	82.9 ± 1.8[c]
07-November	14.31 ± 0.43	–	–	11 ± 0.2	–	–	76.9 ± 1.2	–	–
19-November	18.03 ± 0.43	–	–	14.2 ± 0.45	–	–	78.8 ± 4.2	–	–
05-December	20.8 ± 0.51	–	–	17.12 ± 0.4	–	–	82.2 ± 1.1	–	–

[a, b] Values followed by different letters are significantly different ($P = 0.05$)

Ethephon (N): treatment at fruit set, ethephon (V): treatment at veraison

Table 7 Influence of ethephon on the total polyphenol contents of the extracted oils and their diffusion rates during ripening

Sample	Total polyphenol content of the olives (mg/g olive pulp)			Total polyphenol contents diffusible in extracted oils (mg/kg pulp of olive)			% of diffusion in oils		
	Control	Ethephon (N)	Ethephon (V)	Control	Ethephon (N)	Ethephon (V)	Control	Ethephon (N)	Ethephon (V)
09-September	–	75.2 ± 1.9	–	–	1331.1 ± 13.9	–	–	1.77 ± 0.06	–
21-September	58.8 ± 2[a]	69.6 ± 1[b]	52.7 ± 1.8[a]	804.6 ± 47.9[a]	1291.9 ± 54.2[b]	801.3 ± 10.1[a]	1.37 ± 0.12[a]	1.86 ± 0.1[b]	1.52 ± 0.04[a]
03-October	69.1 ± 2.1[a]	58.8 ± 2.3[b]	71.9 ± 1[b]	685.5 ± 64[a]	1017.4 ± 48.1[b]	1059.7 ± 70.8[b]	1 ± 0.8[a]	1.73 ± 0.07[b]	1.47 ± 0.11[b]
14-October	50.7 ± 2.4[a]	53.9 ± 1.5[b]	56.7 ± 2.4[b]	573.7 ± 54.1[a]	692.2 ± 48[b]	715.7 ± 155[b]	1.14 ± 0.16[a]	1.28 ± 0.05[b]	1.26 ± 0.21[b]
25-October	52.4 ± 1.5[a]	59.1 ± 6.5[b]	52.9 ± 1[a]	446.5 ± 91.1[a]	539.2 ± 84.9[b]	589.3 ± 147.3[b]	0.85 ± 0.15[a]	0.92 ± 0.21[b]	1.11 ± 0.26[b]
07-November	54.2 ± 1.3	–	–	366.3 ± 73.3	–	–	0.67 ± 0.13	–	–
19-November	66.9 ± 1.7	–	–	306.2 ± 12.9	–	–	0.46 ± 0.02	–	–
05-December	69.9 ± 1.4	–	–	216.1 ± 74.9	–	–	0.31 ± 0.11	–	–

[a, b] Values followed by different letters are significantly different ($P = 0.05$)

Ethephon (N): treatment at fruit set; ethephon (V): treatment at veraison

ethylene which is a natural plant hormone. Ethylene influences directly several physiological processes and stimulates the production of endogenous ethylene.

The ethephon application on the olive tree of the Moroccan Picholine variety shows important differences in the growth and ripening of olives according to the treatment applied at fruit set or at veraison. Firstly, a significant influence was observed on the fruit yield of these treated olive trees compared to the control olive trees. This parameter is very important since the fruit tree load conditions the quality and the regularity of the fruit at harvest [25]. In fact, when this chemical product releasing ethylene (phytohormone growth inhibitor) was applied at fruit set, it results in a reduction in the fruit crop estimated at 16% compared to the control and at 12% when applied at veraison. It is currently known that ethephon, which decomposes in cells to produce ethylene, is one of the substances approved in several countries for its role in the chemical thinning of fruit trees [26]. In the pear Harrow Sweet variety, the use of ethephon in 2008 reduced the fruit load by approximately 29, 33, and 9% depending on the stage of application of

Table 8 Influence of ethephon on the contents of various fractions of pectin (µg galacturonic acids/g dry matter) of olives during ripening

Samples	Treatment					
	Soluble pectins (µg galacturonic acids/g dry matter)			Protopectins (µg galacturonic acids/g dry matter)		
	Control	Ethephon (N)	Ethephon (V)	Control	Ethephon (N)	Ethephon (V)
09-September	1846^{a}	1967^{a}	–	18604^{a}	14537^{b}	–
21-September	1928^{a}	2113^{a}	1614^{b}	12888^{a}	9790^{b}	16477^{c}
03-October	1771^{a}	2190^{b}	2081^{b}	11613^{a}	9165^{b}	8996^{b}
14-October	2525^{a}	2442^{a}	2452^{a}	9288^{a}	6014^{b}	5650^{b}
25-October	2934^{a}	2841^{a}	2792^{a}	7150^{a}	3106^{b}	3628^{b}
07-November	3011	–	–	6239	–	–
19-November	3198	–	–	4572	–	–
05-December	–	–	–	–	–	–

$^{a, b}$Values followed by different letters are significantly different ($P = 0.05$)

Ethephon (N): treatment at fruit set; ethephon (V): treatment at veraison

the treatment (i.e., balloon stage, late bloom, and fruit 8–12 mm, respectively) and a reduction of 14, 5, and 17% for the year 2009 [27]. Studies carried out on the peach «Reedhaven» variety [28] show that ethephon applied 30 days after full flowering (fruits 15–20 cm in diameter) results in an exponential reduction in yield estimated at 6–11% for each additional increase of 1 mg/L (between 0 and 400 mg/L). In India, a concentration of 250 mg/L of ethephon causes the thinning of 75% of the peach fruit "July Elberta" variety [29].

Very important morphological, physiological, and biochemical modifications accompany these changes in the fruit yield of olive trees under the influence of ethephon. A similar effect of ethephon on the weight, diameter, moisture, and olive fat content is observed when it is applied at fruit set or veraison: a significant increase in fruit weight and diameter with a larger total lipid accumulation. These results disagree with the work of Rugini and his collaborators [30], which showed that concentrations of ethephon (from 0 to 2000 ppm) do not result in a significant increase in the oil content of olives of the "Giarraffa" variety. Contrariwise, Taheri and his collaborators [28] found that peach weight increases linearly with high concentrations of ethephon, with a 13–15% increase in fruit size for each additional 1 mg/L ethephon increase (between 0 and 400 mg/L).

On the other hand, ethephon results in precocity of the olive harvest estimated at about 45 days compared to the control, both when the treatment is carried out at fruit set or at veraison. Nevertheless, the fruits from olive trees treated at fruit set reached veraison 2 weeks before the control olives. This advancement in fruit ripening

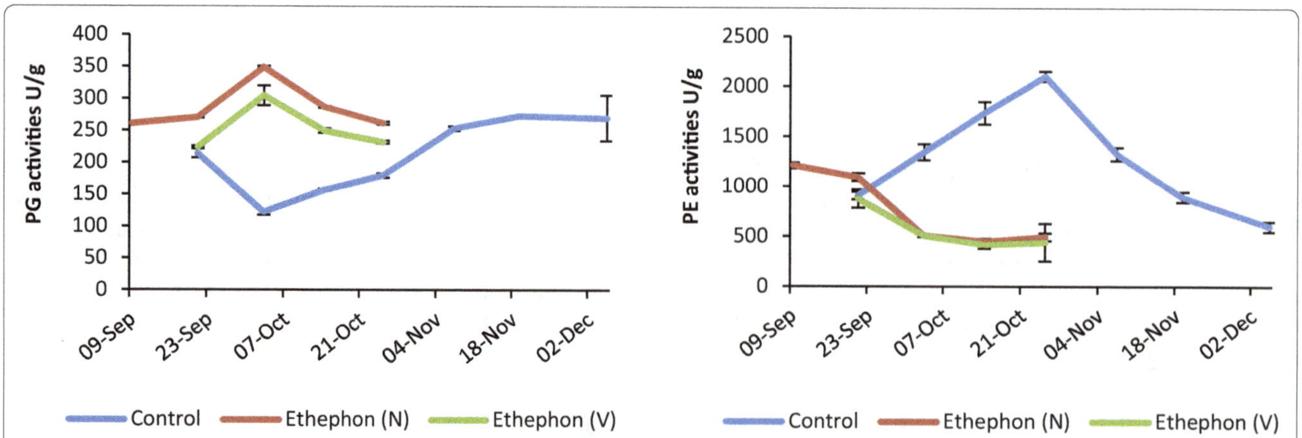

Fig. 2 Influence of ethephon on the evolution of the endogenous enzyme activities contents, the polygalacturonases (PG) and the pectinesterases activities (PE) during the olives ripening. Ethephon (N) treatment at fruit set and ethephon (V) treatment at veraison

is accompanied by a significant and rapid increase in fruit accumulation of anthocyanin. In fruit from treated olive trees at veraison, ethephon leads to an immediate and intense stimulation of the anthocyanins' synthesis which accompanies the precocious softening of the fruits. These results are in agreement with those of Rujini and his collaborators [30] who showed that ethephon applied 30 days after pit hardening at 2000 ppm causes precocity of the olives « Giarraffa » variety with a significant increase in anthocyanin content. The same authors assume that the effect of ethylene is dependent on the respiratory ratio: ethephon treatments will have a marked effect on olive ripening only when applied prior to the appearance of the maximum respiratory ratio. This is why ethylene-releasing chemical treatments, generally applied to olive trees during the pre-harvest period to facilitate fruit abscission, do not appreciably affect fruit ripening. On the other hand, some work on non-climacteric fruits (cherries, figs, and grapes) showed that ethephon causes precocious ripening when applied only a few days before normal maturation [10–13].

The extraction of olive oil is closely related to the cellular maturity state, itself related to the fragility degree of the parietal and membrane structures of the fruit which constitute a physical barrier preventing the release of the cell contents and the effective recovery of the fat. The fragility of the plant cells is estimated by the protopectin contents which ensure the rigidity of the cell walls. In our case, whatever the treatment, these protopectins decrease during the olives ripening but in a very intense and fast way in the fruit of olive trees treated both at the fruit set and at veraison. This promotes the softening of the parietal and membrane structures of the fruit, thus facilitating the fat flow, and ensures an effective recovery of the virgin olive oil. These results disagree with the work of Mínguez-Mosquera and his collaborators [31] on olives cv. Hojiblanca who observed that protopectin levels remained stable during the olives ripening. Moreover, pectin degradation is closely related to the olives levels in endogenous pectolytic activities. Ethephon influences the activity of these endogenous enzymes regardless of whether the treatment is carried out at fruit set or at veraison. These physiological events result in the more or less intense embrittlement of the parietal structures and the release of cellular contents, leading to a flow of fat estimated at 87% for the treatment at fruit set and 83% for the treatment at veraison. The control olives have not yet matured and their percentages of flow have reached only 70%.

Nevertheless, the embrittlement of the cells' parietal structures does not allow an important diffusion of the phenolic compounds in the fat during extraction; a small proportion is found in the oil and this proportion decreases during the olives ripening. The passage of these compounds is related to a modification of the phenolic profile [32]. However, the fruits of olive trees treated by ethylene show an increase in the cellular fragility from the veraison compared to the control following a rapid and intense degradation of the parietal structures. These observations are confirmed by the very important decrease in pectins and the increase in pectolytic activities, mainly the PG activities, which allow the degradation of protopectins. This phenomenon is observed as soon as the olives are treated, both at fruit set and at veraison. These results are in agreement with those of Amrani on grapes [10] and of Agravante and his collaborators [33] and Lohani and his collaborators [34] on the banana ripening. Furthermore, the study of gene expression encoding the parietal hydrolases showed that most of them are induced during ripening and regulated by ethylene [35–38]. These physiological processes allow embrittlement of the olives' cell walls, which has the consequence of facilitating the release of the phenolic compounds and the enrichment of the virgin olive oil by these compounds. Thus, the treatment of olive trees by ethylene at fruit set or veraison makes it possible to extract at total maturity an oil rich in total polyphenols with a gain of these compounds at about 150% in comparison with the phenolic charge of oils of the control olives.

Conclusion

The application effect of ethephon on cell maturity, fat extractability, and diffusion of phenolic compounds in virgin olive oil during the olives ripening of the Moroccan Picholine variety has been studied. In this study, we have shown that the treatment of olive trees by ethephon at fruit set or veraison produces chemical thinning of the fruit and very important precociousness of the harvest due to an increase in the endogenous pectolytic activities which result in a very rapid and intense decrease in the protopectin contents, thus facilitating the fat flow and a very important diffusion of the PCs in the virgin olive oil.

Abbreviations

ABA: abscisic acid; IAA: indole-3-acetic acid; MIA: insoluble material in alcohol; PC: phenolic compound; PE: pectinesterase enzymes; PG: polygalacturonase.

Authors' contributions

The author YM assured the application of the treatments and the samples during the period of ripening of olives and the analyses in the laboratory. In addition, he contributed to the writing of the manuscript. The author ZH assured the processing of the samples (extraction of olive oil, preparations of phenolic and enzymatic extracts) and contributed to the writing of the manuscript. The author KAJ assured statistical treatment of the results obtained and also contributed to the writing of the manuscript. All authors read and approved the final manuscript.

Acknowledgements
Not applicable.

Competing interests
The authors declare that they have no competing interests.

Funding
Nil.

References

1. Abeles FB, Morgan PW, Saltveit ME. Ethylene in plant biology, vol. 15. 2nd ed. San Diego: Academic Press; 1992.
2. Yang SF. Biosynthesis and action of ethylene. HortScience. 1985;20:41–5.
3. DellaPenna D, Giovannoni JJ. Regulation of gene expression in ripening tomatoes. In: Grierson D, editor. Developmental regulation of plant gene expression. Glasgow: Blackie; 1991. p. 182–216.
4. Grierson D, Schuch W. Control of ripening. In: Bevan MW, Harrison BD, Leaver CJ, editors. The production and uses of genetically transformed plants. London: Chapman and Hall; 1994. p. 53–62.
5. Kanellis AK, Chang C, Kende H, Grierson D. Biology and biotechnology of the plant hormone ethylene. Boston: Kluwer Academic Publishers; 1997.
6. Saltveit ME. Relationship between ethylene production and taste panel preference of Starkrimson red delicious apples. Can J Plant Sci. 1983;63:303–6.
7. Blumenfeld A, Epstein E, Bental Y. Ethylene treatment and abscission of olive fruits. HortScience. 1978;13:47–8.
8. Fontanazza G, Rugini E. Ensayos preliminares sobre el uso de fitoreguladores en el aclareo del olivo. Olea: Diciembre; 1979. p. 18–29.
9. Yamaguchi M, Chu CW, Yang SF. The fate of ^{14}C (2-chloroethyl) phosphonic acid in summer squash, cucumber, and tomato. J Am Soc Hort Sci. 1971;96:606–9.
10. Amrani Joutei K. Contribution à l'étude des composés phénoliques des raisins: influence des phytohormones et rôle des activités protéolytiques et pectolytiques purifiées. Thèse de Doctorat d'Etat. Faculté des Sciences et Techniques, Université Sidi Mohamed Ben Abdellah. Fès–Maroc; 2005.
11. Bukovac M, Zucconi JF, Wittenbach VA, Flore JA, Inque H. Effects of (2-chloroethyl)phosphonic acid on development and abscission of maturing sweet cherry (*Prunus avium* L.) fruit. J Am Soc Hort Sci. 1971;96:777–81.
12. Puech AA, Rebeiz CA, Crane JC. The effects of 2-chloroethyl phosphonic acid on pigment changes in the Mission fig fruit. Plant Physiol. 1971;47:89 **(Abstr.)**.
13. Weaver RJ, Pool RM. Effect of (2-chloroethyl)-phosphonic acid (ethephon) on maturation of *Vitis vinifera* L. J Am Soc Hort Sci. 1971;96:725–7.
14. Aenor, Associación Española de Normalización y Certificación. Materias Grasas. Humedad y materias volatiles (Mètodo de la estufa de aire) (Norma UNE 55-020-73, Madrid, España); 1973.
15. Aenor, Associación Española de Normalización y Certificación. Cuerpos Grasos. Determinación del contenido en material grasa total de la aceituna. (Norma UNE 55030, Madrid, España); 1961.
16. Brenes M, Rejano L, Garcia P, Sanchez HA, Garrido A. Biochemical changes in phenolic compounds during Spanish-style green olive processing. J Agric Food Chem. 1995;43:2702–6.
17. Vázquez-Roncero A, Janer del Valle C, Janer del Valle ML. Determinación de polifenoles totales del aceite de oliva. Grasas Aceites. 1973;22:350–5.
18. Catalano L, Franco I, De Nobili M, Leita L. Polyphenols in olive mill waste waters and their depuration plant effluents: a comparison of the Folin-Ciocalteu and HPLC methods. Agrochimica. 1999;43:193–205.
19. Ribereau-Gayon P, Stonestreet E. Le dosage des anthocyanes dans le vin rouge. Bull Soc Chim. 1965;9:2649–52.
20. Saulnier L, Thibault JF. Extraction and characterization of pectic substances from pulp of grape berries. Carbohyd Polym. 1987;7:329–43.
21. Robertson GL. The fractional extraction and quantitative determination of pectic substances in grapes and musts. Am J Enol Vitic. 1979;30(3):182–6.
22. Jesús Tovar M, Paz Romero M, Girona J, Motilva MJ. -Phenylalanine ammonia-lyase activity and concentration of phenolics in developing olive (*Olea europaea* L. cv Arbequina) fruit grown under different irrigation regimes. J Sci Food Agric. 2002;82:892–8.
23. Somogyi M. Notes on sugar determination. J Biol Chem. 1952;195:19.
24. Baron A. Norme française homologuée de détermination des substances pectiques T2. I/MGS/CSM, 0314 et 0607700; 1984.
25. Baab G, Lafer G. Kernobst: Harmonisches Wachstum-optimaler Ertrag. Leopoldsdorf: Österreichischer Agrarverlag; 2005. p. 163.
26. Widmer A, Chriten D. Guide phytosanitaire pour l'arboriculture fruitière 2010/2011. Régulation de la charge par éclaircissage chimique. Revue Suisse Vitic Arboric Hortic. 2010;42(1):56–7.
27. Christen D, Chassot E, Widmer A, Gölles M. Eclaircissage chimique sur poiriers : efficacité et influence sur la production et la qualité de différentes variétés. Revue suisse Vitic Arboric Hortic. 2010;42(3):170–8.
28. Taheri A, Cline JA, Jayasankar S, Pauls PK. Ethephon-induced abscission of 'Redhaven' peach. Am J Plant Sci. 2012;3:295–301. https://doi.org/10.4236/ajps.2012.32035.
29. Sharma MR, Gautam DR. Note on ethephon for thinning peach fruit. Indian J Agric Sci. 1981;51:362–4.
30. Rugini E, Bongi G, Fontanazza G. Effects of ethephon on olive ripening. J Am Soc Hort Sci. 1982;107(5):835–8.
31. Mínguez-Mosquera I, Gallardo-Guerrero L, Roca M. Pectinesterase and polygalacturonase in changes of pectic matter in olives (cv. Hojiblanca) intended for milling. JAOCS. 2002;79(1):93–9.
32. Moustakime Y, Hazzoumi Z, Amrani Joutei K. Effect of proteolytic activities in combination with the pectolytic activities on extractability of the fat and phenolic compounds from olives. SpringerPlus. 2016;5:739.
33. Agravante JU, Matsui T, Kitagawa H. Changes in pectin methylesterase, polygalacturonase and pectic substances of ethanol- and ethylene-treated bananas during ripening. Nippon Shokuhin Kogyo Gakkaishi. 1991;38(6):527–32.
34. Lohani S, Trivedi PK, Nath P. Changes in activities of cell wall hydrolases during ethylene-induced ripening in banana: effect of 1-MCP, ABA and IAA. Postharvest Biol Technol. 2004;31:119–26.
35. Fisher RL, Binnett AB. Role of cell wall hydrolases in fruit ripening. Annu Rev Plant Physiol Plant Mol Biol. 1991;42:675–703.
36. Harriman RW, Tieman DM, Handa AK. Molecular cloning of tomato pectin methylesterase gene and its expression in Rutgers, ripening inhibitor, nonripening, and never ripe tomato fruits. Plant Physiol. 1991;97:80–7.
37. Gonzalez-Bosch C, Brummell DA, Bennett AB. Differential expression of two endo-1,4-beta-glucanase genes in pericarp and locules of wild-type and mutant tomato fruit. Plant Physiol. 1996;111:1313–9.
38. Smith DL, Starrett DA, Gross KC. A gene coding for tomato fruit beta-galactosidase II is expressed during fruit ripening. Cloning, characterization, and expression pattern. Plant Physiol. 1998;117:417–23.

3

MALDI-TOF mass spectroscopy of yeasts and filamentous fungi for research and diagnostics in the agricultural value chain

David Drissner[1] and Florian M. Freimoser[2*]

Abstract

Matrix-assisted laser desorption ionization time-of-flight mass spectrometry (MALDI-TOF MS; MALDI biotyping) has become a standard tool for the accurate, rapid, and economical identification of pathogens in the clinical diagnostics laboratory. The method is continuously being improved, and new applications for distinguishing strains, identifying metabolites or functional characteristics (e.g., antibiotic resistance), and detecting microbes directly in patient samples have been developed. Adopting these methods in other disciplines than clinical diagnostics, for example, in agriculture, food safety and quality testing, or ecology, will open up new opportunities for diagnostics and research. This review focuses on MALDI-TOF MS approaches for the identification of yeasts and filamentous fungi. In contrast to bacterial diagnostics, MALDI biotyping of fungi is more challenging and less established. We thus start by discussing the role of MALDI-TOF MS as a tool for species identification; in particular with respect to DNA-based identification methods. The review then highlights the value of custom-made reference spectra for MALDI biotyping and points out recent advancements of MALDI-TOF MS, mainly from the field of clinical diagnostics that may be adopted and used for fungal diagnostic challenges. The overview ends with a summary of MALDI-TOF MS studies of yeasts and filamentous fungi of agricultural relevance.

Keywords: Agriculture, MALDI-TOF MS, Biotyping, Diagnostics, Filamentous fungi, Yeast

Introduction

Matrix-assisted laser desorption ionization time-of-flight mass spectrometry (MALDI-TOF MS; also called MALDI biotyping) has become increasingly popular for the identification of microorganisms and their functions and is now approved for the routine identification of bacteria and fungi in clinical diagnostic laboratories [1–3]. It is the goal of this review to highlight recent technological advancements of MALDI biotyping and to illustrate applications of this tool for research in agriculture and the agricultural value chain. This outlook focuses on MALDI-TOF MS applications in mycology, because these are usually more demanding than bacterial identifications [4, 5] and will thus greatly benefit from the recent advances in bacterial MALDI-TOF MS diagnostics. In this context, it is important to note that MALDI-TOF MS is not an alternative to DNA sequence-based species identification, but rather a complementary method. In order to make use and realize the potential of MALDI-TOF MS it may thus be helpful bringing to mind the advantages and disadvantages of this method; in particular as compared to DNA-based assays and techniques.

MALDI biotyping and DNA-based identification complement each other

The identification of fungi usually involves a DNA sequence-based analysis that determines the identity of a given isolate based on sequence similarity; often taking advantage of the universal fungal barcode sequence of the nuclear ribosomal internal transcribed spacer (ITS) [6]. Second and third generation sequencing techniques now allow the sequencing of thousands or millions of barcodes, or even entire genomes, in parallel and such

*Correspondence: florian.freimoser@agroscope.admin.ch
[2] Research Division Plant Protection, Agroscope, Schloss 1, 8820 Wädenswil, Switzerland
Full list of author information is available at the end of the article

high-throughput technologies are available to virtually every laboratory. These barcode (amplicon) and metagenome sequencing methods allow identifying hundreds or thousands of species simultaneously and are thus powerful tools to describe the composition of entire microbial communities (Fig. 1a) [7–9]. However, although massive parallel DNA sequencing generates a wealth of data, often more data are generated than needed and storage of the data and extraction of the required information is challenging and often limiting [10]. In addition, preparation and quality control of the template libraries for DNA sequencing is time consuming and costly (Table 1). In specific cases, for the repeated identification of a defined number of species, it may thus be best not to generate the largest amount of data, but rather to generate exactly the amount of data needed in the fastest and most economical way. It is in this realm, where MALDI biotyping shines (Fig. 1a; Table 1).

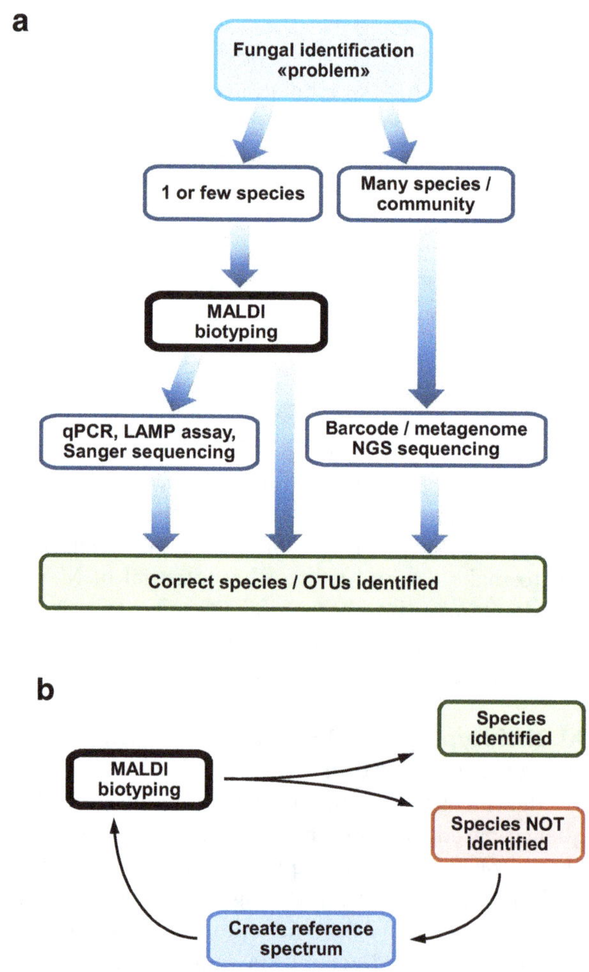

Fig. 1 MALDI biotyping and DNA-based identification complement each other. **a** MALDI biotyping is the method of choice for the identification of a defined number of species. **b** Generating MALDI-TOF MS reference spectra for fungal species identified by DNA-based methods allows the rapid and economical detection of this species by MALDI biotyping in the future

In contrast to DNA sequencing-based approaches, MALDI-TOF MS uses whole cells or crude, acidic extracts, and mass spectra for the identification of individual species [2, 11]. The simple sample preparation, short measurement times, easy preparation of reference spectra, and low costs per sample are important advantages of MALDI-TOF MS, as compared to specific, DNA-based assays that are often laborious to establish and usually require costly reagents (Table 1) [1, 12]. MALDI-TOF MS is also highly flexible and can be combined with additional extraction or processing steps in order to identify specific biomarkers, metabolites, or biochemical functions. Because of these advantages, MALDI biotyping has become a standard in routine clinical diagnostics and MALDI-TOF MS devices have become readily accessible, also for researchers in other disciplines than clinical diagnostics [4, 13, 14]. This opens up new possibilities for functional hypothesis-driven research.

Unless it is the goal to identify entire microbial communities, MALDI-TOF MS is the method of choice for species identification, because it is fast, economical, and many common bacteria and an increasing number of fungi can be identified. In those cases where MALDI-TOF MS does not reveal the identity of a particular organism, classical Sanger sequencing or DNA-based assays (e.g., qPCR or LAMP assays) are used for identification (Fig. 1a; Table 1). Once the organism is identified based on such an assay, the generation of new MALDI-TOF MS reference spectra will allow identifying this species or isolate in the future (Fig. 1b). The iterative procedure comprises DNA-based identification and MALDI-TOF MS reference spectrum generation for each new organism continually increases the number of species that can be identified and thus the benefit of a particular MALDI-TOF MS system (comprises robust software, reliable algorithms, and databases). In particular in large culturomics projects, an integration of MALDI-TOF MS in the identification pipeline results in a comprehensive database of reference spectra that allows identifying a plethora of species and even challenges, or rather complements, DNA sequencing-based analyses of microbiomes [15, 16].

Custom-made reference spectra improve MALDI-TOF MS-based identification

DNA-based identification methods benefit from large, public, DNA sequence repositories, and many openly accessible or downloadable analysis tools. In contrast, MALDI-TOF MS reference spectra, as well as analysis tools, are usually proprietary, only commercially available, and heavily focused on medical applications. Even

Table 1 MALDI-TOF- and DNA sequencing-based approaches complement each other

	MALDI biotyping	qRT-PCR/lamp assay	Sanger sequencing	NGS sequencing
Starting material	*Crude extracts or whole cells*	DNA extract from mixed sample	DNA extract from pure sample	DNA extract from community
Premises	Reference spectrum	Specific assay developed	Clone or PCR product available	*None/PCR product*
Equipment costs (machines)	High	*Low-medium*	High	High
Cost per sample	*Negligible*	Low-medium	Medium	High
Time for sample preparation	*Short*	Short-long	Long	Long
Time for analysis	*Short*	Medium	Medium	Medium-long
Data	Mass spectrum of one sample	Detection of one DNA fragment	DNA sequence of 400–1000 bp	*In total 50 Mbp–1000 Gbp*
Number of species/strains detected	Usually 1 per sample	1 per assay	1 per sample	*Up to many thousands*
Cultivation step required	Usually yes	*No*	Usually yes	*No*

Properties and comparison of MALDI biotyping and DNA sequencing-based approaches (of barcodes/PCR products or genomic DNA) methods for the identification of fungi. References: [100, 101]

The advantages of each method are shown in italics

organisms commonly found in environmental, agricultural, or food samples are thus often not recognized by standard, commercial MALDI-TOF MS systems. In addition, MALDI-TOF MS spectra are not universal, such as a DNA sequence, but depend on the crude cell extract or the physiological state of the fungal cells that are applied onto the target plate and used for recording mass spectra. Consequently, the age and growth conditions of a microbial culture, as well as the settings of the ionizing laser, flight tube, and mass detector, can influence a MALDI biotyping experiment.

Many of these problems can be overcome by generating custom-made, specific MALDI-TOF MS reference spectra for the particular application in question [5]. Since these reference spectra are generated with the same MALD-TOF MS device, settings, and sample preparation as the experimental samples, the scores obtained with these references are usually higher as compared to generic reference databases [5, 17–19]. Custom-made reference spectra thus allow ample flexibility with respect to different sample preparation protocols and experimental set-ups in a particular lab or for specific applications. For example, reference spectra for different cell densities of a particular organism may allow an estimation of cell densities [20]. It is of course also possible to generate reference spectra for fungi grown on different agar plates, different physiological states, different extraction protocols, or directly applied bacterial or yeast cells (as opposed to crude extracts).

The integration of custom-made reference spectra in a commercial MALDI-TOF MS system is only the first step towards a truly open platform for MALDI-TOF MS-based species identification. Although it is easy to generate reference spectra, a comprehensive public repository of MALDI-TOF MS reference spectra is lacking. Such a resource would greatly facilitate the exchange of reference spectra between research groups and thus allow identifying microorganisms for which no reference spectra have been generated in a particular lab [21]. Until now, only few MALDI-TOF MS spectra databases for microbial identification are publicly available. The FoodBIMS database comprises reference mass spectra of food-borne bacteria [22] (http://bioinformatica.isa.cnr.it/Descr_Bact_Dbase.htm). SpectraBank is a freely accessible database that also comprises mainly bacterial reference spectra [23], and the free Spectra site hosts an extended database of MALDI-TOF MS reference spectra for bacteria and fungi (which is made available by the Public Health Agency of Sweden (Folkhälsomyndigheten): http://spectra.folkhalsomyndigheten.se/spectra/welcome.action).

In addition to MALDI-TOF MS data repositories, instrument- and provider-independent analysis tools for microbial biotyping using mass spectra are urgently needed. SPECLUST is an application to perform cluster analyses of MALDI-TOF MS spectra and was for example used to separate the bacterium *Ralstonia solanacearum* into different species [24, 25]. More recently, Starostin et al. [26] developed a tool to use geometric distances between MALDI-TOF MS spectra represented in a multi-dimensional space to distinguish closely related *Bacillus* strains. Mass-up is a comprehensive, open-source tool for the processing and analysis of MALDI-TOF MS data [27]. Besides classification, it has also biomarker discovery, principle component analysis, and clustering functions implemented and was, for example, used to fingerprint bacterial isolates or classify peritoneal dialysis patients by mass spectrometry-based profiling [28, 29]. For baseline correction, normalization, peak detection, and matching, it uses open-source packages such as MALDIquant [30] and MassSpec Wavelet [31], which are also available as separate R packages. Finally, BIOSPEAN (http://software.cr-hana.upol.cz/biospean/login.php) is a web-based application and database that was specifically developed for analyzing whole-cell MALDI-TOF MS data and includes peak picking, generation of MS databases, and data sharing among users [32].

The public deposition of MALDI-TOF MS spectra, together with information concerning culture conditions and experimental details is highly desirable and could largely expand the potential of MALDI-TOF MS for agricultural and food diagnostics, as well as ecological research. Standardized extraction buffers (an acidic extraction using formic acid and acetonitrile seems often used) and particularly matrix solutions (e.g., α-cyano-4-hydroxycinnamic acid, HCCA) would much improve comparability of results among different laboratories. However, since different types of cells and organisms require different extraction buffers and matrices for the best MALDI-TOF MS spectra, including this information in the reference spectra and in biotyping experiments seems necessary.

Clinical MALDI-TOF MS applications benefit fungal diagnosis and research

The low sample preparation costs and fast measurement time, in addition to its accuracy, are highly attractive properties of MALDI-TOF MS; in particular for clinical applications. MALDI biotyping of clinically relevant microorganisms is thus more advanced than of other microbes and new methods are first introduced and tested with clinical samples. Consequently, the majority of MALDI-TOF analyses of fungi so far have dealt with clinical isolates [33]. In particular, the rapid and economical identification of *Candida* species is an important medical application [34–39], but filamentous fungi present in clinical samples have been studied by MALDI biotyping as well [40–43]. Overall, MALDI-TOF MS is an accurate, reliable, and rapid method for the identification of human pathogenic fungi, and new applications for clinically relevant yeasts and filamentous fungi are continuously being developed [43–45]. A comparison of two commercially available systems, VITEK MS (bioMérieux) and MALDI Biotyper (Bruker Daltonics) with their associated databases, has shown similar identification efficiencies of clinically relevant yeasts for the two systems [46, 47], while other studies found differences between biotyping systems [48, 49]. The identification power is improved by including in-house generated reference spectra in the databases [17, 43, 46] (also see above). In addition, it was shown that the sample preparation method and quality of the database are crucial for accurate identification [38]. In general, identification rates for yeasts are above 90% and higher when using acidic extracts, as compared to direct transfer of whole cells [38, 50]. In many cases, clinical diagnostics of human pathogenic bacteria and fungi drives the development of new MALDI biotyping approaches that will also benefit fungal diagnosis and research in other areas. It therefore seems worthwhile discussing new MALDI-TOF MS techniques that are mainly used in the clinical setting in order to highlight the potential and outline opportunities for MALDI-TOF MS of yeasts and filamentous fungi.

Shortening the cultivation time for faster MALDI-TOF MS identification

In clinical diagnostics, as in agricultural and food safety diagnostic, time-to-result and cost are the most important criteria for selecting analytical methods and may, in the most extreme situation, literally be a matter of life or death. The cultivation step, protein extraction, and pure cultures, in many cases required for a successful MALDI-TOF MS identification, may thus prevent this method from being used [51]. Therefore, efforts are undertaken and considerable progress has been made to address these shortcomings. For the diagnosis of bacterial pathogens responsible for bloodstream infections the mean incubation time to identify Gram-positive bacteria was 5.9 h, which dropped to 3.1 h if a crude acidic extract was prepared [52]. In other studies, 97 to 69.5% of bacteria were correctly identified after a short incubation of positive blood cultures on solid media of only 3 to 5 h [53–56], which enabled identification on the same day as the positive blood sample was detected. By increasing the sample concentration (by reducing the sample spot size), creating reference spectra for different cell densities, and immunoaffinity enrichment of bacteria, it was possible to detect as few as 10 to 100 bacterial cells after a blood culture time of only 4 h [20]. These examples illustrate the vastly shortened time-to-result of clinical MALDI-TOF MS applications: the duration of the cultivation step has been reduced to just a few hours and is thus in the same range as DNA amplification steps. In other studies of positive urine or blood samples, or of cerebrospinal fluid, the cultivation step was omitted entirely, and pathogenic bacteria and yeasts were directly identified [57–59].

MALDI-TOF MS applications beyond species identification

MALDI-TOF MS is not only used for taxonomic identification, but also able to draw functional conclusions about clinically relevant properties of a particular isolate [60]. Since MALDI-TOF MS is, in principle, able to identify any ionizable compound [61], it can detect antibiotic resistance (e.g., via the identification of specific proteins or degradation products of antibiotics), but has also been used to reveal recombinant proteins, plasmid insertions in bacteria, or other biomarkers and diagnostic peptides in bacteria [62–66].

Recently, MALDI-TOF MS has been employed to rapidly and simultaneously detect and identify the *Alternaria* mycotoxins alternariol, monomethyl ether, and tentoxin in cereal grains [67]. It may be promising adapting this method for the identification of mycotoxins in further food products in the near future. The identification of clinically relevant anaerobic bacteria has also greatly benefited from MALDI-TOF MS approaches. This slow growing and fastidious group of pathogens has been mainly diagnosed by biochemical tests, but MALDI-TOF MS approaches have successfully identified such anaerobic pathogens and could delineate antibiotic-resistant and antibiotic-susceptible strains within the same species [68–70]. In the case of pathogenic yeasts, the sibling species of the *Cryptococcus gattii/Cryptococcus neoformans* complex, which cannot be discriminated by routine biochemical techniques, have been distinguished by MALDI biotyping [71]. Similarly, MALDI-TOF MS could reliably separate closely related members of the genus *Saccharomyces* (*S. arboricola, S. bayanus, S. cariocanus, S. cerevisiae, S. kudriavzevii, S. mikatae, S. paradoxus*, and *S. pastorianus*) [72]. Finally, MALDI-TOF MS more rapidly identified reduced susceptibility to caspofungin or triazoles in *Candida* and *Aspergillus* species, as compared to classical determination of the minimum inhibitory concentration [39, 73, 74]. With yeast cells from positive blood cultures and prepared by using the Sepsityper kit (Bruker Daltonics), direct MALDI-TOF MS identified the *Candida* species in 62.5% of the samples and antifungals susceptibility results were obtained for 72.7% of the blood samples [35]. Alterations in protein profiles of fungi following the exposure to antifungal compounds could be similarly used to monitor fungicide resistance in agriculturally relevant species.

Sample processing can broaden the use of MALDI biotyping

As for the highly sensitive species identification, discussed above, sample processing (e.g., tryptic digestion, acidic/organic extraction, nano-liquid chromatography) can greatly increase the sensitivity and specificity of a MALDI-TOF MS method and thus enable subspecies identification [66, 75]. A further possibility is the functional modification of the MALDI target plates themselves, which was for example performed with antibodies for direct immunoaffinity MALDI-TOF MS of haptoglobin or by dioxide coating in order to enrich phosphopeptides [76, 77]. In order to increase the number of mass peaks that may be used as biomarkers or the sensitivity of a MALDI-TOF MS analysis, samples have been successfully treated with detergent, sonication, corona plasma discharge, or heat [78–80]. In another study, target bacteria were separated and enriched without prior cultivation by using magnetic nanobeads that were functionalized with specific antibodies [20]. A rapid sample pretreatment consisting of removal of interfering blood cells and centrifugation and washing steps, which can also be performed with a commercially available kit (Sepsityper), is crucial for obtaining high identification rates of microorganisms, including yeasts, directly in positive blood cultures [81, 82]. These examples highlight that sample pretreatments are promising tools that may also benefit applications in agricultural diagnostics.

MALDI-TOF MS of polymicrobial samples and infected material

A standard MALDI biotyping experiment requires pure cultures of the microbial species to be identified. However, the identification of microorganisms from mixed or complex samples derived, for example, from patients, the environment, or infected plant tissue or food by MALDI-TOF MS is an important goal. So far, only a few reports have demonstrated the identification of bacteria and microalgae in mixtures by using biomarkers or correlation coefficients [83–85]. These studies have shown that ion suppression can affect the detection of specific masses of one or the other microorganism and thus complicate species identification in mixed samples [83, 84, 86]. On the other hand, novel mass peaks only observed in the mixed sample were discovered and may serve as biomarkers for mixtures or contaminations [83, 86]. For samples that contain more than two organisms, species identification based on biomarkers that invariably indicate the presence of a particular species performed better than correlation-based methods [84].

Fungal MALDI biotyping as a tool for agricultural diagnostics and research

MALDI biotyping of yeasts and filamentous fungi is more difficult than bacterial identifications, because the former result in less mass peaks and fewer reference spectra are available [4, 5]. In contrast to clinically relevant fungi, only a limited number of fungal plant pathogens, postharvest diseases, or food contaminants have been detected by MALDI biotyping. The non-medical applications of MALDI-TOF MS for fungal diagnostics and research that we are aware of are summarized in Table 2. These studies document that in particular the genera *Fusarium, Trichoderma*, and *Saccharomyces* are being used as models for the development of MALDI-TOF MS applications and for assessing the potential of this technology. In contrast to clinical studies, naturally

Table 2 Overview of MALDI-TOF MS studies of agriculturally relevant fungi

Samples/organisms	Highlights	Reference
Alternaria	Detection of the *Alternaria* mycotoxins alternariol, alternariol monomethyl ether, and tentoxin by MALDI-TOF MS	Sivagnanam et al. [67]
Alternaria	Separation of *A. dauci*, *A. porri*, *A. solani*, and *A. tomatophila* into three clusters by molecular analyses and MALDI-TOF MS	Brun et al. [102]
Alternaria	Identification of 60 isolates of 12 *Alternaria* species by intact cell MALDI-TOF MS with small mycelium samples	Chowdappa et al. [103]
Aspergillus	Optimization of protein extraction for 24 *Aspergillus* species from as few as 10,000 spores and identification of 11 proteins	Sulc et al. [104]
Aspergillus	Identification of 12 *Aspergillus* strains by preparing crude extracts by bead beating	Hettick et al. [105]
Aspergillus	Characterisation of *A. ibericus* strains by MALDI-TOF MS and comparison with related species	Kallow et al. [106]
Aspergillus	Analysis of several aflatoxigenic and non-aflatoxigenic strain belonging to four *Aspergillus* species	Li et al. [107]
Aureobasidium	Analysis of extracellular liamocins (mannitol oils) produced by *A. pullulans*	Price et al. [108]
Aureobasidium	Determination of oil structures of different *A. pullulans* strains	Manitchotpisit et al. [109]
Beer spoilage microorganisms	Detection and distinction of beer spoilage yeasts and bacteria from brewing yeasts	Turvey et al. [19]
Bremia, *Oidium*	Identification of ribosomal proteins and histones as markers for the biotyping of plant pathogens	Beinhauer et al. [97]
Chalara	In vitro and in vivo identification of *C. fraxinea* by secondary metabolites collected in methanol extracts	Pham et al. [110]
Clonostachys	Cluster analysis of MALDI-TOF MS data of 45 *Clonostachys* strains from different substrates	Abreu et al. [111]
Downy and powdery mildews	Identification of the obligate biotrophic mildew fungi *Bremia lactucae* and *Oidium neolycopersici*, also from infected leaves	Chalupova et al. [96]
Fusarium	Identification and characterisation of *F. verticillioides* and fumonisins by MALDI-TOF MS and MALDI-TOF MS/MS	Chang et al. [112]
Fusarium	Differentiation of *Fusarium* subspecies based on spores collected and prepared from isolates	Marchetti-Deschmann et al. [113]
Fusarium	Optimisation of MALDI biotyping of three *Fusarium* species (16 isolates) and identification of proteins following on-target tryptic digestion	Dong et al. [114]
Fusarium	Optimized sample preparation for strongly colored *Fusarium* conidia	Dong et al. [115]
Fusarium	Mixed volume spore preparation for five *Fusarium* species	Kemptner et al. [116]
Fusarium	Differentiation of *Fusarium* species with ferulic acid as the matrix and the dried-droplet technique	Kemptner et al. [117]
Gibberella	Characterisation of *G. zeae* conidia by on-target trypsin digestion	Dong et al. [118]
Metarhizium	Reference spectra for distinguishing 51 isolates of the *M. anisopliae* species complex	Lopes et al. [119]
Monilinia	Identification of *Monilinia* brown rot fungi directly from infect fruits	Freimoser et al. [18]
Monilinia	Identification distinction of four *Monilinia* species cultivated in vitro	Horka et al. [120]
Penicillium	Discrimination of 12 *Penicillium* species based on crude extracts obtained by bead beating	Hettick et al. [121]
Penicillium	Six *Penicillium* species directly detected on citrus and apple fruits	Chen et al. [122]
Puccinia	Identification of different species and pathotypes of *P. triticina* and *P. graminis* by intact spore MALDI-TOF MS	Beinhauer et al. [123]
Rhizopus, *Trichoderma*, *Phanerochaete*	Comparison of sample preparation, matrices, and double-stick tape for collection of fungal material	Valentine et al. [124]
Saccharomyces	Fingerprinting of 33 *Saccharomyces* strains commonly used for wine fermentation	Usbeck et al. [88]
Saccharomyces	Comparison of SAPD-PCR (specifically amplified polymorphic DNA) and MALDI-TOF MS for identifying related *Saccharomyces* species	Blattel et al. [72]
Saccharomyces	Identification of yeasts involved in chichi fermentation	Vallejo et al. [98]
Saccharomyces	MALDI-TOF MS characterization of protein biomarkers desorbed from *S. cerevisiae* by formic acid	Amiri-Eliasi et al. [125]
Sepedonium	Characterisation of mycoparasitic *Sepedonium* species and analysis of low-molecular weight peptides	Neuhof et al. [126]
Spoilage yeasts	Optimization of MALDI-TOF MS assay for *Saccharomyces*, *Wickerhamomyces* and *Debaryomyces* isolated from beverages	Usbeck et al. [87]
Trichoderma	Analysis of 129 *Trichoderma* strains by MALDI-TOF MS as well as ITS and tef1 sequencing	De Respinis et al. [127]

Table 2 continued

Samples/organisms	Highlights	Reference
Trichoderma	Characterisation and clustering of *Trichoderma* strains, their peptaibiotics, and hydrophobins	Degenkolb et al. [128]
Trichoderma	Detection of peptaibols in 28 *Trichoderma* species	Neuhof et al. [129]
Trichoderma	Direct identification of hydrophobins in *Trichoderma* isolates by MALDI-TOF MS	Neuhof et al. [130]
Trichoderma, Rhizoctonia	Visualization of metabolites produced during the antagonistic interaction of *T. atroviride* and *R. solani*	Holzlechner et al. [131]
Verticillium	Identification of six pathogenic *Verticillium* isolates with a protocol involving sonication	Tao et al. [132]
Wood decay fungi	Differentiation of closely related indoor wood decay fungi by MALDI-TOF MS (*Serpula lacrymans, S. himantioides, Coniophora puteana, C. marmorata,* and *Antrodia vaillantii, A. sinuosa*)	Schmidt and Kallow [133]
Yeasts and filamentous fungi	MALDI lipid phenotyping as an alternative method for characterizing and identifying fungi	Stübiger et al. [134]
Yeasts	Identification of food-borne yeasts (≥33 species, 96 isolates) by MALDI-TOF MS and conventional methods	Pavlovic et al. [99]

occurring yeasts have been rarely studied by MALDI-TOF MS, except in the context of fermentations, for example, during winemaking and brewing [19, 72, 87, 88]. In general, the large majority of (non-clinical) fungal MALDI-TOF MS studies were directed towards phytopathogens (e.g., *Aspergillus, Fusarium, Monilinia, Penicillium, Puccinia,* mildews) or potential biocontrol strains (e.g., *Trichoderma, Metarhizium*). MALDI biotyping tools for other important soilborne or phyllosphere plant pathogens have not been established. Examples include *Armillaria* and *Thielaviopsis*, genera of worldwide root pathogens that attack hundreds of plant species [89, 90] or important oomycete pathogens such as *Pythium, Plasmodiophora,* or *Phytophthora* [91]. Also in the phyllosphere, pathogens such as *Venturia inaequalis*, the causative agent of apple scab [92], the apple blotch fungus *Diplocarpon mali* (*Marssonina coronaria*) [93, 94], and many others may be identified by MALDI biotyping and thus facilitate monitoring and studying these diseases. It is also worthwhile realizing that MALDI-TOF MS methods for identifying the majority of the fungal diseases recommended by the European and Mediterranean Plant Protection Organisation (EPPO) to be regulated as quarantine pests (List A1: https://www.eppo.int/QUARANTINE/listA1.htm, List A2: https://www.eppo.int/QUARANTINE/listA2.htm) do not exist. The lack of MALDI-TOF MS protocols for such a wide range of economically important plant pathogens is surprising and highlights avenues for future research.

As in clinical diagnostics, the need to identify fungi often goes along with the goal to identify a particular organism directly in the environment, infected plant tissue, or food, which can harbor complex microbial communities and potentially host tissue. Examples of in situ identifications of microorganisms by MALDI-TOF MS include plant invasive *Rhizobia* [95], obligate biotrophic fungal pathogens [96, 97], or *Monilinia* brown rot fungi [18]. In food science, MALDI biotyping has been mainly used for the identification of food-borne yeasts, which were pre-cultivated prior to MALDI-TOF MS analysis [87, 98, 99]. In own, unpublished work, we have developed an approach for identifying food quality and safety-relevant fungi directly from contaminated apple juices by MALDI biotyping without pre-cultivation of the fungus. MALDI-TOF MS spectra of *Byssochlamys fulva* derived from both approaches—without and with pre-cultivation of the fungus in liquid medium—revealed the same protein profiles and allowed accurate species identification (Fig. 2). These studies and results document the potential of MALDI-TOF MS to identify fungal pathogens directly in infected tissue or food.

In particular for the diagnosis of plant pathogens, food contaminants, and spoilage organisms, biomarker-based identification will permit identifying key species in complex mixtures by MALDI-TOF MS. However, so far the MALDI-TOF MS-based identification of fungal mixtures has not yet been reported.

Conclusions

Although numerous fungi have been identified by MALDI-TOF MS already, this technology is far from being a standard tool for the identification of fungi and their functions. Overall, the examples highlighted here demonstrate that the potential of MALDI-TOF MS for clinical diagnostics, and diagnostic and research applications in general, is far from being exhausted. None of the recent methodological improvements (e.g., direct smear after short cultivation times, sample concentration, immunoaffinity enrichment, subspecies identification, functionalization of MALDI target plates, biomarker-based identification in polymicrobial samples) is inherently limited to clinical samples or only bacteria. Translating these improvements and developments to other areas of diagnostics (e.g., of plant pathogens or

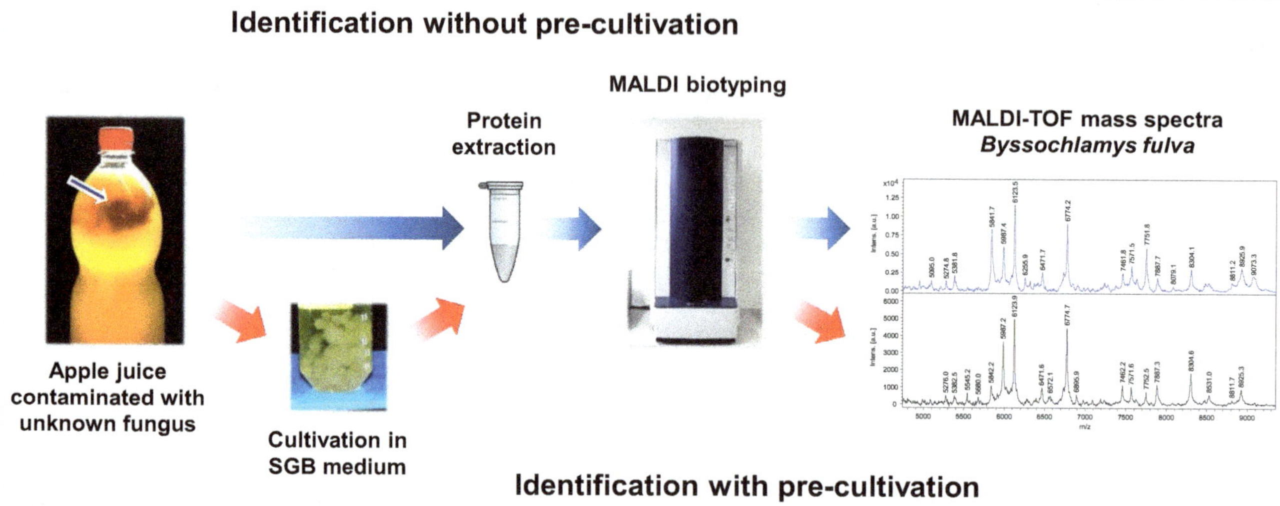

Fig. 2 Direct identification of *Byssochlamys fulva* in contaminated apple juice by MALDI-TOF MS

food spoilages) and to other microorganisms, such as filamentous fungi and yeasts, will thus enable experimental research in bacterial and fungal diagnostics, ecology, physiology, or pathology.

Authors' contributions

Both authors planned and wrote the manuscript together. Both authors read and approved the final manuscript.

Author details

[1] Research Group Microbiology of Plant Foods, Agroscope, Schloss 1, 8820 Wädenswil, Switzerland. [2] Research Division Plant Protection, Agroscope, Schloss 1, 8820 Wädenswil, Switzerland.

Acknowledgements

We are indebted to René Brunisholz (Functional Genomics Center Zurich, FGCZ) for valuable scientific discussions and help with MALDI-TOF MS analysis of *Byssochlamys fulva*.

Competing interests

The authors declare that they have no competing interests.

Funding

The research of both authors is funded by Agroscope.

References

1. Singhal N, Kumar M, Kanaujia PK, Virdi JS. MALDI-TOF mass spectrometry: an emerging technology for microbial identification and diagnosis. Front Microbiol. 2015;6:791.
2. Patel R. MALDI-TOF MS for the diagnosis of infectious diseases. Clin Chem. 2015;61(1):100–11.
3. Posteraro B, Efremov L, Leoncini E, Amore R, Posteraro P, Ricciardi W, et al. Are the conventional commercial yeast identification methods still helpful in the era of new clinical microbiology diagnostics? A meta-analysis of their accuracy. J Clin Microbiol. 2015;53(8):2439–50.
4. Chalupova J, Raus M, Sedlarova M, Sebela M. Identification of fungal microorganisms by MALDI-TOF mass spectrometry. Biotechnol Adv. 2014;32(1):230–41.
5. Rahi P, Prakash O, Shouche YS. Matrix-assisted laser desorption/ionization time-of-flight mass-spectrometry (MALDI-TOF MS) based microbial identifications: challenges and scopes for microbial ecologists. Front Microbiol. 2016;7:1359.
6. Schoch CL, Seifert KA, Huhndorf S, Robert V, Spouge JL, Levesque CA, et al. Nuclear ribosomal internal transcribed spacer (ITS) region as a universal DNA barcode marker for *Fungi*. P Natl Acad Sci USA. 2012;109(16):6241–6.
7. Franzosa EA, Hsu T, Sirota-Madi A, Shafquat A, Abu-Ali G, Morgan XC, et al. Sequencing and beyond: integrating molecular 'omics' for microbial community profiling. Nat Rev Microbiol. 2015;13(6):360–72.
8. Shendure J, Ji H. Next-generation DNA sequencing. Nat Biotechnol. 2008;26(10):1135–45.
9. Niedringhaus TP, Milanova D, Kerby MB, Snyder MP, Barron AE. Landscape of next-generation sequencing technologies. Anal Chem. 2011;83(12):4327–41.
10. Muir P, Li S, Lou S, Wang D, Spakowicz DJ, Salichos L, et al. The real cost of sequencing: scaling computation to keep pace with data generation. Genome Biol. 2016;17:53.
11. Drissner D, Gekenidis MT, Schlapbach R, Brunisholz R. When time-to-result matters: identification of microbes based on MALDI-TOF protein and peptide profiling. Chimia. 2014;68(6):453.
12. Deng J, Fu L, Wang R, Yu N, Ding X, Jiang L, et al. Comparison of MALDI-TOF MS, gene sequencing and the Vitek 2 for identification of seventy-three clinical isolates of enteropathogens. J Thorac Dis. 2014;6(5):539–44.
13. Wenning M, Breitenwieser F, Konrad R, Huber I, Busch U, Scherer S. Identification and differentiation of food-related bacteria: a comparison of FTIR spectroscopy and MALDI-TOF mass spectrometry. J Microbiol Methods. 2014;103:44–52.
14. Quintela-Baluja M, Böhme K, Fernandez-No IC, Morandi S, Alnakip ME, Caamano-Antelo S, et al. Characterization of different food-isolated *Enterococcus* strains by MALDI-TOF mass fingerprinting. Electrophoresis. 2013;34(15):2240–50.
15. Lagier JC, Armougom F, Million M, Hugon P, Pagnier I, Robert C, et al. Microbial culturomics: paradigm shift in the human gut microbiome study. Clin Microbiol Infect. 2012;18(12):1185–93.

16. Lagier JC, Hugon P, Khelaifia S, Fournier PE, La Scola B, Raoult D. The rebirth of culture in microbiology through the example of culturomics to study human gut microbiota. Clin Microbiol Rev. 2015;28(1):237–64.
17. Becker PT, de Bel A, Martiny D, Ranque S, Piarroux R, Cassagne C, et al. Identification of filamentous fungi isolates by MALDI-TOF mass spectrometry: clinical evaluation of an extended reference spectra library. Med Mycol. 2014;52(8):826–34.
18. Freimoser FM, Hilber-Bodmer M, Brunisholz R, Drissner D. Direct identification of *Monilinia* brown rot fungi on infected fruits by matrix-assisted laser desorption/ionization (MALDI) mass spectrometry. Chem Biol Technol Agricult. 2016;3:7.
19. Turvey ME, Weiland F, Meneses J, Sterenberg N, Hoffmann P. Identification of beer spoilage microorganisms using the MALDI Biotyper platform. Appl Microbiol Biotechnol. 2016;100(6):2761–73.
20. Zhu Y, Qiao L, Prudent M, Bondarenko A, Gasilova N, Möller SB, et al. Sensitive and fast identification of bacteria in blood samples by immunoaffinity mass spectrometry for quick BSI diagnosis. Chem Sci. 2016;7(5):2987–95.
21. Del Chierico F, Petrucca A, Vernocchi P, Bracaglia G, Fiscarelli E, Bernaschi P, et al. Proteomics boosts translational and clinical microbiology. J Proteomics. 2014;97:69–87.
22. Mazzeo MF, Sorrentino A, Gaita M, Cacace G, Di Stasio M, Facchiano A, et al. Matrix-assisted laser desorption ionization-time of flight mass spectrometry for the discrimination of food-borne microorganisms. Appl Environ Microbiol. 2006;72(2):1180–9.
23. Böhme K, Fernandez-No IC, Barros-Velazquez J, Gallardo JM, Canas B, Calo-Mata P. SpectraBank: an open access tool for rapid microbial identification by MALDI-TOF MS fingerprinting. Electrophoresis. 2012;33(14):2138–42.
24. Prior P, Ailloud F, Dalsing BL, Remenant B, Sanchez B, Allen C. Genomic and proteomic evidence supporting the division of the plant pathogen *Ralstonia solanacearum* into three species. BMC Genom. 2016;17:90.
25. Alm R, Johansson P, Hjerno K, Emanuelsson C, Ringner M, Hakkinen J. Detection and identification of protein isoforms using cluster analysis of MALDI-MS mass spectra. J Proteome Res. 2006;5(4):785–92.
26. Starostin KV, Demidov EA, Bryanskaya AV, Efimov VM, Rozanov AS, Peltek SE. Identification of *Bacillus* strains by MALDI TOF MS using geometric approach. Sci Rep. 2015;5:16989.
27. Lopez-Fernandez H, Santos HM, Capelo JL, Fdez-Riverola F, Glez-Pena D, Reboiro-Jato M. Mass-Up: an all-in-one open software application for MALDI-TOF mass spectrometry knowledge discovery. BMC Bioinformatics. 2015;16:318.
28. Santos T, Capelo JL, Santos HM, Oliveira I, Marinho C, Goncalves A, et al. Use of MALDI-TOF mass spectrometry fingerprinting to characterize *Enterococcus* spp. and *Escherichia coli* isolates. J Proteomics. 2015;127(Pt B):321–31.
29. Araujo JE, Jorge S, Magrico R, e Costa T, Ramos A, Reboiro-Jato M, et al. Classifying patients in peritoneal dialysis by mass spectrometry-based profiling. Talanta. 2016;152:364–70.
30. Gibb S, Strimmer K. MALDIquant: a versatile R package for the analysis of mass spectrometry data. Bioinformatics. 2012;28(17):2270–1.
31. Du P, Kibbe WA, Lin SM. Improved peak detection in mass spectrum by incorporating continuous wavelet transform-based pattern matching. Bioinformatics. 2006;22(17):2059–65.
32. Raus M, Sebela M. BIOSPEAN: a freeware tool for processing spectra from MALDI intact cell/spore mass spectrometry. J Proteomics Bioinf. 2013;6(12):283–7.
33. Rizzato C, Lombardi L, Zoppo M, Lupetti A. Pushing the limits of MALDI-TOF mass spectrometry: beyond fungal species identification. J Fungi. 2015;1:367–83.
34. De Carolis E, Vella A, Vaccaro L, Torelli R, Posteraro P, Ricciardi W, et al. Development and validation of an in-house database for matrix-assisted laser desorption ionization-time of flight mass spectrometry-based yeast identification using a fast protein extraction procedure. J Clin Microbiol. 2014;52(5):1453–8.
35. Idelevich EA, Grunewald CM, Wüllenweber J, Becker K. Rapid identification and susceptibility testing of *Candida* spp. from positive blood cultures by combination of direct MALDI-TOF mass spectrometry and direct inoculation of Vitek 2. PLoS ONE. 2014;9(12):e114834.
36. Spanu T, Posteraro B, Fiori B, D'Inzeo T, Campoli S, Ruggeri A, et al. Direct MALDI-TOF mass spectrometry assay of blood culture broths for rapid identification of *Candida* species causing bloodstream infections: an observational study in two large microbiology laboratories. J Clin Microbiol. 2012;50(1):176–9.
37. Marklein G, Josten M, Klanke U, Müller E, Horre R, Maier T, et al. Matrix-assisted laser desorption ionization-time of flight mass spectrometry for fast and reliable identification of clinical yeast isolates. J Clin Microbiol. 2009;47(9):2912–7.
38. Vlek A, Kolecka A, Khayhan K, Theelen B, Groenewald M, Boel E, et al. Interlaboratory comparison of sample preparation methods, database expansions, and cutoff values for identification of yeasts by matrix-assisted laser desorption ionization-time of flight mass spectrometry using a yeast test panel. J Clin Microbiol. 2014;52(8):3023–9.
39. Saracli MA, Fothergill AW, Sutton DA, Wiederhold NP. Detection of triazole resistance among Candida species by matrix-assisted laser desorption/ionization-time of flight mass spectrometry (MALDI-TOF MS). Med Mycol. 2015;53(7):736–42.
40. Welham KJ, Domin MA, Johnson K, Jones L, Ashton DS. Characterization of fungal spores by laser desorption/ionization time-of-flight mass spectrometry. Rapid Commun Mass Spectrom. 2000;14(5):307–10.
41. De Carolis E, Posteraro B, Lass-Florl C, Vella A, Florio AR, Torelli R, et al. Species identification of *Aspergillus*, *Fusarium* and *Mucorales* with direct surface analysis by matrix-assisted laser desorption ionization time-of-flight mass spectrometry. Clin Microbiol Infect. 2012;18(5):475–84.
42. Nenoff P, Erhard M, Simon JC, Muylowa GK, Herrmann J, Rataj W, et al. MALDI-TOF mass spectrometry—a rapid method for the identification of dermatophyte species. Med Mycol. 2013;51(1):17–24.
43. L'Ollivier C, Ranque S. MALDI-TOF-based dermatophyte identification. Mycopathologia. 2017;182(1-2):183–92.
44. Ling H, Yuan Z, Shen J, Wang Z, Xu Y. Accuracy of matrix-assisted laser desorption ionization-time of flight mass spectrometry for identification of clinical pathogenic fungi: a meta-analysis. J Clin Microbiol. 2014;52(7):2573–82.
45. Posteraro B, De Carolis E, Vella A, Sanguinetti M. MALDI-TOF mass spectrometry in the clinical mycology laboratory: identification of fungi and beyond. Expert Rev Proteomics. 2013;10(2):151–64.
46. Mancini N, De Carolis E, Infurnari L, Vella A, Clementi N, Vaccaro L, et al. Comparative evaluation of the Bruker Biotyper and Vitek MS matrix-assisted laser desorption ionization-time of flight (MALDI-TOF) mass spectrometry systems for identification of yeasts of medical importance. J Clin Microbiol. 2013;51(7):2453–7.
47. Jamal WY, Ahmad S, Khan ZU, Rotimi VO. Comparative evaluation of two matrix-assisted laser desorption/ionization time-of-flight mass spectrometry (MALDI-TOF MS) systems for the identification of clinically significant yeasts. Int J Infect Dis. 2014;26:167–70.
48. Wang H, Fan YY, Kudinha T, Xu ZP, Xiao M, Zhang L, et al. A comprehensive evaluation of the Bruker Biotyper MS and Vitek MS matrix-assisted laser desorption ionization-time of flight mass spectrometry systems for identification of yeasts, part of the National China Hospital Invasive Fungal Surveillance Net (CHIF-NET) study, 2012 to 2013. J Clin Microbiol. 2016;54(5):1376–80.
49. Cheng JW, Yu SY, Xiao M, Wang H, Kudinha T, Kong F, et al. Identification and antifungal susceptibility profile of *Candida guilliermondii* and *Candida fermentati* from a multicenter study in China. J Clin Microbiol. 2016;54(8):2187–9.
50. McTaggart LR, Lei E, Richardson SE, Hoang L, Fothergill A, Zhang SX. Rapid identification of *Cryptococcus neoformans* and *Cryptococcus gattii* by matrix-assisted laser desorption ionization-time of flight mass spectrometry. J Clin Microbiol. 2011;49(8):3050–3.
51. van Belkum A, Chatellier S, Girard V, Pincus D, Deol P, Dunne WM Jr. Progress in proteomics for clinical microbiology: MALDI-TOF MS for microbial species identification and more. Expert Rev Proteomics. 2015;12(6):595–605.
52. Idelevich EA, Schüle I, Grünastel B, Wüllenweber J, Peters G, Becker K. Rapid identification of microorganisms from positive blood cultures by MALDI-TOF mass spectrometry subsequent to very short-term incubation on solid medium. Clin Microbiol Infect. 2014;20(10):1001–6.

53. Chen Y, Porter V, Mubareka S, Kotowich L, Simor AE. Rapid identification of bacteria directly from positive blood cultures by use of a serum separator tube, smudge plate preparation, and matrix-assisted laser desorption ionization-time of flight mass spectrometry. J Clin Microbiol. 2015;53(10):3349–52.
54. Verroken A, Defourny L, Lechgar L, Magnette A, Delmee M, Glupczynski Y. Reducing time to identification of positive blood cultures with MALDI-TOF MS analysis after a 5-h subculture. Eur J Clin Microbiol Infect Dis. 2015;34(2):405–13.
55. Zabbe JB, Zanardo L, Megraud F, Bessede E. MALDI-TOF mass spectrometry for early identification of bacteria grown in blood culture bottles. J Microbiol Methods. 2015;115:45–6.
56. Kohlmann R, Hoffmann A, Geis G, Gatermann S. MALDI-TOF mass spectrometry following short incubation on a solid medium is a valuable tool for rapid pathogen identification from positive blood cultures. Int J Med Microbiol. 2015;305(4–5):469–79.
57. Ferreira L, Sanchez-Juanes F, Munoz-Bellido JL, Gonzalez-Buitrago JM. Rapid method for direct identification of bacteria in urine and blood culture samples by matrix-assisted laser desorption ionization time-of-flight mass spectrometry: intact cell vs. extraction method. Clin Microbiol Infect. 2011;17(7):1007–12.
58. Ferreira L, Sanchez-Juanes F, Porras-Guerra I, Garcia-Garcia MI, Garcia-Sanchez JE, Gonzalez-Buitrago JM, et al. Microorganisms direct identification from blood culture by matrix-assisted laser desorption/ionization time-of-flight mass spectrometry. Clin Microbiol Infect. 2011;17(4):546–51.
59. Clark AE, Kaleta EJ, Arora A, Wolk DM. Matrix-assisted laser desorption ionization-time of flight mass spectrometry: a fundamental shift in the routine practice of clinical microbiology. Clin Microbiol Rev. 2013;26(3):547–603.
60. Doern CD, Butler-Wu SM. Emerging and future applications of matrix-assisted laser desorption ionization time-of-flight (MALDI-TOF) mass spectrometry in the clinical microbiology laboratory: a report of the association for molecular pathology. J Mol Diagn. 2016;18(6):789–802.
61. Crutchfield CA, Thomas SN, Sokoll LJ, Chan DW. Advances in mass spectrometry-based clinical biomarker discovery. Clin Proteomics. 2016;13:1.
62. Niyompanich S, Srisanga K, Jaresitthikunchai J, Roytrakul S, Tungpradabkul S. Utilization of whole-cell MALDI-TOF mass spectrometry to differentiate *Burkholderia pseudomallei* wild-type and constructed mutants. PLoS ONE. 2015;10(12):e0144128.
63. Penny C, Grothendick B, Zhang L, Borror CM, Barbano D, Cornelius AJ, et al. A designed experiments approach to optimizing MALDI-TOF MS spectrum processing parameters enhances detection of antibiotic resistance in *Campylobacter jejuni*. Front Microbiol. 2016;7:818.
64. Schaumann R, Knoop N, Genzel GH, Losensky K, Rosenkranz C, Stingu CS, et al. A step towards the discrimination of beta-lactamase-producing clinical isolates of *Enterobacteriaceae* and *Pseudomonas aeruginosa* by MALDI-TOF mass spectrometry. Med Sci Monit. 2012;18(9):MT71–7.
65. Pfrunder S, Grossmann J, Hunziker P, Brunisholz R, Gekenidis MT, Drissner D. *Bacillus cereus* group-type strain-specific diagnostic peptides. J Proteome Res. 2016;15(9):3098–107.
66. Cho YT, Su H, Wu WJ, Wu DC, Hou MF, Kuo CH, et al. Biomarker characterization by MALDI-TOF/MS. Adv Clin Chem. 2015;69:209–54.
67. Sivagnanam K, Komatsu E, Rampitsch C, Perreault H, Gräfenhan T. Rapid screening of *Alternaria mycotoxins* using MALDI-TOF mass spectrometry. J Sci Food Agric. 2017;97(1):357–61.
68. Kostrzewa M, Nagy E. How MALDI-TOF mass spectrometry can aid diagnosis of hard-to-identify pathogenic bacteria. Expert Rev Mol Diagn. 2016;16(5):509–11.
69. Nagy E. Matrix-assisted laser desorption/ionization time-of-flight mass spectrometry: a new possibility for the identification and typing of anaerobic bacteria. Future Microbiol. 2014;9(2):217–33.
70. Barreau M, Pagnier I, La Scola B. Improving the identification of anaerobes in the clinical microbiology laboratory through MALDI-TOF mass spectrometry. Anaerobe. 2013;22:123–5.
71. Hagen F, Khayhan K, Theelen B, Kolecka A, Polacheck I, Sionov E, et al. Recognition of seven species in the *Cryptococcus gattii/Cryptococcus neoformans* species complex. Fungal Genet Biol. 2015;78:16–48.
72. Blattel V, Petri A, Rabenstein A, Kuever J, Konig H. Differentiation of species of the genus *Saccharomyces* using biomolecular fingerprinting methods. Appl Microbiol Biotechnol. 2013;97(10):4597–606.
73. De Carolis E, Vella A, Florio AR, Posteraro P, Perlin DS, Sanguinetti M, et al. Use of matrix-assisted laser desorption ionization-time of flight mass spectrometry for caspofungin susceptibility testing of *Candida* and *Aspergillus* species. J Clin Microbiol. 2012;50(7):2479–83.
74. Vella A, De Carolis E, Vaccaro L, Posteraro P, Perlin DS, Kostrzewa M, et al. Rapid antifungal susceptibility testing by matrix-assisted laser desorption ionization-time of flight mass spectrometry analysis. J Clin Microbiol. 2013;51(9):2964–9.
75. Gekenidis MT, Studer P, Wüthrich S, Brunisholz R, Drissner D. Beyond the matrix-assisted laser desorption ionization (MALDI) biotyping workflow: in search of microorganism-specific tryptic peptides enabling discrimination of subspecies. Appl Environ Microb. 2014;80(14):4234–41.
76. Pompach P, Novakova J, Kavan D, Benada O, Ruzicka V, Volny M, et al. Planar functionalized surfaces for direct immunoaffinity desorption/ionization mass spectrometry. Clin Chem. 2016;62(1):270–8.
77. Krasny L, Pompach P, Strohalm M, Obsilova V, Strnadova M, Novak P, et al. *In-situ* enrichment of phosphopeptides on MALDI plates modified by ambient ion landing. J Mass Spectrom. 2012;47(10):1294–302.
78. Horneffer V, Haverkamp J, Janssen HG, ter Steeg PF, Notz R. MALDI-TOF-MS analysis of bacterial spores: wet heat-treatment as a new releasing technique for biomarkers and the influence of different experimental parameters and microbiological handling. J Am Soc Mass Spectrom. 2004;15(10):1444–54.
79. Ryzhov V, Hathout Y, Fenselau C. Rapid characterization of spores of *Bacillus cereus* group bacteria by matrix-assisted laser desorption-ionization time-of-flight mass spectrometry. Appl Environ Microbiol. 2000;66(9):3828–34.
80. Nilsson CL. Fingerprinting of *Helicobacter pylori* strains by matrix-assisted laser desorption/ionization mass spectrometric analysis. Rapid Commun Mass Spectrom. 1999;13(11):1067–71.
81. Yan Y, He Y, Maier T, Quinn C, Shi G, Li H, et al. Improved identification of yeast species directly from positive blood culture media by combining Sepsityper specimen processing and Microflex analysis with the matrix-assisted laser desorption ionization Biotyper system. J Clin Microbiol. 2011;49(7):2528–32.
82. Bidart M, Bonnet I, Hennebique A, Kherraf ZE, Pelloux H, Berger F, et al. An in-house assay is superior to Sepsityper for direct matrix-assisted laser desorption ionization-time of flight (MALDI-TOF) mass spectrometry identification of yeast species in blood cultures. J Clin Microbiol. 2015;53(5):1761–4.
83. Barbano D, Diaz R, Zhang L, Sandrin T, Gerken H, Dempster T. Rapid characterization of microalgae and microalgae mixtures using matrix-assisted laser desorption ionization time-of-flight mass spectrometry (MALDI-TOF MS). PLoS ONE. 2015;10(8):e0135337.
84. Zhang L, Smart S, Sandrin TR. Biomarker- and similarity coefficient-based approaches to bacterial mixture characterization using matrix-assisted laser desorption ionization time-of-flight mass spectrometry (MALDI-TOF MS). Sci Rep. 2015;5:15834.
85. Mahe P, Arsac M, Chatellier S, Monnin V, Perrot N, Mailler S, et al. Automatic identification of mixed bacterial species fingerprints in a MALDI-TOF mass-spectrum. Bioinformatics. 2014;30(9):1280–6.
86. Holland RD, Wilkes JG, Cooper WM, Alusta P, Williams A, Pearce B, et al. Thymol treatment of bacteria prior to matrix-assisted laser desorption/ionization time-of-flight mass spectrometric analysis aids in identifying certain bacteria at the subspecies level. Rapid Commun Mass Spectrom. 2014;28(23):2617–26.
87. Usbeck JC, Kern CC, Vogel RF, Behr J. Optimization of experimental and modelling parameters for the differentiation of beverage spoiling yeasts by matrix-assisted-laser-desorption/ionization-time-of-flight mass spectrometry (MALDI-TOF MS) in response to varying growth conditions. Food Microbiol. 2013;36(2):379–87.
88. Usbeck JC, Wilde C, Bertrand D, Behr J, Vogel RF. Wine yeast typing by MALDI-TOF MS. Appl Microbiol Biotechnol. 2014;98(8):3737–52.
89. Baumgartner K, Coetzee MP, Hoffmeister D. Secrets of the subterranean pathosystem of *Armillaria*. Mol Plant Pathol. 2011;12(6):515–34.
90. Hood ME, Shew HD. Reassessment of the role of saprophytic activity in the ecology of *Thielaviopsis basicola*. Phytopathology. 1997;87(12):1214–9.

91. Fry WE, Grünwald NJ. Introduction to oomycetes. Plant Health Instr. 2010;1207:01.
92. Bowen JK, Mesarich CH, Bus VG, Beresford RM, Plummer KM, Templeton MD. *Venturia inaequalis*: the causal agent of apple scab. Mol Plant Pathol. 2011;12(2):105–22.
93. Harada Y, Sawamura K, Konno K. *Diplocarpon mali* sp. nov., the perfect state of apple blotch fungus *Marssonina coronaria*. Ann Phytopathol Soc Jpn. 1974;40(5):412–8.
94. Lee DH, Back CG, Win NK, Choi KH, Kim KM, Kang IK, et al. Biological characterization of *Marssonina coronaria* associated with apple blotch disease. Mycobiology. 2011;39(3):200–5.
95. Ziegler D, Mariotti A, Pflüger V, Saad M, Vogel G, Tonolla M, et al. *In situ* identification of plant-invasive bacteria with MALDI-TOF mass spectrometry. PLoS ONE. 2012;7(5):e37189.
96. Chalupova J, Sedlarova M, Helmel M, Rehulka P, Marchetti-Deschmann M, Allmaier G, et al. MALDI-based intact spore mass spectrometry of downy and powdery mildews. J Mass Spectrom. 2012;47(8):978–86.
97. Beinhauer J, Lenobel R, Loginov D, Chamrad I, Rehulka P, Sedlarova M, et al. Identification of *Bremia lactucae* and *Oidium neolycopersici* proteins extracted for intact spore MALDI mass spectrometric biotyping. Electrophoresis. 2016;37:2940–52.
98. Vallejo JA, Miranda P, Flores-Felix JD, Sanchez-Juanes F, Ageitos JM, Gonzalez-Buitrago JM, et al. Atypical yeasts identified as *Saccharomyces cerevisiae* by MALDI-TOF MS and gene sequencing are the main responsible of fermentation of chicha, a traditional beverage from Peru. Syst Appl Microbiol. 2013;36(8):560–4.
99. Pavlovic M, Mewes A, Maggipinto M, Schmidt W, Messelhäusser U, Balsliemke J, et al. MALDI-TOF MS based identification of food-borne yeast isolates. J Microbiol Methods. 2014;106:123–8.
100. Knief C. Analysis of plant microbe interactions in the era of next generation sequencing technologies. Front Plant Sci. 2014;5:216.
101. Rhoads A, Au KF. PacBio sequencing and its applications. Genom Proteom Bioinform. 2015;13(5):278–89.
102. Brun S, Madrid H, Gerrits Van Den Ende B, Andersen B, Marinach-Patrice C, Mazier D, et al. Multilocus phylogeny and MALDI-TOF analysis of the plant pathogenic species *Alternaria dauci* and relatives. Fungal Biol. 2013;117(1):32–40.
103. Chowdappa P, Lakshmi MJ, Madhura S. Matrix assisted laser desorption ionization-time of flight (MALDI-TOF) mass spectrometry for identification of plant pathogenic *Alternaria* species. Phytoparasitica. 2013;41(2):169–79.
104. Sulc M, Peslova K, Zabka M, Hajduch M, Havlicek V. Biomarkers of *Aspergillus* spores: strain typing and protein identification. Int J Mass Spectrom. 2009;280:162–8.
105. Hettick JM, Green BJ, Buskirk AD, Kashon ML, Slaven JE, Janotka E, et al. Discrimination of *Aspergillus* isolates at the species and strain level by matrix-assisted laser desorption/ionization time-of-flight mass spectrometry fingerprinting. Anal Biochem. 2008;380(2):276–81.
106. Kallow W, Santos I, Erhard M, Serra R, Venancio A, Lima N, editors. *Aspergillus ibericus*: a new species of section *Nigri* characterised by MALDI-TOF MS. 8th International mycological congress. Cairns: MEDIMOND; 2006.
107. Li TY, Liu BH, Chen YC. Characterization of *Aspergillus* spores by matrix-assisted laser desorption/ionization time-of-flight mass spectrometry. Rapid Commun Mass Spectrom. 2000;14(24):2393–400.
108. Price NP, Manitchotpisit P, Vermillion KE, Bowman MJ, Leathers TD. Structural characterization of novel extracellular liamocins (mannitol oils) produced by *Aureobasidium pullulans* strain NRRL 50380. Carbohydr Res. 2013;370:24–32.
109. Manitchotpisit P, Price NP, Leathers TD, Punnapayak H. Heavy oils produced by *Aureobasidium pullulans*. Biotechnol Lett. 2011;33(6):1151–7.
110. Pham TLH, Zaspel I, Schuemann M, Stephanowitz H, Krause E. Rapid *in vitro* and *in vivo* detection of *Chalara fraxinea* by means of mass spectrometric techniques. Am J Plant Sci. 2013;4(2A):444–53.
111. Abreu LM, Moreira GM, Ferreira D, Rodrigues-Filho E, Pfenning LH. Diversity of *Clonostachys* species assessed by molecular phylogenetics and MALDI-TOF mass spectrometry. Fungal Biol. 2014;118(12):1004–12.
112. Chang S, Porto Carneiro-Leao M, Ferreira de Oliveira B, Souza-Motta C, Lima N, Santos C, et al. Polyphasic approach including MALDI-TOF MS/MS analysis for identification and characterisation of *Fusarium verticillioides* in Brazilian corn kernels. Toxins (Basel). 2016;8(3):54.
113. Marchetti-Deschmann M, Winkler W, Dong HJ, Lohninger H, Kubicek CP, Allmaier G. Using spores for *Fusarium spp.* classification by MALDI-based intact cell/spore mass spectrometry. Food Technol Biotech. 2012;50(3):334–42.
114. Dong H, Marchetti-Deschmann M, Winkler W, Lohninger H, Allmaier G. Intact cell/spore mass spectrometry of *Fusarium* macro conidia for fast isolate and species differentiation. In: Banoub J, editor. Detection of biological agents for the prevention of bioterrorism. NATO Science for Peace and Security Series A: Chemistry and Biology: Springer Netherlands; 2011. p. 47–63.
115. Dong H, Kemptner J, Marchetti-Deschmann M, Kubicek CP, Allmaier G. Development of a MALDI two-layer volume sample preparation technique for analysis of colored conidia spores of *Fusarium* by MALDI linear TOF mass spectrometry. Anal Bioanal Chem. 2009;395(5):1373–83.
116. Kemptner J, Marchetti-Deschmann M, Kubicek CP, Allmaier G. Mixed volume sample preparation method for intact cell mass spectrometry of *Fusarium* spores. J Mass Spectrom. 2009;44(11):1622–4.
117. Kemptner J, Marchetti-Deschmann M, Mach R, Druzhinina IS, Kubicek CP, Allmaier G. Evaluation of matrix-assisted laser desorption/ionization (MALDI) preparation techniques for surface characterization of intact *Fusarium* spores by MALDI linear time-of-flight mass spectrometry. Rapid Commun Mass Spectrom. 2009;23(6):877–84.
118. Dong H, Marchetti-Deschmann M, Allmaier G. Characterization of on-target generated tryptic peptides from *Giberella zeae* conidia spore proteins by means of matrix-assisted laser desorption/ionization mass spectrometry. Mol Cell Probes. 2014;28(2–3):91–8.
119. Lopes RB, Faria M, Souza DA, Bloch C Jr, Silva LP, Humber RA. MALDI-TOF mass spectrometry applied to identifying species of insect-pathogenic fungi from the *Metarhizium anisopliae* complex. Mycologia. 2014;106(4):865–78.
120. Horka M, Kubesova A, Salplachta J, Zapletalova E, Horky J, Slais K. Capillary and gel electromigration techniques and MALDI-TOF MS—suitable tools for identification of filamentous fungi. Anal Chim Acta. 2012;716:155–62.
121. Hettick JM, Green BJ, Buskirk AD, Kashon ML, Slaven JE, Janotka E, et al. Discrimination of *Penicillium* isolates by matrix-assisted laser desorption/ionization time-of-flight mass spectrometry fingerprinting. Rapid Commun Mass Spectrom. 2008;22(16):2555–60.
122. Chen HY, Chen YC. Characterization of intact *Penicillium* spores by matrix-assisted laser desorption/ionization mass spectrometry. Rapid Commun Mass Spectrom. 2005;19(23):3564–8.
123. Beinhauer J, Raus M, Hanzalova A, Horcicka P, Sebela M. Intact spore MALDI-TOF mass spectrometry and proteomic analysis of *Puccinia* pathogenic fungi. Biochim Biophys Acta. 2016;1864(9):1093–103.
124. Valentine NB, Wahl JH, Kingsley MT, Wahl KL. Direct surface analysis of fungal species by matrix-assisted laser desorption/ionization mass spectrometry. Rapid Commun Mass Spectrom. 2002;16(14):1352–7.
125. Amiri-Eliasi B, Fenselau C. Characterization of protein biomarkers desorbed by MALDI from whole fungal cells. Anal Chem. 2001;73(21):5228–31.
126. Neuhof T, Berg A, Besl H, Schwecke T, Dieckmann R, von Döhren H. Peptaibol production by *Sepedonium* strains parasitizing Boletales. Chem Biodivers. 2007;4(6):1103–15.
127. De Respinis S, Vogel G, Benagli C, Tonolla M, Petrini O, Samuels GJ. MALDI-TOF MS of *Trichoderma*: a model system for the identification of microfungi. Mycol Prog. 2010;9(1):79–100.
128. Degenkolb T, Diekmann R, Nielsen KF, Gräfenhan T, Theis C, Zafari D, et al. The *Trichoderma brevicompactum* clade: a separate lineage with new species, new peptaibiotics, and mycotoxins. Mycol Prog. 2008;7:177–219.
129. Neuhof T, Dieckmann R, Druzhinina IS, Kubicek CP, von Döhren H. Intact-cell MALDI-TOF mass spectrometry analysis of peptaibol formation by the genus *Trichoderma/Hypocrea*: can molecular phylogeny of species predict peptaibol structures? Microbiology. 2007;153(Pt 10):3417–37.

130. Neuhof T, Dieckmann R, Druzhinina IS, Kubicek CP, Nakari-Setala T, von Döhren H. Direct identification of hydrophobins and their processing in *Trichoderma* using intact-cell MALDI-TOF MS. FEBS J. 2007;274(3):841–52.
131. Holzlechner M, Reitschmidt S, Gruber S, Zeilinger S, Marchetti-Deschmann M. Visualizing fungal metabolites during mycoparasitic interaction by MALDI mass spectrometry imaging. Proteomics. 2016;16(11–12):1742–6.
132. Tao J, Zhang G, Jiang Z, Cheng Y, Feng J, Chen Z. Detection of pathogenic *Verticillium* spp. using matrix-assisted laser desorption/ionization time-of-flight mass spectrometry. Rapid Commun Mass Spectrom. 2009;23(23):3647–54.
133. Schmidt O, Kallow W. Differentiation of indoor wood decay fungi with MALDI-TOF mass spectrometry. Holzforschung. 2005;59:374–7.
134. Stubiger G, Wuczkowski M, Mancera L, Lopandic K, Sterflinger K, Belgacem O. Characterization of yeasts and filamentous fungi using MALDI lipid phenotyping. J Microbiol Methods. 2016;130:27–37.

4

Water-extractable organic matter and its fluorescence fractions in response to minimum tillage and organic farming in a Cambisol

Han Yin Sun[1,2*], Philipp Koal[3], Georg Gerl[3], Reiner Schroll[2], Rainer Georg Joergensen[4] and Jean Charles Munch[5*]

Abstract

Background: Minimum tillage (MT) and organic farming (OF) are increasingly conducted in agricultural managements from the interest of optimizing soil conditions and developing sustainable agriculture. Our understanding of their effects on water-extractable organic matter (WEOM) is still insufficient.

Methods: To study the effects of MT and OF on WEOM, we analyzed soil materials sampled at two depths (0–8-cm-upper soil and 12–25-cm-deeper soil) from long-term field experiments using different farming and tillage methods. The content, composition, and quality of WEOM were examined.

Results: The results showed organic farming significantly decreased water-extractable organic carbon and nitrogen, but had positive effect on WEOM humic-like components revealed by parallel factor analysis with excitation–emission matrix, soil organic carbon (SOC), total nitrogen (TN), as well as SOC/TN. In addition, organic farming increased the aromaticity and condensation of WEOM as indicated by specific UV absorption and humification index. MT had no effect on WEOM both quantitatively and qualitatively but significantly decreased SOC and TN of the whole investigated soil profile. The depth effect was significant with strong stratification of WEOM, WEOM components as well as SOC and total N in upper soil. Moreover, the WEOM spectroscopic quality showed sharp differences between the upper and deeper soils.

Conclusions: The results indicated that in the combined presence both tillage management and farming management, farming management imposed more influence on WEOM than tillage, and organic farming may facilitate the transformation of WEOM and lead to formation of WEOM with high stability. MT significantly changed the distribution of SOC and WEOM in soil, profile but did not increase the contents of SOC and WEOM in the site of the present study. However, the presence of larger pool of WEOM in MT + OF treatment at upper soil is likely to fuel possibly greater microbial activity and more rapid nutrient cycling in soil which can be favorable practice with potential in improving soil conditions in view of developing a sustainable ecosystem in the studied site

Keywords: Agricultural soil management, Water-extractable organic matter, Minimum tillage, Organic farming, Excitation–emission matrix spectroscopy, Parallel factor analysis

*Correspondence: shyrcy@126.com; munch@tum.de
[1] Chair of Soil Ecology, Technical University of Munich, Ingolstädter Landstr. 1, Neuherberg, 85764 Munich, Germany
[5] Chair of Grassland, Technical University of Munich, WZW, Alte Akademie 12, 85354 Freising, Germany
Full list of author information is available at the end of the article

Background

Agriculture is a potential source of negative impacts on ecosystems. Tillage is basically used to prepare soil for plant growth, while organic farming, as a holistic management practice in agriculture, is conducted for achieving the objectives of protecting human health, conserving natural resources, and preserving the quality of environment while being economically sustainable [1]. Minimum tillage and organic farming practices are reported as having a potential of retaining more soil organic matter and enhancing microbial biomass and activities [2–5]. Water-extractable organic matter (WEOM) is the soluble fraction of organic matter extracted from the soil under various laboratory conditions, the most active and mobile fraction of soil organic matter (SOM). Although it is only a small part of soil organic matter, it has a strong influence on several ecologically relevant processes in soil [6–8]. It is a potential carbon source for soil microorganisms, modulating soil microbial community via the changes of quality and quantity of carbon compounds and their bioavailability [9, 10].

Land-use changes and management activities influence the dynamics of WEOM [6]. Reduced tillage and organic farming as alternative practices of conventional tillage and farming were reported to increase WEOC content by reducing soil disturbance and the addition of organic matter [11–14], while negative or neutral effect also existed [15–18]. However, most of the previous studies focused less on the quality of WEOM. Actually, the concentration, composition, and structural complexity of WEOM affect soil chemical reactions; soil physicochemical and physical degradation in soil ecosystem; and WEOM bioavailability and biodegradability [19, 20]. The characterization of both concentration and composition will contribute to a better understanding of processes in soil and function of soil WEOM in agricultural system. Therefore, the study of WEOM can be of paramount importance to understand the dynamics of soil organic carbon pool and its implication in microbial activities, carbon fluxes, and climate change.

Characterization of WEOM has been performed using specific absorbance [8], the humification index [21], and fluorescence spectroscopy [22]. Furthermore, the advancement in fluorescence spectroscopy enables more detailed characterization on the chemical properties of fluorescing fraction of organic matter. Multidimensional fluorescence spectroscopy is a rapid, nondestructive, and highly sensitive method [7, 19]. More spectral information could be obtained from multidimensional fluorescence spectroscopy than from the traditional fluorescence approaches. The fluorescence index (FI) strongly correlated with the structural conjugation and aromaticity and was used to differentiate source of dissolved organic matter [23]. Biological/freshness index (BIX) or β/α index is an indicator of relative contribution of recently microbially produced dissolved organic matter [24, 25]. With the development of parallel factor analysis, it is also possible to decompose excitation–emission matrix (EEM) data into chemically meaningful components rather than only extract the information by peak picking method [7].

The availability of soils from long-term (21-year) field experiment in adjacent fields with the same soil formation offers an opportunity to evaluate management and tillage effects on SOM and WEOM. Further, the EEM combined with PARAFAC analysis provides a new insight into evaluating WEOM changes in response to tillage. Therefore, we hypothesized that (1) minimum tillage and organic farming increased content and changed the quality of WEOM, and (2) organic farming affects more on WEOM quality compared with minimum tillage.

Methods

Field experiment

Field experiments were carried out since 1992 at the Scheyern Research Farm (TERENO site, http://www.tereno.net) located 40 km north of Munich, Germany (48.50°N, 11.45°E). The altitude of the farm ranges between 445 and 500 masl. The mean annual precipitation and mean annual temperature are 803 mm and 7.4 °C, respectively. The central part of the research station was divided into two parts: organic and integrated, each striving for ecological and economical sustainability. Moreover, detailed studies on management-induced changes were carried out in plots subdivided into integrated and organic farming [26]. The soil types are sandy-to-loamy Cambisols, derived from tertiary sediments and partly covered by loess, and most of the soils have loamy texture [27, 28].

Plot experiments to study tillage-induced [Plow tillage (PL) and minimum tillage (MT)] and management-induced [Integrated farming (IF) and Organic farming (OF)] changes to the systems were set up in two adjacent fields, with the same soil formation and basic soil properties. The soils in both fields have a soil texture of silty loam (USDA), and the soil texture of top soil is composed of 22% sand, 58% silt, and 20% clay. The field trial was arranged in a randomized factor design with three replicate treatments for each factor. The tillage practices in the study represent two possible tillage intensities under the local soil and climatic conditions. PL as a conventional tillage system means to till the soil with moldboard plow (25–30 cm). MT in the study was cultivated soil with chisel mixing in the first 6–8 cm of soil.

For OF system, organic manure (as cattle manure 30 t ha^{-1}, dry weight) was applied instead of synthetic

N-fertilizer, and no pesticides were applied. The OF plots are 12 × 12 m in size, and the crop rotation is a seven-crop rotation: (1) Grass–clover–alfalfa (GCA) (*Lolium perenne* L. + *Trifolium pratense* L. + *Medicago sativa* L.), (2) potatoes (*Solanum tuberosum* L.) + mustard (*Sinapis alba* L.) as cover crop, (3) winter wheat (*Triticum aestivum* L.), (4) sunflower (*Helianthus annuus* L.) + GCA as cover crop, (5) GCA, (6) winter wheat, and (7) winter rye (*Secale cereale* L.) + GCA as cover crop. In IF system, nitrogen fertilization was conducted with UAN (50% urea N + 50% ammonium nitrate N) with a modified boom sprayer with tubes to conduct the solution directly to the soil surface. Pesticide and herbicide were used for pest and weed control. Fertilizer rates were fixed for the cultivated crops (135 kg N ha^{-1} for winter wheat, 105 kg N ha^{-1} for maize, and 100 kg N ha^{-1} for potato), irrespective of credits due to effects of previous crop cultivation or mineralization in the soil. The IF plots are 12 × 12 m in size, and the crop rotation was (1) potatoes + mustard as catch crop, (2) winter wheat, (3) maize (*Zea mays* L.) + mustard as catch crop, and (4) winter wheat.

Therefore, there are four treatments in the present study in total: (1) organic farming + plow tillage (OPL); (2) organic farming + minimum tillage (OMT); (3) integrated farming + plow tillage (IPL); (4) integrated farming + minimum tillage (IMT). The assignments of the treatments to the plots were kept constant since 1992.

Sampling and soil properties

Three replicates of composite samples (each a mixture from 5 cores taken randomly) of top soil were collected from plowing tillage and minimum tillage plots at two soil depths (0–8-cm-upper soil and 12–25-cm-deeper soil) in integrated and organic-farming systems in April 2013. Soil organic carbon (SOC) and total nitrogen were determined using a CN analyzer (EA3000 Eurovector) with an aliquot air-dried soil samples. Soil texture was determined by wet sieving and pipette method [29]. In brief, sand and silt fractions ≥20 μm were measured by sieving after treatment with H_2O_2, silt, and clay <20 μm and using the pipette procedure. The soil pH was measured in a soil suspension with 0.01 M $CaCl_2$ solution (1:5, w/v).

WEOM extraction and spectroscopic characteristics

Soil WEOM was extracted according to the method of Zsolnay [30]. In brief, 2-mm mesh-sieved soil samples were centrifuged with 0.01 M $CaCl_2$ at a ratio of 1:2 (soil: volume) using an overhead shaker for 10 min. After 10-min centrifugation at 3000*g*, the supernatant was filtered through 0.4 μm polycarbonate membrane filter. WEON was quantified by subtracting inorganic N from soluble total N. Soluble total N, NO_3–N, and NH_4–N were quantified using an automated continuous flow analyzer (Skalar).

Specific UV absorption (SUVA), obtained by dividing the absorption at 254 nm by WEOC concentration, provides the information about the aromatic structures of WEOM. Absorption was determined using 1 cm quartz cells with a Varian Cary 50 Bio UV–Visible spectrophotometer [22].

Both pH and molecular concentrations can influence fluorescence [21, 31]. Therefore, to avoid concentration artifacts, dilution was made to have WEOC absorbance <0.1 cm^{-1} at 254 nm [21], and then WEOC extracts were acidified until reaching a constant standard pH of 2 with 2 M HCl [8, 32]. At low pH, most metal complexes disassociate, which should minimize the quenching of fluorescence due to metal complexation [23]. The pH has been used for fluorescence spectra [8, 23, 32]. PARAFAC analysis was also used to decompose the EEMs measurements at a pH range of 2–12.5 [20, 33–35]. In present study, EEMs spectra and fluorescence spectra for humification index calculation were measured separately, and a round of whole measurement procedure including pH adjustment was finished within half an hour, and no precipitation was observed in the samples. All the fluorescence measurements for humification index were performed on a Varian Cary Eclispe fluorescence spectrophotometer using 1-cm quartz cells at 254-nm excitation. The fluorescence spectra were recorded (300–480 nm with 2-nm increment) at 21 ± 1 °C. Even though the optical density was adjusted by dilution in advance, the fluorescence emission spectra were corrected for "inner-filter effect" [36] according to Zsolnay [21] with a simplified formula that measured fluorescence emission multipliers, e^A, where A is the absorbance in cm^{-1} at the 254-nm excitation wavelength [8, 37]. Humification index (HIX) indicating the complexity and condensation (H/C ratios) of WEOM was calculated as the ratio of integrated fluorescence emission peak at longer wavelength region (435–480 nm) over shorter wavelength region (300–345 nm) at the 254-nm excitation wavelength [21].

EEMs were obtained by scanning over excitation wavelengths from 250 to 450 nm with an increment of 5 nm and emission wavelength from 300 to 600 nm with an increment of 5 nm. The slit widths were set as follows: excitation slit: 10 nm, and emission slit: 20 nm. Fluorescence data were corrected for inner-filter effect using the absorption data as suggested by Lakowicz [38]:

$$I_{corr} = I_{obs} \times 10^{0.5(A_{ex}+A_{em})}$$

where I_{corr} and I_{obs} are the corrected and uncorrected fluorescence intensities; and A_{ex} and A_{em} are the absorbance values at the excitation and emission wavelengths of the fluorescence intensity values, respectively. The correction is based on the assumption that the average path length of excitation and emission lights is 50% of the cuvette width, respectively. Although the correction factors of EEMs and fluorescence spectra for humification index calculation were different, this will not affect the results in a comparison study. First, EEMs spectra and fluorescence spectra for humification index calculation were measured separately, the correction for each was done in the same way, respectively. Second, the two correction formulas are both simplified from the same original formula and based on the same theory as reported by Gauthier et al. [36].

The fluorescence index (FI), which strongly correlated with the structural conjugation and aromaticity and was used to differentiate the sources of DOM, was calculated as the ratio of emission intensities at 470 and 520 nm at the fixed excitation wavelength of 370 nm [23]. Biological/freshness index (BIX) or β/α index, an indicator of the relative contribution of the recently microbially produced DOM, was calculated as the ratio of emission intensity at 380 nm (β) to the maximum emission intensity observed between 420 and 435 nm (α) for an excitation wavelength of 310 nm [24, 25].

PARAFAC modeling

PARAFAC provides a way to decompose a dataset of EEM into individual fluorescence components [39, 40]. An EEM can be reduced to trilinear terms and a residual array by

$$X_{ijk} = \sum_{n=1}^{F} a_{in} b_{jn} c_{kn} + \varepsilon_{ijk},$$

where, for the EEM fluorescence data, X_{ijk} is the fluorescence intensity of the ith sample at the kth excitation and at the jth emission wavelength. a_{in} is directly proportional to the concentration (here defined as scores) of the nth fluorophore in the ith sample. b_{jn} and c_{kn} are the estimates of emission and excitation spectra (loadings) of the nth fluorophore at the wavelengths j and k, respectively. F is the number of components, and ε_{ijk} is the residual matrix of the model which represents unexplained variability by the model. For the full description about PARAFAC, reader can refer to Bro [39]. Components extracted by PARAFAC can be ascribed to specific species of organic matter present in liquid samples, but they are more likely to represent groups of organic compounds having similar fluorescence properties. The PARAFAC model identifies the number of components as well quantifies the scores for each component that is directly proportional to the component's (fluorophore) concentration in the sample. This concentration can be converted into actual concentration if specific absorption coefficient or quantum yields associated with excitation and emission of each fluorophore are known [41]. The concentration scores of the PARAFAC components were expressed as maximum fluorescence intensity (F_{max}) (R.U.) [42] for each modeled component in current study. F_{max} gives estimates of the relative concentrations of each component; however, direct comparison of relative concentrations between different components depends on the magnitude of their quantum efficiencies as well as on their individual responses to quenching effects [43].

Before the application of PARAFAC model, zero emission intensities were assigned for excitation wavelengths (λ_{ex}) greater than emission wavelengths (λ_{em}). Raleigh scattering was minimized according to [19, 40]. An EEM of the 0.01 M $CaCl_2$ solution was obtained and subtracted from the EEM of each sample in order to remove most of the Raman scatter peaks; any negative values produced by the subtraction were converted to missing values, and the fluorescence was normalized by dividing the integrated area (Ex 350 nm, Em 371–428 nm) under Raman scatter peak of the corresponding Milli-Q water of each set of measurement. This calibration procedure is universal, and the Raman signal of water can be used irrespective of the sample solvent and matrix, and no spectral changes will occur from applying this method [44] and the fluorescence intensities were reported in Raman units (R.U.). In total, the dataset contained 36 corrected and standardized EEMs. After an initial exploratory analysis (calculating its leverage using DOMFluor), one outlier was identified and removed from the dataset. Therefore, 35 EEMs were performed with PARAFAC analysis in the end. Split-half analysis and examination of residual error plots were applied to get the appropriate number of components. A series of PARAFAC models were generated with Matlab (2009a) by means of DOMfluor toolbox specifically developed for PARAFAC analysis of DOM fluorescence [45].

Statistics

The results presented in the tables are geometric means and expressed on dry-weight basis (for about 24 h at 105 °C). The significance of treatment effects was tested by a two-way ANOVA using tillage and farming management as independent factors and soil depth as repeated measure. ANOVA and Pearson correlation coefficients were calculated using the vegan package of R [46].

Results

Soil total organic matter and water-extractable organic matter

The contents of SOC and total N as well as the SOC/total N and WEOC/WEON were significantly higher in organic farming in comparison with integrated farming system,

Table 1 Effects of tillage and farming practices on soil organic carbon, nitrogen, and water-extractable organic matter

Main effects	SOC (mg g^{-1} soil)	Total N (mg g^{-1} soil)	SOC/total N	WEOC (μg g^{-1} soil)	WEON (μg g^{-1} soil)	WEOC/WEON	NO_3–N (μg g^{-1} soil)	NH_4–N (μg g^{-1} soil)
Organic farming	14.0	1.5	9.1	55	1.6	36.8	10.6	0.5
Integrated	11.9	1.3	8.9	62	5.0	27.4	37.9	0.6
Minimum tillage	12.7	1.4	9.0	59	3.6	32.9	25.5	0.7
Plowing	13.2	1.5	9.1	58	3.0	31.3	23.0	0.4
0–8 cm depth	15.0	1.6	9.2	65	5.3	23.3	40.3	0.7
12–25 cm	10.9	1.2	8.9	52	1.3	40.9	8.1	0.4
Probability values								
Farming	<0.01	<0.01	0.05	0.09	<0.01	0.07	<0.01	NS
Tillage	0.02	0.03	NS	NS	NS	NS	0.02	0.07
Depth	<0.01	<0.01	0.01	<0.01	<0.01	0.01	<0.01	0.09
Farming × tillage	<0.01	0.01	NS	0.09	NS	NS	NS	0.06
Farming × depth	<0.01	<0.01	0.03	NS	<0.01	<0.01	<0.01	0.04
Tillage × depth	<0.01	<0.01	0.02	<0.01	NS	<0.01	<0.01	NS
CV (±%)	2	3	2	14	17	24	9	20

The results presented in the table are geometric means of all plots within each main factor treatments

CV mean coefficient of variation between replicate plots ($n = 4$)

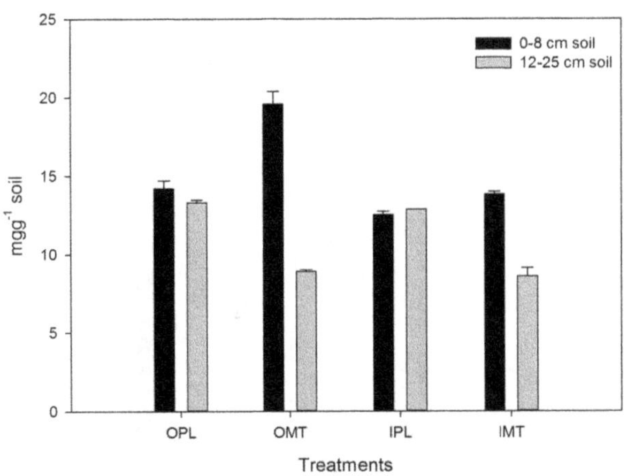

Fig. 1 Soil organic carbon contents at 0–8-cm and 12–25-cm soil depths for all treatments. *OPL* organic farming + plowing tillage, *OMT* organic farming + minimum tillage, *IPL* integrated farming + plowing tillage, *IMT* integrated farming + minimum tillage

while the contents WEOC, WEON, NO_3–N as well as NH_4–N were lower (Table 1). The contents of WEOC and WEON as well as the SOC/total N and WEOC/WEON ratios did not differ between the tillage treatments, while MT significantly decreased the contents of SOC and total N but increased those of NO_3–N and NH_4–N. The accumulation of SOC, total N, WEOC, WEON, NO_3–N and NH_4–N in the 0–8 cm layer was significant in comparison with 12–25 cm layer, usually leading to significant tillage × soil-depth interactions, except WEON and NH_4–N. The ratio of SOC/total N was significantly higher, while WEOC/WEON ratio was significantly lower at 0–8 cm in comparison with 12–25 cm. The significant interactions of farming × tillage on SOC (Fig. 1), but also those on total N, SOC/total N, WEON, NO_3–N, and NH_4–N were caused by the much stronger tillage effects in the organic farming than in the integrated system.

Spectroscopic properties of WEOM

SUVA and HIX were significantly higher, while BIX was significantly lower in organic farming in comparison with integrated farming (Table 2). The FI did not differ between farming systems. Tillage had no effect on any WEOM spectroscopic property. The SUVA values were significantly higher at 0–8 cm than at 12–25 cm, which was especially true for the organic-farming system (Fig. 2). The HIX values were significantly lower at 0–8 cm than at 12–25 cm, which was again most pronounced in the organic-farming system (Fig. 3). In contrast, the BIX values did not change with depth and farming system (Table 2).

PARAFAC components

Three EEM-PARAFAC components (C1, C2, C3), which account for 99.28% of the fluorescence variability, were extracted from the dataset (Fig. 4). The C1 component had two excitation maxima at 240 and 325 nm and one emission maxima at 435 nm, similar to the humic-like substance identified by [47–49]. The C2 component had two excitation maxima at <240 and 260 nm and one emission maxima at 360 nm, similar to the component 5 in [50–52] or peak N in [53] and C4 component in [49].

Table 2 Effects of tillage and farming practices on spectral characteristics and spectrally identified compositions of water-extractable organic matter of all treatments

Main effects	SUVA (m^{-1} mg^{-1} C l)	HIX	C1 (R.U.)	C2 (R.U.)	C3 (R.U.)	FI	BIX
Organic farming	1.18	4.61	1.90	0.94	0.75	1.60	0.74
Integrated	0.79	3.25	1.53	0.93	0.60	1.62	0.77
Minimum tillage	0.95	3.94	1.76	0.93	0.69	1.60	0.75
Plowing	1.02	3.92	1.67	0.94	0.66	1.61	0.76
0–8-cm depth	1.14	3.53	1.98	1.18	0.80	1.58	0.76
12–25 cm	0.82	4.33	1.45	0.69	0.55	1.64	0.75
Probability values							
Farming	<0.01	<0.01	0.01	NS	0.01	NS	0.06
Tillage	NS	NS	NS	NS	NS	NS	NS
Depth	<0.01	<0.01	<0.01	<0.01	<0.01	0.03	NS
Farming × tillage	0.08	NS	NS	NS	NS	NS	NS
Farming × depth	NS	NS	NS	NS	NS	NS	NS
Tillage × depth	NS	0.07	<0.01	<0.01	<0.01	0.06	0.09
CV (±%)	9	6	8	7	9	1	2

The results presented in the table are geometric means of all plots within each main factor treatments

CV mean coefficient of variation between replicate plots ($n = 3$), *SUVA* specific absorption, *HIX* humification index, *FI* fluorescence index, *BIX* freshness index

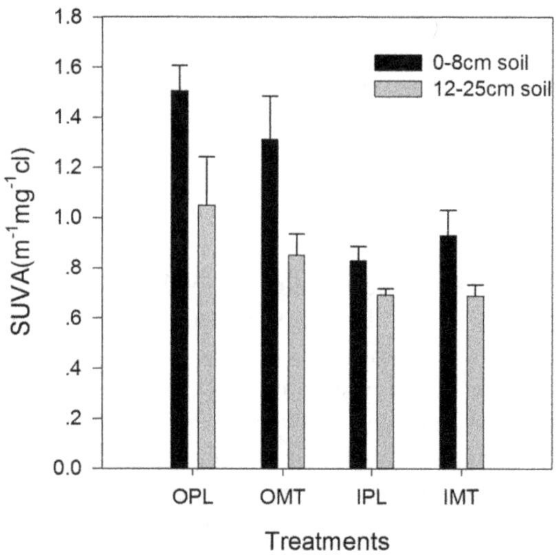

Fig. 2 Specific UV absorptions of water-extractable organic matter at 0–8-cm and 12–25-cm soil of all treatments. *OPL* organic farming + plowing tillage, *OMT* organic farming + minimum tillage, *IPL* integrated farming + plowing tillage, *IMT* integrated farming + minimum tillage

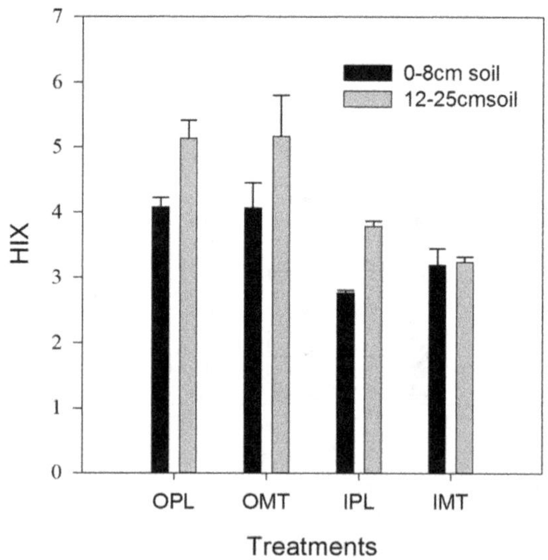

Fig. 3 Humification indexes of water-extractable organic matter at 0–8-cm and 12–25-cm soil of all treatments. *OPL* organic farming + plowing tillage, *OMT* organic farming + minimum tillage, *IPL* integrated farming + plowing tillage, *IMT* integrated farming + minimum tillage

The C2 component is characterized by tryptophan-like substance associated with biological production, and one shoulder could be found on the emission spectrum due to different fluorescence characteristics [54]. The C3 component had one excitation maxima at 265 nm and one emission maxima at 505 nm. The C3 component is similar to humic-like substance reported before [41, 49, 53, 54].

Concentrations of three components as indicated by the concentration scores followed an order of C1 > C2 > C3 (Table 2). The two humic-like substances (C1 and C3) were significantly higher in organic farming in comparison with integrated farming, while the tryptophan-like substance (C2) did not differ between farming systems. The effect of tillage on the three

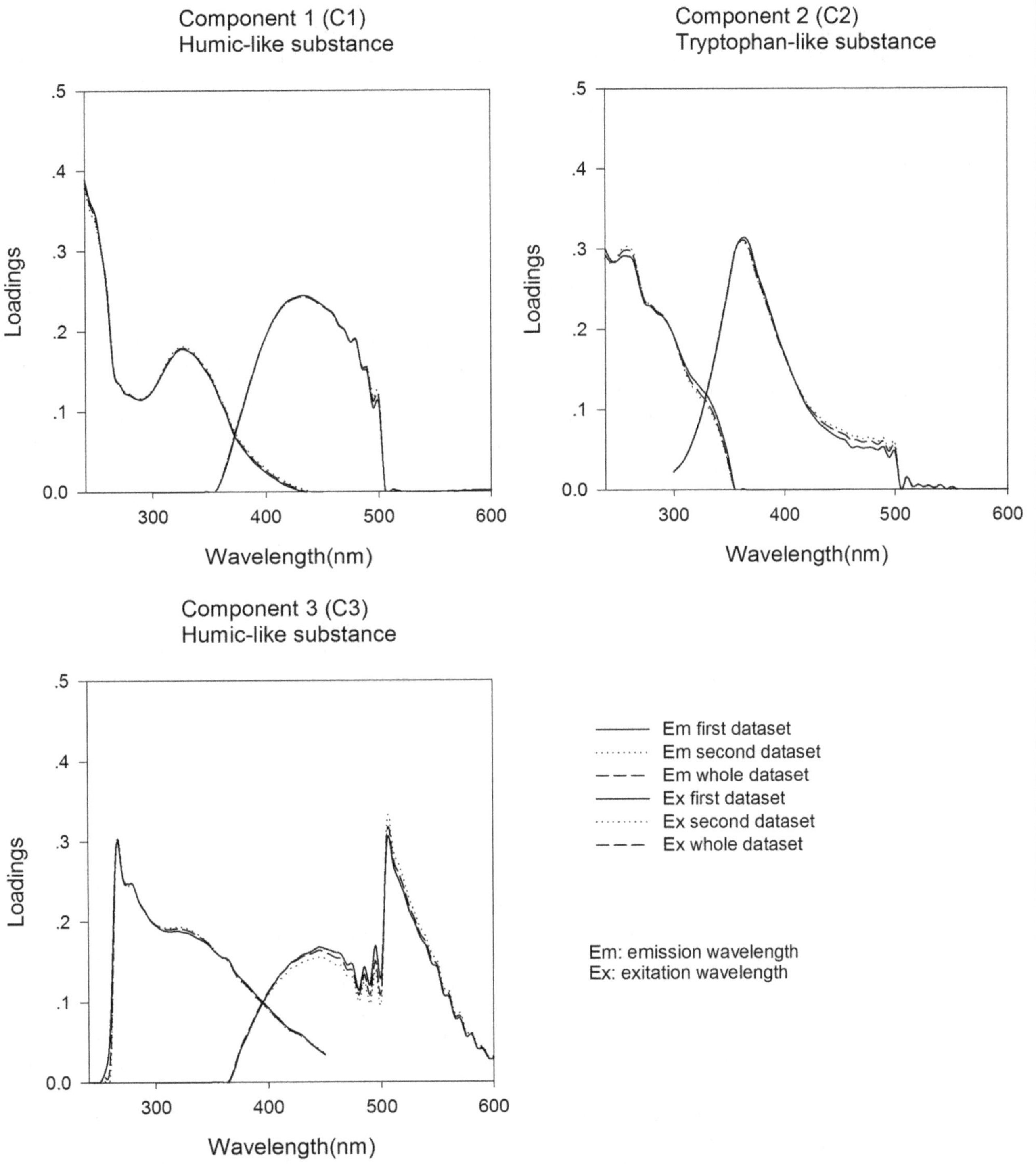

Fig. 4 Split-half analysis of the three components of WEOM extracted by PARAFAC model for all treatments

EEM-PARAFAC components was insignificant. The three EEM-PARAFAC components were significantly higher at 0–8 cm than those at 12–25 cm. All three EEM-PARAFAC components were significantly correlated to SOC and total N at both depths (Table 3), although the relationships were always closer at 12–25 cm. This was also true for the relationships to WEOC, which were already insignificant for C1 and C2 at 0–8 cm. The relationships to WEON were only significant for C1 and C2 at 12–25 cm.

Discussion

Soil organic carbon and nitrogen

In the present study, our results showed MT and organic farming significantly increased SOC and TN in upper soil (Table 1). The results agreed with the results of several previous studies [11, 12, 55]. Higher contents of SOC and TN in MT at upper soil could be attributed to the fact that MT reduces the mineralization of SOM [14] and improves the physical protection of SOM via reducing the disruption of soil aggregates and avoiding physical access of SOM by microbes [56] compared with conventional tillage. Furthermore, crop residues accumulate in the upper layer of soil, while plowing tillage incorporates crop residues into the whole tillage horizon of soil more uniformly resulting in dilution effects of C and N [57, 58]. When the two soil depths were considered, the effect of MT was negative, this being perhaps because of dilution effect in deeper soil part. The results agreed with the findings of Luo et al. [59] that no-tillage changed distribution of C in the soil profile significantly, but did not increase the total SOC, as the increase of SOC in surface soil compensates for the loss of SOC in deeper soil. In terms of organic farming, similar findings [60, 61] were reported, although some inconsistent findings occurred where no obvious improvement of SOC was observed [2, 62]. The difference between studies can be attributed to the difference in the climatic conditions of the studied sites. In the study of Parras-Alcantara et al. [62], their trial site (Los Pedroches Valley) is characterized by cold winters and warm, dry summers with temperatures ranging from −2 to 40 °C (extreme temperatures) and an average annual rainfall of 600 mm, while the climate of our study is milder with mean precipitation of 803 mm and mean temperature of 7.4 °C. Therefore, different degradations and turnovers of organic matter might occur in the two study sites. Further, their soils were sandier. This supports the hypothesis that the tillage-induced changes of soil are soil and site specific [63]. The beneficial effect of organic farming could be associated with the continuous inputs of organic matter by manure, plants, and legume catch-crop-based crop rotations [60]. It is also recognized that the addition of organic manure increases the soil water repellency, thus increasing the aggregate stability [64] and carbon sequestration [65]. In addition, the diverse microbial community under organic farming facilitates the degradation of plant materials, thus increasing the SOM from the decayed plant debris [64].

WEOM components

WEOM is believed to originate from plant roots, litter, and soil humus and is a labile substrate for microbial support and transformations [11]. In the present study, organic farming significantly decreased the contents of WEOC and MT resulting in significant accumulation of WEOC in upper soil. The decreased WEOM in organic-farming treatments could be attributed to the sampling time. Bausenwein et al. [9] reported significant effect of sampling time on WEOC, and the higher WEOC was observed in October in the same research station as ours but with samples from different fields. Our samples were collected in 2013 after a long winter when the snow ended in May. In addition, we also observed increased microbial biomass compared with integrated farming [3]. Therefore, due to more WEOC consumption than its production, the production and consumption of WEOC became unbalanced. Organic farming significantly decreased the content of WEON. One explanation could be that microbial communities in organic farming are more N-depleted, and the N fertilization in integrated farming changed the microbial community's structure thus resulting in the alteration of organic N mineralization [66]; this agreed with Embacher et al.'s report that the fertilizer itself is not the direct source of WEOM but affects WEOM via the biota [32]. The increased WEOC in upper soil under MT treatments was consistent with other findings [11, 12], although some authors also reported tillage had no significant effect at any soil depth [15]. The reason behind this might be that, on the one hand, the MT reduced the mineralization of WEOM [13], and that, on the other hand, the diverse microbial community in MT [67] increased the degradations of SOM, and the plant residues accumulated in upper soil contributed to WEOM.

Organic farming significantly increased the components 1 and 3 (humic-like substance components). This is in line with the result of Zhang et al. [16] who observed that the application of manure increased the humic-like substance identified by EEM-PARAFAC; but contrary to the result reported by Ohno et al. [68] that tillage and cropping management did not affect the humic-like substance components 1 and 5 and amino acid-like component 4. But they found N sources (NH_4NO_3 and poultry litter) significantly affected the identified components of WEOM. In the present study, organic farming caused significantly lower NO_3–N in comparison with integrated farming. Thus, the increased humic-like substance in organic-farming treatments indicated that WEOM in organic farming was highly microbial processed which can also be reflected by SUVA and HIX of WEOM (Table 2). Besides, EEM-PARAFAC components of upper soil were significantly higher than those of deeper soil. This can be explained by the accumulation and degradation of accumulated organic matter by MT in top soil which can be proved by the close association of EEM-PARAFAC components with SOC and total N (Table 3). Compared with humic-like substance, C2 as

Table 3 Correlations between PARAFAC components of WEOM and soil characteristics for all treatments at varying soil depths ($n = 12$)

	SOC	Total N	WEOC	WEON	Clay
0–8 cm					
C1	0.82[b]	0.74[b]	NS	NS	−0.72[b]
C2	0.69[a]	0.64[a]	0.60[a]	NS	−0.86[b]
C3	0.81[b]	0.73[b]	NS	NS	−0.74[b]
12–25 cm					
C1	0.94[b]	0.94[b]	0.85[b]	0.61[a]	0.68[a]
C2	0.94[b]	0.91[b]	0.95[b]	NS	0.93[b]
C3	0.95[b]	0.95[b]	0.84[b]	0.60[a]	0.68[a]

SOC soil organic carbon, *WEOC* water-extractable organic carbon, *WEON* water-extractable organic nitrogen, *C1, C2, C3* EEM-PARAFAC components with C1 and C3 are humic-like substance and C2 is tryptophan-like substance, *NO_3–N* nitrate, *NH_4–N* ammonium

[a] Correlation is significant at the 0.05 level (2-tailed)

[b] Correlation is significant at the 0.01 level (2-tailed)

tryptophan-like substance with simpler chemical structure can be degraded more easily by microbes and will return to background level; this can be a reason for the tillage and organic-farming practices having insignificant effects on C2.

Spectroscopic properties of WEOM and relations to soil characteristics

In comparison with tillage effect, farming practices are the main factor that influenced the quality of WEOM. Increased values of SUVA and HIX in organic farming in the present study implied that WEOM in organic farming is more aromatic and condensed compared with that in integrated farming. This might be because of the altered diverse microbial community's structure in organic farming [69], which might promote the degradation of WEOM in comparison with integrated farming. What's more, the different crop rotations which affected WEOM by crop root exudates, and the return of crop residues might be another reason to explain the different SUVA and HIX values of WEOM. Our result suggested that complex crop rotations release WEOM with higher aromaticity and humification. This is opposite to that of Xu et al. [18] who reported the inclusion of a perennial crop alfalfa in a crop rotation, which releases WEOM with lower aromaticity and lower extent of humification. This may be attributed to simpler crop rotations in their study and different crop species they planted compared with our study. Although FI values of WEOM at upper and deeper soils were statistically different, according to McKnight et al. [23], FI value of WEOM around 1.4 indicates WEOM derives from terrestrial source, while FI around 1.9 indicates WEOM derives from microbial source. Therefore, FI values in the present study were around 1.6 indicating the influence of both terrestrial and microbial source on WEOM. The lower BIX (freshness index) in organic farming indicated that less recently microbial-derived organic matter contributed to WEOM in organic farming, while the higher BIX in MT implied that more freshly derived WEOM contributed to WEOM in MT.

Conclusions

In the present agricultural long-term field study, soil WEOM responded diversely to the long-term applications of MT and OF. MT resulted in negative depth gradients of soil properties and SOC, total N as well as WEOM and its components accumulated in upper soil. In contrast, MT had no effect on the quality and WEOM components. OF had significant effects on SOM and WEOM. Specifically, OF increased SOC, total N content as well as the humic-like components and the complexity of WEOM, while the content of WEOM was lowered by OF. The results indicated that the minimum tillage changed the distribution of SOM in soil profile rather than increasing SOM in the field-trial region of the present study, and organic farming dominated the quality change of WEOM. The combined organic farming plus minimum tillage would be a better option of agricultural practice to improve soil conditions as indicated by the improved soil WEOM at upper soil and is likely to fuel possibly greater microbial activity, favoring the development of nutrient cycling in view of sustainable agricultural practice at the site of present study. Minimum tillage may be considered in the concerned landscape as a more acceptable way in realizing the balance between the yield loss and improved soil conservation [70].

Abbreviations

MT: minimum tillage; PL: plowing tillage; OF: organic farming; IF: integrated farming; WEOM: water-extractable organic matter; WEOC: water-extractable organic carbon; WEON: water-extractable organic nitrogen; EEM: excitation–emission matrix; PARAFAC: parallel factor analysis; C1 and C3: humic-like substance in WEOM; C2: tryptophan-like substance in WEOM; $SUVA_{254}$: specific ultraviolet visual absorption at 254 nm; HIX: humification index; FI: fluorescence index; BIX: freshness index; F_{max}: maximum fluorescence intensity.

Authors' contributions

HYS: through writing—original draft preparation, conceptualization, project management, and realization. PK: through co-field work, advice and help on lab work, and draft preparation; GR: through field-trial management; RS: through advice on lab work and original draft preparation, and supervision; RGJ: through writing—review and editing, and supervision; JCM: through writing—conceptualization, review and editing, and supervision. All the authors read and approved the final manuscript.

Author details

[1] Chair of Soil Ecology, Technical University of Munich, Ingolstädter Landstr. 1, Neuherberg, 85764 Munich, Germany. [2] Institute of Network Biology, Helmholtz-Zentrum München, Munich, Germany. [3] Institute of Biochemical Plant Pathology, Helmholtz-Zentrum München, Ingolstädter Landstr. 1, Neuherberg, 85764 Munich, Germany. [4] Soil Biology and Plant Nutrition, University of Kassel,

Nordbahnhofstr. 1a, 37213 Witzenhausen, Germany. [5] Chair of Grassland, Technical University of Munich, WZW, Alte Akademie 12, 85354 Freising, Germany.

Acknowledgements
The authors thank Franz Buegger and Gudrun Hufnagel for excellent sample analysis. Han Yin Sun was funded by Chinese Scholarship Council (CSC) and Helmholtz Zentrum München.

Competing interests
The authors declare that they have no competing interests.

References

1. Escobar M, Hue N, Pandalai S. Current developments in organic farming. Recent Res Dev Soil Sci. 2007;2:29–62.
2. Gattinger A, Muller A, Haeni M, Skinner C, Fliessbach A, Buchmann N, et al. Enhanced top soil carbon stocks under organic farming. Proc Natl Acad Sci USA. 2012;109(44):18226–31.
3. Sun H, Koal P, Liu D, Gerl G, Schroll R, Gattinger A, et al. Soil microbial community and microbial residues respond positively to minimum tillage under organic farming in Southern Germany. Appl Soil Ecol. 2016;108:16–24.
4. Kandeler E, Tscherko D, Spiegel H. Long-term monitoring of microbial biomass, N mineralisation and enzyme activities of a Chernozem under different tillage management. Biol Fertil Soils. 1999;28(4):343–51.
5. Ghimire R, Adhikari KR, Chen ZS, Shah SC, Dahal KR. Soil organic carbon sequestration as affected by tillage, crop residue, and nitrogen application in rice–wheat rotation system. Paddy Water Environ. 2012;10(2):95–102.
6. Kalbitz K, Solinger S, Park JH, Michalzik B, Matzner E. Controls on the dynamics of dissolved organic matter in soils: a review. Soil Sci. 2000;165(4):277–304.
7. Ohno T, Bro R. Dissolved organic matter characterization using multiway spectral decomposition of fluorescence landscapes. Soil Sci Soc Am J. 2006;70(6):2028–37.
8. Embacher A, Zsolnay A, Gattinger A, Munch JC. The dynamics of water extractable organic matter (WEOM) in common arable topsoils: I. Quantity, quality and function over a three year period. Geoderma. 2007;139(1–2):11–22.
9. Bausenwein U, Gattinger A, Langer U, Embacher A, Hartmann HP, Sommer M, et al. Exploring soil microbial communities and soil organic matter: variability and interactions in arable soils under minimum tillage practice. Appl Soil Ecol. 2008;40(1):67–77.
10. Grosso F, Temussi F, De Nicola F. Water-extractable organic matter and enzyme activity in three forest soils of the Mediterranean area. Eur J Soil Biol. 2014;64:15–22.
11. Dong WX, Hu CS, Chen SY, Zhang YM. Tillage and residue management effects on soil carbon and CO_2 emission in a wheat–corn double-cropping system. Nutr Cycl Agroecosyst. 2009;83(1):27–37.
12. Sun BH, Hallett PD, Caul S, Daniell TJ, Hopkins DW. Distribution of soil carbon and microbial biomass in arable soils under different tillage regimes. Plant Soil. 2011;338(1–2):17–25.
13. Chantigny MH. Dissolved and water-extractable organic matter in soils: a review on the influence of land use and management practices. Geoderma. 2003;113(3–4):357–80.
14. Melero S, Lopez-Garrido R, Madejon E, Murillo JM, Vanderlinden K, Ordonez R, et al. Long-term effects of conservation tillage on organic fractions in two soils in southwest of Spain. Agric Ecosyst Environ. 2009;133(1–2):68–74.
15. Cookson WR, Murphy DV, Roper MM. Characterizing the relationships between soil organic matter components and microbial function and composition along a tillage disturbance gradient. Soil Biol Biochem. 2008;40(3):763–77.
16. Zhang M, He Z, Zhao A, Zhang H, Endale DM, Schomberg HH. Water-extractable soil organic carbon and nitrogen affected by tillage and manure application. Soil Sci. 2011;176(6):307–12.
17. Marinari S, Liburdi K, Fliessbach A, Kalbitz K. Effects of organic management on water-extractable organic matter and C mineralization in European arable soils. Soil Tillage Res. 2010;106(2):211–7.
18. Xu N, Wilson HF, Saiers JE, Entz M. Effects of crop rotation and management system on water-extractable organic matter concentration, structure, and bioavailability in a chernozemic agricultural soil. J Environ Qual. 2013;42(1):179–90.
19. Borisover M, Lordian A, Levy GJ. Water-extractable soil organic matter characterization by chromophoric indicators: effects of soil type and irrigation water quality. Geoderma. 2012;179–180:28–37.
20. Fellman J, D'Amore D, Hood E, Boone R. Fluorescence characteristics and biodegradability of dissolved organic matter in forest and wetland soils from coastal temperate watersheds in southeast Alaska. Biogeochemistry. 2008;88(2):169–84.
21. Zsolnay A, Baigar E, Jimenez M, Steinweg B, Saccomandi F. Differentiating with fluorescence spectroscopy the sources of dissolved organic matter in soils subjected to drying. Chemosphere. 1999;38(1):45–50.
22. Corvasce M, Zsolnay A, D'Orazio V, Lopez R, Miano TM. Characterization of water extractable organic matter in a deep soil profile. Chemosphere. 2006;62(10):1583–90.
23. McKnight DM, Boyer EW, Westerhoff PK, Doran PT, Kulbe T, Andersen DT. Spectrofluorometric characterization of dissolved organic matter for indication of precursor organic material and aromaticity. Limnol Oceanogr. 2001;46(1):38–48.
24. Huguet A, Vacher L, Relexans S, Saubusse S, Froidefond JM, Parlanti E. Properties of fluorescent dissolved organic matter in the Gironde Estuary. Org Geochem. 2009;40(6):706–19.
25. Wilson HF, Xenopoulos MA. Effects of agricultural land use on the composition of fluvial dissolved organic matter. Nat Geosci. 2009;2(1):37–41.
26. Schröder P, Huber B, Olazabal U, Kammerer A, Munch JC. Land use and sustainability: FAM research network on agroecosystems. Geoderma. 2002;105(3–4):155–66.
27. Kölbl A, Kögel-Knabner I. Content and composition of free and occluded particulate organic matter in a differently textured arable Cambisol as revealed by solid-state (13)C NMR spectroscopy. J Plant Nutr Soil Sci. 2004;167(1):45–53.
28. Flessa H, Ruser R, Schilling R, Loftfield N, Munch JC, Kaiser EA, et al. N_2O and CH_4 fluxes in potato fields: automated measurement, management effects and temporal variation. Geoderma. 2002;105(3–4):307–25.
29. Gee G, Bauder J. Particle-size analysis. Methods Soil Anal Part I. 1986;9:383–411.
30. Zsolnay A. Dissolved Humus in Soil Waters. In: Piccolo A, editor. Humic substances in terrestrial ecosystems. Amsterdam: Elsevier Science B.V; 1996. p. 171–223.
31. Zsolnay A. Dissolved organic matter: artefacts, definitions, and functions. Geoderma. 2003;113(3–4):187–209.
32. Embacher A, Zsolnay A, Gattinger A, Munch JC. The dynamics of water extractable organic matter (WEOM) in common arable topsoils: II. Influence of mineral and combined mineral and manure fertilization in a Haplic Chernozem. Geoderma. 2008;148(1):63–9.
33. Yang L, Hur J. Critical evaluation of spectroscopic indices for organic matter source tracing via end member mixing analysis based on two contrasting sources. Water Res. 2014;59:80–9.
34. Cuss CW, Shi YX, McConnell SM, Guéguen C. Changes in the fluorescence composition of multiple DOM sources over pH gradients assessed by combining parallel factor analysis and self-organizing maps. J Geophys Res Biogeosci. 2014;119(9):1850–60.
35. Chen H, Kenny JE. A study of pH effects on humic substances using chemometric analysis of excitation–emission matrices. AES. 2007;1:1–9.
36. Gauthier TD, Shane EC, Guerin WF, Seitz WR, Grant CL. Fluorescence quenching method for determining equilibrium constants for polycyclic aromatic hydrocarbons binding to dissolved humic materials. Environ Sci Technol. 1986;20(11):1162–6.
37. Akagi J, Zsoinay A. Effects of long-term de-vegetation on the quantity and quality of water extractable organic matter (WEOM): biogeochemical implications. Chemosphere. 2008;72(10):1462–6.
38. Lakowicz JR. Principles of fluorescence spectroscopy. Berlin: Springer, US; 2006.
39. Bro R. PARAFAC. Tutorial and applications. Chemom Intell Lab Syst. 1997;38(2):149–71.

40. Andersen CM, Bro R. Practical aspects of PARAFAC modeling of fluorescence excitation–emission data. J Chemom. 2003;17(4):200–15.
41. Singh S, D'Sa EJ, Swenson EM. Chromophoric dissolved organic matter (CDOM) variability in Barataria Basin using excitation–emission matrix (EEM) fluorescence and parallel factor analysis (PARAFAC). Sci Total Environ. 2010;408(16):3211–22.
42. Murphy KR, Stedmon CA, Graeber D, Bro R. Fluorescence spectroscopy and multi-way techniques. PARAFAC. Anal Methods. 2013;5(23):6557–66.
43. Baghoth SA, Sharma SK, Amy GL. Tracking natural organic matter (NOM) in a drinking water treatment plant using fluorescence excitation–emission matrices and PARAFAC. Water Res. 2011;45(2):797–809.
44. Lawaetz AJ, Stedmon CA. Fluorescence intensity calibration using the Raman scatter peak of water. Appl Spectrosc. 2009;63(8):936–40.
45. Stedmon CA, Bro R. Characterizing dissolved organic matter fluorescence with parallel factor analysis: a tutorial. Limnol Oceanogr Methods. 2008;6:572–9.
46. R Core Team. R: a language and environment for statistical computing. Vienna: R Foundation for Statistical Computing; 2015. http://www.R-project.org/. Accessed 18 Oct 2015.
47. Murphy KR, Stedmon CA, Waite TD, Ruiz GM. Distinguishing between terrestrial and autochthonous organic matter sources in marine environments using fluorescence spectroscopy. Mar Chem. 2008;108(1–2):40–58.
48. Yamashita Y, Scinto LJ, Maie N, Jaffe R. Dissolved organic matter characteristics across a subtropical wetland's landscape: application of optical properties in the assessment of environmental dynamics. Ecosystems. 2010;13(7):1006–19.
49. Huang W, McDowell WH, Zou XM, Ruan HH, Wang JS, Ma ZL. Qualitative differences in headwater stream dissolved organic matter and riparian water-extractable soil organic matter under four different vegetation types along an altitudinal gradient in the Wuyi Mountains of China. Appl Geochem. 2015;52:67–75.
50. Stedmon CA, Markager S, Bro R. Tracing dissolved organic matter in aquatic environments using a new approach to fluorescence spectroscopy. Mar Chem. 2003;82(3–4):239–54.
51. Williams CJ, Yamashita Y, Wilson HF, Jaffe R, Xenopoulos MA. Unraveling the role of land use and microbial activity in shaping dissolved organic matter characteristics in stream ecosystems. Limnol Oceanogr. 2010;55(3):1159–71.
52. Hong HS, Yang LY, Guo WD, Wang FL, Yu XX. Characterization of dissolved organic matter under contrasting hydrologic regimes in a subtropical watershed using PARAFAC model. Biogeochemistry. 2012;109(1–3):163–74.
53. Coble PG. Characterization of marine and terrestrial DOM in seawater using excitation–emission matrix spectroscopy. Mar Chem. 1996;51(4):325–46.
54. Stedmon CA, Markager S. Resolving the variability in dissolved organic matter fluorescence in a temperate estuary and its catchment using PARAFAC analysis. Limnol Oceanogr. 2005;50(2):686–97.
55. Sun HY, Wang CX, Wang XD, Rees RM. Changes in soil organic carbon and its chemical fractions under different tillage practices on loess soils of the Guanzhong Plain in north-west China. Soil Use Manag. 2013;29(3):344–53.
56. Six J, Elliott ET, Paustian K. Soil macroaggregate turnover and microaggregate formation: a mechanism for C sequestration under no-tillage agriculture. Soil Biol Biochem. 2000;32(14):2099–103.
57. Chivenge PP, Murwira HK, Giller KE, Mapfumo P, Six J. Long-term impact of reduced tillage and residue management on soil carbon stabilization: implications for conservation agriculture on contrasting soils. Soil Tillage Res. 2007;94(2):328–37.
58. Kaiser M, Piegholdt C, Andruschkewitsch R, Linsler D, Koch HJ, Ludwig B. Impact of tillage intensity on carbon and nitrogen pools in surface and sub-surface soils of three long-term field experiments. Eur J Soil Sci. 2014;65(4):499–509.
59. Luo Z, Wang E, Sun OJ. Can no-tillage stimulate carbon sequestration in agricultural soils? A meta-analysis of paired experiments. Agric Ecosyst Environ. 2010;139(1–2):224–31.
60. Marriott EE, Wander MM. Total and labile soil organic matter in organic and conventional farming systems. Soil Sci Soc Am J. 2006;70(3):950–9.
61. Parras-Alcantara L, Lozano-Garcia B. Conventional tillage versus organic farming in relation to soil organic carbon stock in olive groves in Mediterranean rangelands (southern Spain). Solid Earth. 2014;5(1):299–311.
62. Parras-Alcantara L, Diaz-Jaimes L, Lozano-Garcia B, Rebollo PF, Elcure FM, Munoz MDC. Organic farming has little effect on carbon stock in a Mediterranean dehesa (southern Spain). Catena. 2014;113:9–17.
63. Chatterjee A, Lal R. On farm assessment of tillage impact on soil carbon and associated soil quality parameters. Soil Tillage Res. 2009;104(2):270–7.
64. Mäder P, Fliessbach A, Dubois D, Gunst L, Fried P, Niggli U. Soil fertility and biodiversity in organic farming. Science. 2002;296(5573):1694–7.
65. Blanco-Canqui H, Lal R. Extent of soil water repellency under long-term no-till soils. Geoderma. 2009;149(1–2):171–80.
66. Sun L, Chang SX, Feng YS, Dyck MF, Puurveen D. Nitrogen fertilization and tillage reversal affected water-extractable organic carbon and nitrogen differentially in a Black Chernozem and a Gray Luvisol. Soil Tillage Res. 2015;146:253–60.
67. Kuntz M, Berner A, Gattinger A, Scholberg JM, Mader P, Pfiffner L. Influence of reduced tillage on earthworm and microbial communities under organic arable farming. Pedobiologia. 2013;56(4–6):251–60.
68. Ohno T, He Z, Tazisong IA, Senwo ZN. Influence of tillage, cropping, and nitrogen source on the chemical characteristics of humic acid, fulvic acid, and water-soluble soil organic matter fractions of a long-term cropping system study. Soil Sci. 2009;174(12):652–60.
69. Esperschütz J, Gattinger A, Mader P, Schloter M, Fliessbach A. Response of soil microbial biomass and community structures to conventional and organic farming systems under identical crop rotations. FEMS Microbiol Ecol. 2007;61(1):26–37.
70. Küstermann B, Munch JC, Hülsbergen K-J. Effects of soil tillage and fertilization on resource efficiency and greenhouse gas emissions in a long-term field experiment in Southern Germany. Eur J Agron. 2013;49:61–73.

5

Portable NMR in food analysis

Donatella Capitani[1], Anatoly P. Sobolev[1], Valeria Di Tullio[1], Luisa Mannina[1,2] and Noemi Proietti[1*]

Abstract

Low-field ^{1}H NMR relaxometry is an important tool used to investigate on the most abundant components of intact foodstuffs based on relaxation parameters and amplitude of the NMR signals. In particular, information on water compartments, diffusion and movement can be obtained by detecting proton signals prevalently ascribable to H_2O contained in foodstuffs. The main advantage of this technique is that it does not require any pretreatment of the sample and once developed, standard protocols based on rapid measurements can be easily transferred to quality control applications. An actual breakthrough for the low-field NMR application to food science has certainly been the development of unilateral NMR sensors. These devices present an attractive option for non-invasive assessment of compositional and microstructure of food materials. They are portable and allow easy sample access, which makes them attractive for quality control in industrial environments and directly on sealed packaged foods. A price to pay is the inhomogeneity of the magnetic field, which compromises the sensitivity of these devices. However, recent developments in unilateral magnets design offer larger sensitive volume, higher sensitivity and shorter dead times. In this brief review, we highlight the different uses of unilateral NMR devices in food analysis.

Keywords: Portable NMR, Food science, Relaxation times, Water mobility, Self-diffusion coefficient, Quality control process

Background

Pulsed low-resolution NMR is an appreciated and powerful technique in food science and technology. NMR instruments operating in the time domain (TD) at low field are widely used for investigating on relationships between food properties and its composition and structure. In fact foodstuffs are structurally and chemically heterogeneous systems whose constituents, due to the different molecular mobility, show different relaxation and diffusive properties. Pulsed low-resolution NMR technique is suitable to study the dynamic properties of matter like self-diffusion, flow and relaxation. The two major NMR relaxation phenomena are longitudinal relaxation characterized by a relaxation time T_1, and transverse relaxation, characterized by T_2.

Hahn echo decay experiment can be used to measure the transverse relaxation time. The amplitude of the echo is recorded for different spacings of the two pulses. Echo amplitude is influenced by molecular diffusion (Brownian motion) because of the fluctuations of the local magnetic field. Thus, the amplitude of the echo at the time t is related to the transverse relaxation T_2, gradient G, self-diffusion coefficient D, gyromagnetic ratio γ, and the half echo spacing τ through the following equation:

$$M_t = \exp\left(-\frac{1}{T_2} - \frac{\gamma^2 G^2 D \tau^2 t}{3}\right) \quad (1)$$

In the case of most conventional NMR magnets, the field inhomogeneity is sufficiently small that at short τ values the second term of the equation becomes negligible even for faster diffusing molecules such as water. Instead, in the case of a strong inhomogeneous magnetic field the second term of Eq. (1) cannot be neglected.

The pulsed field gradient NMR method is one of the main NMR methods for obtaining dynamic information, such as the self-diffusion coefficient. Self-diffusion data provide detailed information about molecular organization and phase structure. Self-diffusion rates are quite sensitive to structural changes and to binding and association phenomena. Experimental self-diffusion values are directly related to molecular displacement.

*Correspondence: noemi.proietti@cnr.it

[1] Istituto di Metodologie Chimiche, Laboratorio di Risonanza Magnetica, "Annalaura Segre", CNR, Monterotondo, 00015 Rome, Italy

Full list of author information is available at the end of the article

Many analytical protocols have been established for rapid and precise measurements of compositional properties [1–4]. These protocols are particularly appreciated in the food supply chain for quality and process control. In many applications, differences in self-diffusion coefficient and relaxation times have been exploited to differentiate different phases in food matrices [1–3]. For example, ^{1}H pulsed low-resolution NMR has been applied to study the effect of different physiological conditions on kiwifruit [5, 6], apple [7–10], carrot [11] and strawberry [3, 12].

The effect induced by ripening, drying, freezing, high-pressure processing, bruising, browning, aging and osmotic drying has been studied [1, 2, 13–15].

The relationships between ^{1}H NMR relaxation data and some technological parameters of meat have been investigated [16–18].

NMR is a well-accepted analytical tool for evaluation of oil and moisture content in food [1, 19]. The presence of non-freezable water, water mobility as well as the phase compositional and hydration phenomena have been investigated in many food matrices [1].

A breakthrough in the application of TD NMR to food science has been represented by the conceiving of compact and portable NMR sensors which enable the analysis of intact food matrices in a non-destructive, non-invasive way in situ. Sensors based on permanent magnets have been developed during the last years covering applications involving relaxometry, imaging and even spectroscopy with hand-held systems [20–23]. These devices require low maintenance costs and allow easy sample access, which makes them attractive for quality and process control in industrial environments.

Portable single-sided NMR sensors with open geometry

In conventional NMR instruments, the geometry of the magnet is referred to as closed geometry and the sample under study is placed in a homogeneous magnetic field. While this facilitates high signal-to-noise, geometrically correct, spatially resolved magnetic resonance imaging (MRI), it limits the range of samples that can be examined. In recent years, this limitation has been addressed by development of portable or unilateral NMR [22–24]. The geometry of portable NMR sensors, in which the object is exposed to the stray field of the magnet, is referred to as open geometry. In this case, the magnet may be placed to one side of the object fully preserving the integrity and the dimension of the sample under investigation and also allowing the measurement to whole packed products.

Nowadays, the unilateral NMR sensors available can be categorized into two classes. The first group operates in a strong magnetic field gradient [24], whereas the second one operates in region under more or less homogeneous magnetic field [22, 23]. Both approaches have their strengths and weaknesses.

Unilateral NMR developed by Blumich et al. [25] is characterized by a high magnetic field strength operating in the proton frequency range of 13–18 MHz with a strong magnetic field gradient. This sensor requires very short (in the range of the microseconds) and very powerful RF pulses to achieve the desired frequency bandwidth. An example of such a sensor is sketched in Fig. 1a. Two antiparallel magnets are mounted on an iron yoke and the radio-frequency coil is positioned in the gap between the two magnets. The magnet may be designed either parallel or perpendicular to the sensor surface. Using this sensor, the depth of measurement into the investigated object may be varied by changing the RF coil [26]. The signal-to-noise ratio decreases by increasing the depth of measurement due to the B_1 stray field.

Another class of unilateral NMR instrument has been developed by Marble et al. [22, 27]. In these sensors, the unilateral magnet array consisted of three block magnets all magnetized along the same directions, see Fig. 1b. The sensitive spot (c) is located ~1 cm above the face of the magnet. The magnet spacings are optimized to produce a locally homogeneous field in this region creating a relatively large MR sensitive volume above the surface RF coil (b). The resonance frequency of the NMR system is 4.68 MHz and all pulse lengths are about of 8 μs with 6 dB attenuation for the 90° pulse and no attenuation for the 180° pulse [22, 28].

Another portable NMR sensor with a novel one-sided access magnet design, termed NMR-MOLE (mobile lateral explorer), has been developed by Manz et al. [23]. This sensor is very efficient in terms of sensitivity and penetration depth. The magnet array is based on a barrel magnet operating at 3.3 MHz with a central magnet positioned to provide a homogeneous region from 4 to 16 mm away from the probe, with maximum sensitivity at a depth of 10 mm. Because of the lower diffusion attenuation in homogeneous field sensors, they are better suited for studying liquid samples such as aqueous solutions and biological tissue where unilateral or portable access is required.

A recent evolution of NMR sensors with an open geometry is the profile NMR MOUSE [29]. The magnet geometry consists of four permanent magnet blocks positioned on an iron yoke, see Fig. 1c. Two magnets are polarized along y and two along $-y$. Magnets with the same polarization are separated by a small gap d_S whereas magnets with opposite polarization are separated by a gap d_B. The sensor is placed on a lift which moves the sensitive volume constituted of a thin slice at a well-defined position through the object by varying the distance between the sensor and the object see Fig. 1d. The slice thickness depends on the experimental conditions and typically ranges between a few micrometers and 100 μm. The

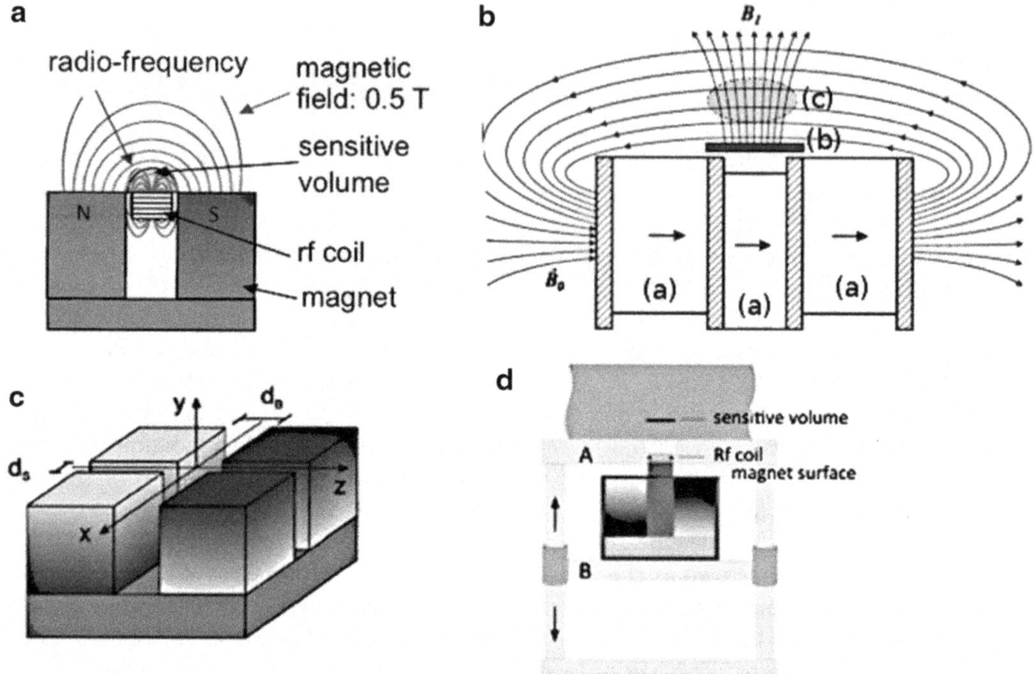

Fig. 1 **a** Skecth of a single-sided NMR sensor with open geometry (reproduced with permission from [25]). **b** Sketch of a single-sided magnet array with a sweet-spot (reproduced with permission from [41]). **c** Sketch of the geometry of the magnet of the profiler NMR MOUSE. **d** Sketch of the lift used to reposition the sensitive slice across the sample with a micrometric precision. The object is placed on *top* of the *plate A*, which is parallel to the movable *plate B* where the magnet is mounted (reproduced with permission from [28])

magnetic field varies linearly with the space generating a constant magnetic field gradient of about 20 T/m [29].

The amplitude values recorded by a modulated displacing of the sensor may provide a profile of proton spin densities as a function of the depth into the sample. However, not only the signal amplitude but also other parameters (i.e. relaxation times) can be reported against the depth scanned. Presently, a penetration depth up to 2.5 cm may be achieved with these sensors.

Due to strongly inhomogeneous magnetic field, the applications of portable NMR are limited as compared to those performed by benchtop instrumentation. FID cannot be detected, because NMR signal decays very quickly and must be recovered as an echo. The measurement of transverse relaxation time suffers for artifacts resulting from molecular diffusion in the strong magnetic field gradient. These artifacts are more pronounced in the case of small molecules that diffuse rapidly. In the presence of diffusion process, the transverse relaxation time values are shortened with respect to those measured in homogeneous field. Furthermore, the decay of the observed transverse magnetization in inhomogeneous field is a mixture of T_1 and T_2 and the early echoes oscillate in amplitude thus leading to an affective echo train decay time T_{2eff} instead only T_2 [30–32]. As well known, it is possible to reduce the effect of molecular diffusion on the NMR signal collected in the magnetic field gradient by generating a series of spin echoes making the interpulse delay τ between $\pi/2$ and π pulses as short as possible.

Another limit in working with permanent magnets is represented by the temperature sensitivity. The NMR MOUSE is not temperature stabilized due to the high gradient of the sensor; a temperature change of the magnet causes a slight shift of the position of the sensitive volume [33]. Consequently, when working at room temperature, the applications are limited as compared to those performed by benchtop instrumentation and are mainly focused on the development of protocols applied in online inspections and for monitoring processes such as ripening, growth rate of fruits, drying process, and so on. Since T_2 tightly depends on temperature, in case of applications in an outdoor environment characterized by a temperature close to that of room temperature, a system to control the temperature during the measurements is strongly recommended to guarantee a good reproducibility. Anferova et al. [34] described two different types of mobile temperature control units operating with the NMR-MOUSE. They reported that in the case of elastomers, by the temperature stabilization the relative error of T_2 decreased by a factor of four on the short time scale. Without temperature stabilization, it may reach 30% in the long time range because of temperature changes in the laboratory.

Application of single-sided NMR sensors to food matrices

Interesting applications of portable single-sided NMR sensors to food matrices have been reported in literature [24, 35]. Despite the inhomogeneous field generated by these open geometry sensors, their potentiality for food analysis and control has been explored. An overview of food examined by unilateral devices is reported in Table 1. The advantage of this methodology is the possibility to analyze foods just placing the sensor on the food surface without needing to cut the sample into pieces suitable to fit into an NMR tube.

The determination of solid fat content (SFC) in food based on pulsed low-resolution NMR is an internationally accepted procedure [1]. Basically, the difference in transverse relaxation properties of solid and liquid fats was exploited to determine the solid and liquid contents. Although this technique is widely used even in industrial processes it cannot be applied on line.

Guthausen et al. investigated the applicability of a single-sided NMR sensor to measure the fat content in packaged food products [36, 37]. At this aim, two methods [35] were applied. One is based on differences in relaxation times which depend on molecular mobility and, consequently on the molecular structure (relaxation weighted method), whereas the other one exploits the fact that the diffusion coefficients of water and fat differ by more than an order of magnitude (diffusion-weighted method). A pulse sequence was purposely developed for measuring fat and water content in packaged coffee creams. The packing material does not influence significantly the measure because it is relatively thin and the polymer material exhibits a short T_{2eff} relaxation time. The measurement time was about 1 min using the surface resonance frequency, whereas measurements at 3 mm depth resulted in considerable longer acquisition times (about 5 min). The ratio of the final echoes and the first few echoes was the relevant NMR parameter, named NMR ratio. The NMR ratio exhibits a linear dependence with fat content (linear regression with a correlation coefficient of 0.996) thus indicating that a reasonable determination of the fat content is possible, see Fig. 2a.

The relaxation method was used for calibrating the fat content in packaged mayonnaises and margarines. In fact, the relaxation time differences of fat and water protons may be exploited to generate a contrast by allowing a more accurate content determination. Again, a linear correlation was found between the fat percentage measured by NMR and the reference fat content, see Fig. 2b. In this case, data were analyzed by a chemometric approach obtaining a correlation coefficient of 0.991. Such a linear correlation confirms the reliability of single-sided NMR for fat measurements even in packaged food. The chemometric approach was applied to the NMR signal which can be analyzed using supervised or unsupervised pattern recognition methods [38, 39]. These methods may be applied in food analysis for both quantitative and qualitative assessments. The knowledge of reference values allows for the calibration of the NMR response.

In another paper, Martini et al. [40] compared SFC determination obtained using conventional offline NMR instrument with determination obtained by single-sided NMR sensor. Authors explored the use of single-sided NMR to measure variation of SFC online during the crystallization of a fat product. Interesterified hydrogenated palm oil was added to canola oil in different proportions to obtain blends with different SFC values. Two different experiments were carried out to study whether the

Table 1 Reviews of portable NMR technologies and their applications in food science

Reference	Food	Focus	Year
Guthausen et al. [36]	Margarine; mayonnaise	Measurements of fat content	2004
Guthausen et al. [37]	Coffee cream	Water moisture and fat analysis in packaged products	2006
Martini et al. [40]	Vegetable oils	Solid fat content determination	2005
Pedersen et al. [41]	Vegetable oils	Measurements of fat content	2003
Haiduc et al. [42]	Fat emulsions	Microstructural quality of packaged food emulsion	2007
Stork et al. [44]	Bottled beverages	Determination of dissolved oxygen in unopened table waters	2001
Xu et al. [46]	Adulterated virgin olive oil	Detection of the adulteration of virgin olive oil in sealed bottles	2014
Pinter et al. [48]	Tomato paste	Monitoring of spoilage in tomato paste processing industry	2014
Veliyulin et al. [43]	Salmon	Fat content in live fish	2005
Veliyulin et al. [28]	Dairy products	Measurements of fat content in packaged dairy products	2008
Capitani et al. [49]	Kiwifruit	Water status as a function of season	2010
Capitani et al. [50]	Kiwifruit	Monitoring of development and ripening	2013
Capitani et al. [51]	Blueberry	Monitoring of shelf life	2014
Adiletta et al. [52]	Pear	Monitoring of drying process	2015

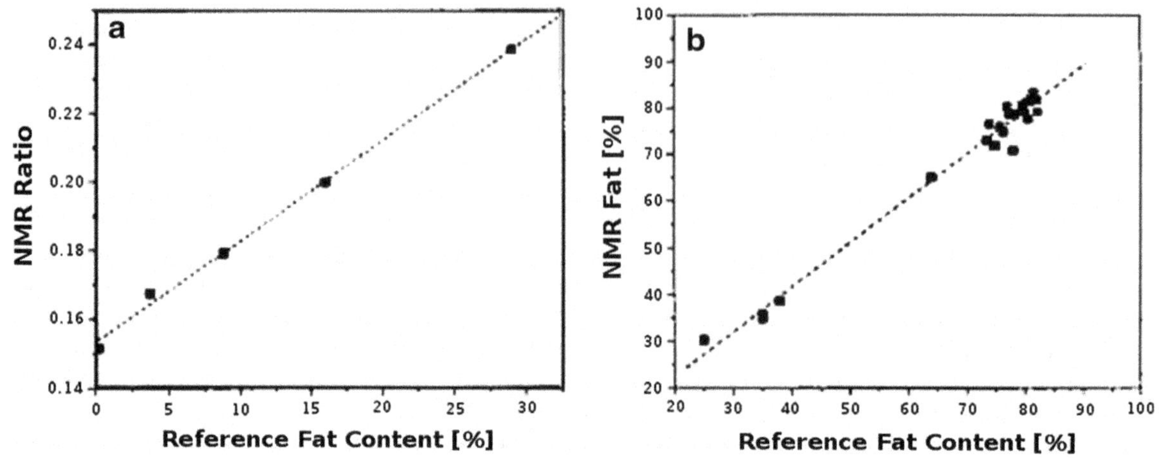

Fig. 2 **a** Correlation of the NMR ratio with the reference fat content of coffee creams. **b** Correlation diagram for measurements on mayonnaises and margarines with the reference fat content. Reproduced with permission from [34]

motion during the crystallization affected the measurements. In one experiment, agitation was stopped during the measurement and then restarted to let the crystallization continue until the next measurement. In the other experiments, measurements were carried out under agitation. To eliminate the temperature effect on measurements carried out by single-sided sensor, a proper correction was applied to the collected data. In the former case, the corrected data were close to those collected by a conventional offline NMR instrument particularly at intermediate SFC values, whereas in the latter case data were found to be significantly different from those determined by the off-line instrument.

The performance of conventional benchtop NMR and single-sided NMR applied to the study of sunflower oil-in-water emulsions was investigated by Pedersen et al. [41]. With conventional NMR, the decay rate of pure water obtained after applying a CPMG sequence is much longer than that of oil, which means that the relaxation behavior of the two components is very different. With single-sided NMR, the presence of a strong magnetic field gradient heavily modifies the decay of magnetization; in fact, the decay rate of water is much faster than that of oil because water is much more affected by diffusion than oil. Nevertheless, this effect makes it possible to unequivocally distinguish the signal of water from oil. The trend in relaxation behavior with increasing oil content measured by single-sided NMR was found to be the reverse of that obtained by conventional benchtop NMR. The reverse of the trend in the decay rates with decreasing the oil content is also foreseeable as a result of the strong magnetic field gradient which affects the apparent transverse relaxation time of water due to its rapid diffusion rate. Authors demonstrated that NMR decays obtained by both conventional benchtop NMR and single-sided NMR can be readily deconvoluted in two components in the case of oil-in-water emulsions with oil content ranging from 10 to 67%.

Hiaduc et al. [42] reported a feasibility study on the use of single-sided NMR for the non-invasive 'through-package' assessment of the microstructural quality of food material. Authors investigated model systems consisting of oil-in-water emulsion gels stabilized by proteins. These systems form complex structures made of fat droplets and a protein aggregate network where water is dispersed in pores with different size. An important quality parameter of such systems is the water exudation (WE). In a classical approach, WE is determined by measuring the amount of water lost from the system when subject to gravitational forces. Multivariate calibration techniques were applied to establish a relation between signals obtained by single-sided NMR and WE. Specifically, to obtain a calibration model allowing for a physical interpretation, an approach based on multilinear regression (MLR) was applied to decays collected with a benchtop instrument and with a single-sided sensor as well. Decays were transformed from the continuous domain to the discrete domain of T_{2eff} distributions and amplitudes using the non-negative least squares (NNLS) algorithm. The decays were averaged and the obtained data fitted without any initial guess on the number of components or T_{2eff} values. To summarize, the MLR model was built using the NNLS amplitudes as responses, and the functional parameters as predictors. The goodness of fits of NNLS_MLR model based on conventional benchtop NMR decays (a) and decays measured with single-sided NMR (b) is observable in Fig. 3. The quality of MLR model applied to both set of data is comparable, thus indicating that single-sided NMR

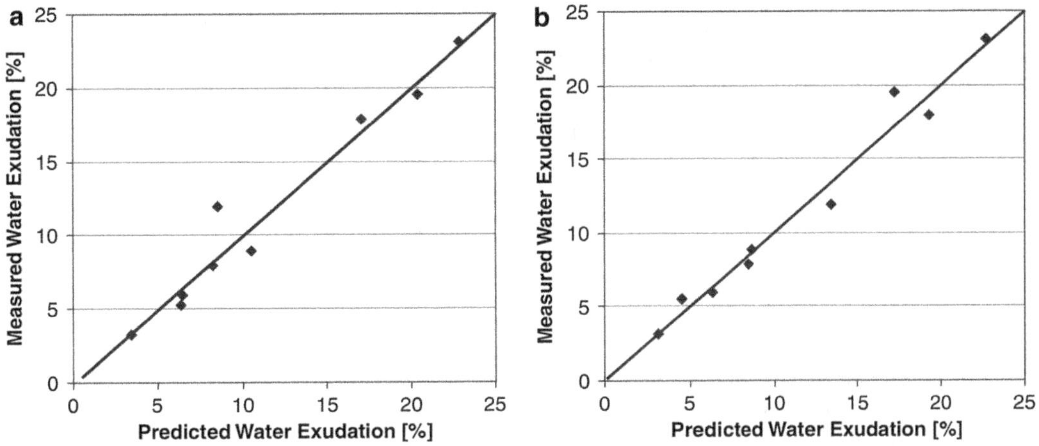

Fig. 3 Fits of NNLS–MLR models for water exudation based on (**a**) decays collected by a conventional benchtop NMR, and (**b**) by single-sided NMR. Reproduced with permission from [38]

may also allow for the assessment of WE in a through package mode. It must be taken into account that due to slow T_1-relaxation, the delay between the individual averages was set to 4 s, which yields an experimental time of 8 min for one MOUSE measurement.

A new procedure for rapid and non-destructive determination of the fat content in dairy products has been proposed by Veliyulin et al. [28]. In this work, a novel design of unilateral NMR instrument with larger measurement volume and higher signal to noise ratio [22, 27] has allowed to reduce the time measurement of the fat content in packaged dairy products about up to few seconds. Two methods were tested: a diffusion-edited CPMG technique and a strongly T_1-weighted conventional CPMG method requiring total measurement times of about 39 and 7 s, respectively. The latter technique has a potential for being implemented for industrial monitoring of the fat content in both packaged products and larger batches of dairy products. Linear relationship between the transverse relaxation time of the fat component and the inverse fat content was observed.

The same unilateral NMR device was tested by Veliyulin et al. [43] to set up rapid non-destructive measurement of the fat content in live or slaughtered Atlantic salmon. The total analysis time per fish was about 20 s. The distribution of transverse relaxation times when measured with a conventional benchtop NMR instrument shows three peaks. The shortest component is usually ascribed to water closely associated with macromolecules, the intermediate one to intracellular water or water within the myofibrillar structure, whereas the longest one accounts for both lipids and water. The overlapping of the fat signal with that of extra-myofibrillar water (both contributing to the longest component) makes very difficult to differentiate fat and water content based on the quantitative T_{2eff} measurements. When measurements are performed by a single-sided sensor, the effective transverse relaxation time is affected by strong field gradient generated by the sensor which influences the self-diffusion coefficient. Because the average self-diffusion coefficient of water in fish muscle is considerably faster than that of fat, this difference may be exploited to make accurate quantification of the fat component from the transverse relaxation response. Figure 4 compares the transverse relaxation curves of pure fish oil and Atlantic salmon white muscle obtained applying a CPMG sequence optimized to ensure the best separation between water and fat components. Prior to perform measurements, the instrument was calibrated using samples with known fat content (mixtures of fish oil and agarose) and characterized by NMR relaxation properties similar to those of fish fat. A significant correlation was

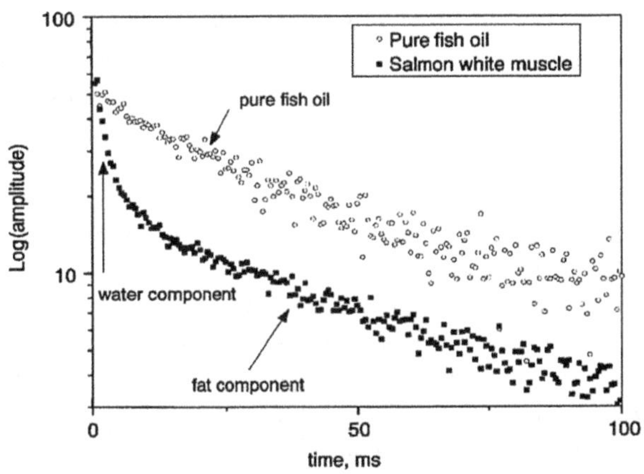

Fig. 4 Transverse relaxation curves of pure fish oil and Atlantic salmon white muscle measured at 4 °C using a single-sided NMR sensor. Reproduced with permission from [42]

found between the fat content measured by single-sided NMR and that measured by chemical analysis. Furthermore, the mobile single-sided sensor was used to map the fat distribution over the whole fish surface. The NMR signal measured on about 50 regions was processed to obtain a post-mortem fat distribution image in the white muscle, red muscle and belly flap of Atlantic salmon.

It is well known that the application of conventional single-sided NMR to highly mobile, watery phases in foods and beverages is hampered by the strong magnetic field. In their paper, Stork et al. used both a single-sided NMR sensor with a reduced magnetic field gradient and another ("semisingle-sided") sensor design with an open sample bay for determining the oxygen content in unopened bottles with superoxygenated table water, and compared the obtained results with those from conventional NMR [44]. In the semisingle-sided sensor, the distance between the two arrays of magnet bricks was chosen in a way that 1.5-L bottles fit in the gap of the magnet system. The latter design allows much better sensitivity without sacrificing the necessary open access needed for measurements on entire food packages such as bottle. For the NMR determination of dissolved oxygen in table waters, the longitudinal relaxation time of the water was measured. It was related to the oxygen concentration C_{Ox} according to the following relationship:

$$\frac{1}{T_1} = \frac{1}{T_{1\mathrm{o}}} + r_{1,\mathrm{ox}} C_{\mathrm{Ox}} \tag{2}$$

where $T_{1\mathrm{o}}$ denoting the relaxation time of the oxygen-free water.

As the relaxivity $r_{1,\mathrm{ox}}$ of oxygen in water is field dependent [45] and no sufficiently precise relaxation data at a proton resonance frequencies of 5–6 MHz are available in the literature, authors performed calibration measurements against samples with a known (supersaturated) oxygen concentration.

Figure 5a shows the comparison between the spin–lattice relaxation rate measured with a single-sided NMR sensor and that measured by conventional NMR. The two sets of data were found to be in a very good agreement. The semisingle-sided sensor was also used for determining the oxygen concentration in a commercial bottle before and after opening the bottle by measuring the water relaxation time; data are reported in Fig. 5b. The sensor was capable of monitoring the oxygen concentration that was constant before opening, and progressively decreased after opening the bottle, thus indicating the possible use of this sensor for online application after a suitable calibration procedure. To achieve a better signal/noise ratio, the measuring times for each relaxation time in these curves were several hours per data point in the single-sided setup and about 40 min in

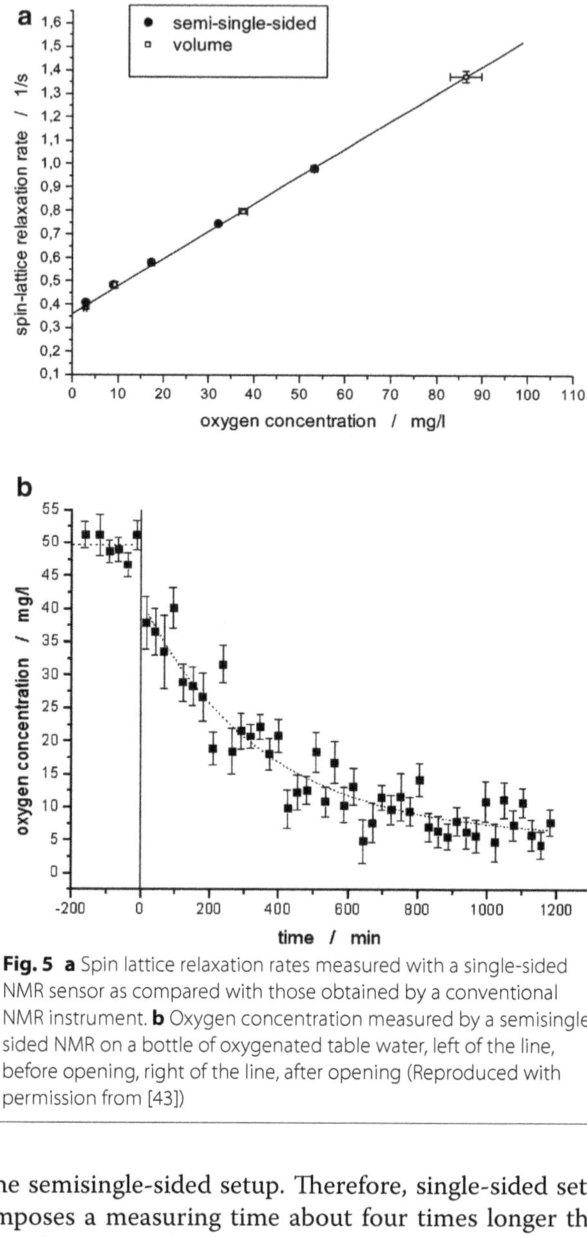

Fig. 5 **a** Spin lattice relaxation rates measured with a single-sided NMR sensor as compared with those obtained by a conventional NMR instrument. **b** Oxygen concentration measured by a semisingle-sided NMR on a bottle of oxygenated table water, left of the line, before opening, right of the line, after opening (Reproduced with permission from [43])

the semisingle-sided setup. Therefore, single-sided setup imposes a measuring time about four times longer than that for semisingle-sided.

Another paper reports a method based on the use of the profile NMR MOUSE [29] to detect adulteration of virgin olive oil through sealed bottles [46]. With this method, the transverse relaxation time and the self-diffusion coefficient of virgin olive oils adulterated with different percentages of sunflower oil and red palm oil were investigated. Authors developed a two-dimensional Laplace inversion according to the algorithm by Venkataramanan [47] to reconstruct the two-dimensional probability density distribution of the transverse relaxation time and self-diffusion coefficient. The two-dimensional D–T_2 distributions were projected to the D axis and T_2 axis, as shown in Fig. 6a, b.

In Fig. 6a, from right to left, there are four peaks representing the pure red palm oil, 10% red palm and 90% olive mixed oil, 20% red palm and 80% olive mixed oil and the pure extra virgin olive oil. The self-diffusion coefficient increases as the percentage of red palm oil increases. While in Fig. 6b, the T_2 decreases as the percentage of red palm oil increases, giving 67.5, 64.1, 60.8 and 54.9 ms, respectively. Comparing with the pure extra virgin olive oil, the red palm mixed oil has a larger self-diffusion coefficient but smaller T_2. These data indicate that the adulteration of extra virgin olive oils with sunflower oil may be readily detected from the self-diffusion coefficient behavior, whereas the adulteration with red palm oil can be detected from both diffusion and transverse relaxation behavior.

The tomato paste processing industry is very interested in developing methods to detect tomato spoilage in 1000-L non-ferrous, metal-lined containers without violating the seal. An early spoilage detection would eliminate shipping costs and disposal costs when spoiled tomato paste arrives at destination. Pinter et al. [48] explored the relaxation properties of sterile and unsterile tomato paste samples as a function of time at both low and room temperature. As long as the storage temperature was maintained below 5 °C, T_1 and T_2 values remain constant over the course of at least 1 month, suggesting that the low temperature significantly slows the bacteria-mediated tomato paste spoilage. At room temperature where tomato processing, packaging and storage occur, the relaxation times for sterile and unsterile samples were different. Although the room temperature T_2 value was constant as a function of time for these samples, the corresponding T_1 value increased exponentially in the first days and tended to a limit value, indicated as T_1^{∞}, for long time of observations. Specifically, the steady-state limit value of spin lattice relaxation time T_1^{∞} obtained by the regression of T_1 values measured as a function of time was the parameter chosen by authors to differentiate between spoiled and unspoiled tomato paste. This result prompted them to extend the study to 1000-L non-ferrous, metal-lined totes using single-sided NMR. To perform measurements directly on the metal container,

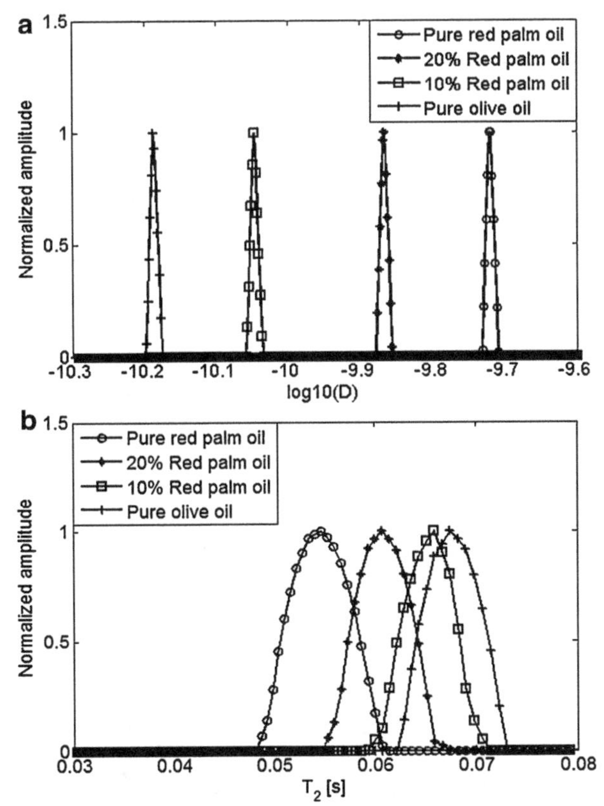

Fig. 6 D-T_2 distribution maps of oils. From the top to the bottom (**a**) pure extra virgin olive oil, extra virgin olive oil adulterated with 10 and 20% of sunflower oil, pure sunflower oil. **b** Pure red palm oil, extra virgin olive oil adulterated with 10 and 20% of red palm oil, and pure extra virgin olive oil (Reproduced with permission from [45])

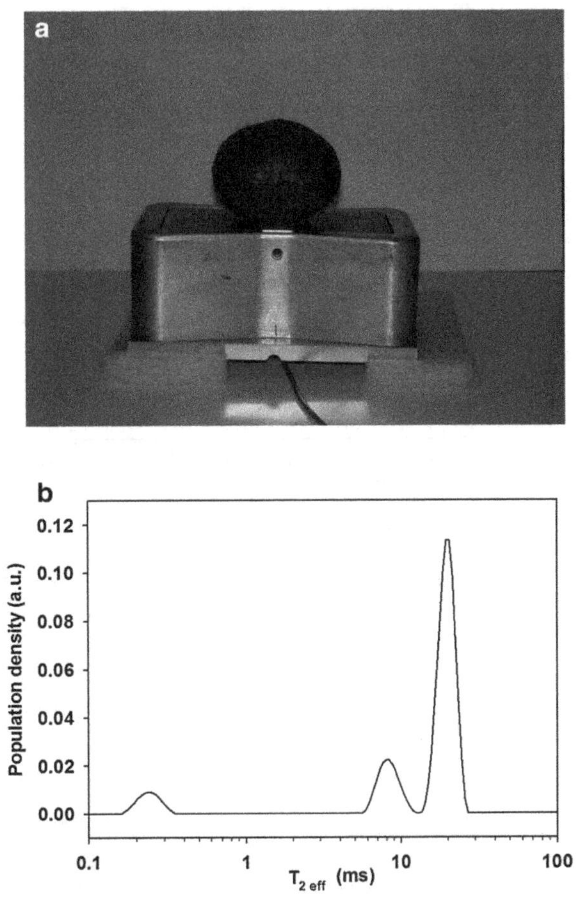

Fig. 7 **a** Measurement of intact kiwifruit with a single-sided NMR sensor. **b** Transverse relaxation times distribution measured on a ripened kiwifruit

its effect on the circuit tuning and impedance matching properties was properly compensated. A modified saturation recovery sequence was used to measure the longitudinal relaxation with single-sided NMR. The NMR signal and T_1 values obtained from the large format container with the single-sided sensor suggested that this device can be used to study tomato paste spoilage in factory process environments.

Capitani et al. monitored the water status of kiwifruits as a function of season using a single-sided NMR sensor [49, 50]. Using this sensor, it was possible to measure the entire fruit at a depth of about 0.5 cm from the peel surface without cutting it, see Fig. 7a. The $T_{2\text{eff}}$ distribution of a mature kiwifruit measured by single-sided NMR shows three peaks, see Fig. 7b. According to the literature, the longest $T_{2\text{eff}}$ component was assigned to

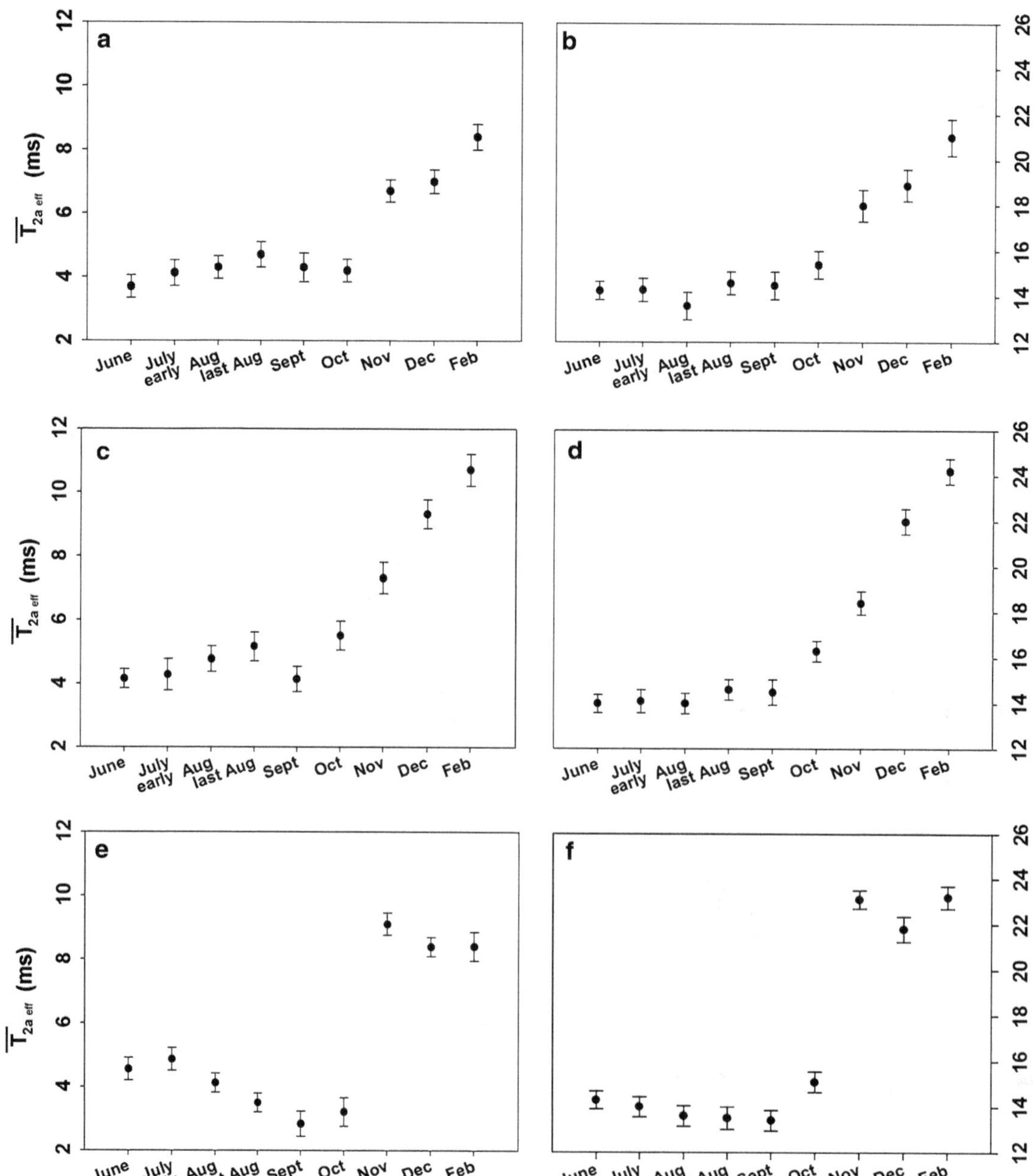

Fig. 8 Average T_2 values for the intermediate ($\bar{T}_{2a}$) and the longest ($\bar{T}_{2b}$) components measured on Hayward (**a**, **b**), CI.GI (**c**, **d**), and Zespri (**e**, **f**) kiwifruits as a function of their developmental stage (Reproduced with permission from [49])

protons in vacuole, the intermediate one to proton in cytoplasm and extra-cellular space, and the shortest one to proton in cell walls. It must be noted that the presence of the strong magnetic field gradient of the single-sided sensor heavily shortens $T_{2\mathrm{eff}}$ values. In fact in the case of ripened kiwifruit measured in homogeneous fields, literature data report average $T_{2\mathrm{eff}}$ values of 800–1000 ms for the longest component, 200–400 ms for the intermediate component, and 20–80 ms for the shortest one. With single-sided NMR, the longest component is as short as 20 ms. Nevertheless, single-sided NMR was suitable to monitor the growth of kiwifruit. Because the shortest component was very poorly affected by the season, only the intermediate and longest components were taken into account. Figure 8 reports the average values of the intermediate component ($\bar{T}_{2a\mathrm{eff}}$) and the longest component ($\bar{T}_{2b\mathrm{eff}}$) measured on nine kiwifruits of three cultivar, namely Hayward (a, b), CI.GI (c, d), and Zespri (e, f) at different stages of development. In all cultivars, $T_{2\mathrm{eff}}$ values were found to be rather constant until October, thereafter they increased. The tendency toward longer $T_{2\mathrm{eff}}$ relaxation times later in the season is consistent with a change in the fruit texture occurring during fruit development. However, whereas in Hayward and CI.GI a gradual lengthening of $T_{2\mathrm{eff}}$ was observed, in Zespri, a net and sharp lengthening of both components occurred between October and November, indicating the earlier maturation of Zespri.

Fig. 9 NMR measurements in field on entire kiwifruits attached to the plant (Reproduced with permission from [49])

The ripening of kiwifruits was also monitored in field on fruits attached to the plant during three campaigns of measurement carried out in October, November and December, see Fig. 9. Figure 10 shows the transverse relaxation decays measured in field. A preliminary optimization of conditions was necessary to achieve a good signal to noise ratio and minimize the measuring times (within 8 min). In all cultivars, a lengthening of the decays with the season was observed; however, the process was complete already in November for Zespri as shown by the perfect overlapping of decays measured in November and December. In agreement with data collected in laboratory on intact kiwifruits, these data confirmed the earlier maturation of Zespri.

The profile NMR MOUSE was used to get preliminary results on three fresh intact blueberries (belonging to the same cultivar) and evaluate the effects of withering outside the fridge for 3 and 6 days [51], see Fig. 11. With this sensor NMR depth profiles were collected, these profiles encode the amplitude of the ^{1}H NMR signal as a function of the depth scanned. The amplitude of profiles measured on withered blueberries was found to be

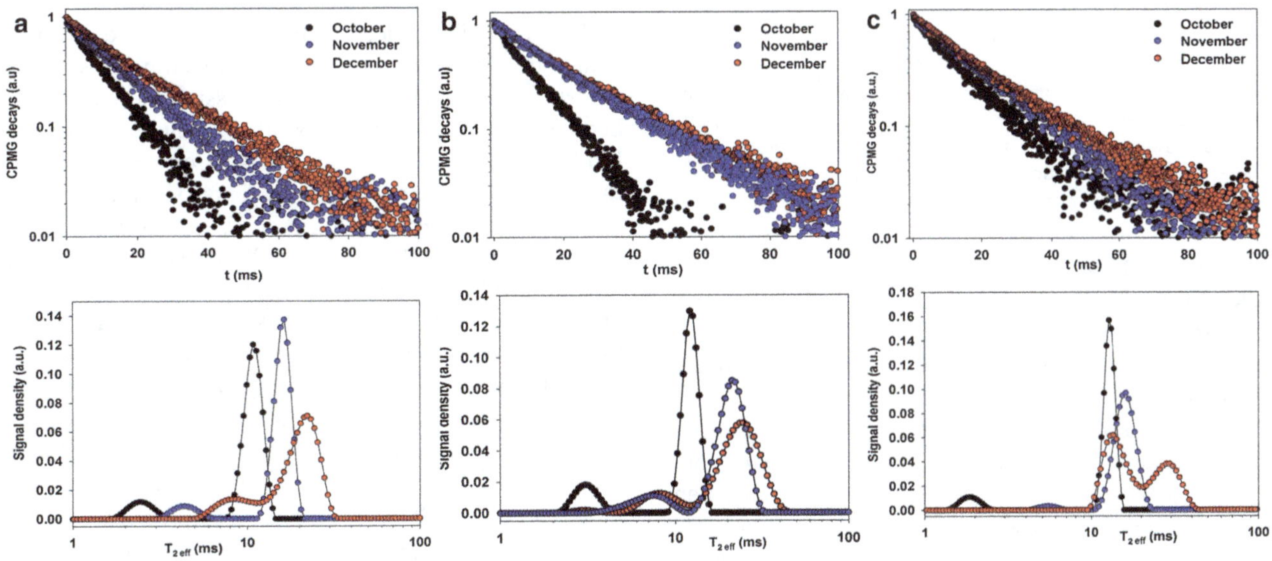

Fig. 10 Transverse relaxation curves measured in field on Hayward (**a**), CI.GI (**b**), and Zespri (**c**) with the corresponding relaxation times distributions

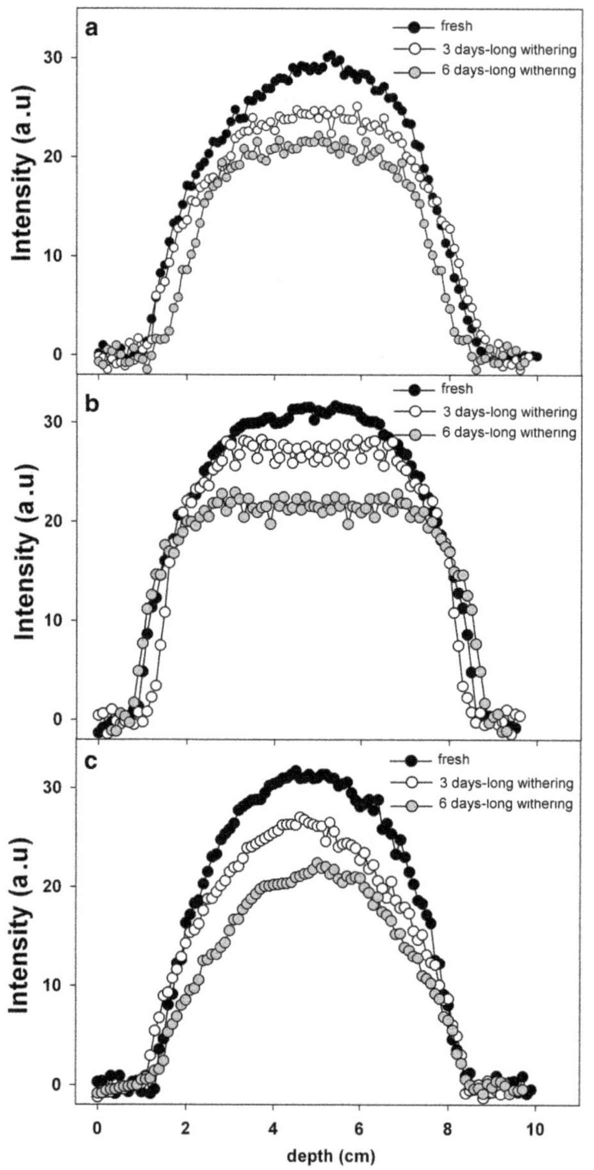

Fig. 11 Depth profiles of three fresh blueberries (**a–c**) and profiles of the same blueberries let to wither for 3 and 6 days. The data reported in **a–c** are related to biological replicates of blueberries belonging to the same cultivar (Reproduced with permission from [51])

lower than that measured on fresh blueberries, indicating a loss of water that was quantified by integrating the profiles. After 3 days of withering, a loss of water of 13% (a), 11% (b), and 16% (c) was measured, whereas after 6 days the loss of water was found to be 30% (a), 20% (b), and 34% (c). Therefore, the integral of the profile is a suitable index of water loss. The same index may be used to monitor changes in foodstuff texture due to maturation, ripening, water and osmotic stress, for monitoring the effect of different types of processing on food matrices, and the effect of storage.

Adiletta et al. used the profile NMR MOUSE to investigate the drying process on pears [52]. Drying is an important process for conservation and marketing of fruits due to their high water activity which makes them perishable. Therefore, the knowledge and optimization of drying process are very important to minimize thermal damage and quality loss. Information on drying kinetics was obtained by measuring the intensity of ^{1}H NMR signal as a function of the thickness of the sample. Figure 12 shows the comparison among profiles of fresh pear and pears dried for 3, 6, 15, 20, 29, and 48 h. The amplitude of profiles measured in dried samples progressively lowered and also the thickness of the profiles progressively decreased, thus indicating a loss of water with increasing of drying time which was accompanied by shrinkage. An evaluation of the loss of water in the outer, intermediate and central regions of pears samples was obtained simply by integrating the profiles in three corresponding regions (without cutting the samples); the results obtained are reported in Fig. 13a. As expected, the water content of the outer region decreases more quickly than that of the intermediate and central ones. Such differences become less pronounced at relatively long drying times. Figure 13b shows the good agreement between values of moisture content (M_t/M_0), obtained by gravimetric method and values of hydrogen content (I_t/I_0) obtained by integrating NMR depth profiles, with a regression coefficient of 0.978.

Concluding remarks

Research applications described above illustrate that unilateral NMR sensors present an attractive option for non-invasive assessment of compositional and microstructure of food materials. The first applications of the unilateral NMR devices in foods were aimed at measuring the water/oil contents in food emulsions and to determine solid fat content. Successively their use has been extended to other foodstuffs as well as at packaged foods. Several applications focused on assessment of food quality and monitoring processes such as ripening, growth rate of fruits, drying process, and so on. The fact that single-sided NMR devices allow easy sample access and portability makes them attractive for quality and process control in the food industry. A main handicap of these devices is magnet field inhomogeneity, which significantly compromises sensitivity and robustness under practical circumstances. The use of echo techniques is typically necessary since FIDs obtained by such devices are also compromised by very short $T_{2\text{eff}}$ values. Other unilateral magnets have been developed with more homogeneous magnetic field, larger sensitive volume and higher sensitivity. These sensors allowed to shorten

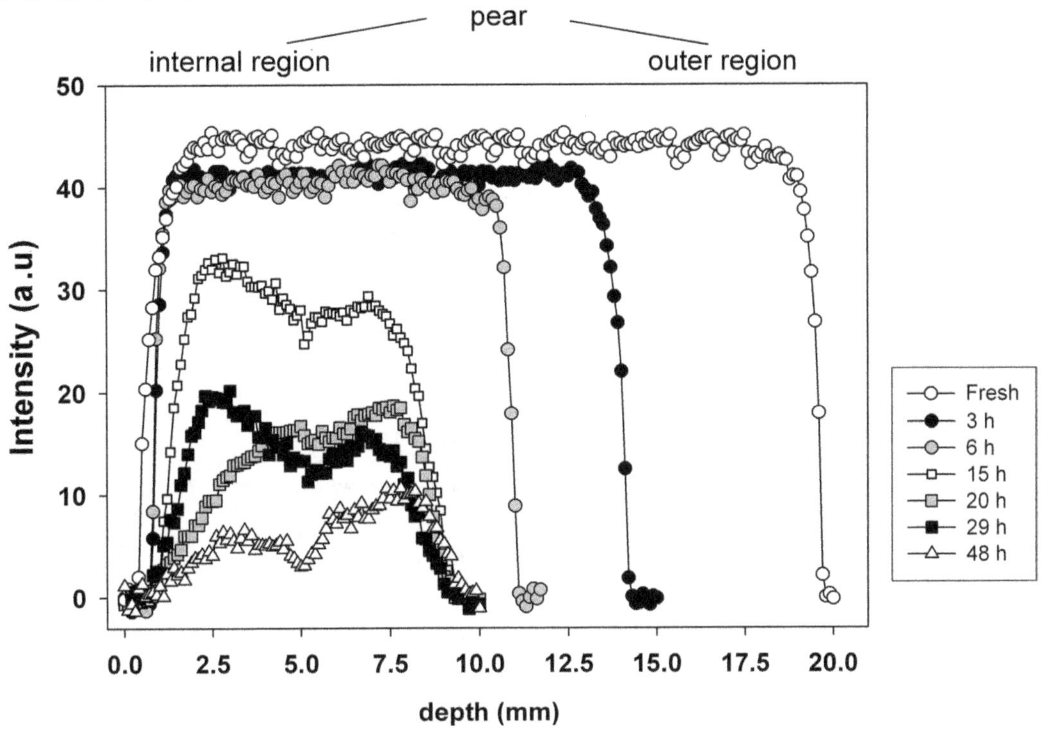

Fig. 12 NMR depth profiles of fresh pear and pear samples dried at 50 °C for 3, 6, 15, 20, 29 and 48 h (Reproduced with permission from [51])

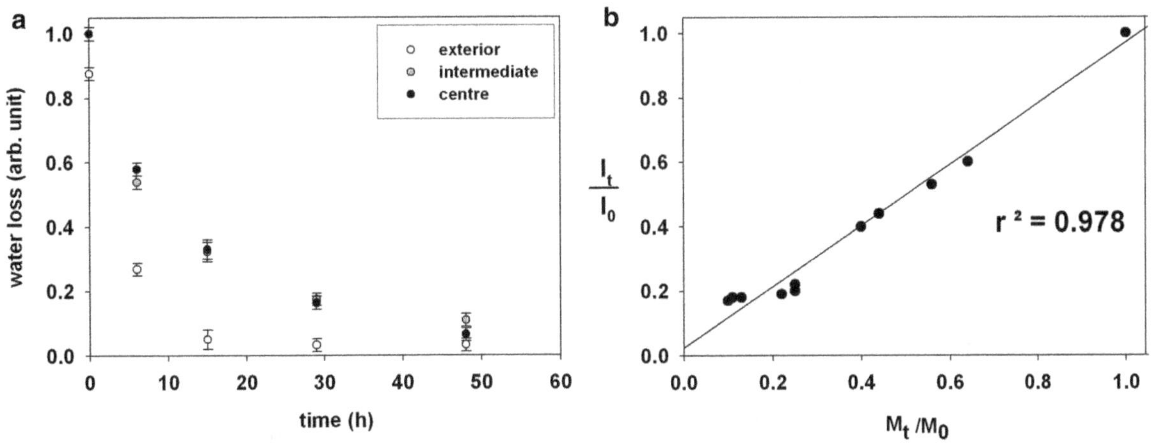

Fig. 13 **a** Water loss in exterior, intermediate, and central part of pear samples measured by single-sided NMR. **b** Relationship between the water loss measured by gravimetric method (M_t/M_0) and single-sided NMR (I_t/I_0) (Reproduced with permission from [51])

measurement times thus paving the way to their future employment in a quality control context.

Nevertheless, use of unilateral NMR devices is still limited. Further studies are needed to expand their utilization to other food matrices and develop protocols to be applied in online inspections for monitoring quality and food processes.

Authors' contributions

All authors contributed equally to this review. All authors read and approved the final manuscript.

Author details

[1] Istituto di Metodologie Chimiche, Laboratorio di Risonanza Magnetica, "Annalaura Segre", CNR, Monterotondo, 00015 Rome, Italy. [2] Dipartimento di Chimica e Tecnologie del Farmaco, Sapienza Università di Roma, Piazzale Aldo Moro 5, 00185 Rome, Italy.

Acknowledgements

This work was supported by "e-ALIERB: un OPEN LAB per caratterizzare e valorizzare i prodotti alimentari ed erboristici del territorio laziale" Project (Regione Lazio LR13/2008).

Competing interests

The authors declare that they have no competing interests.

Funding

This work was supported by "e-ALIERB: un OPEN LAB per caratterizzare e valorizzare i prodotti alimentari ed erboristici del territorio laziale" Project (Regione Lazio LR13/2008).

References

1. van Duynhoven J, Voda A, Witek M, van As H. Time domain NMR applied to food products. Annu Rep NMR Spectrosc. 2010;69:145–97.
2. Hills BP. Applications of low field NMR to food science. Annu Rep NMR Spectrosc. 2006;58:179.
3. Hills BP, Clark CJ. Quality assessment of horticultural products by NMR. Annu Rep NMR Spectrosc. 2003;50:75.
4. Belton PS. Using magnetic resonance to explore food microstructures in food microstructures microscopy, measurements and modelling. In: Morris VJ, Groves K, editors. Cambridge: Woodhead Publishing.
5. Tylewicz U, Panarese V, Laghi L, Rocculi P, Nowacka M, Placucci G, Dalla Rosa M. NMR and DSC water study during osmotic dehydration of *Actinidia deliciosa* and *Actinidia chinensis* Kiwifruit. Food Biophys. 2011;6:327–33.
6. Santagapita P, Laghi L, Panarese V, Tylewicz U, Rocculi P, DallaRosa M. Modification of transverse NMR relaxation times and water diffusion coefficients of kiwifruit pericarp tissue subjected to osmotic dehydration. Food Bioprocess Technol. 2013;6:1434–43.
7. Hills BP, Remigereau B. NMR studies of changes in subcellular water compartmentation in parenchyma apple tissue during drying and freezing. Int J Food Sci Technol. 1997;32:51–61.
8. Marigheto M, Venturi L, Hills BP. Postharvest Biol Technol. 2008;48:331–40.
9. Hills BP. NMR relaxation and diffusion studies of horticultural products in modern magnetic resonance. Netherlands: Springer; 2008. p. 1721–7.
10. Barreiro P, Moya A, Correa E, Ruiz-Altisent M, Fernandez-Valle M, Peirs A, Wright KM, Hills BP. Appl Magn Reson. 2002;22:387.
11. Hills BP, Nott KP. NMR studies of water compartmentation in carrot parenchyma tissue during drying and freezing. Appl Magn Reson. 1999;17:521–35.
12. Marigheto N, Vial A, Wright K, Hills B. A combined NMR and microstructural study of the effect of high-pressure processing on strawberries. Appl Magn Reson. 2004;26:521–31.
13. Marques JPM, Rutledge DN, Ducauze CJ. Low resolution pulse nuclear magnetic resonance study of water equilibration in dried carrots. Int J Food Sci Technol. 1991;26:173.
14. Keener KM, Stroshine RL, Nyenhuis JA. J Am Soc Hortic Sci. 1999;124:289.
15. Cho BK, Chayaprasert W, Stroshine RL. Postharvest Biol Technol. 2008;47:81.
16. Brown RJ, Capozzi F, Cavani C, Cremonini MA, Petracci M, Placucci G. Relationships between ^{1}H NMR relaxation data and some technological parameters of meat: a chemometric approach. J Magn Reson. 2000;147:89–94.
17. Bertram HC, Schafer A, Rosenvold K, Andersen HJ. Physical changes of significance for early post mortem water distribution in porcine *M. longissimus*. Meat Sci. 2004;66:915.
18. Bertram HC, Kristensen M, Andersen HJ. Functionality of myofibrillar proteins as affected by pH, ionic strength and heat treatment—a low-field NMR study. Meat Sci. 2004;68:249.
19. Todt H, Guthausen G, Burk W, Schmalbein D, Kamlowski W. Water/moisture and fat analysis by time-domain NMR. Food Chem. 2006;96:436–40.
20. Zalesskiy SS, Danieli E, Blümich B, Ananikov VP. Miniaturization of NMR systems: desktop spectrometers, microcoil spectroscopy, and "NMR on a chip" for chemistry, biochemistry, and industry. Chem Rev. 2014;114:5641–94.
21. Blümich B. Introduction to compact NMR: a review of methods. Trends Anal Chem. 2016;83:2–11.
22. Marble AE, Mastikhin IV, Colpitts BG, Balcom BJ. A compact permanent magnet array with a remote homogeneous field. J Magn Reson. 2007;186:100–4.
23. Manz B, Coy A, Dykstra R, Eccles CD, Hunter MW, Parkinson BJ, Callaghan PT. J Magn Reson. 2006;183:25–31.
24. Blümich B, Perlo J, Casanova F. Mobile single-side NMR. Progr Nucl Magn Reson Spectrosc. 2008;52:197–269.
25. Eidmann G, Savelsberg R, Blumler P, Blu¨mich B. The NMR mMOUSE: a mobile universal surface explorer. J Magn Reson. 1996;A122:104–9.
26. Blumich B, Anferova S, Sharma S, Segre AL, Federici C. Degradation of historical paper: nondestructive analysis by the NMR-MOUSE. J Magn Reson. 2003;161:204–9.
27. Marble AE, Mastikhin IV, Colpitts BG, Balcom BJ. An analytical methodology for magnetic field control in unilateral NMR. J Magn Reson. 2005;174:78–87.
28. Veliyulin E, Mastikhin IV, Marble AE, Balcom BJ. Rapid determination of the fat content in packaged dairy products by unilateral NMR. J Sci Food Agric. 2008;88:2563–7.
29. Perlo J, Casanova F, Blümich B. Profiles with microscopic resolution by single-sided NMR. J Magn Reson. 2005;176:64–70.
30. Goelman G, Prammer MG. The CPMG pulse sequence in strong magnetic field gradients with applications to oil-well logging. J Magn Reson. 1995;113:11–8.
31. Hurlimann MD, Griffin DD. Spin dynamics of Carr–Purcell–Meiboom–Gill-like sequences in grossly inhomogeneous B0 and B1 fields and application to NMR well logging. J Magn Reson. 2000;143:120–53.
32. Hurlimann MD. Diffusion and relaxation effects in general stray field NMR experiments. J Magn Reson. 2001;148:367–78.
33. Blümich B, Haber-Pohlmeier S, Zia W. Compact NMR. Walter de Gruyter; 2014.
34. Anferova S, Anferov V, Adams M, Fechete R, Schroeder G, Blümich B. Thermo-oxidative aging of elastomers: a temperature control unit for operation with the NMR-MOUSE. Appl Magn Reson. 2004;27(3):361–70.
35. Mitchell J, Gladden LF, Chandrasekera TC, Fordham EJ. Low-field permanent magnets for industrial process and quality control. Progr Nucl Magn Reson Spectrosc. 2014;76:1–60.
36. Guthausen A, Guthausen G, Kamlowski A, Todt H, Burk W, Schmalbein D. Measurement of fat content of food with single-sided NMR. JAOCS. 2004;81:727–31.
37. Guthausen G, Todt H, Burk W, Schmalbein D, Guthausen A, Kamloski A. Single sided NMR in food. In: Modern magnetic resonance. Berlin: Springer; 2006. p.1895–7.
38. Vandeginste RGM, Massart DL, Buydens LMC, De Jong S, Lewi PJ, Smeyers-Verbeke J. Handbook of chemometrics and qualimetrics. Amsterdam: Elsevier Science Publisher; 1998.
39. Guthausen G. Analysis of food and emulsions. Trends Anal Chem. 2016;83:103–6.
40. Martini S, Herrera ML, Marangoni A. New Technologies to determine solid fat content on-line. JAOCS. 2005;82:313–3179.
41. Pedersen HT, Ablett S, Martin DR, Mallet MJD, Engelsen SB. Application of the NMR-MOUSE to food emulsions. J Magn Reson. 2003;165:49–58.
42. Haiduc AM, Trezza EE, van Dusschoten D, Reszka AA, van Duynhoven JPM. Non-invasive through-package assessment of the microstructural quality of a model food emulsion by the NMR MOUSE. Food Sci Technol. 2007;40:737–43.
43. Veliyulin E, van der Zwaag C, Burk W, Erikson U. In vivo determination of fat content in Atlantic salmon (Salmo salar) with a mobile NMR spectrometer. J Sci Food Agric. 2005;85:1299–304.
44. Stork H, Gädke A, Nestle N. Single-sided and semisingle-sided NMR sensors for highly diffusive samples: application to bottled beverages. JAFC. 2006;54:5247–55252.
45. Teng CL, Hong H, Kühne S, Bryant RG. Molecular oxygen spin-lattice relaxation in solutions measured by proton magnetic relaxation dispersion. J Magn Reson. 2001;148:31–4.
46. Xu Z, Morris RH, Bencsik M, Newton MI. Detection of virgin olive oil adulteration using low field unilateral NMR. Sensors. 2014;14:2028–35.

47. Venkataramanan L, Song YQ, Hurlimann MD. Solving Fredholm integrals of the first kind with tensor product structure in 2 and 2.5 dimensions. IEEE Trans Sign Process. 2002;50:1017–26.
48. Pinter MD, Harter T, McCarthy MJ, Augustine MP. Towards using NMR to screen for spoiled tomatoes store in 1,000 L aseptically sealed. Metal-Lined Totes Sensors. 2014;14:4167–76.
49. Capitani D, Mannina L, Proietti N, Sobolev AP, Tomassini A, Miccheli A, Di Cocco ME, Capuani G, De Salvador FR. Monitoring of metabolic profiling and water status of Hayward kiwifruits by nuclear magnetic resonance. Talanta. 2010;82:1826–38.
50. Capitani D, Mannina L, Proietti N, Sobolev AP, Tomassini A, Di Cocco ME, Capuani G, De Salvador FR, Delfini M. metabolic profiling and outer pericarp water state in Zespri, CIGI, and Hayward Kiwifruits. JAFC. 2013;61:1727–40.
51. Capitani D, Sobolev AP, Delfini M, Vista S, Antiochia R, Proietti N, Bubici S, Ferrante G, Carradori S, De Salvador FR, Mannina L. NMR methodologies in the analysis of blueberries. Electrophoresis. 2014;35:1615–26.
52. Adiletta G, Russo P, Proietti N, Capitani D, Mannina L, Crescitelli A, Di Matteo M. Characterization of pears during drying by conventional technique and portable non invasive NMR. Chem Eng Trans. 2015;44:151–6.

6

Panel test and chemical analyses of commercial olive oils: a comparative study

Simone Circi[1], Donatella Capitani[2], Antonio Randazzo[3], Cinzia Ingallina[1], Luisa Mannina[1,2*] and Anatoly P. Sobolev[2]

Abstract

Background: The quality grade of an olive oil is defined according to the results of analytical and organoleptic examinations.The increasing attention towards both olive oil quality and quality verification methods prompted us to undertake a "critical" analysis of analytical and sensory data supplied by an International Certificated Body (ICB), relative to commercial olive oils produced in Mediterranean areas and purchased in Italy and in USA.

Methods: ICB data included chemical analyses namely free acidity, peroxide index, spectrophotometric UV evaluation, fatty acid ethyl esters and stigmadiens content and organoleptic evaluations carried out by nine official International Olive Council labs according to EEC Regulation 2568/91.

Results: The results of the chemical analyses, except the fatty acid ethyl ester content, obtained from the nine labs were consistent giving rise to the same quality grade. In nearly all samples, the fatty acid ethyl ester content was close to the threshold established for extra virgin olive oils indicating a non-excellent quality of the olive oils. Organoleptic evaluations, commonly called panel test, given by the nine labs were not consistent.

Conclusions: The EEC Regulation 2568/91 does not give any indication on the way to report the uncertainty of the results, and in the case of extra virgin olive oils with a borderline value, the way to report the fatty acid ethyl ester content, with or without the uncertainty, can create confusion in defining the olive oil quality grade. Panel test seemed to work well only in the case of extremely good olive oils, whereas, in commercial extra virgin olive oils with borderline value of fatty acid ethyl ester content, a different sensory sensibility seems to be in the different IOC labs.

Keywords: Olive oil, Panel test (organoleptic evaluation), Chemical analyses, Fatty acid ethyl ester content

Background

EEC Regulation 2568/91 [1] and International Olive Council (IOC) [2] have established both analytical and organoleptic criteria to define the quality grade of an olive oil. According to the results of chemical and sensory analyses, an olive oil can be classified as *extra virgin olive oil (EVOO), virgin olive oil or lampante olive oil.* Each category has a completely different commercial and nutritional value. Extra virgin olive oil (EVOO) is a natural fruit juice with peculiar nutritional, healthy [3] and sensory [4] qualities as well as it represents a fundamental component of Mediterranean diet which is very rich in mono-unsaturated fatty acids [5] and polyphenols [6]. On the contrary, lampante olive oils have a distinctly unpleasant smell, are not edible and are used for the production of refined olive oils. It is well known that selling virgin olive oils or lampante olive oils as EVOOs is a fraud. However, another important aspect to consider is that the downgrading of an EVOO to a lower category implies a big economic loss for the producers. So, it is fundamental that both the chemical analyses and the organoleptic evaluations used to define quality grade are "sufficiently" objective and reproducible.

In this paper, a pilot analysis of data supplied by International Certificated Body, relative to 16 olive oils produced in the Mediterranean area, was carried out to verify reproducibility and consistency of chemical analyses and organoleptic evaluations and therefore to

*Correspondence: luisa.mannina@uniroma1.it
[1] Dipartimento di Chimica e Tecnologie del Farmaco, Sapienza Università di Roma, Rome, Italy
Full list of author information is available at the end of the article

highlight possible problems relative to quality grade classification. ICB data, including chemical analyses and organoleptic evaluations, were produced by nine official IOC labs according to EEC Regulation 2568/91. The recent scandals involving Italian olive oils, such as the one revealed in the 'New York Times' [7], prompted us to include data from olive oils purchased not only in Italy, the largest consumer of EVOOs, but also in non-Mediterranean countries such as USA.

Methods

Olive oils (16 samples) of 2014/2015 harvest year were purchased in Italy and Miami (USA) by an International Certificated Body, a recognized EU company that deals with services of certification. ICB also verified the preparation of samples carried out according to EN ISO 5555 reported in EEC Regulation 2568/91, shipped olive oils to laboratories and released a report on all the activity. Samples were collected and labelled with an alphanumeric code without any reference to the origin country. Brands and the packaging places of olive oils were made purposely anonymous to avoid damage from a possible negative or defamatory propaganda.

Chemical analyses and organoleptic evaluation were carried out by nine recognized IOC laboratories, selected from the lists of IOC-recognized laboratories [8, 9], see Additional file 1: Table S1, located in countries with a well-recognized tradition in the production of olive oils such as Italy (5 labs in Abruzzo, Lazio, Liguria, Sicily and Veneto regions), Spain (2 labs), and Greece (1 lab). A Slovenian lab was also included as a reference for the Eastern countries.

Chemical analyses included the determination of free acidity, peroxide index, UV spectrophotometric evaluation (K 232, K 270, ΔK), fatty acid ethyl esters (FAEE) and stigmadiens content. Stigmadiens content, a parameter closely related to the presence of refined oils rather than a quality marker, was also considered as genuineness index. All the analyses were carried out according to the EEC Regulation 2568/91 (Annex II, III, IX, XVII, XX). For the determination of free acidity and peroxide index, the Slovenian lab used IOC method (ISO 660:2009 and ISO 3960:2010, respectively).

According to the official method (EEC Regulation 2568/91, Annex XII), in each IOC lab, organoleptic evaluations were carried out by a group of 8–12 professional tasters trained to recognize, describe and quantify basic taste and odour properties. Olive oils were described through positive ("fruity", "bitter" and "pungent") and negative (for instance "rancid", "fusty", "musty" and "winey") attributes.

Quality grade classifications of olive oils purchased in Italy and in USA were determined by our laboratory according to EEC Regulation 2568/91.

Results and discussion

Chemical analysis results regarding olive oils from Italy and USA are reported in Tables 1 and 2, respectively.

Quality grade classifications of olive oils purchased in Italy and in USA, reported in Tables 3 and 4, respectively, were determined according to EEC Regulation 2568/91 and will be discussed separately. Defects detected by panel test of olive oils purchased in Italy and Miami are reported in Additional files 2: Tables S2 and 3: Table S3, respectively, whereas positive attributes are reported in Additional file 4: Table S4. Classification percentages of the olive oils purchased in Italy and in USA are reported in Additional files 5: Tables S5 and 6: Table S6, respectively.

Samples purchased in Italy

In the case of olive oils purchased in Italy, samples turned out to be EVOOs or virgin olive oils from the chemical point of view (Tables 1, 3 and Additional file 5: Table S5). Five samples (O_i 1, O_i 3, O_i 6, O_i 7 and O_i 8) were judged as EVOOs whereas samples O_i 2, O_i 4 and O_i 5 turned out to be EVOOs or virgin olive oils according to the specific lab. In particular, sample O_i 2 turned out to be EVOO according to the chemical analysis results provided by Sicilian and Venetian labs and virgin olive oil for Abruzzo, Slovenian and Ligurian labs, whereas sample O_i 4 turned out to be EVOO according to Abruzzo and Sicilian labs and virgin olive oil for Slovenian, Ligurian and Venetian labs. Sample O_i 5 was classified virgin olive oil according to all the labs except the Venetian one. The parameter responsible for the declassing of O_i 2, O_i 4 and O_i 5 samples from EVOOs to not EVOOs was the content of FAEE. The importance of FAEE content to define the quality of an olive oil is well known [10, 11]: fatty acid ethyl esters are the products of the trans-esterification reaction between fatty acids and ethanol produced by bacteria fermentation that occurs when olives are of poor quality. Because of this narrow relationship between olive oil quality and FAEE content, the threshold for FAEE content in EVOOs has been decreased along the time, from 40 mg/kg in 2013/2014 harvest year down to 35 mg/kg in 2014/2015. Although this limit has been further reduced to ≤30 mg/kg in 2015, it has been set again to 35 mg/kg [12, 13] on July 2016 by IOC and on September 2016 by EEC.

It is interesting to note, see Table 1, that the FAEE content was reported by different IOC labs with or without uncertainty. For instance, IOC lab located in Veneto (Italy) never provided the uncertainty related to FAEE measurement. Sample O_i 4 with a FAEE content of 39.7 mg/kg reported without any uncertainty has to be classified as virgin olive oil, whereas O_i 1 with a FAEE content of 28 mg/kg has to be classified as extra

Table 1 Chemical analysis results performed by IOC laboratories on eight olive oils purchased in Italy

	Free acidity (% oleic acid)	Peroxide index (meq O_2/kg)	K 232	K 270	Δ*K*	FAEE (mg/kg)	Stigmadienes (mg/kg)
O_i 1							
Greece	/*	/	/	/	/	/	/
Slovenia	0.26 ± 0.03[1]	9.8 ± 1.8[2]	1.80 ± 0.04[3]	0.147 ± 0.009[4]	−0.002 ± 0.001[5]	30.2 ± 4.2[6]	0.0491 ± 0.0280
Spain 1	/	/	/	/	/	/	/
Abruzzo (IT)	0.23 ± 0.02	11.1 ± 0.1	1.79 ± 0.04	0.151 ± 0.040	−0.001 ± 0.001	29 ± 1	0.04 ± 0.01
Lazio (IT)	/	/	/	/	/	/	/
Liguria (IT)	0.2 ± 0.1	9 ± 2	1.88 ± 0.16	0.20 ± 0.06	0.00 ± 0.01	32 ± 8	0.06 ± 0.01
Sicily (IT)	0.24 ± 0.05	11 ± 3	1.77 ± 0.05	0.14 ± 0.02	0.002 ± 0.006	28 ± 8	0.04 ± 0.02
Veneto (IT)	0.28 ± 0.06	10.0 ± 2.3	2.12 ± 0.39	0.16 ± 0.04	0.002 ± 0.002	28	0.04
O_i 2							
Greece	/	/	/	/	/	/	/
Slovenia	0.35 ± 0.30	12.8 ± 2.3	2.07 ± 0.04	0.176 ± 0.009	0.001 ± 0.001	36.8 ± 5.1	0.0376 ± 0.0088
Spain 1	/	/	/	/	/	/	/
Abruzzo (IT)	0.34 ± 0.02	13.5 ± 0.1	1.891 ± 0.004	0.160 ± 0.004	0.0040 ± 0.0001	37 ± 1	0.03 ± 0.01
Lazio (IT)	/	/	/	/	/	/	/
Liguria (IT)	0.3 ± 0.1	13 ± 2	1.96 ± 0.17	0.19 ± 0.06	0.00 ± 0.01	38 ± 10	0.03 ± 0.01
Sicily (IT)	0.36 ± 0.05	13 ± 3	1.95 ± 0.05	0.21 ± 0.02	0.0000 ± 0.0006	30 ± 9	0.05 ± 0.02
Veneto (IT)	0.35 ± 0.06	13.0 ± 2.3	2.04 ± 0.39	0.17 ± 0.04	0.001 ± 0.002	33	0.02
O_i 3							
Greece	/	/	/	/	/	/	/
Slovenia	0.24 ± 0.03	7.3 ± 1.3	1.78 ± 0.04	0.201 ± 0.009	0.004 ± 0.001	24.6 ± 3.4	0.0238 ± 0.0068
Spain 1	/	/	/	/	/	/	/
Abruzzo (IT)	0.23 ± 0.02	9.4 ± 0.1	1.615 ± 0.004	0.168 ± 0.004	0.0090 ± 0.0001	25 ± 1	0.01 ± 0.01
Lazio (IT)	/	/	/	/	/	/	/
Liguria (IT)	0.2 ± 0.1	8 ± 2	1.74 ± 0.15	0.20 ± 0.06	0.00 ± 0.01	27 ± 7	0.01 ± 0.01
Sicily (IT)	0.23 ± 0.05	8 ± 3	1.60 ± 0.05	0.17 ± 0.02	0.004 ± 0.006	20 ± 6	0.05 ± 0.02
Veneto (IT)	0.24 ± 0.06	8.0 ± 2.3	1.82 ± 0.39	0.18 ± 0.04	0.004 ± 0.002	23.7	0.03
O_i 4							
Greece	/	/	/	/	/	/	/
Slovenia	0.28 ± 0.03	6.2 ± 1.1	1.79 ± 0.04	0.122 ± 0.009	0.000 ± 0.001	35.1 ± 4.9	< 0.01
Spain 1	/	/	/	/	/	/	/
Abruzzo (IT)	0.28 ± 0.02	7.9 ± 0.1	1.760 ± 0.004	0.117 ± 0.004	0.000 ± 0.001	35 ± 1	0.03 ± 0.01
Lazio (IT)	/	/	/	/	/	/	/
Liguria (IT)	0.3 ± 0.1	8 ± 2	1.84 ± 0.16	0.13 ± 0.04	0.00 ± 0.01	40 ± 12	0.01 ± 0.01
Sicily (IT)	0.27 ± 0.05	8 ± 3	1.78 ± 0.05	0.12 ± 0.02	0.004 ± 0.006	33 ± 9	0.05 ± 0.02
Veneto (IT)	0.26 ± 0.06	7.0 ± 2.3	1.98 ± 0.39	0.13 ± 0.04	0.001 ± 0.002	39.7	0.02
O_i 5							
Greece	/	/	/	/	/	/	/
Slovenia	0.39 ± 0.03	8.2 ± 1.5	1.86 ± 0.04	0.140 ± 0.009	−0.002 ± 0.001	35.5 ± 5.0	0.0136 ± 0.0025
Spain 1	/	/	/	/	/	/	/
Abruzzo (IT)	0.36 ± 0.02	9.1 ± 0.1	1.812 ± 0.004	0.138 ± 0.004	−0.001 ± 0.001	36 ± 1	0.02 ± 0.01
Lazio (IT)	/	/	/	/	/	/	/
Liguria (IT)	0.4 ± 0.1	8 ± 2	1.96 ± 0.17	0.16 ± 0.05	0.00 ± 0.01	39 ± 10	0.01 ± 0.01
Sicily (IT)	0.36 ± 0.05	9 ± 3	1.69 ± 0.05	0.12 ± 0.02	0.001 ± 0.006	36 ± 10	0.04 ± 0.02
Veneto (IT)	0.38 ± 0.06	8.0 ± 2.3	1.96 ± 0.39	0.16 ± 0.04	0.002 ± 0.002	35	0.04
O_i 6							
Greece	/	/	/	/	/	/	/
Slovenia	0.19 ± 0.03	8.9 ± 1.6	2.21 ± 0.04	0.149 ± 0.009	−0.001 ± 0.001	11.7 ± 1.6	0.0129 ± 0.0023
Spain 1	/	/	/	/	/	/	/

Table 1 continued

	Free acidity (% oleic acid)	Peroxide index (meq O_2/kg)	K 232	K 270	ΔK	FAEE (mg/kg)	Stigmadienes (mg/kg)
Abruzzo (IT)	0.16 ± 0.02	10.2 ± 0.1	2.089 ± 0.004	0.136 ± 0.004	−0.001 ± 0.001	14 ± 1	0.02 ± 0.01
Lazio (IT)	/	/	/	/	/	/	/
Liguria (IT)	0.2 ± 0.1	9 ± 2	2.16 ± 0.19	0.17 ± 0.06	0.00 ± 0.01	10 ± 3	< 0.01 ± 0.01
Sicily (IT)	0.16 ± 0.05	11 ± 3	2.12 ± 0.05	0.16 ± 0.02	0.000 ± 0.006	10 ± 4	0.03 ± 0.02
Veneto (IT)	0.18 ± 0.06	9.0 ± 2.3	2.26 ± 0.39	0.16 ± 0.04	0.002 ± 0.002	11	0.02
O_i 7							
Greece	/	/	/	/	/	/	/
Slovenia	0.23 ± 0.03	7.9 ± 1.4	1.86 ± 0.04	0.126 ± 0.009	−0.003 ± 0.001	24.4 ± 3.4	0.0176 ± 0.0032
Spain 1	/	/	/	/	/	/	/
Abruzzo (IT)	0.23 ± 0.02	9.7 ± 1.0	1.807 ± 0.004	0.126 ± 0.004	−0.003 ± 0.001	30 ± 1	0.02 ± 0.01
Lazio (IT)	/	/	/	/	/	/	/
Liguria (IT)	0.2 ± 0.1	8 ± 2	1.87 ± 0.16	0.15 ± 0.05	0.00 ± 0.01	32 ± 8	0.01 ± 0.01
Sicily (IT)	0.22 ± 0.05	8 ± 3	1.87 ± 0.05	0.15 ± 0.02	0.003 ± 0.006	20 ± 6	0.02 ± 0.02
Veneto (IT)	0.25 ± 0.06	9.0 ± 2.3	1.99 ± 0.39	0.14 ± 0.04	0.004 ± 0.002	22	0.04
O_i 8							
Greece	/	/	/	/	/	/	/
Slovenia	0.31 ± 0.03	11.6 ± 2.1	2.38 ± 0.04	0.152 ± 0.009	−0.002 ± 0.001	29.0 ± 4.1	0.0161 ± 0.0300
Spain 1	/	/	/	/	/	/	/
Abruzzo (IT)	0.32 ± 0.01	13.1 ± 0.1	2.049 ± 0.004	0.143 ± 0.004	−0.002 ± 0.001	33 ± 1	0.03 ± 0.01
Lazio (IT)	/	/	/	/	/	/	/
Liguria (IT)	0.3 ± 0.1	11 ± 2	2.57 ± 0.22	0.18 ± 0.06	0.00 ± 0.01	31 ± 8	0.01 ± 0.01
Sicily (IT)	0.32 ± 0.05	13 ± 3	2.35 ± 0.05	0.15 ± 0.02	0.002 ± 0.006	25 ± 7	0.04 ± 0.02
Veneto (IT)	0.33 ± 0.06	13.0 ± 2.3	2.49 ± 0.39	0.16 ± 0.04	0.003 ± 0.002	27	0.04

/* Analysis was not performed

The limits of the parameters of quality and genuineness of an EVOO are (1) ≤0.8, (2) ≤20, (3) ≤2.50, (4) ≤0.22, (5) ≤0.01, (6) ≤35 (for crop year 2014/2015)

The limits of the parameters of a virgin olive oil are 2≤ (1) ≤0.8, (2) ≤20, (3) ≤2.60, (4) ≤0.25, (5) ≤0.01

virgin olive oil. On the other hand, some labs reported a value higher than 35 mg/kg (the threshold value for year 2014/2015) with the threshold value within the uncertainty, see for example the results of Slovenian lab reported for sample O_i 2 with a FAEE content of 36.8 ± 5.1 mg/kg. Finally, other labs reported a value lower than 35 mg/kg with the threshold within the uncertainty, see for example the results of Ligurian lab for sample O_i 1 with a FAEE content of 32 ± 8 mg/kg. How to use these data to classify olive oils? A strict interpretation of EEC Regulation 2568/91 that does not give any indication on the way to report the uncertainty of the results implies that any olive oil with a FAEE content value above the threshold has to be declassified from extra virgin olive oil to not extra virgin olive. Therefore, O_i 2 with a FAEE content of 36.8 ± 5.1 mg/kg has to be classified as not extra virgin olive oil, whereas O_i 1 with a FAEE content of 32 ± 8 mg/kg has to be classified as extra virgin olive oil. Whereas an analytical point of view to report a result with the uncertainty is a correct way, to consider the uncertainty for the olive oil classification can give rise to confusion. In fact, if the uncertainty is considered, these two olive oils should have the same grade, i.e. extra virgin olive oil. On the other hand, taking into account the upper limit of uncertainty, an olive oil with a value of 32 ± 8 mg/kg could be considered not extra virgin. Therefore, the way to report the FAEE content with the uncertainty can create confusion, when the olive oil commercial classification is required.

Organoleptic evaluation carried out from the different labs were extremely controversial. For instance, all the olive oils were judged as EVOOs according to the results of the Sicilian lab, whereas no sample was judged as EVOO according to the results of the Greek lab. These results clearly show that the same olive oil can be judged EVOO, virgin olive oil or lampante olive oil according to the the lab. This is the case of samples O_i 2, O_i 3, O_i 4, O_i 5, whereas samples O_i 1, O_i 6, O_i 7, O_i 8 were judged virgin olive oils or EVOOs. Only sample O_i 6 was judged EVOO according to the results of all the labs except the Greek one. It is important to note that this olive oil has a FAEE concentration <15 mg/kg well below the threshold

Table 2 Chemical analysis results performed by IOC laboratories on eight olive oils purchased in USA (Miami)

	Free acidity (% oleic acid)	Peroxide index (meq O_2/kg)	K 232	K 270	ΔK	FAEE (mg/kg)	Stigmadienes (mg/kg)
O_u 1							
Greece	/*	/	/	/	/	/	/
Slovenia	0.41 ± 0.03	9.1 ± 1.6	2.17 ± 0.04	0.130 ± 0.009	−0.002 ± 0.001	34.5 ± 4.8	0.0117 ± 0.0021
Spain 1	/	/	/	/	/	/	/
Spain 2	0.39	10.60	2.27	0.13	<0.01	38.80	<0.05
Lazio (IT)	/	/	/	/	/	/	/
Liguria (IT)	0.4 ± 0.1	10 ± 2	2.32 ± 0.20	0.14 ± 0.04	0.00 ± 0.01	35 ± 9	<0.01 ± 0.01
Sicily (IT)	0.40 ± 0.05	10 ± 3	2.04 ± 0.05	0.09 ± 0.02	0.002 ± 0.006	37 ± 10	0.01 ± 0.02
Veneto (IT)	0.40 ± 0.06	10.0 ± 2.3	2.45 ± 0.39	0.15 ± 0.04	0.003 ± 0.002	34	0.02
O_u 2							
Greece	/	/	/	/	/	/	/
Slovenia	0.27 ± 0.03	16.0 ± 2.9	2.79 ± 0.04	0.225 ± 0.009	0.004 ± 0.001	<15	0.089 ± 0.016
Spain 1	/	/	/	/	/	/	/
Spain 2	/	/	/	/	/	/	/
Lazio (IT)	/	/	/	/	/	/	/
Liguria (IT)	0.3 ± 0.1	18 ± 2	2.98 ± 0.26	0.23 ± 0.07	0.00 ± 0.01	22 ± 6	0.10 ± 0.02
Sicily (IT)	0.28 ± 0.05	20 ± 3	2.76 ± 0.05	0.24 ± 0.02	0.002 ± 0.006	22 ± 7	0.10 ± 0.03
Veneto (IT)	0.27 ± 0.06	17.8 ± 2.3	3.08 ± 0.39	0.24 ± 0.04	0.002 ± 0.002	15	0.06
O_u 3							
Greece	/	/	/	/	/	/	/
Slovenia	0.41 ± 0.03	12.0 ± 2.2	2.84 ± 0.04	0.266 ± 0.009	0.006 ± 0.001	18.6 ± 2.6	0.0392 ± 0.0071
Spain 1	/	/	/	/	/	/	/
Spain 2	0.41	16.90	2.71	0.28	0.01	27.60	<0.05
Lazio (IT)	/	/	/	/	/	/	/
Liguria (IT)	0.4 ± 0.1	14 ± 2	2.98 ± 0.26	0.28 ± 0.09	0.00 ± 0.01	25 ± 6	0.05 ± 0.01
Sicily (IT)	0.43 ± 0.05	16 ± 3	2.79 ± 0.05	0.29 ± 0.02	0.006 ± 0.006	28 ± 8	0.05 ± 0.02
Veneto (IT)	0.41 ± 0.06	14.6 ± 2.3	3.02 ± 0.39	0.30 ± 0.04	0.005 ± 0.002	28	0.06
O_u 4							
Greece	/	/	/	/	/	/	/
Slovenia	0.21 ± 0.03	7.9 ± 1.4	1.90 ± 0.04	0.140 ± 0.009	−0.001 ± 0.001	39.0 ± 5.5	0.0162 ± 0.0029
Spain 1	/	/	/	/	/	/	/
Spain 2	0.21	8.80	2.05	0.14	<0.01	47.80	<0.05
Lazio (IT)	/	/	/	/	/	/	/
Liguria (IT)	0.2 ± 0.1	8 ± 2	2.00 ± 0.17	0.16 ± 0.05	0.00 ± 0.01	45 ± 11	0.01 ± 0.01
Sicily (IT)	0.21 ± 0.05	10 ± 3	1.85 ± 0.05	0.14 ± 0.02	0.007 ± 0.006	49 ± 13	0.02 ± 0.02
Veneto (IT)	0.21 ± 0.06	8.3 ± 2.3	2.12 ± 0.39	0.18 ± 0.04	0.002 ± 0.002	46	0.05
O_u 5							
Greece	/	/	/	/	/	/	/
Slovenia	0.21 ± 0.03	8.6 ± 1.6	1.92 ± 0.04	0.238 ± 0.009	0.007 ± 0.001	39.5 ± 5.5	0.0331 ± 0.0060
Spain 1	/	/	/	/	/	/	/
Spain 2	0.20	9.60	2.04	0.24	0.01	43.20	<0.05
Lazio (IT)	/	/	/	/	/	/	/

Table 2 continued

	Free acidity (% oleic acid)	Peroxide index (meq O_2/kg)	K 232	K 270	Δ*K*	FAEE (mg/kg)	Stigmadienes (mg/kg)
Liguria (IT)	0.3 ± 0.1	9 ± 2	2.16 ± 0.19	0.27 ± 0.08	0.01 ± 0.01	43 ± 11	0.03 ± 0.01
Sicily (IT)	0.22 ± 0.05	10 ± 3	1.99 ± 0.05	0.26 ± 0.02	0.007 ± 0.006	48 ± 13	0.03 ± 0.02
Veneto (IT)	0.22 ± 0.06	8.7 ± 2.3	2.10 ± 0.39	0.25 ± 0.04	0.006 ± 0.002	38	0.03
O_u 6							
Greece	/	/	/	/	/	/	/
Slovenia	0.25 ± 0.03	9.9 ± 1.8	2.00 ± 0.04	0.129 ± 0.009	−0.003 ± 0.001	29.7 ± 4.2	0.0084 ± 0.0015
Spain 1	/	/	/	/	/	/	/
Spain 2	0.23	13.20	2.04	0.13	<0.01	35.00	<0.05
Lazio (IT)	/	/	/	/	/	/	/
Liguria (IT)	0.30 ± 0.01	11 ± 2	2.03 ± 0.18	0.14 ± 0.05	0.00 ± 0.01	32 ± 8	0.01 ± 0.01
Sicily (IT)	0.39 ± 0.05	9 ± 3	1.93 ± 0.05	0.13 ± 0.02	0.003 ± 0.006	34 ± 9	0.01 ± 0.02
Veneto (IT)	0.24 ± 0.06	10.9 ± 2.3	2.14 ± 0.39	0.15 ± 0.04	0.004 ± 0.002	29	0.05
O_u 7							
Greece	/	/	/	/	/	/	/
Slovenia	0.27 ± 0.03	10.0 ± 1.8	1.95 ± 0.04	0.148 ± 0.009	−0.001 ± 0.001	63.8 ± 8.9	0.0257 ± 0.0046
Spain 1	/	/	/	/	/	/	/
Spain 2	0.26	11.80	1.99	0.14	<0.01	67.40	<0.05
Lazio (IT)	/	/	/	/	/	/	/
Liguria (IT)	0.3 ± 0.1	10 ± 2	2.07 ± 0.18	0.14 ± 0.05	0.00 ± 0.01	65 ± 17	0.01 ± 0.01
Sicily (IT)	0.27 ± 0.05	12 ± 3	1.95 ± 0.05	0.15 ± 0.02	0.001 ± 0.006	72 ± 18	0.01 ± 0.02
Veneto (IT)	0.26 ± 0.06	10.7 ± 2.3	2.17 ± 0.39	0.16 ± 0.04	0.002 ± 0.002	60	0.02
O_u 8							
Greece	/	/	/	/	/	/	/
Slovenia	0.32 ± 0.03	8.0 ± 1.4	2.01 ± 0.04	0.145 ± 0.009	−0.002 ± 0.001	29.8 ± 4.2	<0.01
Spain 1	/	/	/	/	/	/	/
Spain 2	0.30	12.00	2.16	0.15	<0.01	33.20	<0.05
Lazio (IT)	/	/	/	/	/	/	/
Liguria (IT)	0.3 ± 0.1	9 ± 2	2.15 ± 0.19	0.15 ± 0.04	0.00 ± 0.01	29 ± 8	0.01 ± 0.01
Sicily (IT)	0.31 ± 0.05	8 ± 3	2.03 ± 0.05	0.16 ± 0.02	0.002 ± 0.006	32 ± 9	0.01 ± 0.02
Veneto (IT)	0.29 ± 0.06	10.1 ± 2.3	2.20 ± 0.39	0.16 ± 0.04	0.003 ± 0.002	28	0.04

/* Analysis was not performed

of 35 mg/kg, whereas the other olive oils exhibit a FAEE content which is slightly less or higher than the threshold value. The strict connection between FAEE content and the positive attributes by organoleptic evaluation of an olive oil has been previously reported [10, 14].

Therefore, all the obtained results seem to suggest that only in the case of olive oils with extremely "good" chemical analyses, including a low FAEE content (neither borderline nor higher than the threshold value), the organoleptic evaluation can be reproducible.

Samples purchased in USA

Olive oils purchased in Miami (USA) were judged, see Tables 2 and 4 and Additional file 6: Table S6, EVOOs (O_u 6 and O_u 8), virgin olive oils (O_u 4, O_u 5 and O_u 7) and lampante olive oils (O_u 2 and O_u 3) according to the chemical analysis results. The judgment given by the different labs for these five samples was univocal: samples O_u 2 and O_u 3 turned out to be lampante olive oils because of their high K 232 value (>2.60) whereas samples O_u 4 and O_u 7 were classified as virgin olive oil due to

Table 3 Olive oils purchased in Italy: commercial categories as determined according to EEC Regulation 2568/91

Sample	Origin		LAB-Greece	LAB-Slovenia	LAB-Spain 1	LAB-Abruzzo (IT)	LAB-Lazio (IT)	LAB-Liguria (IT)	LAB-Sicily (IT)	LAB-Veneto (IT)
O_i 1	U.E. (Greece/Italy)	Organoleptic evaluation	Virgin	Virgin	Virgin	Virgin	Virgin	Extra	Extra	Extra
		Chemical analysis	–	Extra	–	Extra	–	Extra	Extra	Extra
O_i 2	U.E.	Organoleptic evaluation	Lampante	Virgin	Lampante	Virgin	Virgin	Virgin	Extra	Virgin
		Chemical analysis	–	Virgin	–	Virgin	–	Virgin	Extra	Extra
O_i 3	U.E.	Organoleptic evaluation	Lampante	Virgin	Lampante	Virgin	Virgin	Virgin	Extra	Virgin
		Chemical analysis	–	Extra	–	Extra	–	Extra	Extra	Extra
O_i 4	U.E.	Organoleptic evaluation	Lampante	Virgin	Virgin	Virgin	Virgin	Virgin	Extra	Virgin
		Chemical analysis	–	Virgin	–	Extra	–	Virgin	Extra	Virgin
O_i 5	U.E.	Organoleptic evaluation	Lampante	Virgin	Virgin	Virgin	Virgin	Extra	Extra	Extra
		Chemical analysis	–	Virgin	–	Virgin	–	Virgin	Virgin	Extra
O_i 6	U.E.	Organoleptic evaluation	Virgin	Extra	Extra	Extra	Extra	Extra	Extra	Extra
		Chemical analysis	–	Extra	–	Extra	–	Extra	Extra	Extra
O_i 7	U.E.	Organoleptic evaluation	Virgin	Virgin	Virgin	Virgin	Virgin	Extra	Extra	Extra
		Chemical analysis	–	Extra	–	Extra	–	Extra	Extra	Extra
O_i 8	U.E. and not U.E. (Italy/Spain/Tunisia)	Organoleptic evaluation	Virgin	Virgin	Virgin	Virgin	Extra	Extra	Extra	Virgin
		Chemical analysis	–	Extra	–	Extra	–	Extra	Extra	Extra

Extra extra virgin olive oil, *Virgin* virgin olive oil, *Lampante* lampante olive oil

Table 4 Olive oils purchased in USA (Miami): commercial categories as determined according to EEC Regulation 2568/91

Sample	Origin		LAB-Greece	LAB-Slovenia	LAB-Spain 1	LAB-Spain 2	LAB-Lazio (IT)	LAB-Liguria (IT)	LAB-Sicily (IT)	LAB-Veneto (IT)
O_u 1	Spain	Organoleptic evaluation	Lampante	Lampante	Virgin	Virgin	Virgin	Virgin	Extra	Virgin
		Chemical analysis	–	Extra	–	Virgin	–	Extra	Virgin	Extra
O_u 2	Italy/Greece/Spain/Turkey/Tunisia/Morocco	Organoleptic evaluation	Lampante	Virgin	Lampante	–	Virgin	Virgin	Extra	Lampante
		Chemical analysis	–	Lampante	–	–	–	Lampante	Lampante	Lampante
O_u 3	Tunisia/Spain	Organoleptic evaluation	Lampante	Virgin	Lampante	Lampante	Virgin	Virgin	Extra	Lampante
		Chemical analysis	–	Lampante	–	Lampante	–	Lampante	Lampante	Lampante
O_u 4	Italy/Greece/Spain/Turkey/Tunisia/Morocco	Organoleptic evaluation	Virgin	Virgin	Virgin	Virgin	Virgin	Extra	Extra	Virgin
		Chemical analysis	–	Virgin	–	Virgin	–	Virgin	Virgin	Virgin
O_u 5	Spain	Organoleptic evaluation	Lampante	Virgin	Lampante	Virgin	Virgin	Virgin	Extra	Virgin
		Chemical analysis	–	Virgin		Virgin	–	Virgin	Virgin	Virgin
O_u 6	Italy/Spain/Greece/Tunisia	Organoleptic evaluation	Extra	Virgin	Extra	Virgin	Virgin	Extra	Extra	Virgin
		Chemical analysis	–	Extra	–	Extra	–	Extra	Extra	Extra
O_u 7	Spain	Organoleptic evaluation	Lampante	Lampante	Lampante	Lampante	Virgin	Virgin	Virgin	Virgin
		Chemical analysis	–	Virgin	–	Virgin	–	Virgin	Virgin	Virgin
O_u 8	Spain/Morocco/Chile/Greece	Organoleptic evaluation	Virgin	Virgin	Virgin	Virgin	Virgin	Virgin	Extra	Virgin
		Chemical analysis	–	Extra	–	Extra	–	Extra	Extra	Extra

Extra extra virgin olive oil, *Virgin* virgin olive oil, *Lampante* lampante olive oil

the high FAEE content (>35 mg/kg). Sample O_u 5 turned out to be virgin olive oil because of the high content of FAEE (>35 mg/kg) in the case of Ligurian and Venetian labs or for both K 270 value (0.22 < K 270 < 0.25) and FAEE content in the case of Slovenian, Sicilian and Spain 2 labs.

Finally, samples O_u 1 and O_u 4 were judged virgin olive oils or EVOOs according to the labs results: the critical point was again the FAEE content given without error or with a high error.

The results of organoleptic evaluation given by the different laboratories turned out to be in disagreement. For instance, all samples were judged virgin olive oils due to a median of defects >0 according to the Lazio lab results, whereas all samples, except one, were judged EVOOs according to the Sicilian lab results. Samples O_u 1, O_u 2, O_u 3 and O_u 5 turned out to be lampante olive oils, virgin olive oils or EVOOs according to the lab, whereas sample O_u 7 was judged lampante olive oil or virgin olive oil. Finally, samples O_u 4, O_u 6 and O_u 8 were judged virgin olive oils or EVOOs according to the labs.

Again, this disagreement of sensory evaluation can be related to some borderline value in the chemical parameters. It is interesting to note that the only two samples, O_u 6 and O_u 8, judged EVOOs by the chemical analysis, with a slightly less FAEE content than the threshold value were classified as EVOOs or virgin olive oils by the organoleptic evaluation. A similar case is also represented by the sample O_u 1 which, through chemical analyses, was judged EVOO in three cases out of five with a borderline FAEE content, whereas, through organoleptic evaluation, it was classified as EVOO, virgin olive oil or even lampante olive oil.

Conclusions

The results reported in this paper highlight some important aspects to be considered in the olive oil commercial classification.

First of all, the non-uniform way to report the FAEE content with or without uncertainty can create confusion in the quality grade classification, especially in case of borderline values of FAEE content. It is important to underline that the EEC Regulation 2568/91 does not give any indication on the way to report the uncertainty of the results.

Our data analysis underlines that it is crucial to find out what causes the variability of judgement in the organoleptic evaluation. Sensory panel test seems to work well in the case of extremely good olive oils, whereas in the case of common commercial EVOOs can give discordant results. A different sensory sensibility seems to be present in the different IOC panel labs, especially in case of EVOOs characterized by a FAEE content very close to the threshold. For a given olive oil, the question to be or not to be classified as EVOO becomes a question strictly linked to the lab.

In our opinion, organoleptic evaluation is extremely important for a global picture of the sensory properties of olive oil and it is particularly precious and not replaceable with other analysis to give add values to PDO or other peculiar EVOOs. On the other hand, this analysis seems to be not enough reproducible in the case of common commercial EVOOs probably due to different sensory sensibility in the different IOC panel labs and therefore it is not suitable for a legally accepted official evaluation.

Additional files

Additional file 1: Table S1. IOC laboratories chosen to carry out chemical analyses and organoleptic evaluation.

Additional file 2: Table S2. Defects, resulting by organoleptic evaluation performed by the eight IOC laboratories, of the eight olive oils purchased in Italy. The median values are reported for each defect. In certain cases, some laboratories provided only the predominant defect of an olive oil despite the detection of more defects. Labs of Spain 1 and Greece just supplied the median of the predominant defect without further details.

Additional file 3: Table S3. Defects, resulting by organoleptic evaluation performed by 8 IOC laboratories, of the eight olive oils purchased in USA (Miami). The median values are reported for each defect. In certain cases, some laboratories provided only the predominant defect of an olive oil despite the detection of more defects. Labs of Spain 2 and Greece just supplied the median of the predominant defect without further details. *Sample O_u 2 was not analyzed by Spain 2 lab.

Additional file 4: Table S4. Positive attributes, resulting by organoleptic evaluation performed by 8 IOC laboratories, of the eight olive oils purchased in Italy and the eight olive oils purchased in USA (Miami). The median values are reported for each defect. Results of "Bitter" and "Pungent" attributes were not provided by all laboratories. *Sample O_u 2 was not analyzed by Spain 2 lab.

Additional file 5: Table S5. Percentage of classification of the eight olive oils purchased in Italy using results obtained by Organoleptic Evaluation and Chemical Analysis performed on these olive oils by 8 IOC laboratories. * In brackets the percentages of classification according to the chemical analysis without the ethyl esters content are reported.

Additional file 6: Table S6. Percentage of classification of the eight olive oils purchased in USA using results obtained by Panel Test and Chemical Analysis performed on these olive oils by 8 IOC laboratories. *In brackets the percentages of classification according to the chemical analysis without the ethyl esters content are reported.

Abbreviations

EVOO: extra virgin olive oil; IOC: International Olive Council; ICB: International Certificated Body; FAEE: fatty acid ethyl esters.

Authors' contributions

All research has been done by the authors. All authors read and approved the final manuscript.

Author details

[1] Dipartimento di Chimica e Tecnologie del Farmaco, Sapienza Università di Roma, Rome, Italy. [2] Istituto di Metodologie Chimiche, CNR, Laboratorio di Risonanza Magnetica "Annalaura Segre", Monterotondo, Rome, Italy. [3] Dipartimento di Farmacia, Università di Napoli "Federico II", Naples, Italy.

Acknowledgements
This work was supported by "e-ALIERB: un OPEN LAB per caratterizzare e valorizzare i prodotti alimentari ed erboristici del territorio laziale" Project (Regione Lazio LR13/2008, CUP B82I15003570002).

Competing interests
The authors declare that they have no competing interests.

Funding
This work was supported by "e-ALIERB: un OPEN LAB per caratterizzare e valorizzare i prodotti alimentari ed erboristici del territorio laziale" Project (Regione Lazio LR13/2008, CUP B82I15003570002).

References

1. The Commission of the European Communities. Commission Regulation (EEC) No. 2568/91 on the characteristics of olive oil and olive-residue oil and on the relevant methods of analysis. Off J Eur Union. 1991;L248:1–102.
2. International Olive Oil Council. Sensory analysis of olive oil, method for the organoleptic assessment of virgin olive oil. 2015. IOC/T.20/Doc. No. 15/Rev. 8. http://www.internationaloliveoil.org/estaticos/view/224-testing-methods. Accessed 22 Jan 2017.
3. Frankel EN. Nutritional and biological properties of extra virgin olive oil. J Agric Food Chem. 2011;59:785–92.
4. Angerosa F. Sensory quality of olive oils. In: Aparicio R, Harwood J, editors. Handbook of olive oil, analysis and properties. Gaithersburg: Aspen Publishers; 2000. p. 355–92.
5. Salas JJ, Harwood JL, Martínez-Force E. Lipid metabolism in olive: biosynthesis of triacylglycerols and aroma components. In: Aparicio R, Harwood J, editors. Handbook of olive oil, analysis and properties. New York: Springer; 2013. p. 97–127.
6. The European Commission. Commission Regulation (EU) No. 432/2012 establishing a list of permitted health claims made on foods, other than those referring to the reduction of disease risk and to children's development and health. Off J Eur Union. 2012;L136:1–40.
7. Camin F, Pavone A, Bontempo L, Wehrens R, Paolini M, Faberi A, Marianella RM, Capitani D, Vista S, Mannina L. The use of IRMS, ^{1}H NMR and chemical analysis to characterise Italian and imported Tunisian olive oils. Food Chem. 2016;19:98–105.
8. International Olive Oil Council. List of chemical testing laboratories recognised by the International Olive Council for the period from 1.12.2014 to 30.11.2015. 2014. IOC/T.21/Doc. No. 13/Rev. 17.
9. International Olive Oil Council. List of laboratories undertaking the sensory analysis of virgin olive oils recognised by the International Olive Council for the period from 1.12.2014 to 30.11.2015. 2014. IOC/T.28/Doc. No. 3/Rev. 17.
10. Gómez-Coca RB, Moreda W, Pérez-Camino MC. Fatty acid alkyl esters presence in olive oil versus organoleptic assessment. Food Chem. 2012;135:1205–9.
11. Conte L, Mariani C, Gallina Toschi T, Tagliabue S. Alchil esteri e composti correlati in oli d'oliva vergini: loro evoluzione nel tempo. RISG. 2014;91:21–9.
12. International Olive Oil Council. Trade standard applying to olive oils and olive pomace oils. 2016. IOC/T.15/NC No. 3/Rev. 11.
13. The European Commission. Commission Delegated Regulation (EU) No. 2016/2095 of 26 September 2016 amending Regulation (EEC) No. 2568/91 on the characteristics of olive oil and olive-residue oil and on the relevant methods of analysis. Off J Eur Union. 2016;L326:1–6.
14. Loreto G, Giansante L, Alfei B, Giacinto L. Alchil esteri ed altri indicatori per la tutela della qualità e della genuinità degli oli extra vergini italiani. RISG. 2014;91:35–45.

7

Food additives and technologies used in Chinese traditional staple foods

Xiaolong Wang, Zhen Ma, Xiaoping Li, Liu Liu, Xiuxiu Yin, Kailong Zhang, Yu Liu and Xinzhong Hu*

Abstract

Noodles and Chinese steamed bread (CSB) represent 70% wheat flour consumption in China. However, fresh noodles and CSB are much produced in small workshop and on family basis. The relatively poor and unstable qualities of wheat flour, as well as the short shelf-life of fresh noodles and CSB have significantly retarded the efficient production of traditional staple foods at large scale and industrial level. This review summarizes the food additives, such as salts, vital wheat gluten, hydrocolloids, esters, enzymes, acids and lots of natural products, used to enhance flour quality and retard staling and microbe growth of noodles and CSB. In addition, recent advances focus on the physical treatments and packaging technologies applied in the production of fresh noodles and CSB were also introduced. The findings in this review would provide reference for further explorations toward the industrialization of traditional staple foods.

Keywords: Noodles, CSB, Quality improvement, Shelf-life extension, Food additives, Package technologies, Industrialization

Introduction

Noodle and Chinese steamed bread (CSB) are two major staple foods in China for many centuries due to their convenience, nutrients and palatability [1–3]. It has been reported that noodles and CSB, respectively, represented 40 and 30% wheat flour consumption in China [4, 5]. However, only 14–20% of wheat flour is applied in the industrial production, and fresh noodle and CSB are much produced in small workshop and on family basis. Currently, due to the socio-economic changes caused by rapid urbanization in China, producing fresh noodles and CSB on large and industrial scale is becoming an inevitable trend.

There are two crucial challenges for the industrialization of fresh noodles and CSB. On the one hand, the production of different noodles and CSB require wheat flours with medium to strong gluten strength, appropriate viscoelasticity and stable hot pasting property [6–8]. But the wheat cultivars in China cannot satisfy the requirements completely due to their weak gluten strength and quality fluctuation caused by genetic and environmental diversity [7, 9, 10]. On the other hand, during storage, some complicated physicochemical and biochemical changes occurred in fresh noodles and CSB, attributing to a shorter shelf-life [11–13]. It is well known that microbial proliferation and staling are the two critical reasons that result in the deterioration of fresh noodles and CSB [14–16]. Previous studies have figured out that mold, bacteria and yeast are the key microbes involved in the microbial proliferation of fresh noodles and CSB [17, 18]. The staling is attributed to both amylose and amylopectin of starch. Specifically, amylose re-association is related to the short-term retrogradation during the initial few hours, while amylopectin re-crystallization is associated with product firming in the longer term [19]. CSB possesses much high moisture content, around 39–44 and 42–47% in crumb and crust, respectively [20–22], so staling increasingly occurred with the water loss during storage, leading to the firmness increase and freshness loss of CSB. In contrast, fresh noodles with lower water content showed slower hardening rate although similar staling trend was observed [23]. In addition, enzymatic browning induced by polyphenol oxidase (PPO) had

*Correspondence: hxinzhong@126.com
College of Food Engineering and Nutrition Science, Shaanxi Normal University, Xi'an 710119, China

been implicated as a leading cause for short shelf-life of fresh noodle too [24].

To improve the quality of wheat flour and extend the shelf-life of fresh noodle and CSB, a variety of additives has been incorporated in the formulations. Salts, vital wheat gluten, hydrocolloids, esters, enzymes, acids and lots of natural products have been widely used as gluten fortifier, emulsifier, thickener, stabilizer, oxidants, reductant and pH regulator in individual or complex ways for improving flour-processing quality, as well as retarding staling and microbe growth of fresh noodle and CSB [25–32]. The quality of fresh noodle and CSB are not only dependent on the quality of the ingredients but also on the processing and storage conditions [3, 13, 21]. Manipulating production process and storage conditions, such as physical treatments with ozone, radiation, microwave, photons and vacuum mixing as well as packaging methods including vacuum packaging, modified atmosphere packaging (MAP), active packaging and hot packaging, showed promising results for enhancing the shelf life of fresh noodles and CSB [2, 12, 21, 33–35].

In this review, a significant portion of the literature is drawn from the Chinese journals and theses where a rather large amount of research was published. We summarized the recent advances in food additives, as well as processing and storage technologies towards quality improvement and shelf-life extension for fresh noodles and CSB, with an aim to provide reference for further research to support the industrialization of traditional staple foods.

Food additives used for quality improvement of noodles and CSB

Inorganic salt

Salt and alkaline (kansui)

In addition to wheat flour and water, salt, alkaline reagents, and combinations of salt and alkaline reagents (mixtures) are also incorporated in noodle manufacturing. Based on the ingredients, wheat flour noodles can be divided into two categories, one is white salted noodles (WSN) made from flour, sodium chloride and water, and another is yellow alkaline noodles (YAN) made from flour, sodium chloride, alkaline reagents (such as sodium and potassium carbonate) and water [36]. It has been proven that WSN originated in northern China represent 90% of total noodle production in China, the rest being mainly YAN [37].

Sodium chloride is an important component in noodles. The addition of sodium chloride at 2–3% level in Asian noodles could enhance the texture of noodles by strengthening and tightening the gluten network to increase the viscoelasticity [38]. It is known that adding sodium chloride decreases water absorption but it increases the dough development time. Although increase in the number of disulphide bond was observed when the dosage of sodium ranged from 1 to 5% [39], more research suggested that the optimum adding amount of sodium chloride was around 2% [32, 40], where the overall quality of noodles, including texture, brightness, cooking properties and flavor, was enhanced significantly [41, 42].

Alkaline reagents are preferred, especially by Cantonese people, for making alkaline noodles [32]. It is believed that the original purpose of adding alkaline in noodles is to prolong the noodle shelf life by retarding the mold growth [37]. Sodium carbonates, or potassium carbonates, or the mixture of the two (typically 9:1, kansui) are the most common alkaline salts used in alkaline noodles [43]. The addition rate of alkaline salts depends on the noodle types. Specifically, the optimum dosage is 1.0–1.5% for fresh alkaline noodles and 0.3–0.5% for steamed alkaline noodles [37]. The quantity of sodium hydroxide is usually 0.3–1.0% for the partially boiled alkaline noodles. Alkaline–flour interactions are responsible for several effects, including strengthening of the dough texture [44]; reducing the dough development time and dough stability; improvement in both the firmness and chewiness of noodles [45]; retardation of starch gelatinization and increase in starch paste viscosity [44, 46]; inhibition of enzyme activity and enzymatic browning [43, 47]; and also contribution to the bright yellow color, aroma and flavor [47].

Phosphates

In some cases, calcium carbonate or sodium/potassium phosphates with or without sodium chloride are also used in cereal foods. The food-grade phosphates (sodium, calcium, and potassium salts of phosphoric acid) impart numerous physicochemical properties in foods, including moisture adsorption, pH adjustment, buffering, protein dispersion, emulsification, ion exchange, sequestration of minerals, flavor, improved whipping, foam stability, cryoprotection, and texture development [48–50]. According to the national hygienic standard of food additives in China, 14 phosphates have been allowed to use as food additives. Sodium tripolyphosphate (STPP), sodium hexametaphosphate (SHMP), sodium pyrophosphate (SPP) and sodium dihydrogen phosphate (SDHP) are commonly used in a single or complex way in the processing of Chinese traditional cereal foods.

Wu et al. [51] found that the moisture-retention capacity of composite phosphate contributed to the full hydration and swelling of gluten protein, which subsequently enhanced the gluten network, dough stability, viscoelasticity and cooking resistance of noodles. It was reported that tripolyphosphate could improve the whiteness of noodle, while hexaphosphate contributed to a delayed browning of noodle [52]. The possible explanation for this finding is that phosphate provided a stable pH and prevented oxidation via sequestration cation in the dough, which both contributed to the good color retention of noodle [53]. Besides, phosphate could enhance the interaction between protein and starch, which not only conducted to a good viscoelasticity but also prevented the starch leaching during cooking [54].

Niu et al. [28] investigated the impacts of four inorganic phosphates on the thermodynamic, pasting and noodle-making properties of whole wheat flour. The Results showed that the gelatinization temperature and enthalpy of melting of starch in whole wheat flour were increased with the addition of phosphates. All phosphates significantly increased peak viscosity and final viscosity of whole wheat flour. Moreover, the whole wheat noodles incorporated with disodium phosphate, trisodium phosphate, and STPP showed brighter appearance, and the use of STPP and SHMP reduced the cooking loss of whole wheat noodles. Texture profile analysis of cooked noodles indicated that adding phosphates significantly decreased the firmness, while the springiness, cohesiveness, and resilience were slightly reduced. The microstructure of whole wheat noodles showed a close connectivity of the protein network and coverage of starch granules in the presence of inorganic phosphates. The results suggested that inorganic phosphates exhibited substantial effects on enhancing the quality of whole wheat noodles. Of all the flour inorganic phosphates, STPP seems to be the most effective one in improving the overall quality of whole wheat noodles.

The moisture retention capacity of phosphates could improve the weight and volume of CSB significantly, while the hardness of steamed bread was reduced as the water absorption increased [55]. Besides, the leaven function was enhanced due to the sequestration of minerals and enhanced protein–starch interactions caused by the addition of phosphates [56]. It has been reported that the phosphate also contributed to the whiteness of steamed bread, which mainly due to a more intense and uniform crumb was formed after the addition of phosphate [57]. Phosphates are commonly used in the quality improvement of quick frozen foods. Quick frozen dumplings accounts for 33% of the total quick frozen foods in China [58], however, the common problems, such as frost cracking, outflow of stuffing juice, sticky texture, poor elasticity, browning and cooking loss, appeared in the practical production retarded its industrialization [59]. Therefore, phosphates were used as additives to alleviate the above-mentioned challenges in production. Phosphates added in frozen dumplings gave cracking resistance and good appearance for frozen dumplings due to their water-holding capacity, in which STPP showed the best performance in all the applied phosphates [59]. For the browning inhibiting, the function of phosphates can be explained in two ways. First, the enzymatic browning was delayed by the sequestering of metal ions [60]. Second, the water-holding rate was increased by the phosphates, which prevented the browning reaction caused by the outflow of stuffing [61].

Vital wheat gluten

Gluten, the dough-forming protein of wheat flour, is the key to the unique ability of wheat suit to a wide range of products. Given the unique viscoelastic properties of wheat gluten, the dry vital wheat gluten, a commodity separated from wheat flour, was widely applied in Chinese traditional cereal foods. Vital wheat gluten usually contains approximately 75% protein, up to 8% moisture, and a small amount of starch, lipid and fiber [26]. It is commonly incorporated into the wheat flour or coarse grain powder to confer a stronger dough strength required for the noodle or CSB making.

Park et al. [62] investigated the impact of gluten on the white salt noodle making with flour from hard and soft wheat. The results showed that gluten incorporation (6%) reduced water absorption of noodle dough by 3%, shortened the length of the dough sheet by 15 and 18%, and increased the thickness of the dough sheet by 18 and 20% in soft and hard wheat flour, respectively. Noodles imbibed less water and imbibed water more slowly during cooking with gluten incorporation, which resulted in a 3-min increase in cooking time for both soft and hard wheat noodles. Despite the extended cooking time of 3 min, noodles incorporated with 6% gluten exhibited decreases in cooking loss by 15% in soft wheat. In hard wheat flour, cooking loss of noodles was lowest with 2% incorporation of gluten. For multi-grain noodles, buckwheat noodle making with 70% wheat flour and 30% buckwheat flour and addition of 5% vital wheat gluten led to the best viscoelasticity and sensory characteristics of cooked noodle [63]. The research conducted by Wang et al. [64] indicated that, when using the combination of vital wheat gluten and transglutaminase (TG) at appropriate formula, oat noodle quality was significantly improved. As reported by a variety of studies, vital wheat gluten played an important role in the quality improvement of CSB. It conferred enhanced specific volume, appearance and texture to CSB, the optimum dosage of vital wheat gluten for CSB making with wheat, wheat–oat mixture, and wheat–corn mixture are 2, 5 and 1%, respectively [2, 65, 66]. It is clearly reflected by the above results that the optimum amount of vital wheat gluten in the formulations of noodles and CSB was depended on the protein quantity and quality in flour, more vital wheat gluten was required for the flour protein with low content and poor quality.

Ester

Esters, acting as thickeners and emulsifiers, are widely used in the formulas of noodles and CSB. Numerous studies have reported that propylene glycol alginate (PGA) can be used for improving noodle quality. Yang et al. [67] carried out a research that PGA influences the acidity of wet noodles, property and texture. The water absorption, dough viscoelasticity, noodle hardness and elasticity were enhanced with the addition of PGA, its optimum adding amount was 0.3%. This finding was confirmed by the research of Yang et al. [68] who also found that the optimum dosage of PGA in fresh noodle was 0.3%. Liu et al. [69] investigated the effects of low-ester PGA on noodle quality. The results showed that low-ester PGA used in the noodles can effectively improve noodle quality; and when the content of low-ester PGA was between 0.2 and 0.3%, the hardness, chewiness and glue viscosity of noodles were increased greatly; meanwhile, the water absorption and dissolution rate of starch were reduced; noodles had the best chewiness, smoothness, and taste. The fatty acid sucrose ester was also involved in noodle making, which showed that the cooking quality of middling noodles improved greatly when the noodles were added 1.5% fatty acid sucrose ester [70].

Regarding CSB, calcium stearoyl lactylate (CSL), sodium stearoyl lactylate (SSL) and glyceryl monostearate (GMS) are widely applied in individual or composite ways for quality improvement. The optimum formulation for the best anti-aging was sodium stearyl lactate 0.15%, glyceryl monostearate 0.05%, sucrose ester of fatty acids 0.15% [27]. Besides, the effects of diacetytartaric esters of monoglycerides (DATEM), CSL and SSL on CSB was investigated, which indicated that CSL–SL was beneficial to dough viscoelasticity; mixture of DATEM, CSL–SSL in flour decreased the dough extensibility while noodle whiteness was increased [71].

Hydrocolloid

Gums consist of a number of water-soluble polysaccharides that come with different chemical structures and provide diverse functional properties such as gelling, thickening, stabilizing, foaming, emulsifying, as well as inhibiting syneresis during a freeze–thaw cycle, water-retention and texture-enhancing properties, by controlling the water molecule mobility [72]. Based on the previous studies, the impacts of gums on dough-functional performance and food quality are depending upon the nature, origin, particle size, molecular structure and ionic charges of the gums, and also on the dosages of gums added to the dough formulations [73]. Alginate, k-carrageenan, xanthan, guar gum, pectin, hydroxypropylmethylcellulose (HPMC) and some modified starches are extensively used in the formulations in the Chinese traditional cereal foods, especially for noodles and CSB. The positive functions of these hydrocolloids are summarized as following [74]: (1) improve food texture, (2) retard the retrogradation of the starch, (3) increase moisture retention, (4) extend the overall quality of the product during time, and also (5) as gluten substitutes in the formulation of gluten-free breads since gums could act as polymeric substances that mimic the viscoelastic properties of gluten in bread dough.

There are a number of studies focussing on the effects of hydrocolloids on dough-rheological characteristics. It has been reported that the hydration property of wheat flour–hydrocolloids mixtures was greatly increased due to the high affinity of hydrocolloids to water, which would affect the dough-rheological properties and food quality significantly [75]. Rosell et al. [72] investigated the effect of four widely used hydrocolloids, including sodium alginate, k-carrageenan, xanthan gum and HPMC, on the rheological properties of the wheat flour dough. The results showed that xanthan and alginate had the most pronounced effect on dough properties yielding strengthened dough. A great improvement in dough stability during fermentation was achieved by adding hydrocolloids. Liu et al. [75] investigated the effects of hydrocolloids (HPMC, carboxymethylcellulose (CMC), xanthan gum (XG), and apple pectin (AP)) at different concentrations on potato flour dough-rheological properties and found that HPMC remarkably increased the development time; potato flour dough containing CMC or 0.5% HPMC and AP exhibited lower stability than control. Conversely, the stability was increased by the addition of XG (1.0 and 2.0%) and HPMC (1.0%). Moreover, the effect of hydrocolloids on the pasting properties and gelling behavior of cereal four was also investigated. The influence of the selected hydrocolloids (guar gum, pectin, alginate, k-carrageenan, XG and HPMC) on wheat four pasting was tested by Rojas et al. [74] using amylograph and differential scanning calorimetry (DSC). The results showed that hydrocolloid addition largely modified the amylograph parameters of wheat four at the low levels tested [0.5 and 1% (w/w), four basis] and the extent of their influence depended upon the chemical structure of the added hydrocolloid. The greatest effect on pasting temperature was observed when 1% alginate was added, which produced a decrease of 3 °C. Xanthan and pectin increased the cooking stability while k-carrageenan and alginate did not modify it. The setback was augmented by guar gum and HPMC but alginate, xanthan and k-carrageenan showed the opposite effect. The bump area related to the formation of amylose–lipid complex, was favored by k-carrageenan, alginate and pectin, and slightly affected by xanthan and HPMC. Amritpal et al. [76] found that the addition of xanthan and guar gum improved peak viscosity, hot-paste viscosity and final viscosity for mung and corn starches; while for potato starch, guar gum increased peak and final viscosities and decreased hot paste viscosity while xanthan gum increased hot-paste- and final viscosities and decreased peak viscosity. All those results about the functional properties of hydrocolloids concluded that a unique trend did not exist and the impact of hydrocolloids on dough thermo-mechanical properties greatly depend on the nature of the flour components, the chemical structure of specific hydrocolloid and the quantity of added hydrocolloids [75].

Based on their unique physicochemical properties, hydrocolloids are increasingly incorporated in the formulations of Chinese traditional cereal foods in individual or composite ways, especially for noodles and CSB. The application status of different hydrocolloids is summarized in the next section.

Alginates are a group of viscous polysaccharides extracted from the cell wall of brown seaweeds or produced as an extracellular matrix by some bacterial species. As one of the popular hydrocolloids, alginate is an unbranched polymer composed of (1-4)-linked β-D-mannuronic acid and α-L-guluronic acid that are arranged in homopolymerically or heteropolymerically alternating sequence. Alginates have been widely used as thickeners, emulsifiers, water-holding agents and stabilizing agents in a number of food applications.

According to a number of reports from China, alginates were widely involved in the formulations of Chinese dry noodles and fresh noodles. Generally, noodles with optimum quantity of alginates usually exhibited an increase in the cooked weight, a decrease in the cooking loss and significant increase in the cutting and tensile forces [77, 78]. Moreover, the noodles containing more alginate showed lower lightness and redness, while yellowness was increased. The optimum additive amount of alginates for different noodles was also investigated. Zhao et al. [78] found that the sodium alginate with the viscosity of 300 mPa. s was suitable for noodle making, and the optimum concentration was 0.2–0.25%. Yang et al. [77] reported that the oat flour containing 0.3% sodium alginate exhibited best quality in oat noodle making. As for the buckwheat flour with high dietary fiber (5%), 4% sodium alginate contributed to the best performance of buckwheat noodles [79]. According to the quantity variation of sodium alginate incorporated in different noodles, it is supposed that the optimum quantity of sodium alginate is partially depended on the gluten strength of mixed flour. Wheat flour had stronger gluten strength comparing with oat and buckwheat flour, hence lower quantity of sodium alginate was used in the noodle making. Conversely, relatively the higher quantity of sodium alginate was required for the noodle making using oat flour or buckwheat flour due to their weak gluten caused by lack of gluten proteins. Apart from application in noodle making, alginates, acting as an anti-staling agent, were also increasingly involved in the making of CSB. The results from the study of Sim et al. [30] suggest that at 0.2% addition level, alginate is better than konjac glucomannan in delaying staling of CSB though slight reduction in spread ratio and specific volume were evident.

Apart from alginates, other hydrocolloids, such as CMC, carrageenan, xanthan, guar gum, pectin, konjac gum, artemisia sphaerocephala krasch gum, edible gelatin and some modified starches, usually play a similar role as alginates in improving the overall quality of fresh noodles, dry noodle or CSB. However, variations on detailed performance among these hydrocolloids were observed as reflected by the specific quality indicator.

To select the optimum thickening agent for fresh noodle making, five kinds of hydrocolloids, CMC, sodium alginates, xanthan gum, guar gum and konjac gum, were incorporated in the making of fresh noodle at different concentrations [80]. The results indicated that overall quality of noodle was improved by the addition of all the five gums. The overall quality score increased to the maximum and then decreased with the quantity increasing of hydrocolloids. It is notable that the optimum concentration of hydrocolloids appeared between 0.3 and 0.4%, and the performances of xanthan gum and konjac gum were much better than that of the other three hydrocolloids.

Modified plant starch is an alternative source for hydrocolloids applied in fresh noodle, noodle containing it showed a much springier, smoother, more elastic, and chewier texture than those without it [81]. Yan et al. [82] investigated the effects of three modified starches (waxy potato starch, hydroxypropyl starch, hydroxypropyl distarch phosphate) and artemisia sphaerocephala krasch gum on the dough- and noodle-quality improvement. The results reflected that, when applied individually, artemisia sphaerocephala krasch gum at the concentration of 1%, hydroxypropyl starch at 3 or 7% resulted a better dough quality, while composite hydrocolloids consist of 1% artemisia sphaerocephala krasch gum and 5% hydroxypropyl starch or 1.25% artemisia sphaerocephala krasch gum and 7% hydroxypropyl starch contributed to a better quality of noodle.

The effect of different hydrocolloids on the quality of coarse cereal noodles was also studied. Chen et al. [83] found that the composite thickener showed better performance than single thickener on the quality improvement of oats noodles, and the optimum formula of the composite thickener was: konjac flour 0.2%, edible gelatin

4%, sodium polyacrylate 0.1% and CMC-Na 0.4%. Meanwhile, the functional property of five hydrocolloids in sorghum noodles was evaluated by Kou et al. [84], which concluded that the contribution of the tested hydrocolloids was ranked as: sodium alginate > CMC > konjac gum > guar gum > XG. Low-protein sorghum noodle got the best cooking quality and sensory quality when the sodium alginate was added at 1% level.

Comparing with noodles, few researches on the hydrocolloids application in CSB were reported so far. Sim et al. [30] found that when wheat flour was blended with 0.2% sodium alginates and 0.8% konjac gum to prepare CSB, the quality evaluation in terms of spread ratio, specific volume and staling behavior indicated that, at 0.2% addition level sodium alginates is better than konjac gum in delaying staling of CSB though the slight reduction in spread ratio and specific volume were evident, at 0.8% level, however, konjac gum seems to be better than sodium alginates in enhancing CSB properties. Liu et al. [75] investigated the effects of hydrocolloids (HPMC, CMC, XG, and AP) at different concentrations on the quality of gluten-free potato-steamed bread. The results showed that steamed breads with hydrocolloids presented higher specific volume and lower hardness, and the rapidly digestible starch and estimated glycemic index significantly decreased from 45.51 to 20.64 and from 69.54 to 55.17, respectively. They suggested that HPMC and XG could be used as improvers in the gluten-free potato-steamed bread.

The results described above together reflected that different hydrocolloids show different effects on wheat dough, noodle and CSB quality properties due, probably to their distinctly different molecular structures. It is supposed that hydrocolloids may form higher molecular weight aggregation via the interactions with protein and starch, and which were closely related to the type, ratio and content of the incorporated hydrocolloids. The incorporation of hydrocolloids always shows positive effects on the overall quality of noodles. In noodle making, higher dosage of hydrocolloids is required for the dough with weaker gluten strength, and composite hydrocolloids shows better performance in noodle quality improvement. The effects of hydrocolloids on CSB quality are largely dependant on their molecular structure and addition quantity. In dough matrix, entanglement coupling between glutenin molecules is responsible to maintain the elasticity of the dough, while the extensibility of the dough are governed by the breaking of secondary valence bonds and slippage of entangled chains under deformation. Therefore, low resistance and high extensibility are observed when the rate of chain slippage is much greater than the rate of elongation of the chain. However, if the rate of chain slippage is relatively low, meaning chains slippage occur insufficiently rapidly in response to the applied stress, the chains will break resulting very short distance of elongation. The hydrocolloids differ in structures and additional levels are supposed to change the rate ratio of the chain slippage and chain elongation uncertainly due to their different interaction patterns with the glutenin entanglement. These different interactions are prone to affect the gas-holding capacity of dough and lead to a good or poor CSB quality consequently. As for the anti-staling activity of hydrocolloids, it can be explained by their interactions with starch. The hydrocolloids present in CSB or noodles could hinder the development of macromolecular entanglement and retard the starch recrystallization, thus delaying the staling of CSB and noodles.

Enzymes

Enzymes, a group of "green" food additives with the advantages of low application dosage, high reaction rate, mild reaction conditions and instinctive safety, are commonly used in food processing [29, 85]. As reported in the previous studies, α-amylase, glucose oxidase, xylanase, lipase, lipoxidase and TG are increasingly involved into the production of noodles and CSB (Table 1).

Alpha amylase

Alpha amylase could hydrolyze the starch to dextrin, which provide carbon source to the yeast to produce enough carbon dioxide in dough fermentation [86]. Moreover, the amylose–amylopectin ratio that is closely related to the pasting stability could be changed after the incorporation of α-amylase. It is well known that the starch-pasting property is closely related to the noodle quality. Shi et al. [87] reported that shorter cooking time and lower water absorption of noodle were observed with the increased addition of α-amylase. The adding quantity of α-amylase was negatively related to L^* and a^* value of fresh noodle, while positively related to the b^* value. They concluded that the amylose–amylopectin ratio has been modified by the α-amylase addition, leading to the quality variation of fresh noodle. The study conducted by He et al. [88] showed that the development time of fresh noodle was decreased by the incorporation of α-amylase, and the α-amylase also contributed to the enhancement of texture and sensory properties of fresh noodle. In the CSB making, Ma et al. [89] reported that, when the addition quantity of α-amylase is below 30 mg/kg, the specific volume, internal structure, crumb texture and overall evaluation score of CSB were increasingly improved with the concentration growth of α-amylase.

Table 1 Application of enzymes in the production of fresh noodles and CSB

Enzymes	Function properties		Dosage		References
	Fresh noodles	CSB	Fresh noodles	CSB	
Alpha amylase	Enhance dough fermentation, texture and sensory properties	Improving specific volume, internal structure and crumb texture	60 mg/kg 24 mL/100 kg	< 30 mg/kg	[87–89]
Glucose	Enhance appearance, chewiness, smoothness, cooking tolerance and whiteness	Soft crumb, high whiteness and delayed staling	20–40 mg/kg	20 mg/kg	[91–94]
Lipase	Enhance appearance, whiteness, and elasticity; flavor generation		60 mg/kg		[92]
Xylanase (XynA)		Improving specific volume and hardness		120 U/kg	[98]
Transglutaminase	Reducing water absorption and cooking loss		0.01% (buckwheat noodle)		[102, 103]
	Enhancing hardness and tensile force		2% (YAN)		

Glucose oxidase and lipoxidase

Glucose oxidase and lipoxidase had similar effects on flour-dough development, the hydrogen peroxide produced after the addition of these two oxidases would promote the formation of additional disulphide bond in dough matrix, enhancing the viscoelasticity of dough [90]. Besides, flour whiteness could be improved because of the interaction between carotene and lipoxidase [90]. Researches on the effects of glucose oxidase on noodle quality indicated that the noodle appearance, chewiness, smoothness, cooking tolerance and whiteness were improved by the addition of glucose oxidase and the recommended dosage was at 20–40 mg/kg [91, 92]. Glucose oxidase was also used in the formulation of CSB produced by wheat flour or multi-grain flour, it conferred soft crumb, high whiteness and delayed staling to the CSB and the optimum dosage was determined at 20 mg/kg [93, 94]. As for the lipoxidase, Cato et al. [95] investigated the effects of lipoxidase on WSN and found that whiter noodle sheets were obtained when a soybean lipoxidase was added to the formulation. Textural and structural properties of WSN were not adversely affected by enzyme addition, giving firm, elastic and non-sticky products. It is concluded that the incorporation of the lipoxidase offered prospects for color enhancement of white salted noodles. These findings were in line with the study of Zhang [96], who elucidating the contribution of a recombinant lipoxidase (ana-rLox) to the noodle quality improvement.

Lipase

The lipase could hydrolyze the lipid in flour into a mixture of fatty acids, monoglycerides and diglycerides. It is supposed that the monoglycerides are prone to form complex with amylose and protein, which reducing the starch staling and cooking loss of cereal foods [90]. Lipase was seldom used individually in noodle making. Shan et al. [92] studied the impacts of lipase and glucose oxidase on the noodle quality and found that the appearance, whiteness, and elasticity of noodles were improved at most when the formulation containing lipase at 60 mg/kg and glucose oxidase at 40 mg/kg simultaneously. Relationships between lipase, lipoxidase and peroxidase activity, along with quantity of individual free fatty acids and levels of headspace volatile compounds of boiled buckwheat noodles, were investigated using 12 different buckwheat varieties by Suzuki et al. [97]. The results indicated that lipase and peroxidase in buckwheat flour was involved into the flavor generation of boiled buckwheat noodles. This finding was important for increasing desirable flavor of buckwheat products as well as for selecting varieties with improved flavor.

Xylanase

Arabinoxylans (AX) constitutes the major non-starch polysaccharides in wheat, water-extractable AX (WE-AX) and solubilized high-molecular-mass AX (S-AX) have a positive effect on dough stability and mechanical properties, while water-unextractable AX (WU-AX) has a negative effect [98]. Therefore, xylanases that preferentially hydrolyze WU-AX and leave WE-AX and S-AX unharmed have the greatest beneficial effect on flour products, when added at the appropriate quantities [99, 100]. Yang et al. [98] have incorporated a recombinant xylanase named XynA to the formulation of CSB. The results showed that, when using XynA at the optimal amount in the preparation of CSB, both the specific volume and hardness of the CSB were improved greatly.

Transglutaminase

Transglutaminase (TG) catalyzes acyl-transfer reactions, creating inter- and intra-molecular cross-links between lysine residues and glutamines residues, forming the so-called e(c-glutamyl)-lysine isopeptide bonds [101]. The cross-linking action of TG leads to the formation of very high molecular weight gluten proteins, reinforcing the gluten protein network and enhancing the mechanical properties and hot stability of dough [102]. The investigation for effects of TG on quality of buckwheat noodle showed that, addition of TG significantly decreased the water absorption and cooking loss of noodles, increased the hardness and tensile force. When adding 0.01% TG, the lowest cooking loss of buckwheat noodle was observed [103]. Bellido et al. [102] investigated the effects of TG on the protein composition, mechanical properties and microstructure of yellow alkaline noodles. The results indicated that, in the presence of TG, the unextractable glutenin was greater than 225–300% over that of corresponding flour from three different varieties. Comparison of the mechanical properties of YAN formulated with and without TG showed that the firmness and elastic-like behavior of the noodle increased with the TG supplementation.

Composite enzyme

Although enzymes play important roles in the quality improvement of noodles and CSB, individual utilization of enzyme which catalyzing specific reaction always limits the overall quality improvement. As so, the explorations of composite enzymes aiming at the overall quality improvement of noodles and CSB have been conducted. A Box–Behnken and Response Surface Methodology were used to analyze the effect of enzymes on noodle quality. The independent variables were xylanase, glucose oxidase and lipase and their interactions between each other were also taken into consideration. The model was optimized by response of sensory evaluation score and L^*. The optimized formula of enzyme was xylanase 213 mg/kg, glucose oxidase 67.43 mg/kg and lipase 5 mg/kg [104]. Regarding the CSB, Ren et al. [105] found that the composite enzymes consisted of amylase 3 mg/kg, xylanase 10 mg/kg, lipase 4 mg/kg and glucose oxidase 18 mg/kg were appropriate for the CSB making, while Meng et al. [94] recommended the formula including glucose oxidase, lipase and α-amylase, respectively, at 20, 50 and 5 mg/kg for the CSB supplementation. It seems that the addition amount of α-amylase, glucose oxidase in the formula for the CSB making have remained relatively consistent. However, some enzymes, such as lipase and xylanase used in individual or composite ways for noodle making showed great variations on their optimum dosages. It is supposed that the differences in the ingredients, enzyme source and activity as well as processing conditions all lead to the optimum dosage variations of enzymes.

Although some enzymes, esters and hydrocolloids show the anti-staling property in the production and storage of CSB and noodles, but they interact with starch in different patterns. In terms of esters, they can form complexion with the helical amylose via hydrophobic bonds, thus reducing the recrystallization of starch. Meanwhile, the interaction between external branch of amylopectin and esters via hydrogen bonds also impedes the starch staling. The anti-staling mechanism of hydrocolloids is as follows: (1) the good film-forming property of hydrocolloids reduces the moisture loss; (2) the hydrogen bonds formed between hydrocolloids and starch impeding the recrystallization of starch; (3) the high water-holding capacity of hydrocolloids improves the moisture of noodles and CSB. As for the enzymes, α-amylase could hydrolyse the starch to dextrin, which interfering the crystallization of starch; lipase could hydrolyse the lipid into a mixture of fatty acids, monoglycerides and diglycerides, which reducing the starch staling like the role of esters.

Comparing with other food additives, enzymes possess unique properties such as high specificity, high rate of catalysis, high thermo-stability and safety. Most of the enzymes applied in noodles or CSB are intended to catalyze the reaction of polysaccharide or proteins. The enzymes (especially amylases of various sources) aimed at the modification of starch, show great potential for the quality improvement of noodles. However, the transglutaminase, oxidases and protease, which focusing on the gluten network modification, might be of more benefit to the quality of CSB. Optimum dosage of enzymes in noodles or CSB is essential to maximize their function. However, the optimum dosage is susceptible to the food ingredients and processing conditions. As so, on the one hand, the enzyme suitable for specific noodle or CSB is required in future. On the other hand, the enzyme act at wide pH and temperature should be also developed. Comparing to the enzyme application in bread, more enzymes specific for Chinese traditional foods should be developed. For example, enzymes used for the flavor enhancement and anti-staling of CSB or noodles should be taken into consideration.

Preservative applied in the production of noodles and CSB

Because of high moisture content, the shelf life of fresh noodles and CSB is largely shortened by the microbial growth. Therefore, chemical and natural preservatives were commonly used to prolong the shelf time of traditional cereal products. Varieties of preservatives with different properties were involved into the production of fresh noodles. Inorganic acids, such as benzoic acid, sorbic acid and propionate as well as organic acids including glycine, lactic acid and citric acid were widely used as antimicrobial agents due to their pH-regulating capacity [31]. A complex preservative, consisting of ε-polylysine 120 mg/kg, citric acid 0.5%, potassium sorbate 0.08%, sodium diacetate 0.04% and calcium propionate 0.015%, developed by Zhang et al. [106] has been proved for extending the shelf time of fresh noodle to 60 h at 20 °C. Some oxidants were used as antimicrobial agents in fresh noodles. Chlorine dioxide and hydrogen peroxide at 0.04 and 0.03% in wheat flour has prolonged noodle shelf time to 4 and 6 days, respectively [107]. Composite preservatives consisting of natural products, such as nisin, lysozyme, natamycin, chitosan and Chinese herb medicine extracts, were also incorporated in fresh noodles. The appropriate mixture of dimethyl fumarate, nisin and natamycin has prolonged the shelf life of fresh noodle to more than 85 days at 37 °C [108]. While Zhou et al. [109] found that the combination of lysozyme and Chinese herb medicine extracts showed great potential to inhibit the microbial growth in noodles.

As for the preservatives application in CSB, fewer researches were conducted than that of fresh noodles. Zhao et al. [110] investigated the effects of different preservatives on CSB shelf time. The results showed that the shelf time of CSB reached over 78 h when incorporating the complex of vinegar and calcium propionate, while it felt below 12 h when vitamin c and ethanol were used. Besides, it is notable that, adding complex spicy essential oil diluted by pure ethanol (0.3 μl/L) together with cold storage condition can prolong the shelf time of CSB significantly [25]. Moreover, the impact of sourdough (10, 20, 30%) on quality and microbial shelf life of CSB was investigated. Results showed that the addition of sourdough (10, 20 and 30%) decreased the pH of CSB to 4.5, 4.3, 4.0, and extended the microbial shelf life to 8, 30 and 40 days, respectively [111].

According to the existed researches, lots of chemicals and natural products have been used for the shelf-life extending of fresh noodles, the potential of natural products especially Chinese herb medicine need to be further explored to meet the requirements of specific product. In contrast, fewer studies about the preservatives in CSB have been carried out, suggesting more works are needed in the future.

Technologies for shelf-life extension of noodles and CSB

It is well known that microbial proliferation and staling are the two critical reasons resulting in the deterioration of fresh noodles and CSB [2, 15]. The staling is attributed to both amylose and amylopectin of starch. Specifically, amylose re-association is related to the short-term retrogradation during the initial few hours, while amylopectin re-crystallization is associated with product firming in the longer term [19]. Fresh noodles and CSB have exhibited the similar staling trend during storage, although a slower hardening rate was observed in fresh noodles. As for the microbial growth, mold and bacteria are supposed to be the two crucial microbes determining the microbiological shelf life of fresh noodles and CSB [31, 111].

As mentioned in the previous sections, varieties of anti-staling agents including esters, hydrocolloids and enzymes, etc., as well as chemical and natural preservatives have been utilized to enhance the shelf life of fresh noodles and CSB. However, the increase in shelf life only depending on the food additives cannot satisfy the requirement of efficient production of fresh noodles and CSB at large scale and industrial level; hence, extending shelf life via manipulating production process and storage conditions has already become an irresistible trend [2, 12, 21, 33–35]. In the following section, the research progress focus on the physical treatment and packaging technology of fresh noodles and CSB would be introduced.

New technologies used in fresh noodle

Physical treatments

Treatment with ozone, radiation, microwave and vacuum mixing were reported as the new technologies enhancing shelf life or quality of fresh noodles. Li et al. [2] investigated the effects of ozone treatment on the microorganism mortality in wheat flour and shelf life of fresh noodles. The results showed that the total plate count (TPC) can be largely reduced in wheat flour exposed to ozone gas for 30 and 60 min. Microbial growth and darkening rate of fresh noodles made from ozone-treated flour were delayed significantly. While Yang et al. [35] found that the optimal ozone treatment parameters for fresh fish noodle were ventilation of 6 L/min for 20 min then sterilizing for 10 min. Under the conditions, the shelf life of noodles treated by 100% CO_2 in the package was 22 days at − 0.8 °C. Cai et al. [112] reported that the shelf life of fresh noodle increased to 10 days with the Co60 γray treatment at 10 kGy, while significant extending of shelf life of fresh noodle was also observed under the treatment at 4 kGy [113]. Moreover, Xie and Li [114] found that the fresh noodle can be stored over 28 days with the 20 s microwave (700 W) treatment at 4 °C. As for the vacuum mixing, − 0.06 MPa was considered as the best degree for noodle production. In addition, vacuum mixing could promote the formation of bound water in the dough and effectively reduce water activity at certain moisture content, resulting in the extended shelf life of fresh noodle [33].

Packaging technology

Vacuum packaging, modified atmosphere packaging (MAP) and active packaging are widely used in food package. Vacuum packaging is not suitable for fresh noodles due to the surface adhesion and deformation of noodles caused by the squeeze. The effect of reduced O_2- and elevated CO_2-modified atmospheres and abuse temperature (15 °C) on the growth of *Penicillium aurantiogriseum* were evaluated by Zardetto [115]. The results indicated that, no fungal growth was observed for up to 550 h of incubation at 15 °C when samples were packaged in atmospheres with CO_2 concentrations higher than 70%. Nobile et al. [116] explored the combined effects of chitosan and MAP to improve the microbiological quality of amaranth-based homemade fresh pasta. The results suggested that there was a combined effect between MAP and chitosan in delaying the microbial quality loss of pasta during storage. Moreover, it was also found that, among the tested MAP conditions, the ratio of N_2/CO_2 at 30/70 is the most efficient combination, which promoting an extension of the microbial acceptability limit beyond 2 months.

Generally, the individual physical treatment or packaging technology usually has limitation for extending shelf life. Exploration on the synergistic interactions among preservatives, temperature, physical treatment and packaging technologies should be highly regarded for long-term storage of noodles and CSB.

New technologies used in CSB

Physical treatments

Similar to noodles, a series of physical treatments have been applied to increase the shelf life of CSB. Han and Wang [117] studied the effects of irradiation on the quality of CSB, the results showed that ^{60}Co at 6 kGy companied with 10 min steaming sterilization (100 °C) pre-treatment can effectively extended the shelf life of CSB for 6 months. The sterilization of steamed bread made by yeast and special flour were investigated by Xiong et al. [118]. The results indicated that the shelf life of CSB was extended to 3–6 months when the sterilization condition was microwave 150 °C for 70 s. Liu et al. [34] explored the effects of photos on the quality of CSB, the results reflected that, after the samples were stored at 4 °C for 2, 4, 7, and 12 days, the sensory scores of the photon-treated group were all higher than those of control group, and mold spots appeared on the control samples on day 12. The differences on the retrogradation degree and protein conformation between CSB with and without photon treatment revealed that the preservation effects of photons on steamed bread could be achieved by affecting the starch retrogradation and protein-secondary structure of steamed bread.

High temperature (above 100 °C) companied with appropriate irradiation or microwave treatment showed excellent performance on the shelf-life extension of CSB. However, the greatly extended shelf life was mainly ascribed to the inhibited microbial growth in CSB. Increasing hardness and taste variation were observed with the prolonging of storage time. Different from irradiation or microwave treatment, photon treatment was supposed to prolong the shelf life of CSB due to its anti-staling activity, but much faster microbial growth occurred during storage comparing to irradiation or microwave treatment. Because the limitation of different physical treatments, composite treatments or physical treatments combined with food additives should be considered for the shelf-life extension of CSB.

Packaging technology

Several packaging methods towards extending shelf life of CSB have been developed. The effects of MAP on the quality of CSB showed that, the shelf life of CSB can be increased significantly when packing without oxygen. Extended shelf life of CSB was observed with the increasing content of CO_2, and the shelf life was prolonged to 12 days when packing with the 100% CO_2. In addition, the shelf life of CSB was extended to 18 days when the ultraviolet radiation and MAP packaging were applied simultaneously [119].

The active packaging [oxygen absorber (OA) and ethanol emitter (EE)] has been applied to extending the shelf life of CSB [21]. Microbiological shelf life of CSB (mold, bacteria and yeast) was extended up to 11 days [(by oxygen absorber combined with ethanol emitter (OA + EE)]. Textural and DSC analysis showed that active packaging decreased the CSB staling. This may be due to the packaging which prevented the moisture loss during storage. CSB packaged with (OA + EE) had lower hardness development in the first 7 days of storage than that with OA [21]. Ethanol might act as plasticizers for the protein fractions, lowering the protein–starch associations during storage. Though ethanol had no effect on the recrystallization of amylopectin, it might interact with amylose fraction for the formation of complexes. The amylose–ethanol complexes may reduce the amylose–amylose interactions for retrogradation.

The moisture migration from crust to crumb was related to the staling of CSB [120]. Hot packaging is a novel technology that can be used for extending CSB shelf life. As shown in Fig. 1 [120], the moisture migration in CSB was ended after storage in room temperature for 20 days. Two challenges need to be faced in hot packaging: the water vapor will condense inside the package and the packaging bag will expand because of heat. But, the unique advantage is that it can protect against

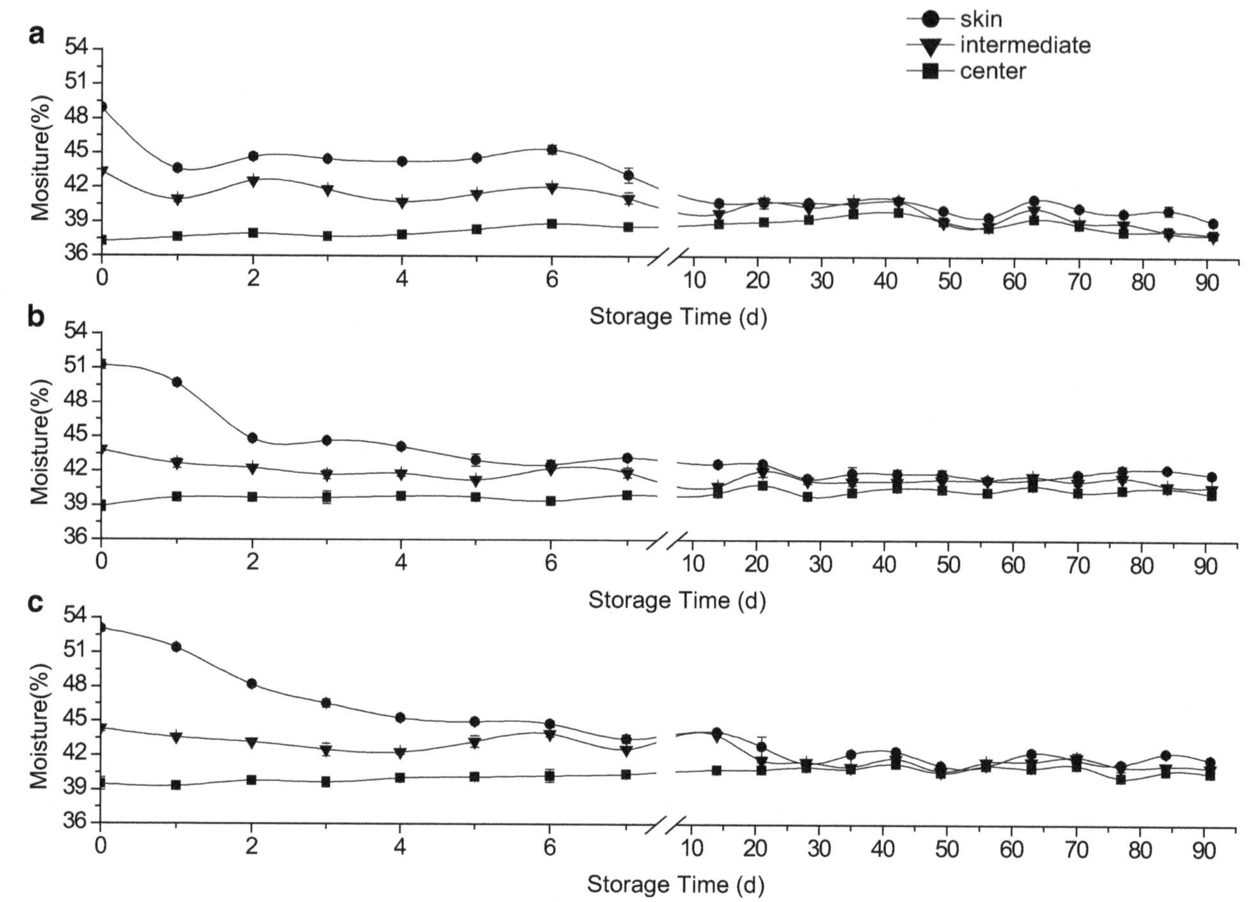

Fig. 1 The changes of moisture of different zones of steamed buns during storage at room temperature: water addition of dough is 40% (**a**), 42% (**b**) and 44% (**c**) (adopted from [120])

contamination, and save space without cooling. For CSB manufacturers, hot package technology can significantly increase processing efficiency and reduce costs. To solve the problems of the short storage time, mold growth at room temperature storage, retrogradation in steamed bread production, and also to meet the requirements of safety and convenience, as well as industrialization, there is promising data from the technology using traditional yeast for fermentation, combined with online hot-packing technology to produce CSB. The results showed that the microorganism index, including *Staphylococcus aureus* and *E. coli*, met the requirements of the national standard. In addition, there was no significant change in protein content, moisture content, fat content, a\nd starch retrogradation, as measured using X-ray diffraction [120]. Utilizing this technology, water vapor disappeared after 3 days of storage, without leading to water condensation on the surface, thus having a similar effect as vacuum packaging. This technology saved energy consumption and minimized secondary contamination during transportation and retailing. The CSB packaged by this technology can be maintained fresh up to 90 days without any additives, giving good taste along with least starch retrogradation [120].

As described above, to achieve the prolonged shelf life, physical treatment and high concentration of CO_2 were required for the MAP, while oxygen absorber and ethanol emitter had to be prepared for the active package. In contrast, hot packaging, which showing additional advantages such as low secondary contamination, self-sterilization, self-vacuum, low energy input and simple operation, contributed to the much longer shelf life than that of MAP or active packaging. It is supposed that the hot packaging technology would be widely applied in the storage of Chinese traditional staple foods in future.

Conclusion and future perspective

Traditionally, fresh noodles and CSB are much produced in small workshop and on family basis in China. Developing noodles and CSB industry requires efficient strategies to solve the problems appeared at processing and storage stages. To date, there are great advances in the development of food additives towards good quality and prolonged shelf life of fresh noodles and CSB. On the one hand, salts, vital wheat gluten, hydrocolloids, esters and enzymes have been incorporated in individual or composite ways to enhance the processing quality or to retard the staling of fresh noodles and CSB. On the other hand, antimicrobial agents, such as acids, oxidants and some natural products were utilized to extend the microbiological shelf life of fresh noodles and CSB. Apart from additives, manipulating production process and storage conditions also showed promising results for enhancing quality and shelf life of fresh noodles and CSB. Treatment with ozone, radiation, microwave, photons and vacuum mixing, as well as packaging methods such as vacuum packaging, MAP, active packaging and hot-online packaging have been employed to manufacture fresh noodles or CSB with significantly extended shelf life.

Quality improvement and shelf-life extending are still two major challenges for the production of fresh noodles and CSB at large scale and industrial levels. Compared to the research on bread, pasta and instant noodles, little is done in fresh noodles and CSB. To further promote the industrialization of fresh noodles and CSB, following topics should be better addressed in future:

(1) To ensure the excellent and stable quality performance of wheat flour, wheat cultivars suitable to the production of specific noodle or CSB should be screened out.
(2) Sodium salts, such as sodium chloride, sodium alginate, sodium hydrates and sodium phosphates were involved in the formulations of noodles and CSB for the quality improvement. However, high intake of sodium deviated from the principle of low sodium diet. As so, sodium in noodles and CSB need to be partially substituted by potassium, enzyme or some organic products with the similar property.
(3) Comparing with other additives, enzymes have inherent advantages in the production of noodles and CSB. Although some enzymes have been applied as the substitutes of traditional additives, for example lipase has been used to replace some emulsifiers, while TG is the substitute of vital gluten, further investigation about the food-grade enzymes is still required to support the large scale production of noodles and CSB in China. First, more accurate formula of composite enzyme aiming at specific noodle or steamed bread need to be developed. Second, based on the technology of recombinant DNA, protein engineering, etc., novel recombinant enzymes suitable for the specific processing conditions should be manufactured. Finally, although enzymes used in food processing have historically been considered non-toxic, some characteristics, such as allergenicity, activity-related toxicity, residual microbiological activity and chemical toxicity arising from their chemical nature and source, are of high concern. Therefore, safety evaluation of the food-grade recombinant enzymes including those produced by the genetically modified microorganisms is still essential to assure consumer safety.
(4) Hydrocolloids and enzymes usually act as emulsifier, thickener, gluten fortifier, anti-staling agent even antimicrobial agent in the processing and storage of noodles and CSB. The function properties of these additives mainly attributed to their interactions with protein, starch, lipid and other components in foods, hence further exploration on the interaction mechanisms need to be conducted.
(5) Multi-grain noodle and CSB is being a research focus due to their nutritional value and high consumer acceptance. However, oat, buckwheat or other cereals usually show poor viscoelasticity in dough development due to the lack of gluten proteins, thus more formulas of composite additive consisting of vital wheat gluten, hydrocolloids, TG and other gluten fortifiers should be developed to enhance the dough stability of multi-grain flour. Moreover, complex additive aiming at specific products has played important role in the noodle and CSB making, the mechanism of cooperation and antagonism of different additives such as salts, hydrocolloids, enzymes and preservatives in the mixture should be clarified to support the development of new composite additive with accurate formula.
(6) Physical treatments and packaging methods have showed much higher efficiency in shelf-life extending for noodles and CSB. On the one hand, novel technologies such as high pressure, ultrasound, pulsed electric field, and plasma treatments can be employed for their potential anti-staling impact. On the other hand, the function mechanism for the physical treatments and packaging methods are worth of further investigation.

Abbreviations

CSB: Chinese steamed bread; MAP: modified atmosphere packaging; WSN: white salted noodles; YAN: yellow alkaline noodles; STPP: sodium tripolyphosphate; SHMP: sodium hexametaphosphate; SPP: sodium pyrophosphate; SDHP: sodium dihydrogen phosphate; TG: transglutaminase; PGA: propylene glycol alginate; CSL: calcium stearoyl lactylate; SSL: sodium stearoyl lactylate; GMS: glyceryl monostearate; DATEM: diacetytartaric esters of monoglycerides; HPMC: hydroxypropylmethylcellulose; CMC: carboxymethylcellulose; XG: xanthan gum; AP: apple pectin; DSC: differential scanning calorimetry; AX: arabinoxylans; WE-AX: water-extractable arabinoxylans; S-AX: solubilized arabinoxylans; WU-AX: water-unextractable arabinoxylans; TPC: total plate count; OA: oxygen absorber; EE: ethanol emitter.

Authors' contributions

XH and XW conceived the paper. XW, XH, ZM participated in the literature view and selection and in drafting of the manuscript. XL and LL participated in the literature selection and revised the manuscript critically. XY, KZ and YL participated in the literature selection, paper writing and revising. All authors read and approved the final manuscript.

Acknowledgements

The authors are grateful to colleagues for providing kind technical assistance and advices in papering writing.

Competing interests

The authors declare that they have no competing interests.

Funding

This work was supported by the National Natural Science Foundation Project (31671889) from Ministry of Science & Technology of the People's Republic of China, Shaanxi Science and Technology as a key project (2015KTZDNY01-07), National Oat and Buckwheat Industrial Technology System (CARS-D-08), and Research team project of Shaanxi Normal University (GK201601002) supported by the Fundamental Research Funds for the Central Universities.

References

1. Gan CY, Ong WH, Wong LM, Easa AM. Effects of ribose, microbial transglutaminase and soy protein isolate on physical properties and in vitro starch digestibility of yellow noodles. LWT Food Sci Technol. 2009;42:174–9.
2. Li M, Zhu KX, Wang BW, Guo XN, Peng W, Zhou HM. Evaluation the quality characteristics of wheat flour and shelf-life of fresh noodles as affected by ozone treatment. Food Chem. 2012;135:2163–9.
3. Zhu F. Influence of ingredients and chemical components on the quality of Chinese steamed bread. Food Chem. 2014;163:154–62.
4. Yan J, Zhang Y, He Z. Investiagition on paste property of Chinese wheat. Sci Agric Sin. 2001;34:1–4 **(In Chinese with English abstract)**.
5. Yangsoo K, Huang W, Zhu H, Patricia RD. Spontaneous sourdough processing of Chinese Northern-style steamed breads and their volatile compounds. Food Chem. 2009;114:685–92.
6. He ZH, Liu L, Xia XC, Liu JJ, Pena RJ. Composition of HMW and LMW glutenin subunits and their effects on dough properties, pan bread, and noodle quality of Chinese bread wheats. Cereal Chem. 2005;82:345–50.
7. Huang S, Yun SH, Quail K, Moss R. Establishment of flour quality guidelines for northern style Chinese steamed bread. J Cereal Sci. 1996;24:179–85.
8. Kovacs MIP, Fu BX, Woods SM, Khan K. Thermal stability of wheat gluten protein: its effect on dough properties and noodle texture. J Cereal Sci. 2004;39:9–19.
9. He ZH, Liu AH, Pena RJ, Rajaram S. Suitability of Chinese wheat cultivars for production of northern style Chinese steamed bread. Euphytica. 2003;131:155–63.
10. Zhang PP, He ZH, Zhang Y, Xia XC, Liu JJ, Yan J, et al. Pan bread and Chinese white salted noodle qualities of Chinese winter wheat cultivars and their relationship with gluten protein fractions. Cereal Chem. 2007;84:370–8.
11. Jh HITV. Microbial and biochemical spoilage of foods: an overview. Int J Food Microbiol. 1996;33:1–18.
12. Sha K, Qian P, Wang LJ, Lu ZH, Li LT. Effect of storage time on the physicochemical and sensory properties of man-tou (Chinese Steamed Bread): international Journal of Food Engineering. Int J Food Eng. 2007;3:1–17 **(In Chinese with English abstract)**.
13. Sha K, Zhang ZJ, Liu HJ, Li-Te LI. Effects of storage temperature on firming and retrogradation of amylopectin in steamed bread. Food Sci Technol. 2009;45:767–74 **(In Chinese with English abstract)**.
14. Fang XB, Su DM, Hu LH. Research advances on antiretrogradation technology of Chinese steamed bread. Storage Process. 2011;11:47–9 **(In Chinese with English abstract)**.
15. Hu X, Sheng X, Liu L, Ma Z, Li X, Zhao W. Food system advances towards more nutritious and sustainable mantou production in China. Asia Pac J Clin Nutr. 2015;24:199–205.
16. Wu P, Zhou T, Dong HZ, Zhang XS, Du HY, Jiang XL. Factors effecting mantou quality. J Chin Cereals Oils Assoc. 2012;27:5107–17 **(In Chinese with English abstract)**.
17. Li M, Ma M, Zhu KX, Guo XN, Zhou HM. Delineating the physico-chemical, structural, and water characteristic changes during the deterioration of fresh noodles: understanding the deterioration mechanisms of fresh noodles. Food Chem. 2017;216:374–81.
18. Ray B. Fundamental food microbiology. [M]. Boca Raton: CRC Press; 2013.
19. Ronda F, Caballero PA, Quilez J, Roos YH. Staling of frozen partly and fully baked breads. Study of the combined effect of amylopectin recrystallization and water content on bread firmness. J Cereal Sci. 2011;53:97–103.
20. Sha K, Qian P, Liu HJ, Li-Te LI. Effect of process conditions on specific volume and extent of firming in steamed bread. Food Sci Technol. 2007;12:54–7 **(In Chinese with English abstract)**.
21. Sheng Q, Guo XN, Zhu KX. The effect of active packaging on microbial stability and quality of Chinese steamed bread. Packag Technol Sci. 2015;28:775–87.
22. Wang S, Khamchanxana P, Zhu F, Zhu C, Pan J. Textural and sensory attributes of steamed bread fortified with high-amylose maize starch. J Texture Stud. 2016;48:3–8 **(In Chinese with English abstract)**.
23. Ding W, Wang Y, Xia W. Mechanism and measuring method of starch retrogradation. Cereal Feed Ind. 2004;12:28–30 **(In Chinese with English abstract)**.
24. Kruger JE, Hatcher DW, Depauw R. A whole seed assay for polyphenol oxidase in Canadian prairie spring wheats and its usefulness as a measure of noodle darkening. Cereal Chem. 1994;71:324–6.
25. Chai XH. Studies on antisepsis of steamed bread in the atmosphere of spice essential oil. Food Mach. 2011;27:126–8 **(In Chinese with English abstract)**.
26. Day L, Augustin MA, Batey IL, Wrigley CW. Wheat-gluten uses and industry needs. Trends Food Sci Technol. 2006;17:82–90.
27. He C, Lin X. Anti-aging effect of emulsifier on steamed bread. Acad Period Farm Prod Process. 2010;208:20–2 **(In Chinese with English abstract)**.
28. Niu M, Li X, Wang L, Chen Z, Hou GG. Effects of inorganic phosphates on the thermodynamic, pasting, and Asian noodle-making properties of whole wheat flour. Cereal Chem. 2014;91:1–7.
29. Olempska-Beer ZS, Merker RI, Ditto MD, Dinovi MJ. Food-processing enzymes from recombinant microorganisms—a review. Regul Toxicol Pharmacol. 2006;45:144–58.
30. Sim SY, Aziah AAN, Cheng LH. Characteristics of wheat dough and Chinese steamed bread added with sodium alginates or konjac glucomannan. Food Hydrocoll. 2011;25:951–7.
31. Wang XW, Cui YK, Zhao JL, Zhao RY. A review of preservation technology for fresh noodles. Food Ind. 2013;12:405–14 **(In Chinese with English abstract)**.
32. Wu J, Beta T, Corke H. Effects of salt and alkaline reagents on dynamic rheological properties of raw oriental wheat noodles. Cereal Chem. 2006;83:211–7.
33. Li M, Luo LJ, Zhu KX, Guo XN, Peng W, Zhou HM. Effect of vacuum mixing on the quality characteristics of fresh noodles. J Food Eng. 2012;110:525–31.
34. Liu XL, Zhen-Jing LI, Zhou QL. Effect of photons on the preservation of steamed bread. Mod Food Sci Technol. 2017;33:229–36 **(In Chinese with English abstract)**.
35. Yang JJ, Xian LI, Liu YM, Zhao SM, Xiong SB. Research on the preservation technology of the fresh fish noodles. Sci Technol Food Ind. 2012;33:336–9 **(In Chinese with English abstract)**.
36. Asenstorfer RE, Wang Y, Mares DJ. Chemical structure of flavonoid compounds in wheat (*Triticum aestivum* L.) flour that contribute to the yellow colour of Asian alkaline noodles. J Cereal Sci. 2006;43:108–19.
37. Fu BX. Asian noodles: history, classification, raw materials, and processing. Food Res Int. 2008;41:888–902.

38. Hou G. Oriental noodles. Adv Food Nutr Res. 2001;43:141–93.
39. Chen J, Shi L, Wang L, Bian K. The effects of sodium chloride on the dough extensibility of reman. Food Feed Ind. 2015;12:35–8 **(In Chinese with English abstract)**.
40. Jing P, Zheng X, Ding X, Liu C, Han X. Effects of salt on the quality of dough pieces and noodle. Cereal Feed Ind. 2014;09:32–5 **(In Chinese with English abstract)**.
41. Kubomura K. Instant noodles in Japan. Cereal Foods World. 1998;43:194–7.
42. Rho KL, Seib PA, Chung OK, Deyoe CW. Noodles. VII. Investigating the surface firmness of cooked oriental dry noodles made from hard wheat flours. Cereal Chem. 1988;65:320–6.
43. Moss HJ, Miskelly DM, Moss R. The effect of alkaline conditions on the properties of wheat flour dough and Cantonese-style noodles. J Cereal Sci. 1986;4:261–8.
44. Terada M, Minami J, Yamamoto T. Rheological properties of dough made from flour exposed to gaseous ammonia. Cereal Chem. 1981;58:101–5.
45. Shin SY. Cooking properties of dry noodles prepared from HRW-WW and HRW-ASW wheat flour blends. Korean J Food Sci Technol. 1993;25:232–7.
46. Bean MM, Nimmo CC, Fullington JG, Keagy PM, Mecham DK. Dried Japanese noodles. II. Effect of amylase, protease, salts, and pH on noodle doughs. Cereal Chem. 1974;51:427–33.
47. Miskelly DM, Moss HJ. Flour quality requirements for Chinese noodle manufacture. J Cereal Sci. 1985;3:379–87.
48. Deman JM, Melnychyn P. Symposium: phosphates in food processing. Ontario: University of Guelph; 1970.
49. Furia TE. EDTA in foods, a technical review. Food Technol. 1964;18:1874–82.
50. National Academy of Sciences. Chemicals used in food processing. Publication No. 1274. Washington, D. C: National Academy of Sciences—National Research Council; 1965.
51. Wu X, Li L. Improving effects of compound phosphate on noodle quality. Cereal Feed Ind. 1998;12:45–6 **(In Chinese with English abstract)**.
52. Bao Y. The application of phosphates in noodle production. J Henan Univ Technol Nat Sci Ed. 2009;30:69–72 **(In Chinese with English abstract)**.
53. Young LL, Lyon CE. Effects of rigor state and addition of polyphosphate on the color of cooked turkey meat. Poult Sci. 1994;73:1149–52.
54. Zhou Y, Hou GG. Effects of phosphate salts on the pH values and rapid visco analyser (RVA) pasting parameters of wheat flour suspensions. Cereal Chem. 2012;89:38–43.
55. Li A, Ou S, Wu J, Liang Z. The application of phosphates in frozen steamed bread. J Food Sci Technol. 2000;21:58–60 **(In Chinese with English abstract)**.
56. Chang Y, Liu C, Zhang X. The effects of phosphates on the quality of steamed bread. Grain Process. 2011;36:31–2 **(In Chinese with English abstract)**.
57. Zhang Z, Cao X, Li G. The processing technology of quick frozen and freeze-dried foods, vol. 102–108. Beijing: Chemical Industry Press; 2008 **(In Chinese)**.
58. Hou HR. Effect of mixed additives on the crack ratio of quick-freezing dumpling. J Food Sci Technol. 2008;01:111–3 **(In Chinese with English abstract)**.
59. Ding L, Chen J, Liu FL, Wang C. The effects of phosphates on frozen dumpling. Food Ferment Ind. 2007;33:87–90 **(In Chinese with English abstract)**.
60. Zhang LH, Zhang GD, Sun CC, Yang WZ, Chen HN. Effect of mixed additives on the cracking rate and browning of deep-frozen dumpling. Food Res Dev. 2010;31:24–6 **(In Chinese with English abstract)**.
61. Xue S, Zou Y, Chen X, Yang H, Xing T, Xu X, et al. Effects of sodium tripolyphosphate on functional properties of low-salt single step high pressure processed chicken breast sausage. Int J Food Sci Technol. 2016;51:2106–13.
62. Park SJ, Baik BK. Quantitative and qualitative role of added gluten on white salted noodles. Cereal Chem. 2009;86:646–52.
63. Li L, Wang S. The effects of vital wheat gluten on the quality of noodle with high quantity of buckwheat. T Chin Soc Agric Mach. 2011;67–69. **(In Chinese)**.
64. Wang F, Huang W, Kim Y, Liu R, Tilley M. Effects of transglutaminase on the rheological and noodle-making characteristics of oat dough containing vital wheat gluten or egg albumin. J Cereal Sci. 2011;54:53–9.
65. Liu Y, Zhang Z, Yang J, Liu M, Zhuo W, Song R. Effects of vital wheat gluten on the processing quality of wheat cultivar Pubing 9946 meal. J Chin Inst Food Sci Technol. 2014;14:164–9 **(In Chinese with English abstract)**.
66. Zhang Z, Cao X. Technology of buckwheat steamed bread production. J Chin Cereals Oils Assoc. 2006;21:60–2 **(In Chinese)**.
67. Yang Y, Gong-Ming YU, Wang CZ. Effect of PGA on nature construction of acidic wet noodles. Cereals Oils. 2009;5:16–9 **(In Chinese with English abstract)**.
68. Yang AH, Wang CZ, Yang Y. Effect of PGA on the nature construction of noodles. Sci Technol Food Ind. 2010;31:323–5 **(In Chinese with English abstract)**.
69. Liu R, Jiang J, Yang Y, Xie S, Liu H, Wang X, et al. Effect of low-esterification PGA with different concentrations on the quality of noodles. China Food Addit. 2016;148–52. **(In Chinese with English abstract)**.
70. Dou KN, Wang F, Xie JH. Study on the cooking quality of the wheat-middling noodles. Food Ind. 2013;34:77–78 **(In Chinese with English abstract)**.
71. He Y, Ma T, Wang F. The effects of emulsifiers on the quality of Chinese northern steamed bread. Grain Process. 2009;34:46–9 **(In Chinese with English abstract)**.
72. Rosell CM, Collar C, Haros M. Assessment of hydrocolloid effects on the thermo-mechanical properties of wheat using the Mixolab. Food Hydrocoll. 2007;21:452–62.
73. Collar C, Andreu P, MartíNez JC, Armero E. Optimization of hydrocolloid addition to improve wheat bread dough functionality: a response surface methodology study. Food Hydrocoll. 1999;13:467–75.
74. Rojas JA, Rosell CM, Barber CBD. Pasting properties of different wheat flour-hydrocolloid systems. Food Hydrocoll. 1999;13:27–33.
75. Liu X, Mu T, Sun H, Zhang M, Chen J, Fauconnier ML. Influence of different hydrocolloids on dough thermo-mechanical properties and in vitro, starch digestibility of gluten-free steamed bread based on potato flour. Food Chem. 2018;239:1064–74.
76. Kaur A, Shevkani K, Singh N, Sharma P, Kaur S. Effect of guar gum and xanthan gum on pasting and noodle-making properties of potato, corn and mung bean starches. J Food Sci Technol. 2015;52:8113 **(In Chinese with English abstract)**.
77. Yang Y, Wang C. Study on effect of sodium alginate on oats noodle quality. Cereal Food Ind. 2010;17:21–5 **(In Chinese with English abstract)**.
78. Zhao Z, Yu G, Liu H, Ma D. The effects of sodium alginate on the noodle texture. Grain Process. 2008;33:78–81 **(In Chinese with English abstract)**.
79. Shao P, Xu M. Optimization of processing conditions of buckwheat noodles by adding wheat bran dietary fibers. J Agric Sci. 2006;27:38–40 **(In Chinese with English abstract)**.
80. Zhang J, Li M, Fan Y, Gong X. Effects of thickeners on the quality improvement of fresh noodles. Sci Technol Food Ind. 2007;28:185–186. (In Chinese with English abstract).
81. Liu LY, Zhao K, Yang CH, Zhang N, Miao M. Application of starch and modified starch in quality improvement of noodles. J Harbin Univ Commer. 2011;27:51–3 **(In Chinese with English abstract)**.
82. Yan SQ, Zhou YH, Shen Q. The effects of four hydrocolloids on wheat starch, gluten protein properties and noodle quality. Food Res Dev. 2011;32:63–7 **(In Chinese with English abstract)**.
83. Chen RQ, Dong WB, Xiu XH. Effect of different thickeners on quality of oat noodles. Cereals Oils. 2013;26:21–4 **(In Chinese with English abstract)**.
84. Kou X, Du F, Xu T, Li W, Liu Z, Liu Z, et al. Effect of different thickeners on quality of low protein sorghum noodles. J Food Sci Technol. 2015;40:150–4 **(In Chinese with English abstract)**.
85. Beilen JBV, Li Z. Enzyme technology: an overview. Curr Opin Biotechnol. 2002;13:338–44.
86. Chen LM. The application of fungal α- amylase to the processing of steamed bread (mantou). Sci Technol Food Ind. 2008;08:308–308 **(In Chinese with English abstract)**.

87. Shi Y, Xu, C, Zhang Y. The effects of α-amylase on the quality of fresh noodles. T Chin Soc Agric Mach. 2013;57–61 **(In Chinese)**.
88. He Y, Song J, Li G, Shi Y, Xiao M. The application of a α-amylase in wet-fresh noodles making. Food Res Dev. 2017;38:104–8 **(In Chinese)**.
89. Ma T, He Y, Wang F. The effects of four enzymes on the quality of Chinese northern steamed bread. Cereals and Oils Process. 2009;86–90 **(In Chinese with English abstract)**.
90. Zhang JZ, Zhao XW. Application of enzyme in flour product. Food Sci Technol. 2006;07:185–8 **(In Chinese with English abstract)**.
91. Feng W. The application of glucose oxidase in food processing. Sci Technol Food Ind. 2000;21:67–8 **(In Chinese with English abstract)**.
92. Shan CJ, Zhou JZ. Effects of glucose oxidase and lipase on quality of noodle. Acta Agric Jiangxi. 2008;20:65–7 **(In Chinese with English abstract)**.
93. Lin J, Zhu K. The quality improvement of multi-grain steamed bread by the incorporation of glucose oxidase. Cereal Food Ind. 2011;18:17–9 **(In Chinese)**.
94. Meng-Qin LI, Zhang J, Lei N, Feng ZQ, Wang CX. Study on flour quality improvers for better Steamed-bread. Food Sci. 2008;29:150–3 **(In Chinese with English abstract)**.
95. Cato L, Halmos AL, Small DM. Measurement of lipoxygenase in Australian white wheat flour: the effect of lipoxygenase on the quality properties of white salted noodles. J Sci Food Agric. 2010;86:1670–8.
96. Zhang C, Zhang S, Bie X, Zhao H, Lu F, Lu Z. Effects of recombinant lipoxygenase on the rheological properties of dough and the quality of noodles. J Sci Food Agric. 2015;96:3249–55.
97. Suzuki T, Sunju K, Mukasa Y, Morishita T, Noda T, Takigawa S, et al. Effects of lipase, lipoxygenase, peroxidase and free fatty acids on volatile compound found in boiled buckwheat noodles. J Sci Food Agric. 2010;90:1232–7.
98. Yang X, Jing W, Zheng K, Dan W. A xylanase from *Streptomyces* sp. FA1: heterologous expression, characterization, and its application in Chinese steamed bread. J Ind Microbiol Biotechnol. 2016;43:663–70.
99. Courtin CM, Delcour JA. Arabinoxylans and endoxylanases in wheat flour bread-making. J Cereal Sci. 2002;35:225–43.
100. Courtin CM, Gelders GG, Delcour JA. Use of two endoxylanases with different substrate selectivity for understanding arabinoxylan functionality in wheat flour breadmaking. Cereal Chem. 2001;78:564–71.
101. Autio K, Kruus K, Knaapila A, Gerber N, Flander L, Buchert J. Kinetics of transglutaminase-induced cross-linking of wheat proteins in dough. J Agric Food Chem. 2005;53:1039–45.
102. Bellido GG, Hatcher DW. Effects of a cross-linking enzyme on the protein composition, mechanical properties, and microstructure of Chinese-style noodles. Food Chem. 2011;125:813–22.
103. Wei XM, Guo XN, Zhu KX, Peng W, Zhou HM. Effects of transglutaminase on qualities of buckwheat noodles. Food Mach. 2016;32:188–92 **(In Chinese with English abstract)**.
104. Lv Y, Zhang FF, Chen J. Response surface analysis the effect of enzymes on noodle quality. China Food Addit. 2013;1:175–80 **(In Chinese with English abstract)**.
105. Ren H, Chen L. The effects of composite enzyme on the quality of Chinese steamed bread. Mod Flour Mill Ind. 2007;62 **(In Chinese)**.
106. Zhang C, Dou H, Jia R, Yue X. The application and research on ε-polylysine in antiseptic humid noodle. Food Ind. 2008;02:60–2 **(In Chinese with English abstract)**.
107. Liu ZG, Xue-Ming XU, Jin ZY. Study on effect of chlorine dioxide and hydrogen peroxide on the preservation of wet raw noodles. Sci Technol Food Ind. 2008;29:96–102 **(In Chinese with English abstract)**.
108. Wang H, Huang H. The effects of preservation on the shelf life of fresh noodles. Light Ind Sci Technol. 2009;25:10–1 **(In Chinese)**.
109. Zhou B, Shi L, Meng T, Wang H, Shan F, Bian J, et al. Reseach on the preservation of coarse fresh noodle. J Chin Cereals Oils Assoc. 2008;23:42–4 **(In Chinese)**.
110. Zhao YL, Liu CH, Zhi-Gang MA, Zhu-Sheng LI. Effects of food antimicrobial additives on quality and storage life of steamed bread. Sci Technol Cereals Oils Foods. 2005;13:4–6 **(In Chinese)**.
111. Sheng Q. Study on preservation of Chinese steamed bread store at room temperature. Jiangsu: Jiangnan University; 2015 **(In Chinese with English abstract)**.
112. Cai J. Preservation of fresh noodles by irradiation. Radiat Phys Chem. 1998;52:35–8.
113. Man L, Zhu K, Xu G, Wei P, Zhou H. Effect of water activity (aw) and irradiation on the shelf-life of fresh noodles. Innov Food Sci Emerg Technol. 2011;12:526–30 **(In Chinese with English abstract)**.
114. Xie X, Li X. The study on the preservation technology of fresh noodles. T Chin Soc Agric Mach. 2012;65–7 **(In Chinese)**.
115. Zardetto S. Effect of modified atmosphere packaging at abuse temperature on the growth of *Penicillium aurantiogriseum* isolated from fresh filled pasta. Food Microbiol. 2005;22:367–71.
116. Nobile MAD, Benedetto ND, Suriano N, Conte A, Corbo MR, Sinigaglia M. Combined effects of chitosan and MAP to improve the microbial quality of amaranth homemade fresh pasta. Food Microbiol. 2009;26:587–91.
117. Han ZH. Study on sterilization technology of steamed bread. Food Mach. 2013;29:209–11 **(In Chinese with English abstract)**.
118. Xiong L, Fan-Fei LI, Sun QJ. Studies on processing technology and quality properties of long life steamed bread. Cereals Oils. 2009;01:24–6 **(In Chinese with English abstract)**.
119. Li X. Research on the circulation preservation of steamed bread. Tianjiang: Tianjing Universit y of Science and Technology; 2013 **(In Chinese with English abstract)**.
120. Sheng X, Ma Z, Li X, Liu L, Hu X. Effect of water migration on the thermal-vacuum packaged steamed buns under room temperature storage. J Cereal Sci. 2016;72:117–23.

8

Efficiency and effectiveness of vitamin C-substrate organo-mineral straight fertilizer in lettuce (*Lactuca sativa* L.)

Hyo-Seok Chae[1*], Hye-Ji Noh[2], Woo Seung Song[1] and Hyang-Hyun Cho[1]

Abstract

Mineral fertilizers are expensive and last only a relatively short time. Proper handling of fertilizers has a direct impact on crop quality and profitability. Currently, straight fertilizers are divided into types, A and B, based on the fertilizer's sedimentation properties when mixed. Therefore, to solve these flaws, we developed a vitamin C-substrate organo-mineral fertilizer, AGH-Ringer. This study consisted of a fertilizer equivalency test, which lasted 15 weeks of soil cultivation in a farm field, using lettuce (*Lactuca sativa* L., 'Jeok Chi Ma') with AGH-Ringer as a test fertilizer. This fertilizer was based on powder compound-type fertilizer r-Ringer and manufactured by chelating essential minerals with a vitamin C substrate to have NPK contents in a ratio of 18.9:14:4.0. In the onsite evaluation, the superiority of AGH-Ringer was demonstrated through an effect of growth stimulation in both aerial and rhizosphere parts of the plants at suitable soil temperatures. This fertilizer also increased yield, even in inadequate low-temperature conditions. In addition, AGH-Ringer increased the absorption and bioavailability of minerals in lettuce leaves at harvest. The mineral absorption ability showed a uniform pattern during the entire harvest period. It has been confirmed that AGH-Ringer, manufactured with vitamin C as a substrate, is a fertilizer with excellent efficacy that increases growth and harvest simultaneously, solving the existing problems of chemical fertilizers such as sedimentation and lack of mineral absorptivity in nutrient solution systems.

Keywords: Straight fertilizer, Vitamin C (substrate), Organo-mineral fertilizer, Fertilizer efficiency

Background

Agriculture has a played a major role in the development of human civilization for a long time. In prehistoric times, humans formed clans and started cultivating plants as a food supply. Conventional agricultural systems can cause multifarious negative impacts on the environment such as inefficient use of water, large-scale land requirements, outflow of residual pesticides, and erosion with soil degeneration [1]. Therefore, over time, people have attempted to develop cultivation techniques for more efficient food production and environmental conservation. Representative examples include hydroponic farming and bio-farming using bio-inoculation. Hydroponic cultivation, which is an advanced cultivation technique, has advantages such as increase in yield, continuous production throughout the year, and a need for only a small area for cultivation of many kinds of agricultural products [2]. On the other hand, management of this cultivation system and the supply of culture medium are expensive. Recently, some organic fertilizers have been reported to play an important role in plant growth and development through plant growth-promoting microorganisms that colonize the rhizosphere and interact with the plant [3–5]. However, these microbial inoculations are difficult to use due to the diversity of open-field applications and laboratory results.

The concentration range of inorganic elements that promote optimal growth in crops varies by nutrient type and crop species. Plant growth is optimal when essential elements are supplied properly. When the absorbed inorganic element is either insufficient or excessive, the growth of stems and roots is impaired [6, 7]. In addition,

*Correspondence: iseulii@gmail.com
[1] HANIL BIOMED Inc., Gwangju 62363, Korea
Full list of author information is available at the end of the article

total yield is proportional to the mineral uptake. Therefore, at present, highly concentrated mineral fertilizers are frequently used in the cultivation of all crops. These fertilizers cause serious leaching and infiltration of N and P into soil and surface water, and also are inconvenient to use due to sedimentation [8]. Because of these problems, fertilizer is divided into types A and B, and each can be supplied to the crop only after dissolution through a dedicated supply system.

In plants, the minerals from the soil solution traverse the plant roots via apoplastic and/or symplastic pathways to the stele. In some plant species, several cytotoxic minerals such as Ca, Mo, Na, Cd, and Al are retained in the roots or transported in a chelated form [9]. Organic minerals called chelated or proteinated minerals are formed when the mineral is joined with an organic ligand such as a protein or a specific amino acid [10]. These organic minerals promote the movement of minerals from root to shoot, which is inhibited by the exchange capability of the xylem cell [11]. As the most representative of antioxidants, vitamin C is the most abundant water-soluble oxide in plant cells. It accumulates in chloroplasts and plays an important role in the balance between excess and absent free radicals in the plant cells [12]. In addition, because vitamin C has an –OH group that can bind to minerals, it is expected that the two molecules and mineral ions will have the ability to chelate through coordinate covalent bonding. In Korea, it has been reported that using vitamin C-substrate zinc as a feed additive enhances the intramuscular fat content of bovines, and selenium with chitosan and a fatty acid as a substrate induce an increase in yield and promotion of growth in tomato [13, 14]. These reports suggest that chelated mineral fertilizers or feeds have some utility value. Therefore, in this study, vitamin C was selected as substrate and a chelated mineral fertilizer was generated.

The aim of this study was to make an organic fertilizer using vitamin C as a substrate that maximized the production of high-quality crops. Additionally, this fertilizer was expected to complement declination of body absorptivity, cohesion and sedimentation, necessity of a nutrient solution supply system, and accumulation of soil salt and environmental contamination, which will improve the convenience of use and profitability in farm cultivation.

Methods

Plant material and test site

Seedlings of the lettuce cultivar 'Jeok Chi Ma' (*Lactuca sativa* L.) were obtained from Hampyeung Nursery (Worya-myeon, Hampteong-gun, Korea) about 7 days after sowing. Test site management and evaluation during the whole test period were carried out at Sannaedeul Farm (Daejeon-myeon, Damyang-gun, Korea). Dried and crushed soil samples were used for chemical and physical analyses. The green (organic farming) standard range of soil chemical and physical properties were pH 6.5–7.0, organic matter 20–30 g/kg, phosphate (P) 250–400 mg/kg, potassium (K) 250–400 cmol+/kg, calcium (Ca) 6–7 cmol+/kg, magnesium (Mg) 2.0–2.5 cmol+/kg, and soil electrical conductivity (EC) 0.0–2.0 dS/m. The ground temperature was monitored weekly for the entire test period [15–23].

Preparation of fertilizer

The fertilizer r-Ringer (AceKorea Co., Korea), which is used in agricultural fields, was purchased and used as a control fertilizer treatment. The test fertilizer, AGH-Ringer (Hanil Biomed Co., Korea), was manufactured based on r-Ringer by simply mixing its components with vitamin C (Nam Yung Commercial Co., Ltd., Korea) as a substrate (Table 1). All experiments were performed by using the two fertilizers diluted to different concentrations.

Experimental design and treatments

There were four test groups, each group with 16 lettuce seedlings. The control group was drenched with only water, and the other two groups were drenched with water and 50 mL diluted r-Ringer (diluted 1/2000) or AGH-Ringer (diluted 1/500, 1/1000, or 1/2000) once per week.

Chemical damage

To investigate the damage caused to the plant by the concentration and treatment with each fertilizer, the number of tipburns and leaf deaths per seedling was counted 13 days after transplanting.

Growth analysis before harvest

Twenty-four days after transplanting, growth status and tissue density were examined through the plant height (cm), stem diameter (mm), leaf length (cm), leaf width (cm), leaf number (ea/plant), leaf diameter (mm), chlorophyll concentration (mg/g FW), root length (cm), fresh weight (g/plant), and dried weight (g/plant).

Crop yield

The yield was measured during two periods after seedling transplant. These periods were defined based on ground temperature: suitable low temperature (14.6–18 °C, weeks 1–7) and inadequate low temperature (10.3–12.1 °C, weeks 8–15). Each test group consisted of 16 seedlings and was repeated three times.

Table 1 Component of fertilizers

Feed materials	Control fertilizer (r-Ringer)		Test fertilizer (AGH-Ringer)	
	Mixed amount (g)	Mix ratio (%)	Mixed amount (g)	Mix ratio (%)
NH_4NO_3	633	63.3	633	49.7
KH_2PO_4	176	17.6	176	13.8
$Mg(NO_3)_2 \cdot 6H_2O$	176	17.6	176	13.8
Fe-EDTA (Fe: 13%)	2.11	0.21	2.11	0.17
H_3BO_3	0.70	0.07	0.70	0.05
$CuSO_4 \cdot 5H_2O$	0.05	0.005	0.05	0.003
$ZnSO_4 \cdot 7H_2O$	11.3	1.13	11.3	0.88
$MnSO_4 \cdot H_2O$	0.49	0.05	0.49	0.04
$Na_2MoO_4 \cdot 2H_2O$	0.06	0.006	0.06	0.004
$C_6H_8O_6$ (vitamin C)	–	–	275	21.57
Sum	1000	100	1275	100

Mineral content of leaves

Three weeks after seedling transplantation, when lettuce showed maximum growth and yield, body absorption pattern and bioavailability were measured by ICP analysis. The contents of 16 inorganic ions (Ca, Zn, Fe, Cu, Se, K, Mg, Mn, B, Mo, Na, Co, P, S, Cr) and four heavy metals (As, Cd, Hg, Pb) in the lettuce samples were measured three times and their average values were used for ICP analysis.

Statistical analysis

The data were statistically evaluated using Duncan's multiple range tests after one-way ANOVA using SPSS 12.0 K for Windows (SPSS Inc., Chicago, IL, USA). Statistically significant differences were considered at the $p < 0.05$ level.

Results and discussion

Characterization of test fertilizer

The composite fertilizer AGH-Ringer was manufactured based on the mineral component ratio of the conventional fertilizer r-Ringer, which is commonly used in agricultural fields in Korea. AGH-Ringer, which was made by simply mixing vitamin C with r-Ringer, has an NPK content ratio of 18.9:14:4.0 and has non-sedimentation and water-soluble properties (Fig. 1, Table 2). The powder forms are light-brown and a pale yellow before and after dissolution, respectively. This fertilizer is similar to or the same as the conventional fertilizer in this phenotype, and it is thought that it will be convenient to use because no sediments are formed when it is dissolved in water.

Analysis of soil conditions

To determine whether the basal soil was adequate for plant growth and within the organic farming range before the experiment, pH, organic matter, phosphate, potassium, calcium, magnesium, cation exchange capacity, and soil temperature were measured. The soil condition was: pH 7, organic matter 25 g/kg, phosphate 438 mg/kg, potassium 0.37 cmol+/kg, calcium 6.5 cmol+/kg, magnesium 2.7 cmol+/kg and EC 1.4 ds/m. All except potassium were within the range of green standards (data not shown).

Chemical damage

One typical symptom of crop damage caused by fertilizer is the occurrence of tipburn. In general, lettuce is cultivated at low temperatures and low-light conditions to inhibit tipburn caused by a calcium disorder. Tipburn should be avoided because it significantly reduces the salability of the crop [24, 25]. Therefore, to investigate the chemical damage caused by the test fertilizer, the number of tipburns per transplanted seedling was counted at the 13th day after transplanting. Considerable (29%) tipburn damage was observed in the groups treated with high-concentration (diluted 1/500) r-Ringer, but not in any concentrations of AGH-Ringer or conventional treatments (data not shown). While tipburn is generally considered a calcium deficiency problem, symptoms may appear despite sufficient calcium supply in the soil. The problem lies in the plant's inability to move sufficient calcium to the leaves rapidly. Calcium moves from the root to the leaves of the plant with water drawn by the transpiration process. These results suggest that the reason for the lack of tipburn in the AGH-Ringer treated group is that calcium movement is facilitated by a vitamin C–calcium complex. This means that the test fertilizer is safe for plant growth as it does not cause chemical damage.

Fig. 1 The characteristics of the fertilizers on the third day after dissolving in water. Commonly used fertilizers (**a**, **a′**) used in Korea have sediments but not AGH-Ringer (**b**, **b′**). **a**, **b** Powder type and **a′**, **b′** dissolved type. Arrow indicates the sediment in the control fertilizer (**a′**)

Table 2 Comparison of characteristics between control fertilizer and test fertilizer

Classification	Control fertilizer (r-Ringer)	Test fertilizer (AGH-Ringer)
Fertilizer type	Composite fertilizer	Composite fertilizer
Appearance	White powder	Light-brown powder
Solubility (%)	12.7	100
Sediment rate (%)	87.3	0
NPK content (%)	18.9:14:4.0	18.9:14:4.0
Method of use	Foliar or drench	Foliar or drench
Nutrient solution supply system	Necessary	Unnecessary

Effects on lettuce growth

To investigate the effects of AGH-Ringer on plant growth, growth of the aerial parts (top) and rhizosphere parts (root) were compared. Chlorophyll content was measured to compare the absorptivity and bioavailability of minerals in the plant body. In the r-Ringer groups, growth of stem diameter, leaf length, and leaf width were significantly improved when compared to the water-only control group. Likewise, two AGH-Ringer groups (diluted 1/1000 and 1/500) showed significant increases in growth of the aerial parts and inhibition in the rhizosphere parts (Table 3). These results suggest that even

Table 3 Effects of fertilizers on growth at harvest

Treatment groups	Plant height (cm)	Stem diameter (mm)	Leaf length (cm)	Leaf width (cm)	Leaf number (ea/plant)	Leaf diameter (mm)	Chlorophyll (mg/g FW)	Root length (cm)
Control (water only)	22.1 ± 1.6^{b}	3.0 ± 1.0^{c}	12.6 ± 1.4^{d}	6.5 ± 0.4^{b}	7.7 ± 1.2^{b}	0.2 ± 0.1^{a}	21.2 ± 0.5^{a}	8.3 ± 0.3^{a}
r-Ringer (1/2000)	24.4 ± 0.7^{a}	5.5 ± 0.8^{b}	16.4 ± 0.8^{ab}	9.6 ± 1.2^{a}	9.3 ± 1.5^{ab}	0.4 ± 0.5^{a}	21.3 ± 3.1^{a}	8.0 ± 0.5^{a}
AGH-Ringer (1/2000)		5.6 ± 0.2^{b}	14.5 ± 1.4^{c}	7.2 ± 0.9^{b}	9.0 ± 1.0^{ab}	0.1 ± 0.01^{a}	21.4 ± 2.0^{a}	7.8 ± 0.6^{a}
AGH-Ringer (1/1000)	25.8 ± 0.8^{a}	6.9 ± 0.7^{a}	17.6 ± 0.3^{a}	9.6 ± 0.2^{a}	11.0 ± 1.7^{a}	0.1 ± 0.01^{a}	23.3 ± 7.4^{a}	6.8 ± 0.5^{b}
AGH-Ringer (1/500)	23.7 ± 1.2^{ab}	5.5 ± 0.4^{b}	15.2 ± 0.1^{bc}	7.8 ± 0.8^{b}	10.3 ± 1.5^{ab}	0.2 ± 0.01^{a}	20.6 ± 1.1^{a}	6.5 ± 0.9^{b}

All data are expressed as mean ± standard deviation ($n = 3$). Different letters show a significant difference at $p < 0.05$ as determined by Duncan's multiple range tests

though AGH-Ringer treatment did not induce root growth promotion, it can induce promotion of growth in the aerial parts. In other words, the body mineral absorptivity and bioavailability ability of AGH-Ringer may be better than that of r-Ringer. This was confirmed by the difference in chlorophyll contents in the plants treated by the two fertilizers. The chlorophyll content was increased by 109% by the 1/1000 dilution of AGH-Ringer, but the r-Ringer groups showed similar values to the control group (Table 3). Many studies have reported that nutritional management can regulate the balance between plant reproduction and growth [26, 27]. Plant volume and tissue density are related to yield and storage period at harvest. Therefore, the effects of the AGH-Ringer on biomass and dry weight were investigated. The fresh weight of the r-Ringer groups significantly increased by a factor of more than two in both the aerial and rhizosphere parts when compared to the water-only control group. Leaf dry weight was increased significantly, but root and stem dry weights were decreased or showed no differences (Table 4). In the comparison between r-Ringer and AGH-Ringer treated groups, only the 1/1000-dilution groups showed significantly increased fresh and dry weights, and there were no differences in the other fertilizer concentration groups (Table 4). The chelated minerals promote movement of minerals from root to leaf [11]. Therefore, the improvement of growth in the aerial parts and increased stem weights can be attributed to the easy movement of minerals caused by chelation with vitamin C. Taken together, these results demonstrate that the effectiveness of fertilizer using vitamin C as substrate is excellent in promoting plant growth. AGH-Ringer is expected to increase the yield and production of high-quality lettuce.

Effect on lettuce yield

Lettuce is cultivated at low temperatures to reduce damage, such as tipburn, caused by high temperatures. Therefore, to investigate the effects of AGH-Ringer on lettuce yield, yields were investigated for 15 weeks after transplantation, during both an inadequate low-temperature period (weeks 1–7, 10.3–12.1 °C) and a suitable low-temperature period (weeks 8–15, 14.6–18 °C), based on ground temperature. As a result, the yield during the whole harvest period increased by 127% in the case of r-Ringer and increased by 144–157% in the case of AGH-Ringer (Table 5). At the beginning of the harvest, the yield of lettuce increased by 151–174% compared to

Table 4 Effects of fertilizers on tissue density at harvest

Treatment groups	Root		Stem		Leaf	
	Fresh weight (g/plant)	Dry weight (g/plant)	Fresh weight (g/plant)	Dry weight (g/plant)	Fresh weight (g/plant)	Dry weight (g/plant)
Control (water only)	0.5 ± 0.01^{b}	0.07 ± 0.01^{c}	0.1 ± 0.1^{c}	0.01 ± 0.001^{b}	5.80 ± 0.8^{c}	0.40 ± 0.1^{c}
r-Ringer (1/2000)	1.0 ± 0.2^{a}	0.13 ± 0.03^{ab}	0.3 ± 0.1^{bc}	0.01 ± 0.001^{b}	12.9 ± 5.0^{b}	0.70 ± 0.2^{b}
AGH-Ringer (1/2000)	0.9 ± 0.1^{a}	0.1 ± 0.01^{b}	0.2 ± 0.1^{bc}	0.01 ± 0.001^{b}	10.9 ± 1.6^{bc}	0.60 ± 0.1^{b}
AGH-Ringer (1/1000)	1.1 ± 0.2^{a}	0.2 ± 0.03^{a}	0.6 ± 0.02^{a}	0.04 ± 0.001^{a}	19.2 ± 3.8^{a}	1.20 ± 0.2^{a}
AGH-Ringer (1/500)	1.0 ± 0.1^{a}	0.1 ± 0.02^{ab}	0.3 ± 0.1^{b}	0.02 ± 0.01^{b}	13.0 ± 2.0^{b}	0.90 ± 0.1^{b}

All data are expressed as mean ± standard deviation ($n = 3$). Different letters show a significant difference at $p < 0.05$ as determined by Duncan's multiple range tests

Table 5 Effect of fertilizers on the lettuce yield in harvest periods

Treatment groups	Weeks 1–7 (14.6–18 °C)		Weeks 8–15 (10.3–12.1 °C)		Total	
	Total yield (g/plant)	Yield rate (%)	Total yield (g/plant)	Yield rate (%)	Yields (g/plant)	Yield rate (%)
Control (water only)	89.9	100	47.7	100	137.6	100
r-Ringer (1/2000)	120.8	134	54.5	114	175	127
AGH-Ringer (1/2000)	154.7	172	61.7	129	216.4	157
AGH-Ringer (1/1000)	136.1	151	61.4	129	197.6	144
AGH-Ringer (1/500)	156.5	174	58.2	122	214.7	156

Total crop yield in each soil temperature condition was calculated as the sum of the average (g/plant) of 16 seedlings in every week

the control group and the highest value was 174% at the highest AGH-Ringer concentration (1/500). On the other hand, in the inadequate low-temperature period, the rate of increase was 122–129%, which was much lower than that of the adequate low-temperature period. These results show that treatment with fertilizers has effect on crop yield. The effect of AGH-Ringer is greater than that of r-Ringer in the increase of lettuce harvest in both suitable low-temperature and inadequate low-temperature conditions. This result also can be attributed to increased movement of minerals due to chelation with vitamin C in AGH-Ringer. In addition, AGH-Ringer is expected to have low-temperature resistance and maintain yields even in inadequate cold conditions.

Evaluation of mineral contents of lettuce leaves

To determine whether the AGH-Ringer affected plant mineral absorption and bioavailability, we carried out ICP analysis on the mineral contents in harvested leaves. When the 20 minerals were examined, the total amount detected was the lowest (53,203 ppm) in the r-Ringer treated group, and the AGH-Ringer groups showed significantly higher values than the control group, which increased in a concentration-dependent manner. In case of r-Ringer, Zn, Fe, S, and Cr were increased compared to the control group, and AGH-Ringer (diluted 1/2000) showed an increase in all minerals except Fe and P. Additionally, other concentrations of AGH-Ringer led to similar patterns (Table 6). These results show that r-Ringer exhibits typical inorganic fertilizer characteristics, but AGH-Ringer has greater efficacy than r-Ringer in mineral bioavailability and absorption through its organic functions because chelating minerals with vitamin C improves movement of minerals from root to leaf, and increases absorption into the plant body. Additionally, this result supports the results of previous experiments on growth promotion and increase of yield though AGH-Ringer treatment. Therefore, fertilizers that have organic

Table 6 Mineral contents in harvested leaves

Minerals	Control (water only)	r-Ringer (1/2000)	AGH-Ringer (1/2000)	AGH-Ringer (1/1000)	AGH-Ringer (1/500)
Ca	7143 ± 13^c	7045 ± 21^d	7706 ± 5^b	7032 ± 23^d	7845 ± 9.46^a
Zn	37 ± 0.11^e	41 ± 0.12^d	51 ± 0.16^a	44 ± 0.31^c	45 ± 0.31^b
Fe	255 ± 2.17^c	335 ± 1.21^a	172 ± 0.33^e	196 ± 0.85^d	304 ± 1.48^b
Cu	0.31 ± 0.02^b	0 ± 0^c	1.13 ± 0.06^a	0 ± 0^c	0.25 ± 0.05^b
Se	0 ± 0^a	0 ± 0^a	0 ± 0^a	0 ± 0^a	0 ± 0^a
K	$40{,}953 \pm 146^c$	$35{,}549 \pm 58^d$	$43{,}048 \pm 148^b$	$45{,}822 \pm 127^a$	$45{,}811 \pm 225^a$
Mg	2095 ± 7.06^d	2034 ± 5.93^e	2228 ± 6.99^c	2291 ± 24.2^b	2356 ± 2.69^a
Mn	13 ± 0.02^d	13 ± 0.03^d	17 ± 0.05^c	18 ± 0.07^b	27 ± 0.19^a
B	20 ± 0.12^c	19 ± 0.06^d	25 ± 0.17^b	31 ± 0.22^a	20 ± 0.03^c
Mo	0 ± 0^a	0 ± 0^a	0 ± 0^a	0 ± 0^a	0 ± 0^a
Na	2039 ± 12^d	1439 ± 3^e	2230 ± 0.32^c	2494 ± 4^d	2308 ± 9.0^a
Co	0 ± 0^a	0 ± 0^a	0 ± 0^a	0 ± 0^a	0 ± 0^a
P	6465 ± 28^b	4632 ± 12^e	6364 ± 11^c	5306 ± 17^d	6807 ± 49^a
S	1873 ± 10^c	2051 ± 14^b	2076 ± 11^b	1548 ± 15^d	2119 ± 20^a
Cr	39.6 ± 0.15^e	44.9 ± 0.2^d	53.8 ± 0.14^a	46.6 ± 0.29^c	48.7 ± 0.55^b
Sum	$60{,}933 \pm 219$	$53{,}203 \pm 115$	$63{,}972 \pm 183$	$64{,}829 \pm 213$	$67{,}691 \pm 318$

All data are expressed as mean ± standard deviation ($n = 3$). Different letters show a significant difference at $p < 0.05$ as determined by Duncan's multiple range tests

functions due to vitamin C as a substrate may be better than existing inorganic fertilizers in terms of plant growth, yield, and mineral bioavailability.

Conclusions

Straight fertilizers are very convenient to use when a farm is equipped with proper facilities. However, the preparation and use of these facilities requires a lot of money. In addition, nutrient fertilizers, which are compounds of inorganic ions, form sediments by ionic reaction when dissolved in water, and these sediments cause damage to plant growth and yield. Therefore, we developed a better fertilizer, which is no different in mineral composition than conventional fertilizer, by simply mixing conventional fertilizer with vitamin C. Vitamin C is the richest water-soluble oxidized substance in plant cells, and has ability to chelate and bond with minerals due to the –OH group it contains. The vitamin C-substrate straight fertilizer developed here, AGH-Ringer, has chelated minerals that easily move from root to leaves and had no sedimentation even when stored a long time after it was dissolved (Tables 1, 2, Fig. 1). Therefore, when lettuce was cultivated with this fertilizer, plant growth was enhanced by the minerals chelated with vitamin C (Table 3). It was also demonstrated that its mineral bioavailability and absorption are superior to conventional fertilizers (Tables 4, 6). For these reasons, there was an increase in crop yield (Table 5). Based on these results, AGH-Ringer is expected to increase famers' profits by increasing the yield of high-quality crops and convenience of use.

Authors' contributions

Study conception and design—H-JN, WSS and H-HC. Acquisition of data—H-SC, H-JN, WSS. Analysis and interpretation of data—H-SC, H-JN. Drafting of manuscript: critical revision—H-SC. All authors read and approved the final manuscript.

Author details

[1] HANIL BIOMED Inc., Gwangju 62363, Korea. [2] MEDINUTROL Co. Ltd., Yeonggwang-gun 57024, Korea.

Acknowledgements

This research was supported by Advanced Production Technology Development Program, Ministry of Agriculture, Food and Rural Affairs.

Competing interests

The authors declared that they have no competing interests.

Funding

Advanced Production Technology Development Program, Ministry of Agriculture, Food and Rural Affairs.

References

1. Barbosa GL, Almeida Gadelha FD, Kublik N, Proctor A, Reichelm L, Weissinger E, et al. Comparison of land, water, and energy requirements of lettuce grown using hydroponic vs. conventional agricultural methods. Int J Environ Res Public Health. 2015;12:6879–91.
2. Brechner M, Both AJ. Hydroponic lettuce handbook. Cornell University CEA Program. 1996; 48.
3. Beneduzi A, Ambrosini A, Passaglia LMP. Plant growth-promoting rhizobacteria (PGPR): their potential as antagonists and biocontrol agents. Genet Mol Biol. 2012;35:1044–51.
4. Franche C, Lindström K, Elmerich C. Nitrogen-fixing bacteria associated with leguminous and non-leguminous plants. Plant Soil. 2009;321:35–59.
5. Tuchy L, Chowańska J, Chojnacka K. Seaweed extracts as biostimulants of plant growth: review. Chemik. 2013;67:636–41.
6. Nakano Y, Fujitani K, Kurihara J, Ragan J, Usui-aoki K, Shimoda L, et al. Mutations in the novel membrane protein spinster interfere with programmed cell death and cause neural degeneration in *Drosophila melanogaster*. Mol Cell Biol. 2001;21:3775–88.
7. Bamsey M, Graham T, Thompson C, Berinstain A, Scott A, Dixon M. Ion-specific nutrient management in closed systems: the necessity for ion-selective sensors in terrestrial and space-based agriculture and water management systems. Sensors (Switzerland). 2012;12:13349–92.
8. Voogt W. Fertigation in greenhouse production. Proceedings of International Symposium on Fertigation, Beijing, China. 2005; p. 116–29.
9. Marschner H. Mineral nutrition of higher plants. 2nd ed. London: Academic Press; 1995.
10. Behnia R, Clark DA, Carter AG, Clandinin TR, Desplan C. Processing properties of ON and OFF pathways for Drosophila motion detection. Nature. 2014;512:427–30. https://doi.org/10.1038/nature13427.
11. Zacchini M, Pietrini F, Mugnozza G, Iori V, Pietrosanti L, Massacci A. Metal tolerance, accumulation and translocation in polar and willow clones treated with cadmium in hydroponics. Water Air Soil Pollut. 2009;197:23–34. https://doi.org/10.1007/s11270-008-9788-7.
12. Smirnoff N. The function and metabolism of ascorbic acid in plants. Ann Bot. 1996;78:661–9.
13. Lee CK. Effects of selenium supplying methods on the growth and se uptake of hydroponically grown tomato plants. J Bio-Environ Control. 2005;14(4):284–8.
14. Yoon BY. Development of beef cattle feed additive for stress diminution and enhancement of marbling by using Zn-Vit C chelated compound. IPET Report No. 11-1541000-001487-01:5–15; 2010.
15. Alkhader AMF, Abu Rayyan AM. Improving water use efficiency of lettuce (*Lactuca sativa* L.) using phosphorous fertilizers. Springerplus. 2013; 2:563. http://www.pubmedcentral.nih.gov/articlerender.fcgi?artid=3825228&tool=pmcentrez&rendertype=abstract.
16. Bower CA, Wilcox LV. Soluble salts. In: Balck C, editor. Methods of soil analysis. Part 2. Chemical and microbiological properties Agronomy 9. Madison: American Society of Agronomy; 2010. p. 933–51.
17. Chapman HD. Cation-exchange capacity. In: Balck C, editor. Methods of soil analysis. Part 2. Chemical and microbiological properties. Agronomy 9. Madison: American Society of Agronomy; 1965. p. 891–900.
18. Day TR. Particle size analysis. In: Balck C, editor. Methods of soil analysis. Part 2. Chemical and microbiological properties. Agronomy 9. Madison: American Society of Agronomy; 1965. p. 562–6.
19. Allison LE. Organic carbon. In: Balck C, editor. Methods of soil analysis. Part 2. Chemical and microbiological properties. Agronomy 9. Madison: American Society of Agronomy; 1965. p. 1376–8.
20. Allison LE. Moodie CD Carbonates. In: Balck C, editor. Methods of soil analysis. Part 2. Chemical and microbiological properties Agronomy 9. Madison: American Society of Agronomy; 1965. p. 1379–96.
21. Brechner M, Both AJ (1996) Hydroponic lettuce handbook. Cornell University of CEA Program 48.
22. Olsen SR, Dean LA. Phosphorus. In: Balck C, editor. Methods of soil analysis. Part 2. Chemical and microbiological properties. Agronomy 9. Madison: American Society of Agronomy; 1965. p. 1035–48.
23. Pratt PF. Potassium. In: Balck C, editor. Methods of soil analysis. Part 2. Chemical and microbiological properties. Agronomy 9. Madison: American Society of Agronomy; 1965. p. 1022–30.

9

NMR-based metabolomics of water-buffalo milk after conventional or biological feeding

Pierluigi Mazzei* and Alessandro Piccolo*

Abstract

Background: Biological farming in dairy production is often advocated as one of the most virtuous solutions to the environmental problems of conventional farming while improving the sustainability of production and cattle welfare. However, it is still under debate whether the conversion from conventional to biological farming has an influence on milk composition. In addition, the possible frauds related to biological dairy products call for analytical tools enabling the authentication of products quality and consumers protection. The aim of this work was to determine the composition of milk produced by water-buffaloes and to identify the specific metabolic profiles discriminating a biological from a conventional feeding diet.

Methods: Liquid-state ^{1}H, ^{13}C, and ^{31}P nuclear magnetic resonance (NMR) spectroscopies were used to study milk samples which were supplied during a 2-year-long experimentation by a single dairy farm and sampled from conventionally and biologically fed buffaloes (CFM and BFM, respectively). For each milk sample, we obtained NMR spectra of both raw milk and milk cream fractions comprising neutral lipids and phospholipids.

Results: The elaboration of multinuclear spectroscopic NMR results by the principal component analysis (PCA) enabled the identification of diagnostic differences in the milk composition between CFM and BFM samples. In particular, BFM were characterized by larger content of unsaturated lipids and phosphatidylcholine. Our findings confirmed that the conversion from a conventional to biological feeding regime influenced the buffalo milk composition, with possible implications for sensorial and nutritional properties of dairy products. Finally, the analytical methodology of NMR spectroscopy shown here may be considered as a useful tool to assess the quality and the authenticity of biological milk.

Keywords: Water-buffalo, Milk, Metabolomics, Multinuclear NMR spectroscopy, Principal component analysis, Biological diet

Background

Bovine milk is an essential constituent of the human diet, particularly in the Western world, being it nutrient-rich and well-balanced with respect to the nutritional values of carbohydrates, fats, and proteins [1, 2]. Milk consists in a complex mixture of metabolites including water, lactose, lipids, vitamins, minerals, and proteins, including caseins and immunoglobulins [3]. Milk metabolites derive from various cell types and originate downstream of different metabolic processes occurring in the bovine [4]. Milk is also processed into a variety of dairy products such as cream, butter, yogurt, ice cream, and cheese [5]. The sensorial and nutritional properties of dairy products depend strictly on milk composition, that, in turn, may be influenced by factors such as bovine lactation stage and age, farming methods, milk processing, and bovine diet [6–10]. The latter is particularly critical, since it is strongly involved in the metabolism of the mammalian organism and has a direct qualitative and quantitative influence on the lactation process [1, 11, 12].

*Correspondence: pierluigi.mazzei@unina.it; alessandro.piccolo@unina.it
Centro Interdipartimentale per la Risonanza Magnetica Nucleare per l'Ambiente, l'Agro-Alimentare ed i Nuovi Materiali (CERMANU), Università di Napoli Federico II, Via Università 100, 80055 Portici, Italy

In respect to the conventional farming, the biological approach is increasing in European dairies due to a general reduced impact on the environment and an improvement of bovine welfare [13], but also to income increase of producers for the growing demand of organic milk and dairy products [14]. However, it is still under debate whether the variation from conventional to biological bovine diet influences the molecular composition of milk, and, consequently, the corresponding dairy products [10, 14, 15]. Advanced analytical techniques are thus required to identify metabolites in the complex milk matrix which can be directly attributable to a biological diet. Nuclear magnetic resonance (NMR) spectroscopy is one of the most powerful techniques to evaluate the total molecular composition of milk, since it enables the concomitant detection of lipids, organic acids, nucleotides, amino acids, and carbohydrates [16–18]. Despite its relatively low sensibility, NMR offers a number of unique advantages, including the application of different, complementary, and multidimensional experiments to reach a molecular profiling and perform both semi- and quantitative analyses [19]. In fact, it has been proven that NMR-based metabolomics enables the appraisal of both the authenticity and geographical origin of milk [14, 20–22] and assesses important milk aspects, such as quality-related technological parameters [16, 18, 23, 24], fats composition [25, 26], and nutritional quality [27–29].

While NMR applications were mainly devoted to evaluate bovine rather than buffalo milk (*Bubalus bubalis*) [25, 30], the latter is the essential component for the production of "mozzarella di bufala campana" cheese, that represents an economically valuable PDO (protected denomination of origin) dairy product of the Italian Campania region [31]. The aim of this work was then to apply multinuclear NMR-based metabolomics to differentiate buffalo milk after feeding animals with either biological or conventional fodder. In particular, liquid-state ^{1}H, ^{13}C and ^{31}P nuclear magnetic resonance (NMR) spectroscopies were employed on milk samples supplied by a single dairy farm breeding buffaloes fed by conventionally and biologically fodder (CFM and BFM, respectively) during a 2-year-long experimentation. For each milk sample, NMR spectra were obtained for both raw milk and milk cream fractions (neutral lipids and phospholipids).

Methods

Milk samples

Buffalo milk samples were supplied by the dairy "Allevamento Biologico Bufalino Di Mauro Nicola" that is located in Caserta (Italy). During a 2-year-long experimentation, milk samples were collected from the same 24 buffalo animals. The selection of these specific bovines was based on the facts that they calved in the same period and were characterized by similar age and average productive levels. Half of these animals was fed with a conventional diet, whereas the other half was exclusively fed with a fodder mix from a controlled biological production, consisting in alfalfa hay, straw hay, and a biologic mix prepared with maize meal (40%), barley meal (17%), proteic peas (35%), and flax seeds (8%). Both types of feeding diet were modulated to be isoenergetic and isoproteic. In 2010 and 2011, milk samples from both the conventional (CFM) and biological (BFM) buffalo populations were collected once a month during the lactation period spanning from March to October. During sampling sessions, a milk volume between 150 and 200 mL was collected from each of the 24 selected bovines (12 CFM + 12 BFM), unless the selected animal was unable to produce milk due to specific reasons (i.e., relatively delayed birth, precocious end of lactation stage). In all cases, milk was machine-milked during mornings, immediately stored at 4 °C, and delivered to the NMR laboratory. Subsequently, each sample was divided into two aliquots corresponding to 5 and 95 mL, one of which was stored at − 20 °C until the NMR analysis, while the other one was immediately freeze-dried.

Samples preparation for NMR analysis

The molecular composition of raw milk (RM) was assessed for refrigerated samples, which were prepared by diluting 0.01 mL of milk with 0.99 mL of deuterated (99.8% D_2O/H_2O, Armar Chemicals) phosphate buffer (90 mM, pH 6.5). 0.05 mg of 3-(tri-methylsilyl)propionic-2,2,3,3-d4 acid (TMSPA, EurisoTop, France) was also added to the mixture as an internal standard.

The freeze-dried milk fraction was redissolved in water and centrifuged at 10,000 rpm at 4 °C for 90 min to precipitate the caseins, and concomitantly isolate the milk cream emerging on the supernatant surface. The apolar lipids (AL) were extracted from 5 g of milk cream by shaking twice in 200 mL of acetone (HPLC grade, Merck). The extracted material was then roto-evaporated to dryness and suspended in a 2:1 chloroform/acidic water solution (v/v), and the organic phase separated and again roto-evaporated until dried. Prior ^{13}C NMR analysis, AL extracts (150 μL) were diluted with 700 μL of deuterated chloroform (99.8% D, $H_2O < 0.01\%$, Eurisotop) and added with 23.33 μL mL^{-1} of tetramethylsilane (TMS, 99.9%, Eurisotop) as an internal standard.

The phospholipid (PL) fraction was isolated from the AL extract by suspending this solid residue into 60 mL of a 2:1 chloroform/methanol solution (v/v) (HPLC grade, Merck) and then filtering the supernatant (PVDF filters, 0.45 μm polyvinylidene fluoride, Sigma-Aldrich) to which 60 mL of a Na_4-EDTA (0.01 M) and NaCl (0.1 M) aqueous solution was added to remove metals possibly affecting NMR resolution [32]. The PL-containing organic phase was separated, dried with anhydrous Na_2SO_4, filtered (PVDF filters, 0.45 μm), and roto-evaporated at 37 °C [26, 33]. Prior to ^{31}P NMR analysis, PL were redissolved in 1 mL of a 1:3.3 mixture of triethylamine (> 99%, Aldrich Chemie) and deuterated dimethylformamide (99.5% D/D_2O + H_2O < 0.05%, Eurisotop) (v/v) [26], containing 76.9 mg mL^{-1} of guanidine chloride (Sigma-Aldrich). 0.75 mg mL^{-1} of triphenylphosphate (TPP, > 99%, Sigma-Aldrich) was also added as an internal standard. Each sample was centrifuged at 12,000 rpm and 6 °C for 15 m to remove possible suspensions before loading the NMR tube.

NMR spectroscopy

All NMR experiments were conducted at 25 ± 1 °C on a 400 MHz Avance magnet (Bruker Biospin, Rheinstetten, Germany), equipped with a 5 mm Bruker Broadband Inverse (BBI) probe working at the ^{13}C, ^{31}P, and ^{1}H frequencies of 100, 161, and 400 MHz, respectively.

^{1}H spectra of RM samples (84 replicates per treatment and per year) were acquired through conventional, T_2-edited, and diffusion-edited NMR experiments. The setup for the conventional pulse sequence consisted in a 2 s-long on-resonance presaturation of water signal (54–65 dB for power level attenuation), 32k data point, and 248 scans. T_2-edited spectra were achieved through the spin-echo-based CPMG (Carr–Purcell–Meiboom–Gill) pulse sequence. In particular, the total spin–spin relaxation delay ($2n\tau$), to make the T_2-filter effective, was 1320 ms and was composed by single echo times (τ) of 3 ms. The T_2-edited acquisitions consisted in a spectral width of 12 ppm (5 kHz), 32 K time domain points, and 248 scans. Diffusion-edited spectra were achieved using a stimulated echo pulse sequence based on bipolar gradients and combined with a watergate 3–9–19 pulse train to suppress the water signal. Spectra were acquired by setting 2000 μs-long sine-shaped gradients (δ) at 32.030 G cm^{-1} and selecting a diffusion delay of 0.2 s (Δ) between encoding and decoding gradients. The acquisition implied a recycle delay of 2 s, 16k time domain points, a spectral width of 12 ppm (5 kHz), 96 scans, and 4 dummy scans. The conventional and T_2-edited spectra were Fourier transformed multiplying by a 0.1 Hz exponential function and without zero-filling, whereas diffusion-edited spectra were processed multiplying by a 0.6 Hz exponential function and applying a twofold zero-filling.

AL samples (35 replicates per treatment and per year) were examined through proton-decoupled ^{13}C spectra. Acquisitions were attained through the inverse-gated pulse sequence and consisted in 32k time domain points, a spectral width of 299.34 ppm (30,120.482 Hz), 5 s of recycle delay, 248 scans, and a carbon 45° pulse in the range 8–10 μs. ^{13}C spectra were processed by multiplying by a 2 Hz exponential function and applying a twofold zero-filling.

^{31}P NMR spectroscopy was used to evaluate PL fractions (35 replicates per treatment and per year). A decoupling scheme, based on the inverse-gated sequence, was applied to remove long-range proton-phosphorous couplings and improve spectral resolution. The setup consisted in 7 s of recycle delay, a phosphorous 45° pulse ranging within 7 and 8.5 μs, a spectral width of 250.9 ppm (40,650 Hz), 32k time domain points, and 6500 scans. ^{31}P NMR spectra were Fourier transformed by multiplying by a 3 Hz exponential function and applying a twofold zero-filling.

Structural identification of compounds detected in RM and AL samples was reached on the basis of existing literature [4, 16, 17, 21, 25] and two-dimensional (2D) NMR experiments, such as homonuclear ^{1}H–^{1}H correlation spectroscopy (COSY), total correlation spectroscopy (TOCSY), J-RES, heteronuclear ^{1}H–^{13}C heteronuclear single-quantum correlation (HSQC), and heteronuclear multiple-bond correlation (HMBC). All 2D NMR experiments were acquired with a spectral width of 16 (6410.3 Hz) and 300 (30,186.8 Hz) ppm for ^{1}H and ^{13}C nuclei, respectively, and a time domain of 2048 points (F2) and 256 experiments (F1). Homonuclear 2D spectra were based on 16 dummy scans and 64 total transients. In addition, a mixing time of 80 ms and a trim pulse length of 2500 ms were set for the TOCSY experiment. HSQC and HMBC heteronuclear experiments were acquired with 16 dummy scans, 80 total transients, and 0.5 μs of trim pulse length. The experiments were optimized by considering 145 and 6.5 Hz as the optimal ^{1}H–^{13}C short- and long-range J couplings, respectively.

For all 1D and 2D NMR spectra, the axes of ^{1}H, ^{13}C, and ^{31}P nuclei were referred to the resonance frequencies of respective internal standards (δ ^{13}C and ^{1}H = 0 ppm for both TMS and TMSPA standards; δ ^{31}P = – 17.7 ppm for TPP standard). All spectra were processed using both Bruker Topspin software (version 2.1, Bruker Biospin, Rheinstetten, Germany) and MNOVA software (version 9.0, Mestrelab Research, Santiago de Compostela, Spain).

Multivariate analysis

All detected signals in ^{1}H, ^{13}C and ^{31}P spectra were integrated and divided by the area of the respective internal standard (TMSPA, TMS and TPP, respectively). The standardized data were Pareto-scaled prior to perform principal component analysis (PCA). One-way ANOVA test was used to examine and filter the loading-vectors (variables) mostly represented in the applied PCA axes. In all cases, only the loading-vectors of variables which contributed significantly (p values < 0.05 at a significance α of 0.05) to differentiate the conventional from biological treatments were considered and superimposed in related score-plot. Statistical elaboration was conducted by the XLStat software (version 2012, Addinsoft, Paris, France).

Results and discussion

Buffalo raw milk (RM)

NMR spectra of RM samples provided rapid qualitative and quantitative information on the most abundant milk metabolites. These samples were prepared by diluting a small volume of fresh milk in an aqueous solution buffered at pH 6.5, to minimize the effect of signal broadening due to suspended components (i.e., proteins and liposoluble compounds). Our preliminary experiments have shown that the best compromise to reach maximum NMR detectability and sample stability consisted in a dilution of 1:99 (v/v, milk/aqueous solution). A 90:10 (v/v, milk/deuterated water) dilution was previously proposed to evaluate raw milk samples through liquid-state NMR [17]. However, while the latter dilution was meant to favor the NMR detection of relatively insensitive and low naturally abundant ^{13}C nuclei [17], a greater amount of milk was not desirable in our case, because both an increased time-dependent phases separation and abundance of suspended fraction represent interferences for any reliable metabolomic analysis.

The residual water peak at 4.704 ppm was suppressed during proton NMR acquisitions of RM, thus enabling emergence of overlapped peaks and detection of low-intense signals, otherwise incompatible with the dynamic range of the digitizer. A relevant number of overcrowded signals were detected in the 7–0.5 ppm range of conventional proton spectra of RM samples (Fig. 1a) and included intense signals in the alkyl region (2.4–08 ppm), typically ascribable to lipidic components.

Filtered spectra (also referred to as edited spectra) based on specific NMR pulse sequences enabled selective suppression of signals modulated by intrinsic metabolites properties, such as molecular tumbling rate, diffusivity, hydrodynamic radius, and aggregation extent [31, 34]. T_2-edited and diffusion-edited spectra are most used in NMR-based metabolomics studies, since they are very robust, reliable, and advantageous [35, 36]. In particular, a T_2-filtered NMR acquisition eliminates signals corresponding to slow-moving molecules, including lipid compounds, which are characterized by a relatively shorter spin–spin relaxation. This is due to the inverse correlation existing between the compound dimension and its mobility, so that the larger is the molecular size of a compound, the shorter is the NMR spin–spin relaxation time (T_2) [37]. Conversely, a diffusion-based pulse sequence selectively suppresses signals of metabolites with small hydrodynamic radius and faster diffusivity, thus enhancing visibility and detectability of larger molecules.

Both diffusion-edited (Fig. 1b) and T_2-edited (Fig. 1c) NMR experiments were hence acquired to simplify and enhance readability of proton spectra of RM samples. However, as compared to the conventional spectrum (Fig. 1a), the NMR filters may lead to both a partial reduction of signal intensity and a distortion in J-coupling resolution. Nevertheless, T_2-edited spectra (Fig. 1c) allowed a clearer observation of singlets and multiplets of relatively fast-moving molecules, such as carbohydrates, organic acids, and nitrogen-containing compounds. The most intense signals in T_2-edited spectra were attributed to lactose, choline, creatine, and several organic acids, including fumarate, citrate, and isobutyrate. Remarkably, most signals within the 5.3–3.33 ppm range rose from the magnetically non-equivalent hydroxy-alkyl protons of glucopyranosyl and galactopyranosyl moieties of lactose disaccharide. Moreover, the most deshielded signals at 5.25, 4.7, and 4.48 ppm are assignable to protons bound to lactose anomeric carbons.

Diffusion-edited acquisitions isolated broad signals of slow-diffusing lipidic moieties (Fig. 1b), with short spin–spin relaxation and relatively large molecular weights. In particular, the signal at 5.37 ppm corresponds to protons bound to *sp*2-hybridized carbon atoms in unsaturated lipids, whereas the three multiplets resonating between 5.3 and 4.05 ppm correspond to methylene and methine protons of glycerol moieties (Fig. 1b). The signals at 2.08 and 2.82 are due to methylene protons geminally bound to a single unsaturated carbon (i.e., oleic acyl group) and two unsaturated carbons (i.e., linoleic and linolenic acyl groups), respectively. The 2.3 and 1.63 ppm signals may be assigned to protons bound to C2 and C3 carbons in acyl chains, whose NMR frequencies result from progressively longer distance from the deshielding carboxyl group (C1). Finally, the signals at 0.92 and 1.33 ppm derive from terminal CH_3 and superimposed methylene $-(CH_2)_n-$ protons in acyl chains, respectively (Fig. 1b).

T_2-edited and diffusion-edited NMR spectra of CFM (collected in 2010, during the first year of experimentation) and BFM (collected in 2011, during the second year of experimentation) samples were compared to evaluate the short-medium effect of the two diets on buffalo RM composition. From a qualitative standpoint, BFM samples did not show any new signal directly related to the specific diet or at least at a concentration compatible with the instrumental detection limit. To further evaluate NMR data and attempt to reveal even small diet-related semiquantitative differences, we applied the principal component analysis (PCA). This multivariate statistical procedure is an unsupervised pattern-recognition technique that explores the intrinsic variations within different samples types [38, 39]. It offers the practical advantage to evaluate in a single output, referred to as score-plot, the response induced on the metabolome by numerous variables (described by the loading-plot).

The PCA score-plots based on T_2-edited (Fig. 2a) and diffusion-edited (Fig. 2b) spectra revealed a difference in milk composition attributable to the buffaloes feeding regime. In fact, the relatively distant spread between CFM (2010) and BFM (2011) samples suggests that the biological diet systematically affected the metabolome of buffalo milk involving both small non-lipidic (Fig. 2a) and lipidic (Fig. 2b) components. It is noteworthy that replicates from the same feeding treatment were projected relatively close to each other, thereby revealing a good reproducibility of the proposed method.

The PCA based on T_2-edited spectra (Fig. 2a) showed that most of BFM samples were positioned in the direction of the first quadrant, with a correlation to the lactose loading-vectors. This indicates that BFM samples exhibited a larger abundance than CFM samples of this disaccharide after a 1-year of biological feeding regime. While the partial superimposition of CFM and BFM replicates (Fig. 2a) also suggests that the differences in lactose content between these two types of RM samples were not so large, the ANOVA test confirmed that, though small, the differences were statistically significant (p values < 0.05 at a significance α of 0.05). Besides being the most abundant saccharide in milk (> 98%, ≈ 30% of the caloric value of the whole milk), lactose plays the important biological role of osmotic regulator during lactation, thus generally undergoing only small content variations in milk. Nevertheless, our results indicate that the biological diet contributed to enhance lactose biosynthesis in buffaloes, even though to a small extent. A possible explanation may reside in (i) an enhanced glucose uptake from blood exerted by secretory cells of mammary gland, and (ii) an up-regulation of genes involved in lactose biosynthesis [40].

The PCA score-plot of diffusion-edited spectra (Fig. 2b) again suggested a role played by the administered diet on milk composition. In fact, BFM samples were differentiated from CFM along the first and the third quadrant. On the basis of the loading-plot, the placement of BFM

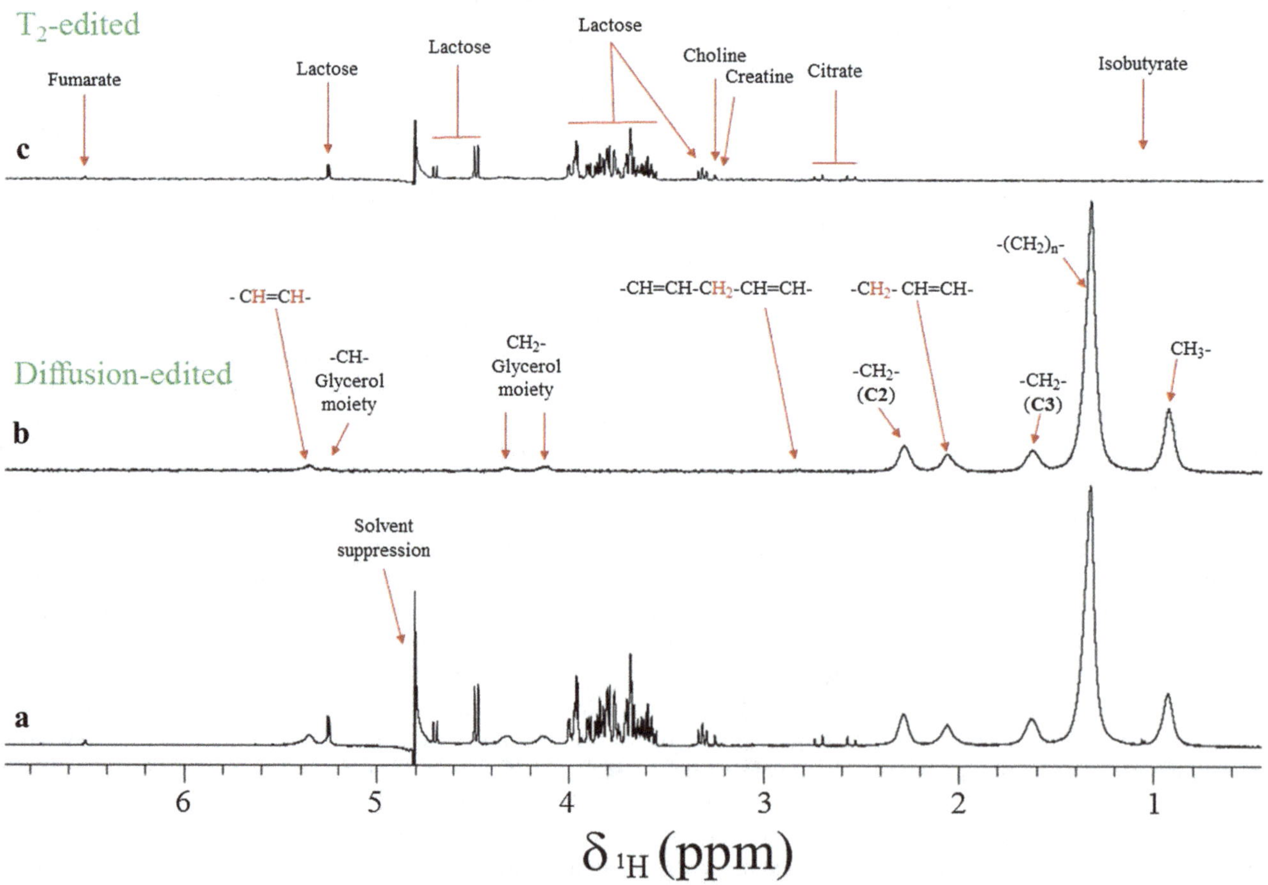

Fig. 1 ^{1}H NMR spectra of raw milk (RM) diluted in a deuterated solution buffered at pH 6.5. Conventional (**a**), diffusion-edited (**b**), and T_2-edited (**c**) proton spectra. The figure reports the assignment of most intense signals

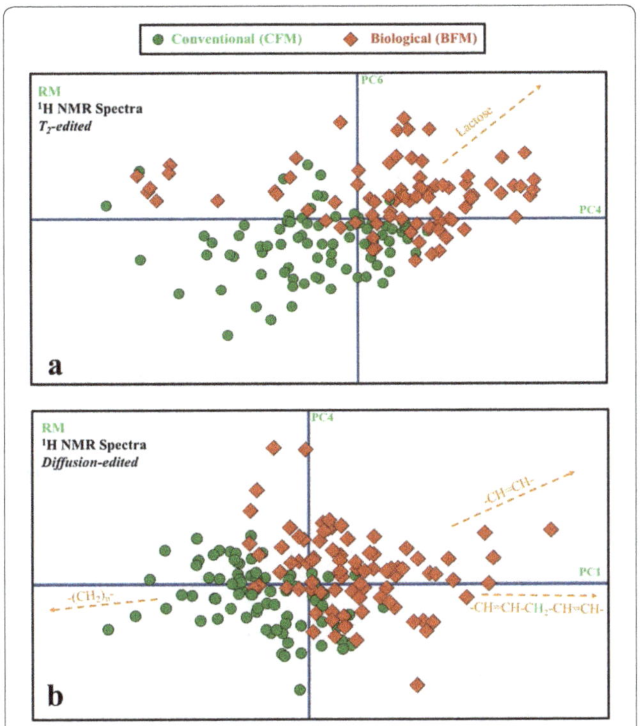

Fig. 2 Score-plots of principal component analysis (PCA) based on T_2-edited (**a**) and diffusion-edited (**b**) NMR proton spectra of CFM (green circles, collected in 2010) and BFM (light brown rhombi, collected in 2011) samples. The figure reports exclusively the name and the direction of loading-vectors which significantly differentiated CFM from BFM samples. The total variances explained for score-plots shown in **a**, **b** correspond to 11.8 and 46.2%, respectively

samples was due to a greater intensity of –CH=CH– and –CH=CH–CH_2–CH=CH– proton signals, as well as by a smaller intensity for saturated –$(CH_2)_n$– methylene groups. This indicates a larger abundance of mono- and polyunsaturated lipids in BFM samples. Remarkably, the behavior of these variables is consistent since the larger the abundance of unsaturated carbons, the lower the number of protons contributing to the –$(CH_2)_n$– methylenes signals in acyl chains. Moreover, such a dissimilarity in polyunsaturated milk components is in line with previous reports on the differentiation between biological and conventional cow milk based on the larger concentration of α linolenic acid [14, 41]. Finally, the ANOVA test confirmed again that, though to a small extent, all these variables showed a statistically significant difference (p values < 0.05 at a significance α of 0.05) between CFM to BFM. Interestingly, the larger content of polyunsaturated lipids also denotes an enhanced quality for biological milk, due to the beneficial properties for human health commonly recognized to these compounds.

The apolar lipids (AL) fraction in buffalo milk

Lipids are present in the milk matrix as globules emulsified in the aqueous phase. In buffalo milk, the total lipid content ranges between 8.3 and 15% [11] and about 98% is represented by triacylglycerides, whereas the rest is composed by phospholipids, diacylglycerols, monoacylglycerols, and free fatty acids [42]. In bovine animals, the long-chain fatty acids (mainly C16 and C18, saturated, monounsaturated and polyunsaturated fatty acids) in milk are of dietary origin, while short-chain fatty acids (C4–C10, mainly saturated) are biosynthesized in the mammary gland [43, 44].

The previous works showed the high potential of ^{13}C NMR spectroscopy in evaluating the fatty acids in bovine milk [25, 42, 43]. Therefore, ^{13}C NMR spectra of the AL fraction extracted from buffalo milk were acquired here and revealed more than 30 signals due to acyl chains and glycerol carbon atoms. In particular, signals common to most of triacylglycerides consisted in terminal methyls (14.5–13.3 ppm), C3 carbons (22.7 ppm), –$(CH_2)_n$– methylenes (30–28.8 ppm), and both methine and methylene of glycerol moiety (68 and 62 ppm, respectively) (Fig. 3). Other signals permitted to distinguish different acyl types and identify the position of acyl esterification on the glycerol backbone. The carboxyl (C1) signals resonating from 173.4 to 172.7 ppm identified the related acyl position (*Sn*-1,3 or *Sn*-2) in the glycerol backbone and differentiated between saturated (S) and unsaturated (U) acyl chains (Fig. 3). The two signals at 34.24 and 34.03 ppm indicated C2 carbons of acyl chains in *Sn*-1,3 and *Sn*-2 positions, respectively. The signals detected in the spectral region of *sp2* hybridized carbons (130.2–129.6 ppm) confirmed the presence of unsaturated acyls in either *Sn*-1,3 or *Sn*-2 positions (Fig. 3). The resonances at 24.9 and 27.2 ppm corresponded to carbons directly bound to either one or two *sp2* hybridized carbons, respectively. Finally, signals at 18.37, 35.9, and 173.1 ppm were assigned, respectively, to C3, C2, and C1 carbons of butyric acid, which is the most abundant short-chain acyl group in buffalo milk triacylglycerides [44].

The PCA score-plot based on ^{13}C NMR spectra of the AL fraction showed an almost neat separation of BFM (2011) from CFM (2010) samples along the horizontal PC2 (15.57% variance explained) (Fig. 4). The related loading-plot revealed that the variables responsible for the differentiation consisted in C9(U) and C10(U) carbons in both glycerol positions, the –CH=CH–CH_2–CH=CH– carbons and the $(CH_2)_n$ methylene signals. Excluding the latter, all variables resulted associated with positive PC2 values. The fact that most of BFM

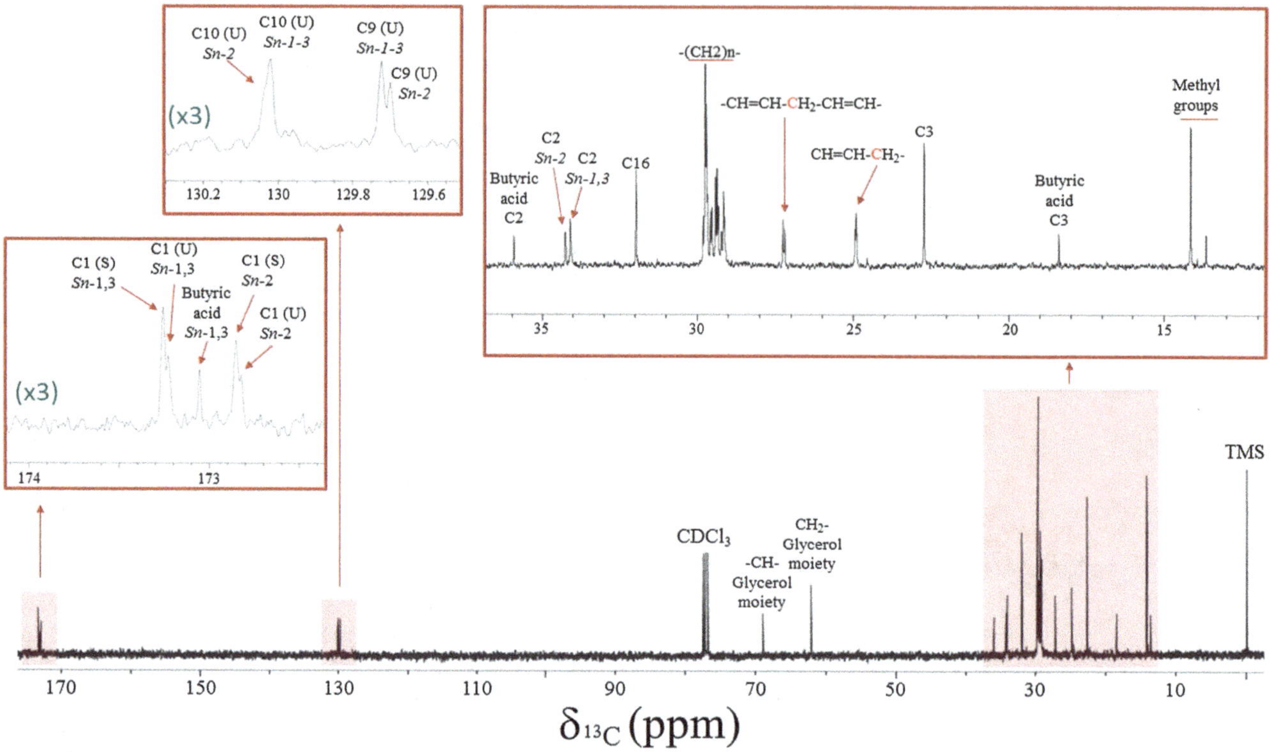

Fig. 3 ^{13}C NMR spectrum of the apolar lipids (AL) fractions extracted from milk cream and dissolved in $CDCl_3$. The spectral regions within the colored rectangles are enlarged, and, in the case of the 130.3–129.5 ppm region, vertically magnified by a factor of 3. The figure reports the assignment of most intense peaks (*Sn*-1,3 and *Sn*-2, position of acyl esterification on the glycerol backbone; U and S mean unsaturated and saturated acyl chains, respectively)

observations were placed in the positive PC2 region suggests a larger content of mono- and polyunsaturated lipids in the milk generated by the biological diet. In agreement with diffusion-edited RM results (Fig. 2), the abundance of mono- and polyunsaturated lipids was accompanied by a lower intensity for $(CH_2)_n$ methylene signals. Our findings are in line with a previous work proving that a different diet in bovines may influence the amount and the composition of mono and polyunsaturated fatty acids in milk [45].

The phospholipids (PL) fraction in buffalo milk

Phosphorous (P) in milk is mainly present as inorganic orthophosphate, occurring as Ca-phosphate salt in casein micelles or in solution. The rest of P is in organic forms as phosphate monoesters, phosphodiesters and, in small amount, in nucleotides [33]. The phosphodiesters are principally represented by PL, whose important role is due to their bioactive functions and capacity to emulsify fat globules in the whole milk. More than 65% of PL in milk is composed by external bilayer of fat globule membranes that surround the lipid droplets secreted by mammary gland cells [46].

Glycerophospholipids and sphingolipids are the two main PL groups. The former consists of fatty acids esterified to glycerol backbone, a phosphate group, and a hydrophilic residue (e.g., choline, ethanolamine, serine, and inositol), while the latter differ from glycerophospholipids because presents a long-chain amino-alcohol sphingosine, rather than the glycerol backbone [46]. Sphingomyelin, that is composed of a phosphorylcholine head group linked to the ceramide, is the dominant Sphingolipid species in milk [47].

^{31}P NMR spectroscopy was applied to examine the principal phospholipids in buffalo milk (Fig. 5). Phosphatidylcholine (0.02 ppm), sphingomyelin (0.82 ppm), phosphatidylethanolamine (PE, 0.55 ppm), and phosphatidylinositol (1.05 ppm) are the most prevalent classes of bovine milk phospholipids, whereas phosphatidylserine (0.48 ppm), phosphatidylglycerol (1.2 ppm), monomethyl P.E., and P.E. plasmalogens (0.57 ppm) are present to a lesser extent [7] (Fig. 5). Signals at 0.22 and 0.18 ppm are attributed to adducts in which one or more molecules of triethylamine are linked, via hydrogen bonds, to the amine protons of the polar head of Plasmalogens and PE, respectively [26] (Fig. 5).

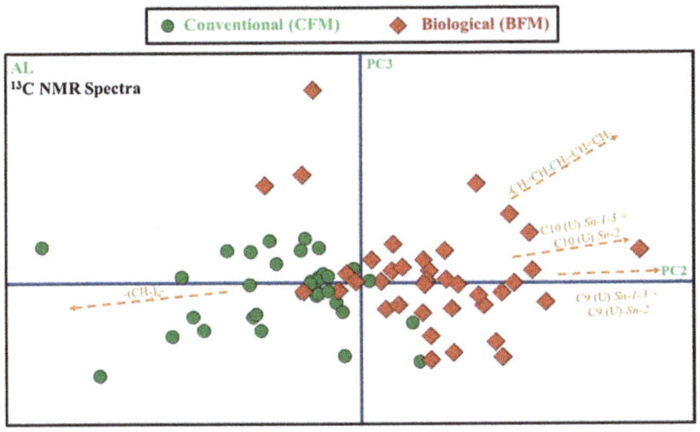

Fig. 4 Score-plot of principal component analysis (PCA) based on ^{13}C NMR spectra of apolar lipids (AL) extracted from milk cream of CFM (green circles, collected in 2010) and BFM (light brown rhombi, collected in 2011) samples (23.1% of total variance explained). The figure reports exclusively the name and the direction of loading-vectors which significantly differentiated CFM from BFM samples

Again, the PCA score-plot revealed an almost neat separation between BFM (2011) and CFM (2010) samples (Fig. 6), whose difference, according to the loading-plot, was due to a diverse content of phosphatidylcholine and sphingomyelin. In particular, BFM showed the largest and the smallest amount of phosphatidylcholine and sphingomyelin, respectively (Fig. 6). This indicates that the diet fed to buffaloes during the 2-year-long experimentation not only influenced the composition of major metabolites in milk, but also their minor components, such as PL. Remarkably, the detection of the largest content of phosphatidylcholine in BFM is related to the quality and beneficial value of dairy products derived from biological feeding regimes, since the assumption of PL has potential implications for human health [48]. In fact, it has been shown that phosphatidylcholine mitigates inflammatory processes, such as rheumatoid arthritis [49], and ulcerative colitis [50], while reducing the risk of cardiovascular diseases [51].

Conclusions

This work revealed that liquid-state multinuclear NMR spectroscopy enables to determine the buffalo milk metabolome and concomitantly differentiate milk deriving from buffaloes fed with either a conventional or a biological diet. The T_2-edited proton spectra of RM indicated that BFM samples were characterized by a slightly larger content of lactose than CFM samples. Moreover, BFM milk showed consistently more abundant mono- and polyunsaturated lipids in both ^{1}H diffusion-edited and ^{13}C spectra. Moreover, the ^{31}P NMR spectra of PL extracts revealed that the biological diet further influenced milk biosynthesis by inducing a larger production

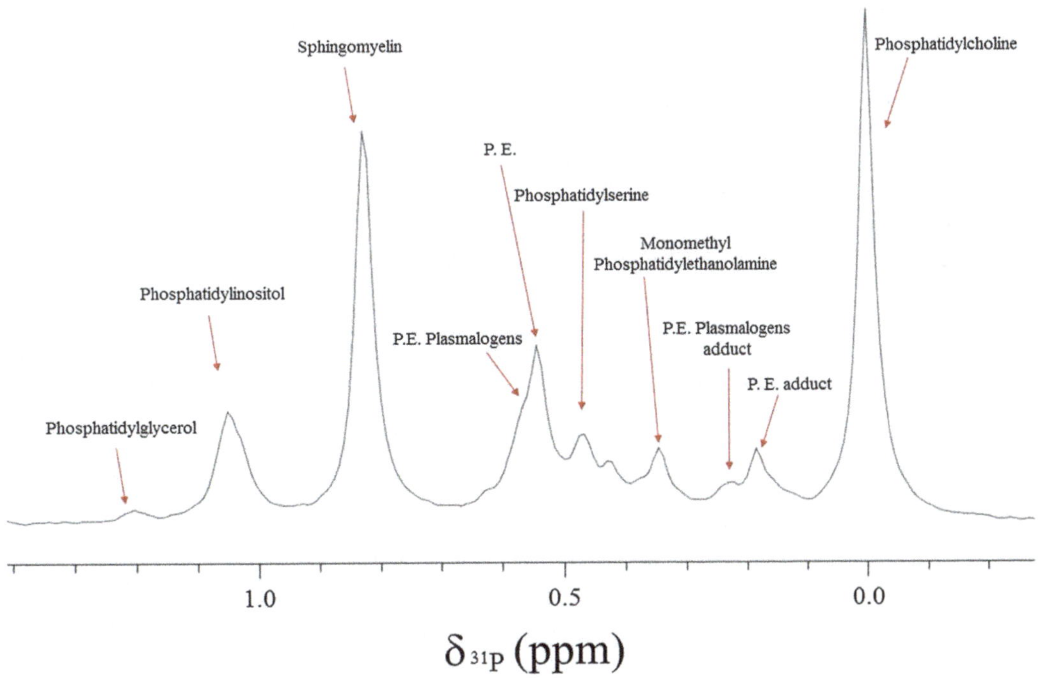

Fig. 5 ^{31}P NMR spectrum of the phospholipids (PL) fraction extracted from milk cream and dissolved in a monophasic solution composed by triethylamine, deuterated dimethylformamide, and guanidine chloride. The figure reports the assignment of the most intense peaks (PE, phosphatidylethanolamine)

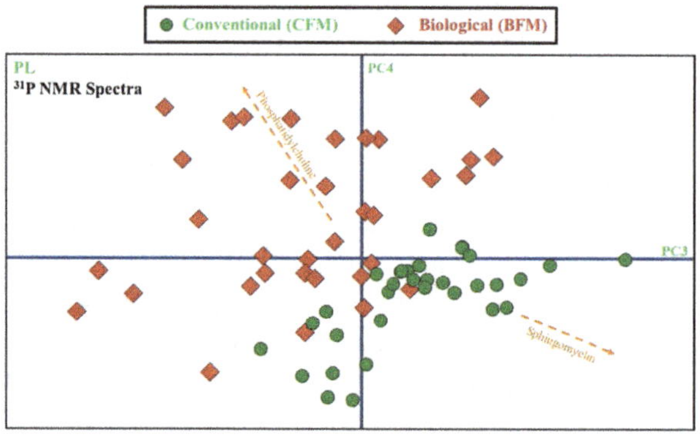

Fig. 6 Score-plot of principal component analysis (PCA) based on ^{13}C NMR spectra of the phospholipids (PL) fraction extracted from milk creams of CFM (green circles, collected in 2010) and BFM (light brown rhombi, collected in 2011) samples (19.5% of total variance explained). The figure reports exclusively the name and the direction of loading-vectors which significantly differentiated CFM from BFM samples

of phosphatidylcholine. It is noteworthy that these differences in milk composition were relatively small and resulted significant only after a 2-year-long application of the biological feeding diet. In fact, the comparison of the 2010 CFM samples with either the 2010 BFM or the 2011 CFM samples did not show any significant difference (data not shown).

Our results indicate that the conversion from a conventional to a biological feeding diet for water-buffalo animals exerts a significant influence on milk composition, but only after a short-medium period (> 1 year), especially if the feeding regimes were carefully modulated to be isoenergetic and isoproteic. Finally, our findings suggest the capacity of liquid-state multinuclear NMR spectroscopy to assess the quality of biological dairy products and its potential contribute to identify and contrast the possible frauds in the realm of biological farming.

Authors' contributions

The manuscript was written by PM and AP. Samples preparation, extraction, and analysis were conducted by PM. Multivariate elaboration of data was performed by PM and AP. Both authors read and approved the final manuscript.

Acknowledgements

This work was supported by the project "ECOBIOBUFFALO-Influenza dell'alimentazione biologica sulla qualità del latte bufalino" within the PSR 2007–2013 program of the MIPAAF (Ministero delle Politiche Agricole, Alimentari e Forestali).

Competing interests

The authors declare that they have no competing interests.

References

1. Solomons NW. Nature's perfect food revisited: recent insights on milk consumption and chronic disease risk. Nutr Rev. 2002;60(6):180–2.
2. DuPuis M. Nature's perfect food: how milk became America's drink. New York: New York University Press; 2002.
3. Fox PF. Milk: an overview. In: Thompson A, Boland M, Singh H, editors. Milk proteins—from expression to food. New York: Academic Press; 2009. p. 1–53.
4. Sundekilde UK, Larsen LB, Bertram HC. NMR-based milk metabolomics. Metabolites. 2013;3:204–22.
5. Boudonck KJ, Mitchell MW, Wulff J, Ryals JA. Characterization of the biochemical variability of bovine milk using metabolomics. Metabolomics. 2009;5:375–86.
6. Auldist MJ, Walsh BJ, Thomson NA. Seasonal and lactational influences on bovine milk composition in New Zealand. J Dairy Res. 1998;65(03):401–11.
7. Jensen RG. The composition of bovine milk lipids: January 1995 to December 2000. J Dairy Sci. 2002;85(2):295–350.
8. Dewhurst RJ, Fisher WJ, Tweed JK, Wilkins RJ. Comparison of grass and legume silages for milk production. 1. Production responses with different levels of concentrate. J Dairy Sci. 2003;86:2598–611.
9. Ellis KA, Innocent G, Grove-White D, Cripps P, McLean WG, Mihm M. Comparing the fatty acid composition of organic and conventional milk. J Dairy Sci. 2006;89:1938–50.
10. Butler G, Nielsen JH, Slots T, Seal C, Eyre MD, Sanderson R, et al. Fatty acid and fat-soluble antioxidant concentrations in milk from high- and low-input conventional and organic systems: seasonal variation. J Sci Food Agric. 2008;88:1431–41.
11. Varricchio ML, Di Francia A, Masucci F, Romano R, Proto V. Fatty acid composition of Mediterranean buffalo milk fat. Ital J Anim Sci. 2007;6(S1):509–11.
12. Kirovski D, Blond B, Katić M, Marković R, Šefer D. Milk yield and composition, body condition, rumen characteristics, and blood metabolites of dairy cows fed diet supplemented with palm oil. Chem Biol Technol Agric. 2015;2(6):1–5.
13. Berentsen PBM, Giesen GWJ, Schneiders MMFH. Conversion from conventional to biological dairy farming: economic and environmental consequences at farm level. Biol Agric Hortic. 1998;16:311–28.
14. Erich S, Schill S, Annweiler E, Waiblinger HU, Kuballa T, Lachenmeier DW, et al. Combined chemometric analysis of ^{1}H NMR, ^{13}C NMR and stable isotope data to differentiate organic and conventional milk. Food Chem. 2015;188:1–7.
15. Bloksma J, Adriaansen-Tennekes R, Huber M, van de Vijver LPL, Baars T, de Wit J. Comparison of organic and conventional raw milk quality in The Netherlands. Biol Agric Hortic. 2008;26:69–83.
16. Belloque J, Ramos M. Application of NMR spectroscopy to milk and dairy products. Trends Food Sci Technol. 1999;10:313–20.
17. Hu F, Furihata K, Ito-hishida M, Kaminogawa S, Tanokura M. Nondestructive observation of bovine milk by NMR spectroscopy: analysis of existing states of compounds and detection of new compounds. J Agric Food Chem. 2004;52:4969–74.
18. Sundekilde UK, Frederiksen PD, Clausen MR, Larsen LB, Bertram HC. Relationship between the metabolite profile and technological properties of bovine milk from two dairy breeds elucidated by NMR-based metabolomics. J Agric Food Chem. 2011;59:7360–7.
19. Larive CK, Barding GA, Dinges MM. NMR spectroscopy for metabolomics and metabolic profiling. Anal Chem. 2015;87:133–46.
20. Sacco D, Brescia MA, Sgaramella A, Casiello G, Buccolieri A, Ogrinc N, et al. Discrimination between Southern Italy and foreign milk samples using spectroscopic and analytical data. Food Chem. 2009;114:1559–63.
21. Brescia MA, Monfreda M, Buccolieri A, Carrino C. Characterisation of the geographical origin of buffalo milk and mozzarella cheese by means of analytical and spectroscopic determinations. Food Chem. 2005;89:139–47.
22. Lamanna R, Braca A, Di Paolo E, Imparato G. Identification of milk mixtures by ^{1}H NMR profiling. Magn Reson Chem. 2011;49:S22–6.
23. Frederiksen PD, Andersen KK, Hammershøj M, Poulsen HD, Sørensen J, Bakman M, et al. Composition and effect of blending of noncoagulating,

poorly coagulating, and well-coagulating bovine milk from individual Danish Holstein cows. J Dairy Sci. 2011;94:4787–99.
24. Zabbia A, Buys EM, De Kock HL. Undesirable sulphur and carbonyl flavor compounds in UHT milk: a review. Crit Rev Food Sci. 2011;52:21–30.
25. Andreotti G, Trivellone E, Lamanna R, Di Luccia A, Motta A. Milk identification of different species: ^{13}C-NMR Spectroscopy of triacylglycerols from cows and buffaloes' milks. J Dairy Sci. 2000;83:2432–7.
26. Murgia S, Mele S, Monduzzi M. Quantitative characterization of phospholipids in milk fat ^{31}P NMR using a monophasic solvent mixture. Lipids. 2003;L9135(38):585–91.
27. Klein MS, Almstetter MF, Schlamberger G, Nurnberger N, Dettmer K, Oefner PJ, et al. Nuclear magnetic resonance and mass spectrometry-based milk metabolomics in dairy cows during early and late lactation. J Dairy Sci. 2010;93:1539–50.
28. Marincola FC, Noto A, Caboni P, Reali A, Barberini L, Lussu M, et al. A metabolomic study of preterm human and formula milk by high resolution NMR and GC/MS analysis: preliminary results. J Matern Fetal Neonatal Med. 2012;25:62–7.
29. Monakhova Y, Kuballa T, Leitz J, Andlauer C, Lachenmeier D. NMR spectroscopy as a screening tool to validate nutrition labeling of milk, lactose-free milk, and milk substitutes based on soy and grains. Dairy Sci Technol. 2012;92:109–20.
30. Brescia MA, Mazzilli V, Sgaramella A, Ghelli S, Fanizzi FP, Sacco A. ^{1}H NMR characterization of milk lipids: a comparison between cow and buffalo milk. J Am Oil Chem Soc. 2004;81(5):431–6.
31. Mazzei P, Piccolo A. ^{1}H HRMAS-NMR metabolomic to assess quality and traceability of mozzarella cheese from Campania buffalo milk. Food Chem. 2012;132:1620–7.
32. Costello AJR, Glonek T, Van Wazer JR. Phosphorus-31 chemical shift variations with countercation and ionic strength for the various ethyl phosphates. Inorg Chem. 1976;15(4):973–4.
33. Andreotti G, Trivellone E, Motta A. Characterization of buffalo milk by ^{31}P-nuclear magnetic resonance spectroscopy. J Food Compos Anal. 2006;19:843–9.
34. Mazzei P, Piccolo A. Interactions between natural organic matter and organic pollutants as revealed by NMR spectroscopy. Magn Reson Chem. 2015;53(9):667–78.
35. Tang H, Wang Y, Nicholson JK, Lindon JC. Use of relaxation-edited one-dimensional and two dimensional nuclear magnetic resonance spectroscopy to improve detection of small metabolites in blood plasma. Anal Biochem. 2004;325:260–72.
36. Mazzei P, Piccolo A. HRMAS NMR spectroscopy applications in agriculture. Chem Biol Technol Agric. 2017;4(11):1–13.
37. Jacobsen NE. NMR spectroscopy explained—simplified theory, applications and examples for organic chemistry and structural biology. Hoboken: Wiley; 2007.
38. Brereton RG. Pattern recognition. Data analysis for the laboratory and chemical plant. In: Eilers P, editor. Chemometrics, chapter 4. Chichester: Wiley; 2003. p. 183–269.
39. Worley B, Powers R. Multivariate analysis in metabolomics. Curr Metabolomics. 2012;1:92–107.
40. Lin Y, Sun X, Hou X, Qu B, Gao X, Li Q. Effects of glucose on lactose synthesis in mammary epithelial cells from dairy cow. BMC Vet Res. 2016;12:81.
41. Molkentin J, Giesemann A. Differentiation of organically and conventionally produced milk by stable isotope and fatty acid analysis. Anal Bioanal Chem. 2007;388:297–305.
42. Andreotti G, Lamanna R, Trivellone E, Motta A. ^{13}C NMR spectra of TAG: an easy way to distinguish milks from different animal species. JAOCS. 2002;79(2):123–7.
43. Mansbridge RJ, Blake JS. Nutritional factors affecting the fatty acid composition of bovine milk. Br J Nutr. 1997;78(S1):S37–47.
44. Yonezawa T, Haga S, Kobayashi Y, Katoh K, Obara Y. Short-chain fatty acid signaling pathways in bovine mammary epithelial cells. Regul Pept. 2009;153:30–6.
45. Renou JP, Deponge C, Gachon P, Bonnefoy JC, Coulon JB, Garel JP, et al. Characterization of animal products according to geographic origin and feeding diet using nuclear magnetic resonance and isotope ratio mass spectrometry: cow milk. Food Chem. 2004;85:63–6.
46. Contarini G, Povolo M. Phospholipids in milk fat: composition, biological and technological significance, and analytical strategies. Int J Mol Sci. 2013;14:2808–31.
47. Potočki S. Potential health benefits of sphingolipids in milk and dairy products. Mljekarstvo. 2016;66(4):251–61.
48. Küllenberg D, Taylor LA, Schneider M, Massing U. Health effects of dietary phospholipids. Lipids Health Dis. 2012;11(3):1–16.
49. Eros G, Ibrahim S, Siebert N, Boros M, Vollmar B. Oral phosphatidylcholine pretreatment alleviates the signs of experimental rheumatoid arthritis. Arthritis Res Ther. 2009;11(2):R43.
50. Stremmel W, Hanemann A, Ehehalt R, Karner M, Braun A. Phosphatidylcholine (lecithin) and the mucus layer: evidence of therapeutic efficacy in ulcerative colitis? Dig Dis. 2010;28(3):490–6.
51. Olthof MR, Brink EJ, Katan MB, Verhoef P. Choline supplemented as phosphatidylcholine decreases fasting and postmethionine-loading plasma homocysteine concentrations in healthy men. Am J Clin Nutr. 2005;82(1):111–7.

10

Genetic diversity, microbiological study and composition of soil associated with wild *Pleurotus ostreatus* from different locations in Ondo and Ekiti States, Nigeria

Taye Victor Familoni[1], Clement Olusola Ogidi[1,2*], Bamidele Juliet Akinyele[1] and Anthony Kayode Onifade[1]

Abstract

Background: Genetic variation among different strains of wild macrofungi can be expressed and documented using different molecular tools. In this study, the genetic diversity and relatedness of *Pleurotus ostreatus* from different locations in Ondo and Ekiti States were investigated.

Methods: Random amplified polymorphic DNA-polymerase chain reaction marker was adopted to assess the genetic diversity of some wild *P. ostreatus*. Microorganisms associated with the wild *P. ostreatus* and their soil (substrate) were enumerated and identified using standard microbiological methods, while composition of soil around the mushrooms was evaluated using the conventional method.

Results: A total of 114 positive DNA bands were observed among wild *P. ostreatus* based on random amplified polymorphic DNA analysis (RAPD) with 10 primers, showing a total number of polymorphic markers of 91 and average polymorphism of 80.24%. *P. ostreatus* from Ala quarters and Igbatoro Road in Akure related to *P. ostreatus* from Ado-Ekiti, while *P. ostreatus* from Ido-Ekiti and Usi-Ekiti are genetically similar. The bacterial count from wild *P. ostreatus* and associated soil ranged from 1.20×10^4 to 4.70×10^5 and 1.70×10^7 to 6.10×10^8 CFU g^{-1}, respectively. The highest fungal count of 3.40×10^4 and 7.40×10^7 SFU g^{-1} were obtained for wild *P. ostreatus* and associated soil at Usi-Ekiti, which possesses the highest organic content (5.90%). Isolated microorganisms were *Pseudomonas putida, Streptomyces* spp., *Trichoderma* spp., *Penicillium italicum* and others. The highest crude fiber (25.04%) and protein content (24.07%) were obtained in wild *P. ostreatus* from Usi-Ekiti.

Conclusion: RAPD revealed the genetic relatedness and genetic diversity among studied wild *P. ostreatus*, indigenous to two sates in Southwestern Nigeria. This will improve the strain selection for further utilization and documentation.

Keywords: RAPD-PCR, Mushroom growth-promoting bacteria (MGPB), Basidiomycetes, Food nutrient, Soil microorganisms

*Correspondence: clementogidi@yahoo.com
[2] Department of Biotechnology, Faculty of Science, Kings University, PMB 555, Odeomu, Nigeria
Full list of author information is available at the end of the article

Background

Pleurotus species are predominant in both temperate and tropical parts of the world; it naturally grows on decaying woody materials as well as above soil [1]. They are devoid of chlorophyllous tissues and, hence, considered as primary decomposer of organic matter. Mushrooms play important ecological niche by decomposing complex organic matter when their mycelial network extend into the soil to absorb nutrients, thereby contributing to the improvement of soil composition [2, 3].

Generally, species of edible and medicinal mushrooms are gaining much attention for advance biotechnological uses, especially in pharmaceutical and agro allied companies. The biodiversity of *Pleurotus* spp. has created potential utilization with profound biological and economical values [4], but their genetic variability and taxonomic controversies still express some limiting factors [5]. However, exact characterization and identification of *Pleurotus* species is fundamental to reveal their true identity for maximum exploitation in food industries and for research documentation. The improvement in the use of deoxyribonucleic acid (DNA) for molecular technology has been proven to be a successful tool to solve taxonomic chaos [6]. Random amplified polymorphic DNA (RAPD) is one of the cheapest and quickest methods for assessing the variability at DNA level, being explicitly useful in intraspecific analysis and displaying the accurate genotypic identity of fungi [7].

Moreover, microbial ecology and substrate composition are other paramount concerns that affect the relatedness and nutrient contents of wild edible mushrooms. Soil microbial communities are extremely diverse; relatively to their organic contents, water activity and soil nutrient, physical and chemical properties [8]. The microbial succession on wild edible mushroom as well as their substrates is crucial for mushroom farmers to enhance productivity, eliminate post-harvest contamination and maintain quality product. Therefore, microbial community associated with wild mushrooms and their substrate (soil) has not been adequately investigated and revealed. Our study, thus, provides information on the molecular variability of some *P. ostreatus* from different locations and reveals the microorganisms associated with *P. ostreatus* as well as composition of soil around them.

Materials

Collection of wild *Pleurotus* species and soil around them

Wild edible mushrooms, *Pleurotus* spp., were picked from the forest in Ala quarters (116.8 g), Igbatoro Road (97.5 g) in Ondo State, Ido-Ekiti (124.2 g), Usi-Ekiti (145.8 g) and Ado-Ekiti (106.6 g) in Ekiti State, Nigeria (Fig. 1). The soil associated with the growth of wild *P. ostreatus* at different locations was collected using hand soil auger from the surface to a depth of 0–20 cm into sterile cellophane bags. The samples were transferred to the laboratory immediately for analysis.

Methods

Extraction of DNA using cetyl trimethyl ammonium bromide (CTAB)

Each of the mushroom samples (5 g) was air dried and ground in mill machine. The prepared mushroom sample (0.5 g) was transferred into Eppendorf tube; DNA extraction buffer (500 μl) consisting of 2% w/v of CTAB diluted in 100 mM Tris–HCl, 20 mM EDTA, 1.4 M of NaCl and β-mercaptoethanol (2% v/v) was added and incubated at 65 °C for 60 min [9]. Thereafter, the sample was allowed to cool and centrifuged at the speed of 4800*g* at 4 °C for 10 min. 0.5 ml of chloroform and isoamyl alcohol (24:1) were added and vortexed for 10 s, followed by spinning at 5400*g* for 10 min. The supernatant was transferred into a new Eppendorf tube and 0.5 ml of cold isopropanol was added to precipitate the DNA. The sample was kept in the freezer for 1 h and later spun at 5400*g* for 10 min. The supernatant was discarded and the pellet was washed with 70% v/v ethanol. The DNA was dried at room temperature (26 ± 1 °C) and resuspended in 100 μl of sterile distilled water. Thereafter, the DNA concentration was measured using a spectrophotometer (Beckman Instruments Inc., Fullerto CA, USA) at 260 nm.

Random amplified polymorphic DNA-polymerase chain reaction (RAPD-PCR) analysis

Ten (10) RAPD primers were used for the generation of polymorphic fragments of *P. ostreatus* strains (Table 1). Amplifications was performed on 25 μl reaction mixture consisting of the following: 2.5 μl of PCR buffer, 2.5 μl of 10 mM dNTP mix, 2.5 μl of $MgCl_2$, 1.0 μl of DNA, 1.0 μl of 10 pM primer (Operon Technologies, USA), 1.0 μl of Taq polymerase (Boehringer, Germany) and 14.5 μl of distilled water. The reaction volume in Eppendorf tube was amplified in a thermal cycler (Applied Biosystems 9700). The cycling program for amplifying RAPD-PCR assays was: 1st cycle of 94 °C for 3 min for initial denaturation, followed by 45 cycles of 94 °C for 20 s for denaturation, annealing of primer at 37 °C for 20 s and extension at 72 °C for 40 s with a final extension at 72 °C for 7 min.

Gel electrophoresis of PCR amplicons

Agarose gel electrophoresis was used to determine the quality and integrity of DNA by size fractionation on agarose gel. Hence, amplification products were resolved by electrophoresis using 1.0% agarose gel in 100 ml, 0.5× TBE buffer: 20 mM Tris, 89 mM boric acid, 20 mM EDTA, pH 8.0. The gel was allowed to cool down, 10 μl of 5 mg ml^{-1} ethidium bromide was added and mixed

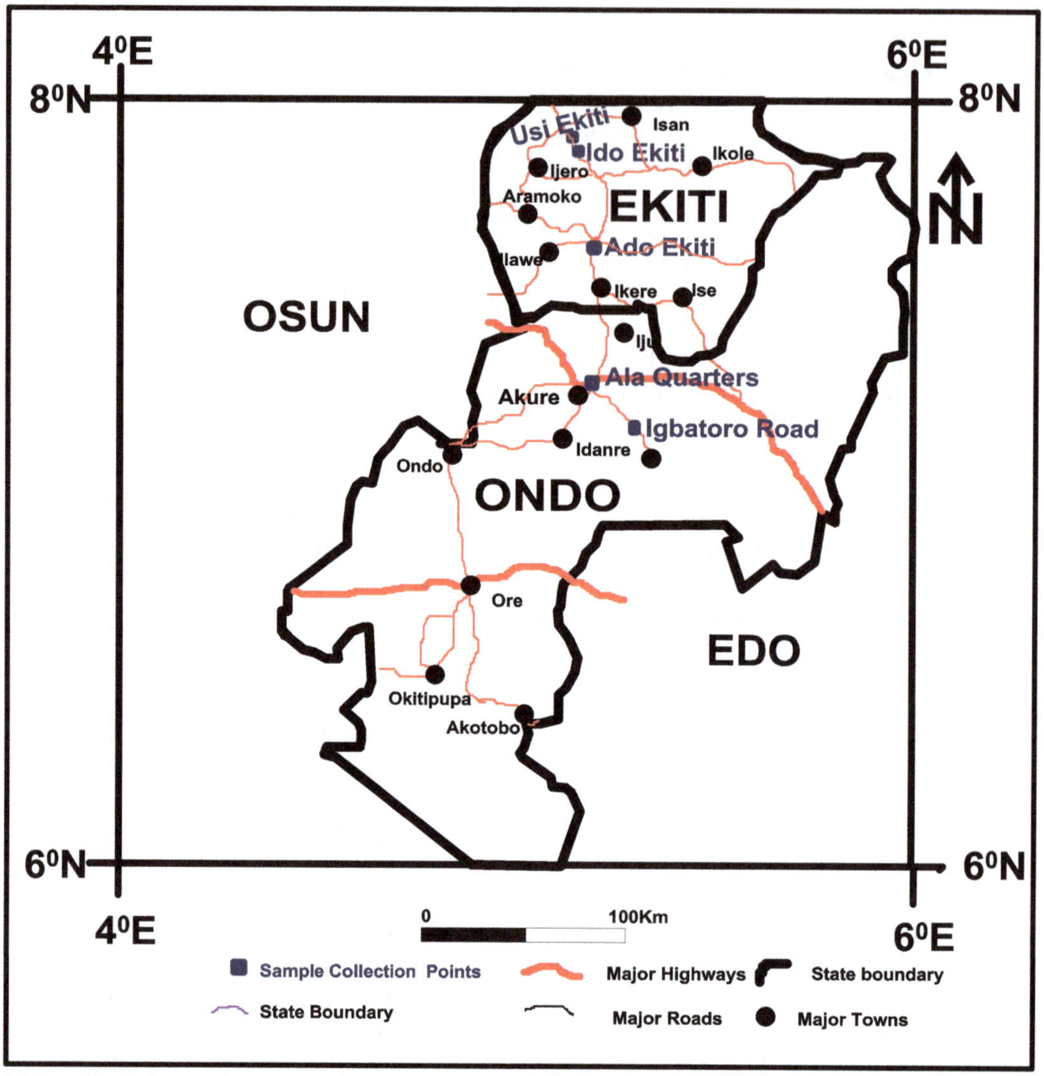

Fig. 1 Map of Ekiti and Ondo States indicating site of mushroom sample collection

together before pouring into an electrophoresis chamber set with the combs inserted. After the gel had solidified, 10 μl of each PCR amplification product was mixed with a 2 μl of 6× loading buffer in the well created on agarose gel. Electrophoresis was done at 100 V for 2 h and the integrity of the DNA was visualized, photographed under UV light and documented with a gel documentation system (UVP, USA).

Cluster analysis

The position of the RAPD bands was scored and transformed into a binary character matrix as "1" stands for the presence, while "0" stands for the absence of an RAPD band at a particular position. Using the NTSYS-pc version 2.02 K package [10], the transformed binary character matrix data are first transferred into the software data collection module, from which pairwise distance matrices are compiled. The genetic diversity parameters were determined using the output data and graphical module of the software, and a phylogenetic tree was created by the unweighted pair-group method arithmetic average (UPGMA) to group individuals into clusters [11].

Isolation of bacteria and fungi from wild *P. ostreatus* and soil

The stock of each sample was prepared by putting 2.0 g of freshly harvested mushroom and 5.0 g of soil into sterilized peptone water of 8 ml and 45 ml, respectively. 1.0 ml of each sample was serially diluted in different test tubes to obtain appropriate dilution. Isolation of bacteria and fungi was done using pour plate method [12]. An aliquot (1.0 ml) of each diluted sample was transferred into a Petri dish with addition of nutrient agar (Lab M Lancashire, UK) for bacteria and potato dextrose agar (Lab

Table 1 RAPD primers used for the analysis of *Pleurotus ostreatus* strains, number of all amplified DNA fragments, number of polymorphic fragments and percentage of polymorphism

Primer	Primer sequence	Number of fragment amplified (bands)	Number of polymorphic markers	% polymorphism
OPI-05	5′-TGTTCCACGG-3′	10	8	80.0
OPI-15	5′TCATCCGAGG-3′	11	10	91.0
OPL-03	5′-CCAGCAGCTT-3′	12	9	75.0
OPL-13	5′-ACCGCCTGCT-3′	10	8	80.0
OPN-10	5′ACAACTGGGG-3′	12	8	67.0
OPP-11	5′ AACGCGTCGG-3′	13	9	69.2
OPP-18	5′-GGCTTGGCCT-3′	14	12	86.0
OPS-13	5′-GTCGTTCCTG-3′	11	9	82.0
OPS-15	5′ CAGTTCACGG-3′	12	10	83.3
OPQ-06	5′-GAGCGCCTTG-3′	9	8	88.9
	Overall	114	91	
	Average			80.24

M, Lancashire, UK) for fungi. The plates were allowed to solidify, incubated at 37 °C for 24 h and 28 ± 2 °C for 48 h for bacteria and fungi, respectively. Thereafter, the discrete colonies formed on growth media were counted and recorded as colony forming units per gram (CFU g^{-1}) for bacteria and spore-forming unit per gram (SFU g^{-1}) for fungi. The microbial mixture was transferred to the edge of an agar plate with an inoculating loop and streaked out over the surface of fresh agar plate. Pure colonies obtained were maintained on agar slants at 4 °C.

Identification of isolated microorganisms from wild mushrooms and soil samples

The isolated bacteria from wild mushrooms and soil samples were stained and subjected to biochemical tests such as catalase, citrate utilization, methyl red, Voges Proskauer, nitrate reduction, sulfur reduction, indole production, motility (SIM), triple sugar iron (TSI) and other sugar fermentation according to Cappuccino and Sherman [13] and Cheesbrough [12]. The results of biochemical tests were interpreted and identification of bacteria was carried out according to Cowan and Steel [14]. Fungi spores were stained with drops of lactophenol and viewed under a microscope. The microscopic characteristics were interpreted according to Barnett et al. [15].

Determination of soil composition around the wild mushrooms

The total organic matter and total organic carbon were determined by Walkley and Black method through chromic acid digestion [16, 17] with slight modification. Briefly, 0.5 g of each air-dried sample was put into a conical flask. Ten milliliter (10 ml) of 0.167 potassium dichromate ($K_2Cr_2O_7$) and 20 ml of concentrated sulfuric acid (H_2SO_4) were added to the soil. The content was stirred to ensure good mixing of the soil sample with the reagents. After 30 min in a fume cupboard, 200 ml of distilled water, 10 ml of concentrated H_3PO_4 and 1 ml of 0.16% diphenylamine (indicator) were added. The excess dichromate that was not reduced in the reaction was determined by volumetric titration with 1 N $FeSO_4$ solution. The percentage of sand (0.06–2.0 mm), silt (0.002–0.06 mm) and clay (less than 0.002 mm) were read on a textural triangle to determine the soil texture.

The mineral contents of the soil and mushroom were determined using the method of AOAC [18]. Briefly, each sample of mushroom and soil (1.0 g) was digested with HNO_3, H_2SO_4 and $HClO_4$ (3:1:1, v:v:v) with addition of hydrogen peroxide (500 µl) at 200 °C. The resulting solutions were cooled with 10 ml of ultrapure water (Milli-Q system, Millipore, USA) and analyzed for magnesium (Mg), iron (Fe), calcium (Ca), and zinc (Zn) with atomic absorption spectrophotometer (Buck 201 VGP). Flame photometer (Jenway PFP 7, Staffordshire, UK) was used to determine the potassium (K) and sodium (Na) contents of the samples. Each of the mineral standard solutions (1000 µg/ml Merck, Germany) was used for the calibration of the atomic absorption spectrophotometer.

Proximate composition of wild mushrooms *Pleurotus ostreatus*

The proximate composition of dried mushrooms was determined according to the method of AOAC [18]. Briefly, the moisture content in the mushrooms was determined by air oven drying 5 g of each sample at 110 °C for 2 h. The ash content was determined by incinerating the mushrooms (5 g) at 550 °C for 4 h, while the crude fiber in the mushroom samples was determined by dilute acid and alkali hydrolysis. The fat content in *Pleurotus* mushrooms was determined by extracting with diethyl ether in a Soxhlet apparatus, and the crude protein in *Pleurotus* mushrooms was determined by using Kjeldahl method and multiplying by 4.38. The total carbohydrates in the mushroom samples were calculated by difference using the formula below:

$$\begin{aligned}\text{Carbohydrates}\ (\%) \\ &= 100 - [\text{moisture}\ (\%) + \text{crude fiber}\ (\%) \\ &\quad + \text{fat}\ (\%) + \text{ash}\ (\%) + \text{crude protein}\ (\%)].\end{aligned}$$

Statistical analysis

Data obtained during the experiment were subjected to analysis of variance. Variables with significant effects were characterized using Duncan's new multiple range test at $p \leq 0.05$ with the aid of SPSS version 17.0, Chicago, Illinois, USA.

Results and discussion

A total of 114 positive DNA bands were observed among the wild *P. ostreatus* based on RAPD analysis, showing total number of polymorphic markers of 91 and accounting for percentage of polymorphism ranging from 67 to 91% (Table 1). Typical RAPD gel photographs with marker bands of wild *P. ostreatus* are shown in Fig. 2. The outcome of Jacquard estimates with NTSYS using neighbor-joining cluster (Fig. 3) grouped the studied *P. ostreatus* into two clusters. *Pleurotus ostreatus* obtained from Ala quarters and Igbatoro Road in Akure labeled as 1 and 2, respectively, belonged to the same clade, showing 100% similarity. The harvested *P. ostreatus* from Ido-Ekiti (3) and Usi-Ekiti (4) are also genetically related. The pairing of *P. ostreatus* from Ado-Ekiti (5) in the same clade with *P. ostreatus* in Ala quarters and Igbatoro Road in Ondo State indicated a genetic relatedness. The RAPD cluster of *Pleurotus* species revealed both genetic relatedness

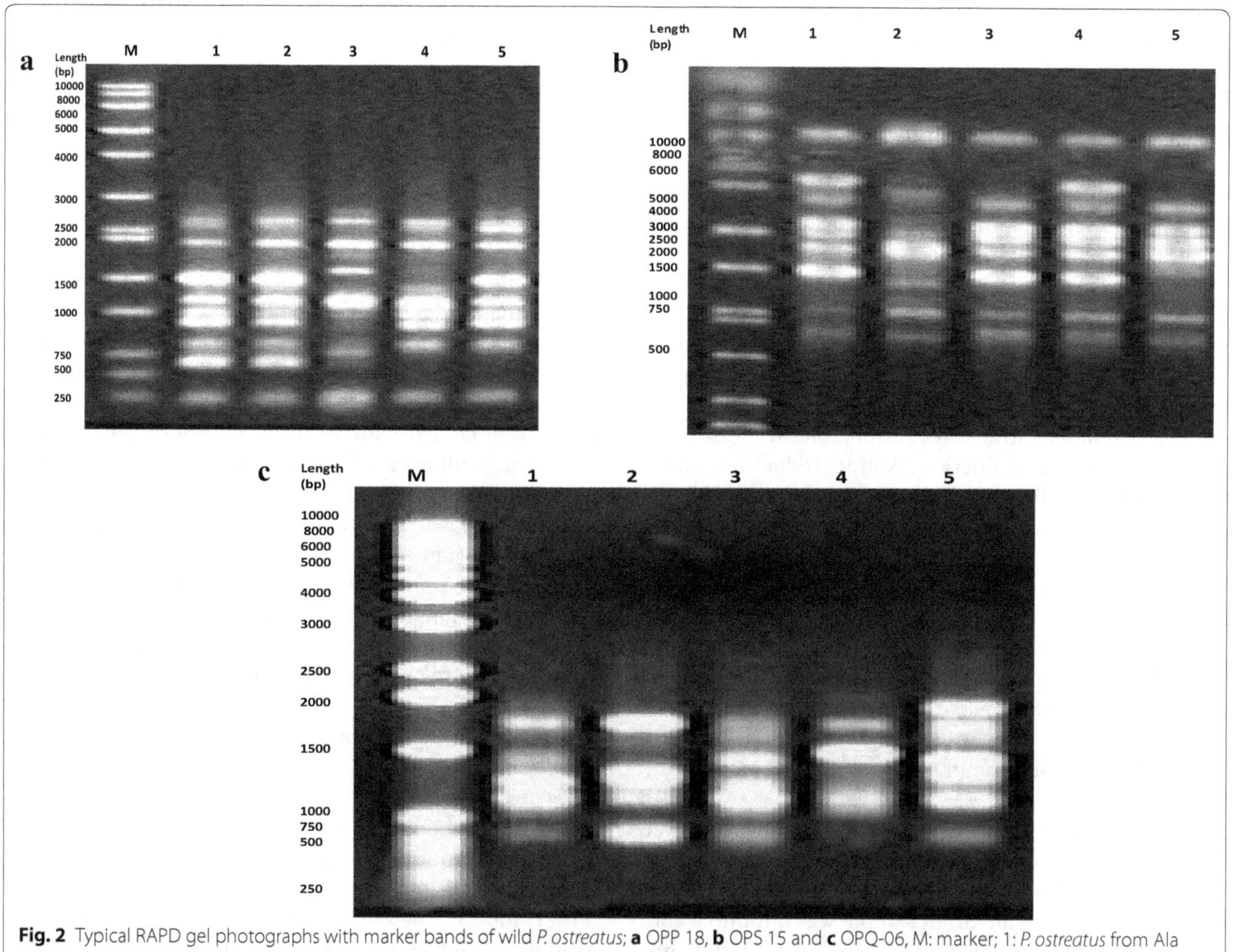

Fig. 2 Typical RAPD gel photographs with marker bands of wild *P. ostreatus*; **a** OPP 18, **b** OPS 15 and **c** OPQ-06, M: marker; 1: *P. ostreatus* from Ala quarters in Akure; 2: *P. ostreatus* from Igbatoro Road in Akure; 3: *P. ostreatus* from Ido-Ekiti; 4: *P. ostreatus* from Usi-Ekiti; and 5: *P. ostreatus* from Ado-Ekiti

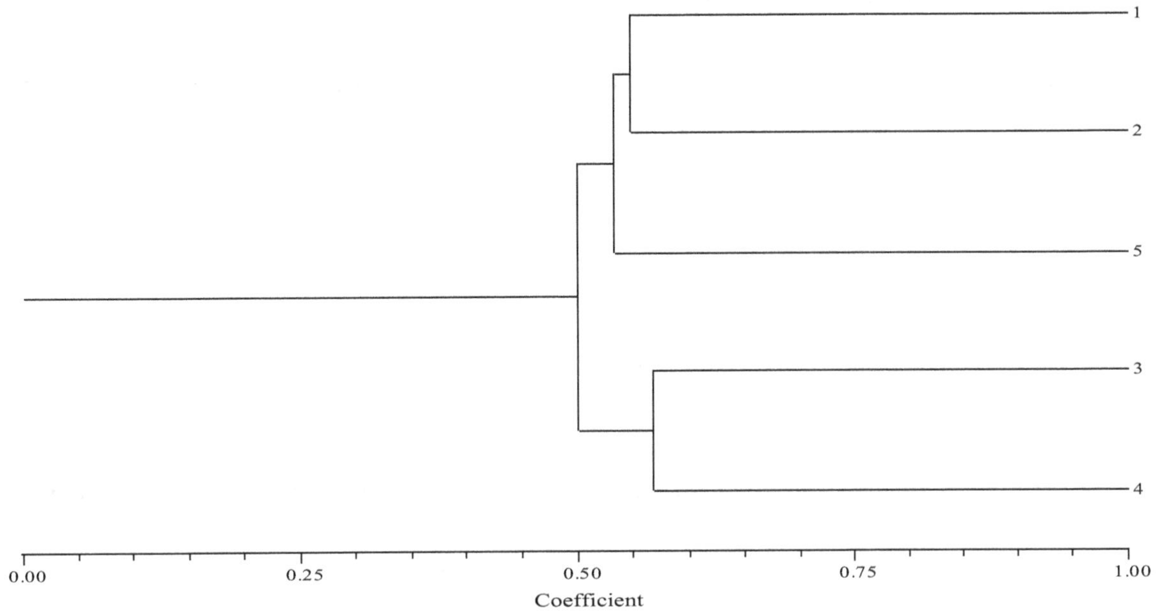

Fig. 3 Dendrogram of wild *P. ostreatus* from different geographic areas and labeled as 1: *P. ostreatus* from Ala quarters in Akure; 2: *P. ostreatus* from Igbatoro Road in Akure; 3: *P. ostreatus* from Ido-Ekiti; 4: *P. ostreatus* from Usi-Ekiti; and 5: *P. ostreatus* from Ado-Ekiti based on polymorphic RAPD markers showing genomic diversity of fungal strains using neighbor-joining cluster analysis method produced from Jacquard estimates with NTSYS Spc Version 2.02

and difference according to their geographical origins. The genetic relatedness could be attributed to the wide dispersal of fungi spore through a short- or long-distance distribution of *Pleurotus* species in the natural population across forest fragmentation, being promoted by human and animal movement [19].

The outcome of phylogenetic relation based on molecular analysis between different or similar populations of *Pleurotus* species provides a better resolution and understanding of their biogeography and speciation [5]. RAPD is a valuable method in distinguishing the different genotypes in mushrooms and evaluating their genetic similarities [7]. Other various biochemical and molecular techniques applied for genetic diversity, relatedness and identification of *Pleurotus* species were highlighted by Maftoun et al. [20] and Correa et al. [4]. These methods have been verified to be efficient in identification, expressed biodiversity in taxonomic system and unique in the selection of fungal strains.

The enumeration of bacteria (1.20×10^4 to 6.10×10^8 CFU g^{-1}) and fungi (1.50×10^3 to 7.40×10^7 SFU g^{-1}) from wild *P. ostreatus* and soil around *P. ostreatus* is presented in Table 2. The microbial count was more correlated to the organic matter than the genetic relationship exhibited by *Pleurotus* mushrooms. It was noted that soil from Usi-Ekiti possesses the highest microbial count: 7.40×10^7 SFU g^{-1} for fungi and 6.10×10^8 CFU g^{-1} for bacteria. The highest organic content (5.90%) in the location may require high microbial communities to improve nutrient cycle. Increase in soil bacteria and fungi creates microbial metabolic function to improve the soil biological characteristics [21]. The occurrence of isolated bacteria and fungi from *P. ostreatus* and their associated soil at different locations is shown in Tables 3 and 4. The presence of *Pseudomonas putida* conformed to the findings of Cho et al. [22] and Young et al. [23]. In their studies, they revealed the bacterium as one of the mushroom growth-promoting bacteria (MGPB) during the growth of *P. ostreatus* and *Agaricus bisporus*. *Pseudomonas putida* initiates primordial development and increases the yield of cultivated mushrooms [24], but Munsch and Alatossava [25] revealed that other members of the genus *Pseudomonas* exhibited saprophytic and pathogenic activities on *A. bisporus* and *Pleurotus* spp.

The species of *Paecilomyces* and *Penicillium* have been associated with white button mushroom in the study of Siyoum et al. [26]. The researchers revealed the microbial succession of compost, casing and mushroom using a plate count technique, denaturing gradient gel electrophoresis (DGGE) and sequencing of 16S and 18S rDNA. *Trichoderma* spp. was isolated with the highest occurrence from the examined *P. ostreatus* (33.30%) and their associated soil (17.10%). The cellulolytic filamentous

Table 2 Total bacterial and fungal count from harvested wild *Pleurotus ostreatus* and soil associated with their growth

Sample	Wild *P. ostreatus*		Soil around wild *P. ostreatus*	
	Bacteria (CFU g^{-1})	Fungi (SFU g^{-1})	Bacteria (CFU g^{-1})	Fungi (SFU g^{-1})
1	$1.20^a \times 10^4$	$1.50^a \times 10^3$	$1.70^a \times 10^7$	$5.00^b \times 10^6$
2	$1.60^b \times 10^4$	$2.60^c \times 10^3$	$2.30^b \times 10^7$	$5.30^c \times 10^7$
3	$2.40^c \times 10^4$	$2.00^b \times 10^3$	$5.40^d \times 10^7$	$6.90^d \times 10^6$
4	$4.70^d \times 10^5$	$3.40^d \times 10^4$	$6.10^e \times 10^8$	$7.40^e \times 10^7$
5	$4.50^d \times 10^5$	$2.50^c \times 10^3$	$4.10^c \times 10^8$	$4.80^a \times 10^7$

Values are mean of replicates ($n = 5$)

Mean values with the same superscript letters along the column are not significantly different ($p \leq 0.05$)

1: *P. ostreatus* and associated soil from Ala quarters; 2: *P. ostreatus* and associated soil from Igbatoro Road; 3: *P. ostreatus* and associated soil from Ido-Ekiti; and 4: *P. ostreatus* and associated soil from Usi-Ekiti. 5: *P. ostreatus* and associated soil from Ado-Ekiti

Table 3 Occurrence (%) of bacteria and fungi isolated from wild *Pleurotus ostreatus* at different habitats in Ondo and Ekiti States

	POAL	POIT	POID	POUS	POAE	N	% occurrence
Bacteria							
Aerococcus viridans	−	−	−	−	+	1	5.55
Actinomyces bovis	+	+	−	−	−	3	16.70
Pseudomonas putida	−	−	+	+	+	3	16.70
Fungi							
Fusarium spp.	+	+	−	−	−	2	11.10
Mucor hiemalis	+	−	−	+	−	3	16.70
Trichoderma spp.	+	+	+	+	+	6	33.30

+: present; −: absent

POAL: *P. ostreatus* from Ala quarters in Akure; POIT: *P. ostreatus* from Igbatoro Road in Akure; POID: *P. ostreatus* from Ido-Ekiti; POUS: *P. ostreatus* from Usi-Ekiti; POAE: *P. ostreatus* from Ado-Ekiti; N: number of isolates

Table 4 Occurrence (%) of bacteria and fungi isolated from soil associated with the growth of wild *Pleurotus* species at different locations in Ondo and Ekiti States

	SOAL	SOIT	SOID	SOUS	SOAE	N	% occurrence
Bacteria							
Klebsiella edwardsii	−	−	+	−	−	1	2.80
Bacillus subtilis	−	−	+	+	−	2	5.70
Actinomyces bovis	−	−	−	+	+	2	5.70
Streptomyces spp.	+	+	+	+	−	4	11.40
Pseudomonas putida	+	+	+	+	+	5	14.30
Fungi							
Phialophora verrucosa	−	−	−	−	+	1	2.80
Paecilomyces victoriae	−	−	−	+	−	2	5.70
Fusarium spp.	+	−	−	+	−	2	5.70
Mucor hiemalis	+	+	−	−	−	2	5.70
Microascus brevicaulis	+	+	−	−	+	3	8.60
Penicillium italicum	+	+	+	+	+	5	14.30
Trichoderma spp.	+	+	+	+	+	6	17.10

+: present; −: absent

SOAL: soil associated with the growth of wild *P. ostreatus* in Ala quarters; SOIT: soil associated with the growth of wild *P. ostreatus* in Igbatoro Road; SOID: soil associated with the growth of wild *P. ostreatus* in Ido-Ekiti; SOUS: soil associated with the growth of wild *P. ostreatus* in Usi-Ekiti; SOAE: soil associated with the growth of wild *P. ostreatus* in Ado-Ekiti; N: number of isolates

Table 5 Composition of soil (%) associated with the growth of wild *Pleurotus ostreatus* from different geographical areas in Ondo and Ekiti States

Soil	Sand	Silt	Clay	TOM	TOC	K	Na	Mg	Ca	Fe	Zn
SOAL	$82.00^{d} \pm 1.73$	$5.66^{b} \pm 0.58$	$10.30^{c} \pm 0.58$	$1.41^{a} \pm 0.03$	$0.82^{a} \pm 0.02$	$651.84^{c} \pm 3.90$	$102.89^{a} \pm 5.90$	$218.01^{a} \pm 11.0$	$319.11^{a} \pm 20.17$	$189.1^{d} \pm 8.12$	$161.90^{a} \pm 8.02$
SOIT	$88.33^{e} \pm 1.00$	$3.33^{a} \pm 0.58$	$3.67^{a} \pm 0.50$	$2.94^{b} \pm 0.04$	$1.70^{b} \pm 0.03$	$573.80^{a} \pm 8.09$	$176.11^{b} \pm 0.80$	$309.77^{c} \pm 14.10$	$500.89^{b} \pm 33.8$	$98.11^{b} \pm 0.96$	$204.14^{b} \pm 0.94$
SOID	$66.30^{a} \pm 1.15$	$11.33^{d} \pm 1.15$	$17.70^{d} \pm 0.60$	$2.95^{b} \pm 0.12$	$1.72^{b} \pm 0.05$	$942.71^{e} \pm 14.7$	$376.09^{d} \pm 4.99$	$285.14^{b} \pm 7.90$	$694.66^{d} \pm 27.9$	$108.80^{c} \pm 2.18$	$215.9^{c}7 \pm 1.89$
SOUS	$75.00^{c} \pm 0.74$	$7.80^{c} \pm 1.15$	$7.00^{b} \pm 0.08$	$5.90^{d} \pm 0.70$	$4.32^{d} \pm 0.45$	$614.18^{b} \pm 11.67$	$402.81^{e} \pm 16.0$	$407.07^{d} \pm 19.45$	$548.09^{c} \pm 23.90$	$79.09^{a} \pm 1.98$	$391.09^{d} \pm 3.90$
SOAE	$68.30^{ab} \pm 1.58$	$11.30^{d} \pm 0.03$	$12.33^{c} \pm 1.05$	$5.17^{cd} \pm 0.03$	$3.00^{c} \pm 0.05$	$701.55^{d} \pm 17.7$	$217.15^{c} \pm 8.90$	$394.11^{d} \pm 11.89$	$715.26^{e} \pm 37.8$	$287.90^{e} \pm 2.09$	$217.09^{c} \pm 8.90$

Values are mean ± SD of replicates ($n = 3$)

Means with the same superscript letters down the column are not significantly different ($p \leq 0.05$)

SOAL: soil associated with the growth of wild *P. ostreatus* in Ala quarters; SOIT: soil associated with the growth of wild *P. ostreatus* in Igbatoro Road; SOID: soil associated with the growth of wild *P. ostreatus* in Ido-Ekiti; SOUS: soil associated with the growth of wild *P. ostreatus* in Usi-Ekiti; SOAE: soil associated with the growth of wild *P. ostreatus* in Ado-Ekiti; TOM: total organic matter and TOC: total organic carbon

Table 6 Proximate composition (on dry basis %) and mineral profile (mg/100 g) of harvested wild *P. ostreatus* from different locations in Ondo and Ekiti States

Fungi	Moisture	Ash	Crude fiber	Fat	Protein	CHO	K	Na	Mg	Ca	Fe	Zn
POAL	$8.10^{b} \pm 0.13$	$6.26^{a} \pm 0.57$	$23.00^{b} \pm 0.25$	$0.95^{a} \pm 0.10$	$17.30^{a} \pm 0.32$	$44.80^{c} \pm 0.54$	$93.80^{b} \pm 8.21$	$0.93^{a} \pm 0.04$	$48.62^{b} \pm 1.30$	$50.30^{c} \pm 1.86$	$22.80^{a} \pm 0.94$	$13.89^{a} \pm 0.58$
POIT	$10.19^{c} \pm 0.16$	$9.17^{c} \pm 0.40$	$24.85^{c} \pm 0.39$	$1.77^{c} \pm 0.67$	$22.65^{c} \pm 0.39$	$33.27^{b} \pm 0.14$	$148.90^{c} \pm 18.70$	$3.01^{b} \pm 0.03$	$79.20^{c} \pm 2.06$	$40.90^{a} \pm 1.00$	$81.03^{e} \pm 1.04$	$43.00^{b} \pm 0.30$
POID	$7.30^{a} \pm 0.05$	$8.45^{b} \pm 0.02$	$19.50^{a} \pm 0.38$	$1.70^{c} \pm 0.05$	$17.90^{a} \pm 0.44$	$44.80^{c} \pm 0.51$	$221.06^{e} \pm 15.11$	$2.91^{b} \pm 0.00$	$111.40^{e} \pm 22.0$	$80.59^{d} \pm 0.81$	$67.77^{d} \pm 0.14$	$25.89^{bc} \pm 0.90$
POUS	$10.05^{c} \pm 0.74$	$7.80^{b} \pm 0.55$	$25.04^{c} \pm 0.09$	$0.85^{a} \pm 0.01$	$24.07^{d} \pm 0.66$	$30.06^{a} \pm 0.33$	$181.50^{d} \pm 23.90$	$1.78^{a} \pm 0.04$	$104.78^{d} \pm 17.0$	$58.80^{b} \pm 4.88$	$26.10^{b} \pm 0.10$	$32.80^{b} \pm 1.45$
POAE	$7.90^{a} \pm 0.03$	$8.31^{b} \pm 0.44$	$19.05^{a} \pm 0.30$	$1.08^{b} \pm 0.25$	$18.36^{ab} \pm 0.23$	$45.10^{c} \pm 0.53$	$71.08^{a} \pm 9.04$	$4.05^{c} \pm 0.01$	$27.00^{a} \pm 2.89$	$101.78^{e} \pm 6.05$	$34.84^{c} \pm 1.09$	$10.71^{a} \pm 1.90$

Values are mean ± standard deviation (SD) of replicates ($n = 3$)

Means with the same superscript letters down the column are not significantly different ($p \leq 0.05$)

POAL: *P. ostreatus* from Ala quarters in Akure; POIT: *P. ostreatus* from Igbatoro Road in Akure; POID: *P. ostreatus* from Ido-Ekiti; POUS: *P. ostreatus* from Usi-Ekiti; and POAE: *P. ostreatus* from Ado-Ekiti; CHO: carbohydrates

fungi are responsible for green mold disease on mushrooms and spoilage of mushroom spawns [27]; the mold causes substantial and economic losses of some edible mushrooms, namely, species of *Pleurotus, Agaricus* and *Lentinula* in South Korea, Italy, Hungary, Canada, USA, Iran, New Zealand and Romania [28, 29]. *Streptomyces* spp., *Bacillus subtilis, Actinomyces bovis, Pseudomonas putida, Penicillium italicum, Fusarium* spp. and others were isolated from the soil around wild *P. ostreatus*. *Streptomyces* spp. produce indole-3-acetic acid (IAA), which acts as biocontrol against soil-borne pathogenic microorganisms. Many microorganisms exert growth-promoting effects by secreting beneficial secondary metabolites and enzymes that support spore germination and fruiting formation of wild mushrooms [30, 31]. The microbial ecology couples with optimal environmental conditions, and available nutrient or substrate composition are also selective parameters for the colonization and growth of mushrooms or other ectomycorrhizal fungi [32, 33].

In addition, soil microorganisms markedly influence soil composition by playing a key role in decomposing organic matter, cycling of carbon, nitrogen, phosphorous and stabilization of soil structure [34]. Therefore, the constituent of soil microorganisms could be a sensitive indicator of the growth of wild mushrooms.

The low occurrence and absence of some microorganisms from the examined mushrooms and soil may be due to shortcomings of traditional methods, that is, routine microbiological method adopted for microbial cultivation. This microbiological technique offers a narrow array for microbial isolation due to the fact that some microorganisms may remain viable but non-cultivable (VBNC). The method is only suitable for fast-growing and non-fastidious microbes, and overgrown fungi capable of producing large amounts of spores suppressing other fungi [35]. Another perception in relation to the low microbial occurrence could be associated with antibacterial, antifungal agents, lytic enzymes and volatile compounds secreted by *Pleurotus* mushrooms and their flora microorganisms, which contribute to the survival strategies of the edible fungi by reducing the occurrence of spoilage microorganisms, since mushrooms are often confronted by different types of bacteria, fungi and viruses [36, 37]. Hence, in associated ecosystem processes, a succession of microbial communities on wild mushrooms and their habitat (soil) involved intra- or interspecific interaction of different associations, which may include antibiosis, competition and mutualism [38].

Soil composition around wild *Pleurotus* mushrooms is presented in Table 5, while the nutrient contents of harvested wild *P. ostreatus* from different geographical locations in Ondo and Ekiti States are presented in Table 6. The sand content in the soil around *P. ostreatus* ranged from 66.30 to 88.33%. The clay contents of 10.30, 12.33 and 17.70% were obtained for soil collected around wild *P. ostreatus* at Ala quarters, Ado-Ekiti and Ido-Ekiti, respectively. They were higher ($p \leq 0.05$) than those of Igbatoro (3.67%) and Usi-Ekiti (7.00%). The highest organic content (5.90%) in the soil around *P. ostreatus* mushroom harvested from Usi-Ekiti could contribute to its highest protein (24.07%) and crude fiber contents (25.04%). Nakalembe et al. [39] reported a high concentration of nutrient in edible mushrooms from the humid zone, which could be as a result of higher organic matter in the soil. Organic substances in the soil support high microbial biomass, metabolic activities and, thus, the assembly of more active nutrients into the food crops [40, 41]. The moisture, ash, dietary fiber, protein and carbohydrate contents in the harvested *P. ostreatus* mushrooms ranged from 7.30 to 10.19%, 6.26 to 8.45%, 19.05 to 25.04%, 0.85 to 1.77%, 17.30 to 24.07% and 30.06 to 45.10%, respectively. These values are within what were reported in the literature for proximate composition of *Pleurotus* spp. by the findings of Maftoun et al. [20]. Essential minerals, potassium (K), magnesium (Mg), calcium (Ca), iron (Fe) and zinc (Zn) in the studied mushrooms ranged from 71.08 to 221.06 mg/100 g, 27.00 to 111.40 mg/100 g, 40.90 to 101.78 mg/100 g, 26.10 to 81.03 mg/100 g and 10.71 to 43.00 mg/100 g respectively. The mineral contents of edible fungi may vary from one species to another based on their varieties, absorption mechanisms from the ecosystem, soil nutrient, mineral concentration in the substrates and environmental conditions [42]. Low fat (0.85–1.77%) and sodium (0.93–4.05 mg/100 g) contents were observed in the *Pleurotus* mushrooms. Several studies have indicated edible mushrooms as a better source of protein, dietary fibers, carbohydrate and essential minerals due to their ability to bioaccumulate nutrients into their fruit body with low or no cholesterol [43].

Conclusion

The study examined *Pleurotus* spp. indigenous to two states in Southwestern Nigeria and revealed their genetic relatedness and diversity. The use of molecular technology needs to be adopted to ascertain proper identification and documentation. The varying degrees in the microbial and nutrient contents of the studied wild *P. ostreatus* could be as a result of environmental factors, substrate constituent and geographical locations.

Authors' contributions

All the authors made substantial contributions to the conception, design and acquisition of the research study. TVF carried out the data collection and OCO did the interpretation. The manuscript was drafted by OCO and revised by all authors to meet up with the intellectual content. All authors read and approved the final manuscript.

Author details
[1] Department of Microbiology, The Federal University of Technology, PMB 704, Akure, Nigeria. [2] Department of Biotechnology, Faculty of Science, Kings University, PMB 555, Odeomu, Nigeria.

Acknowledgements
The authors appreciate the technical support received from the Department of Microbiology, The Federal University of Technology, Akure, Nigeria.

Competing interests
The authors declare that they have no competing interests.

Funding
Not applicable.

References

1. Chang ST, Miles PG. Mushrooms-cultivation, nutritional value, medicinal effect and environmental impact. 2nd ed. Boca Raton: CRC Press; 2004.
2. Cunningham AB, Yang X. Mushrooms in forests and woodlands: resource management, values and local livelihoods. London: Earthscan Ltd; 2011.
3. de Mattos-Shipley KMJ, Ford KL, Alberti F, Banks AM, Bailey AM, Foster GD. The good, the bad and the tasty: the many roles of mushrooms. Stud Mycol. 2016;85:125–57.
4. Correa RCG, Brugnari T, Bracht A, Peralta MR, Ferreira ICFR. Biotechnological, nutritional and therapeutic uses of *Pleurotus* spp. (Oyster mushroom) related with its chemical composition: a review on the past decade findings. Trends Food Sci Technol. 2016;50:103–17.
5. Menolli N, Breternitz BS, Capelari M. The genus *Pleurotus* in Brazil: a molecular and taxonomic overview. Mycosci. 2014;55:378–89.
6. Glen M. The use of DNA techniques to identify fungi. In: Potter K, Rimbawanto A, Beadle C, editors. Heart rot and root rot in tropical Acacia plantations. In: Proceedings of a workshop held in Yogyakarta, Indonesia, 7–9 February 2006. Canberra: ACIAR Proceedings;2006;124:46–54.
7. Park Y-J, Kwon O-C, Son E-S, Yoon D-E, Han W, Nam J-Y, Yoo Y-B, Lee C-S. Genetic diversity analysis of *Ganoderma* species and development of a specific marker for identification of medicinal mushroom *Ganoderma lucidum*. Afri J Microbiol Res. 2012;6(25):5417–25.
8. Cong J, Yang Y, Liu X, Lu H, Liu X, Zhou J, et al. Analyses of soil microbial community compositions and functional genes reveal potential consequences of natural forest succession. Sci Rep. 2015;5:10007–17.
9. Zolan ME, Pukkila PJ. Inheritance of DNA methylation in *Coprinus cinereus*. Mol Cell Biol. 1986;6:195–200.
10. Rohlf FJ. NTSYS-pc numerical taxonomy and multivariate analysis system, version 2.02. New York: Exeter Publication Setauket; 1998.
11. Sneath PHA, Sokal RR. Numerical taxonomy: the principle and practice of numerical classification. San Francisco: Freeman; 1973.
12. Cheesbrough M. District laboratory practice in Tropical Countries, part 2. Cambridge: Cambridge University Press; 2000. p. 12–69.
13. Cappuccino GJ, Sherman N. Microbiology, A laboratory manual; Biochemical activities of microorganism. 5th ed. California: Benjamin/Cumming science publishing; 1999.
14. Cowan ST, Steel KJ. Manual for the identification of medical bacteria. 3rd ed. Cambridge: Cambridge University Press; 1993.
15. Barnett JA, Payne RW, Yarrow O. Yeast characteristics and identification. 3rd ed. London: Cambridge University Press; 2000.
16. Sato JH, de Figueiredo CC, Marchão RL, Madari BE, Benedito LEC, Busato JG, de Souza DM. Methods of soil organic carbon determination in Brazilian savannah soils. Sci Agricol. 2014;71(4):302–8.
17. Walkley A, Black IA. An examination of Degtjareff method for determining soil organic matter and a proposed modification of the chromic acid filtration. Soil Sci. 1934;37:29–38.
18. Association of Official Analytical Chemists A.O.A.C. Official Methods of Analysis. Association of Official Analytical Chemists. 19th Gaithersburg, MD, USA; 2012.
19. Otieno OD, Onyango C, Onguso JM, Matasyoh LG, Bramwel W, Wamalwa M, Harvey JJW. Genetic diversity of Kenyan native oyster mushroom (*Pleurotus*). Mycolgia. 2015;107(1):32–8.
20. Maftoun P, Johari H, Soltani M, Malik R, Othman NZ, El Enshasy HA. The edible mushroom *Pleurotus* spp.: I. biodiversity and nutritional values. Int J Biotechnol Wellness Ind. 2015;4:67–83.
21. Yang Q, Wang X, Shen Y. Comparison of soil microbial community catabolic diversity between rhizosphere and bulk soil induced by tillage or residue retention. J Soil Sci Plant Nutri. 2013;13(1):187–99.
22. Cho Y-S, Kim J-S, Crowley DE, Cho BG. Growth promotion of the edible fungus *Pleurotus ostreatus* by fluorescent pseudomonads. FEMS Microbiol Lett. 2003;218:271–6.
23. Young L-S, Chu J-N, Hameed A, Young C-C. Cultivable mushroom growth-promoting bacteria and their impact on *Agaricus blazei* productivity. Pesq Agropec Bras. 2013;48(6):636–44.
24. Zarenejad F, Yakhchali B, Rasooli I. Evaluation of indigenous potent mushroom growth promoting bacteria (MGPB) on *Agaricus bisporus* production. World J Microbiol Biotechnol. 2012;28:99–104.
25. Munsch P, Alatossava T. Several pseudomonads, associated with the cultivated mushrooms *Agaricus bisporus* or *Pleurotus* sp. are hemolytic. Microbiol Res. 2002;157:311–5.
26. Siyoum NA, Surridge K, van der Linde EJ, Korsten L. Microbial succession in white button mushroom production systems from compost and casing to a marketable packed product. Ann Microbiol. 2016;66:151–64.
27. Kredics L, Jimenez LG, Naeimi S, Czifra D, Urbán P, Manczinger L, Vágvölgyi C, Hatvani L. A challenge to mushroom growers: the green mould disease of cultivated champignons. In: Mendez-Vilas A, editor. Current research, technology and education topics in applied microbiology and microbial biotechnology. Badajoz: Formatex Research Center; 2010. p. 295–305.
28. Colavolpe MB, Mejía SJ, Alberto E. Efficiency of treatments for controlling *Trichoderma* spp. during spawning in cultivation of lignicolous mushrooms. Bra J Microbiol. 2014;45(4):1263–70.
29. Hatvani L, Antal Z, Manczinger L, Szekeres A, Druzhinina IS, Kubicek CP, Nagy A, Nagy E, Vágvölgyi C, Kredics L. Green mold diseases of *Agaricus* and *Pleurotus* spp. are caused by related but phylogenetically different *Trichoderma* species. Phytopathol. 2007;97:532–7.
30. Kim MK, Math RK, Cho KM, Shin KJ, Kim JO, Ryu JS, Lee YH, Yun HD. Effect of *Pseudomonas* sp. P7014 on the growth of edible mushroom *Pleurotus eryngii* in bottle culture for commercial production. Bioresour Technol. 2008;99(8):3306–8.
31. Frey-Klett P, Burlinson P, Deveau A, Barret M, Tarkka M, Sarniguet A. Bacterial-fungal interactions: hyphens between agricultural, clinical, environmental, and food microbiologists. Microbiol Molecul Biol Rev. 2011;75(5):583–609.
32. Wang P, Liu Y, Yin Y, Jin H, Wang S, et al. Diversity of microorganisms isolated from the soil sample surround *Chroogomphus rutilus* in the Beijing region. Intern J Bio Sci. 2011;7(2):209–20.
33. Yamada A, Ogura T, Ohmasa M. Cultivation of mushrooms of edible ectomycorrhizal fungi associated with *Pinus densiflora* by in vitro mycorrhizal synthesis. Myco. 2001;11(2):59–66.
34. Loranger-Merciris G, Barthes L, Gastine A, Leadley P. Rapid effects of plant species diversity and identity on soil microbial communities in experimental grassland ecosystems. Soil Biol Biochem. 2006;38:2336–43.
35. Agrawal PK, Agrawal S, Shrivastava R. Modern molecular approaches for analyzing microbial diversity from mushroom compost ecosystem. 3 Biotech. 2015;5:853–66.
36. Lim Y, Ryu JS, Shi S, Noh W, Kim E, Le QV, Lee H-S, Ro H-S. Isolation of bacteria associated with the King Oyster mushroom, *Pleurotus eryngii*. Mycobiology. 2008;36(1):13–8.
37. Kim S-W, Kim S, Lee H-J, Park J-W, Ro H-S. Isolation of fungal pathogens to an edible mushroom, *Pleurotus eryngii*, and development of specific ITS primers. Mycobiology. 2013;41(4):252–5.
38. Boon E, Meehan CJ, Whidden C, Wong DH-J, Langille MGI, Beiko RG. Interactions in the microbiome: communities of organisms and communities

of genes. FEMS Microbiol Rev. 2014;38:90–118.
39. Nakalembe I, Kabasa JD, Olila D. Comparative nutrient composition of selected wild edible mushrooms from two agro-ecological zones, Uganda. SpringerPlus. 2015;4:433.
40. Ortíz-Castro R, Contreras-Cornejo HA, Macías-Rodríguez L, López-Bucio J. The role of microbial signals in plant growth and development. Plant Signal Behav. 2009;4(8):701–12.
41. Vieira RF, Pazianotto RAA. Microbial activities in soil cultivated with corn and amended with sewage sludge. SpringerPlus. 2016;5:1844.
42. Zhu F, Qu L, Fan W, Qiao M, Hao H, Wang X. Assessment of heavy metals in some wild edible mushrooms collected from Yunnan Province, China. Environ Monit Assess. 2011;179(1–4):191–9.
43. Kalac P. Chemical composition and nutritional value of European species of wild growing mushrooms. In: Andres S, Baumann N, editors. Mushrooms: types, properties and nutrition. New York: Nova Science Publishers; 2012. p. 30–51.

11

Larvicidal effect of compounds isolated from *Maerua siamensis* (Capparidaceae) against *Aedes aegypti* (Diptera: Culicidae) larvae

Saksit Nobsathian[1*], Vasakorn Bullangpoti[2], Nutchaya Kumrungsee[3], Natnicha Wongsa[3] and Dussadee Ruttanakum[3]

Abstract

Background: Dengue is a major problem for humanity. Most people use insecticides to eliminate larvae of *Aedes aegypti*, which requires heavy chemicals use that affects the environment and human health. Therefore, in this research, the focus was on the larvicidal efficacy of pure compounds from the leaves and twigs of *Maerua siamensis* against the larvae of *A. aegypti*.

Results: Larval mortality was observed after a 24-h exposure. The 1H-indole-3 acetonitrile glycosides cappariloside A and cappariloside B and the triterpene lupeol showed strong larvicidal effects (24-h $LC_{50} = 71.14$, 99.79 and 133.03 ppm). After 48 h, cappariloside B caused the most potential mortality with an LC_{50} of 1.56 ppm and lupeol had the highest lethal concentration at $LC_{50} = 158.71$ ppm. Additionally, consistency was observed between the toxicity tests and detoxification enzyme activity. Most compounds, except for lupeol and vanillin, reduce the activity of glutathione-s-transferase, whereas no significant differences were between control and treated groups for carboxylesterase.

Conclusions: Cappariloside A and cappariloside B are good potential larvicide agents. They showed larvicidal activity against *Ae. aegypti* larvae with $LC_{50} = 71.14$ and 99.79 ppm at 24 and 48 h, respectively.

Keywords: *Maerua siamensis*, Larvicidal agent, *Aedes aegypti*, Toxicity, Detoxification enzyme

Background

The yellow fever mosquito *Aedes aegypti* (Diptera: Culicidae) is a primary vector that causes dengue and dengue haemorrhagic fever [1]. In the 20th century, more than 1.8 billion people (more than 70%) in 110 countries were at risk of dengue infection. In the past decade, because of the public health problems [2], most people use insecticides to eliminate the mosquitoes. However, the current use of insecticides has caused inhibition of cholinesterases and chromosomal aberrations in human peripheral leukocytes [3, 4]. Additionally, these insecticides also contaminate air, water, and soil in surrounding areas and therefore cause mortality to animals [5, 6]. Thus, other insecticidal substances with lower toxicity to the environment than the ones in current use are urgently required [7].

Some plant compounds can eliminate larvae. The crude ethyl acetate extract of leaves of *Acalypha fruticosa* shows significant larvicidal activity with lethal concentration values LC_{50} and LC_{90} of 253.08 and 455.40 ppm, respectively [8]. The methanol extract from the leaves of *Ocimum sanctum* against fourth instar larvae of *Ae. aegypti* has an LC_{50} value of 429.54 ppm [9]. Deguelin and tephrosin, rotenoid types isolated from the seeds of *Millettia dura* show potent larvicidal activity with $LC_{50} = 1.6$ and 1.4 μg/ml at 24 h, respectively [10]. The methanol extract of the leaves of *Atalantia monophylla* (Rutaceae) shows larvicidal activity against second stages *Ae. aegypti* with a lethal concentration = 0.002 mg/l [11].

*Correspondence: saksit.nob@mahidol.ac.th
[1] Nakhonsawan Campus, Mahidol University, Nakhonsawan 60130, Thailand
Full list of author information is available at the end of the article

Plants of Capparidaceae are found in tropical and subtropical regions of the world. Most plants of this family are found in Africa with 17 genera and 450 species [12]. Four genera are found in Thailand [13]: *Capparis, Cleome, Crateva,* and *Maerua*. All 90 species in the genus *Maerua* are found in tropical Asia, including the African continent. The crude extracts of plants in this genus have biological activities and ethnomedical applications. The aqueous root extract of *M. oblongifolia* is an anti-diabetic in rats [14]. The crude methanol extract of leaves of *M. angolensis* DC is active against *Streptococcus pyogenes, Escherichia coli,* and *Neisseria gonorrhoeae* [15]. The ethyl acetate fraction of the tuber parts of *M. pseudopetalosa* shows cytotoxic activity in brine shrimp larvae [16].

In this study, we focused on the isolation of larvicidal agents from *Maerua siamensis* (Kurt) Pax, the only species of the genus found in Thailand. The isolation of natural compounds from the leaves and twigs of this plant led to the separation of eight known compounds with promising insecticidal bioactivity.

Methods

General experimental procedure

Column chromatography (CC) used silica gel 60 (70–230 mesh, Merck Millipore, Darmstadt, Germany). Preparative thin-layer chromatography (Prep-TLC) used Kieselgel 60 PF254 (0.5 mm Merck Millipore, Darmstadt Germany). Benzenesulfonyl fluoride hydrochloride (AEBSF), EDTA, 1-chloro-2,4-dinitrobenzene (CDNB), glutathione-*s*-transferase, potassium phosphate buffer (pH 7.2), carboxylesterase and 4-nitrophenyl acetate (pNPA) were purchased from Sigma-Aldrich. The melting point was recorded on an electrothermal instrument. Optical rotations were determined on a JASCO DIP-370 digital polarimeter using a 50 mm microcell (1 ml). UV spectra were measured in EtOH or MeOH with a JASCO 530 spectrometer, and IR spectra were recorded on a Perkin Elmer 2000. The ^{1}H-NMR, ^{13}C-NMR, DEPT, and 2-D NMR spectra were recorded on a Bruker Ascend™ 400 MHz or Bruker AVANCE 500 MHz. in $CDCl_3$ using TMS as an internal standard, unless otherwise stated. Finally, HR-TOF-MS results were recorded on a Micromass model VQ-TOF2.

The leaves and twigs of *M. siamensis* were collected in Nakhonsawan Province, Thailand, in August 2015. A voucher specimen (BKF No. 180668) was deposited in and identified by the Forest Herbarium, Royal Forest Department in Bangkok.

Dried and finely powdered leaves and twigs (1.70 kg) of *M. siamensis* were percolated with MeOH at room temperature to produce a crude MeOH extract (215.00 g). After dissolution in MeOH: EtOAc (1:1) and solvent removal, a crude MeOH: EtOAc (1:1) soluble fraction (105.00 g) was obtained. The active MeOH: EtOAc (1:1) soluble fraction was separated by Si-gel CC (SiO_2, 1.8 kg, CH_2Cl_2–hexane and MeOH–CH_2Cl_2 gradients) to give fractions A_1–A_5. Fraction A_1 (2.57 g) provided glochidone (**3**) (125.20 mg) after two consecutive Si-gel CCs (CH_2Cl_2–hexane gradients), followed by recrystallisation from MeOH–CH_2Cl_2. Fraction A_3 (13.20 g) afforded fractions B_1–B_7 after Si-gel CC (acetone–hexane, and MeOH–acetone gradients). Fraction B_2 (5.20 g) gave fractions C_1–C_5 after Si-gel CC (EtOAc–hexane, and MeOH–EtOAc gradients). Fraction C_1 (1.52 g) gave lupeol (**4**) (121.20 mg) as white needles after recrystallisation from MeOH–CH_2Cl_2. Fraction C_2 (1.07 g) yielded chrysoeriol (**5**) (80.70 mg) after two consecutive CCs (1st CC: Si-gel, EtOAc–hexane gradient; 2nd CC: Sephadex LH-20, MeOH), followed by recrystallisation from ethanol. Fraction C_3 (1.11 g) yielded cappariloside A (**1**) (51.70 mg) after separation by CC (CH_2Cl_2–hexane gradients), followed by recrystallisation from MeOH. Fraction B_3 (5.05 g) provided cappariloside B (**2**) (14.80 mg) after two consecutive Si-gel CCs (1st CC: EtOAc–hexane gradient; 2nd CC: acetone–hexane gradient), followed by prep-TLC (2% MeOH–CH_2Cl_2 as the eluent) and recrystallisation from EtOAc–hexane. Fraction B_5 (2.15 g) after two consecutive Si-gel CCs (1st CC: CH_2Cl_2–hexane and CH_2Cl_2–MeOH gradients; 2nd CC: EtOAc–hexane, and MeOH–EtOAc gradients) gave fractions D_1–D_7. Fraction D_6 (180.80 mg) provided cinnamic acid (**6**) (17.30 mg) after recrystallisation from MeOH–CH_2Cl_2. The residue (98.20 mg) yielded 3,4-dimethoxybenzoic acid (**7**) (21.20 mg) after recrystallisation from CH_2Cl_2–hexane. Fraction B_6 (3.42 g) was further purified by Si-gel CC (acetone–hexane gradient), followed by prep-TLC (2% MeOH–CH_2Cl_2) to provide fractions E_1–E_4. Fraction E_3 (174.00 mg) gave vanillin (**8**) (32.10 mg) following recrystallization from EtOAc. The structures of all pure compounds are shown in Fig. 1.

Cappariloside A (**1**) [17]: amorphous solid; m.p. 228.0–229.0 °C; UV (MeOH) λ_{max} (log ε) 267 (4.79), 278 (4.61), 289 (4.14) nm.; IR (KBr disc) ν_{max} 3525, 3495, 3400, 3359, 1625, 1590, 1508, 1170, 1084; HR-TOF-MS (ESI positive) m/z 357.1069 $[M + Na]^+$ (calcd. for $C_{16}H_{18}N_2O_6Na$, 357.1063).

Cappariloside B (**2**) [17]: amorphous solid; m.p. 230.1–231.3 °C; UV (MeOH) λ_{max} (log ε) 272 (4.00), 279 (3.88), 289 (4.12) nm.; IR (KBr disc) ν_{max} 3390, 2855, 2255, 1625, 1540, 1510, 1120 cm^{-1}; HR-TOF-MS (ESI positive) m/z 519.1589 $[M + Na]^+$ (calcd. for $C_{22}H_{28}N_2O_{11}Na$, 519.1591).

Glochidone (**3**) [18]: colorless needles; m.p. 166.0–166.6 °C; UV (EtOH) λ_{max} (log ε) 228 (3.96), 333 (1.73).; IR (KBr disc) ν_{max} 2945, 2873, 1662 (C=O stretching of conjugated ketone), 1456, 1382, 1284, 1229, 1161, 1103,

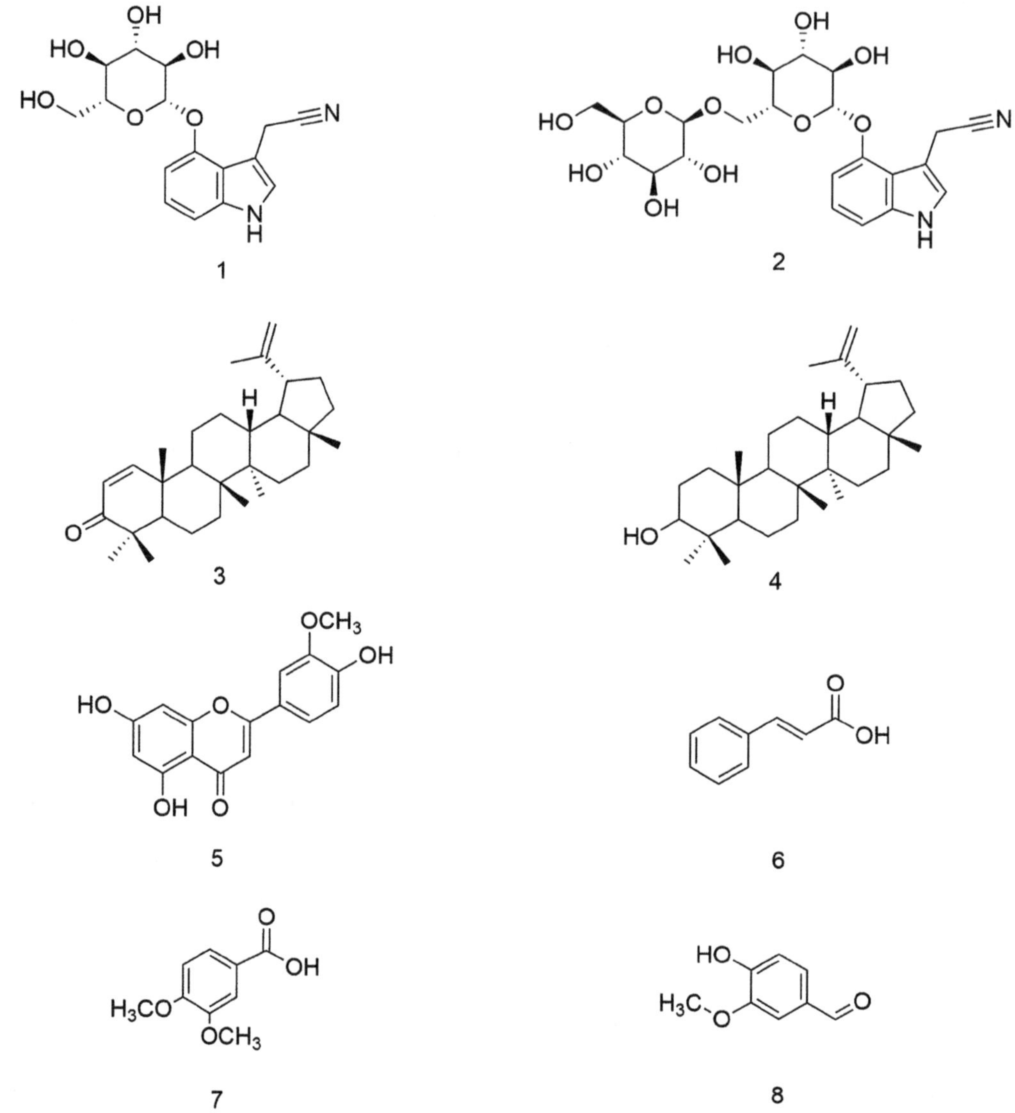

Fig. 1 Compounds isolated from *Maerua siamensis*. All compounds were purified and their structures were elucidated based on spectroscopic data as described in the text

947, 888, 825 cm^{-1}; HR-TOF-MS (ESI positive) m/z 427.2044 [M + Na]$^+$ (calcd. for $C_{30}H_{28}ONa$, 427.2038); Optical rotation: $[\alpha]^{30}_{589}$ +68.6° (*c* 0.1, $CHCl_3$).

Lupeol (**4**) [19]: white powder; m.p. 212.4–213.0 °C; IR ($CHCl_3$) ν_{max} 3486, 2933, 1640, 1473, 1384, 1037, 870 cm^{-1}; HR-TOF-MS (ESI positive) m/z 449.3750 [M + Na]$^+$ (calcd. for $C_{30}H_{50}ONa$, 449.3759).

Chrysoeriol (**5**) [20]: yellow powder; m.p. 325.1–326.2 °C; UV (MeOH) λ_{max} (log ε) 269 nm (3.44), 340 nm (4.52) nm.; IR (KBr disc) 3350, 3088, 2925, 1777, 1719, 1655, 1607, 1561, 1506, 1256 cm^{-1}; HR-TOF-MS (ESI positive) m/z 323.0540 [M + Na]$^+$ (calcd. for $C_{16}H_{12}O_6Na$, 323.0532).

Cinnamic acid (**6**) [21]: white powder; m.p. 193.9–194.3 °C; UV (MeOH) λ_{max} (log ε) 272 (2.85) nm.; IR (KBr disc) ν_{max} 3448, 1711, 1638, 1577, 1551, 1450 cm^{-1}; HR-TOF-MS (ESI positive) showed [M + Na]$^+$ 171.0411 [M + Na] + (calcd. for $C_9H_8O_2Na$, 171.0422).

3,4-Dihydroxybenzoic acid (**7**) [22]: white powder; m.p. 130.9–131.0 °C; UV (MeOH) λ_{max} (log ε) 212 (4.66), 268 (4.56) 313 (4.17) nm.; IR (KBr disc) ν_{max} 3200, 2839, 1674, 1603 cm^{-1}; HR-TOF-MS (ESI positive) m/z 177.0170 [M + Na]$^+$ (calcd. for $C_7H_6O_4Na$, 177.0164).

Vanillin (**8**) [23]: white powder; m.p. 81.0–83.2 °C; UV (MeOH) λ_{max} (log ε) 221 (3.66), 271 (3.56) nm.; IR (KBr disc) ν_{max} 3475, 3444, 3184, 2733, 2669 cm^{-1};

HR-TOF-MS (ESI positive) showed $[M + Na]^+$ 175.0380 [M + Na] + (calcd. for $C_7H_6O_4Na$, 175.0371).

Mosquito strain

Eggs of *Ae. aegypti* (Thailand laboratory strain) were received from the Ministry of Public Health, Thailand. Larvae were reared in 500-ml glass beakers containing water, fed a fish diet and maintained in our culture room at 28 °C and 70% RH with a 14:10 DL photoperiod. The same conditions were used for the pupae. After emergence, adults were maintained in cages, in the same locale.

Larvicidal toxicity assay

The larvicidal bioassay against third instars of *Ae. aegypti* was modified from the WHO (1981) method under laboratory conditions at 28 ± 1 °C and, 70% RH 14:10 DL photoperiod. Twenty-five larvae were placed in a small cup filled with 50 ml of distilled water, to which 0.5 ml of each extract dissolved in 0.5% acetone was added. The final concentration range for each treatment was from 125 to 1000 ppm dissolved in acetone. For all concentrations, 15 replicates were used per concentration. In controls, 0.5 ml of 0.5% acetone was used in each case. During the aqueous dispersion test, mosquito larvae were not provided with food. After 24 and 48 h, mortality of the larvae in each treatment was recorded.

Mode of action studied

The larvae that survived the LC_{50} value treatment at 24 h of exposure were homogenised in 0.5 ml of homogenised buffer [100 mM phosphate buffer (pH 7.2) and 1% Triton X-100]. The homogenate was centrifuged at 10,000×*g* for 15 min at 4 °C, and the supernatant was used as the enzyme source.

To measure in vivo enzymes activities, 24 h LC_{50} value-treated larvae were homogenised in buffer A [100 mM phosphate buffer (pH 7.2) containing 1 M DTT, 100 mM 4-(2-aminoethyl) benzenesulfonyl fluoride hydrochloride (AEBSF) and 0.5 M EDTA]. The homogenate was centrifuged at 10,000×*g* for 5 min at 4 °C, and the resultant supernatant was used for carboxylesterase and glutathione-*s*-transferase activity analyses.

Carboxylesterase (CE) activity was determined by the modified method of Kumrungsee [24]. Enzyme solution 50 μl, was mixed with 4-nitrophenyl acetate (pNPA) (10 mM in DMSO) and phosphate buffer (50 mM, pH 7.4). Enzyme activity was measured at 410 nm and 37 °C for 90 s with a microplate reader in the kinetic mode. The activity of CE was determined using the extinction coefficient of 176.4705 for pNPA.

The method for determining glutathione-*s*-transferase (GST) activity was according to Oppenoorth [25]. The reaction solution contained 100 μl of enzyme solution, 50 mM potassium phosphate buffer (pH 7.3) and 150 mM 1-chloro-2,4-dinitrobenzene (CDNB). Optical density was recorded at intervals of 30 s for 3 min at 37 °C and 340 nm with a microplate reader. The GST activity was determined from the extinction coefficient of 0.0096 for CDNB.

Three biological replicates per treatment were analysed. The protein content of each fraction used as an enzyme source was determined by the Bradford method before measuring enzyme activities.

Statistical analyses

Probit analysis was used to calculate LC_{50} and LC_{90} values were determined using the STATPLUS program (version 2017). The range of detoxification enzyme activity from each treatment was compared using one-way analysis of variance (ANOVA).

Results and discussion

All pure compounds isolated from *M. siamensis* were identified by the comparison of their physical properties and spectroscopic data with those reported in the literature [17–23].

The larvicidal activities in the different periods of exposure to *Ae. aegypti* larvae are presented in Tables 1 and 2. Cappariloside A had the lowest lethal concentration with an LC_{50} of 71.14 ppm, and vanillin had the highest LC_{50} value at = 2846 ppm at 24 h of exposure. After 48 h, cappariloside B caused maximum mortality with an LC_{50} 1.56 ppm and lupeol had the highest lethal concentration $LC_{50} = 158.71$ ppm.

The effect of pure compounds on detoxification enzyme activity is shown in Table 3. Consistency was found between toxicity tests and detoxification enzyme activity. Most compounds except for lupeol and vanillin, reduce glutathione-*s*-transferase activity whereas, differences between the control and treated groups for carboxylesterase were not significant.

For triterpenes, glochidone and lupeol; lupeol ($LC_{50} = 133.03$ ppm) had better activity than betulinic acid [26] ($LC_{50} = 142$ ppm), but the activity of betulinic acid was better than that of glochidone ($LC_{50} = 382.34$ ppm) after 24 h. No reports on the larvicidal activity of indole alkaloids are available. Cappariloside A and B were actively toxic ($LC_{50} = 71.14$ and 99.79 ppm, respectively). Finally, cinnamic acid, 3,4-dimethoxybenzoic and vanillin were weakly toxic. However, after 48 h, cappariloside B, cinnamic acid, lupeol, cappariloside A, 3,4-dimethoxybenzoic acid and chrysoeriol were actively toxic ($LC_{50} = 1.56$, 22.76, 58.87, 71.14, 72.54 and 77.55 ppm, respectively). Vanillin and

Table 1 The LC_{50} and LC_{90} values (ppm) of compounds isolated from *Maerua siamensis* against 3rd instars of *Aedes aegypti* after 24 h of exposure

Compounds	Slope	LC_{50} ± SE (ppm)	LC_{90} ± SE (ppm)	χ^2
Cappariloside A	0.89 ± 0.37	71.14 ± 5.41a	1980.51 ± 182.71a	1.48
Cappariloside B	1.14 ± 0.21	99.79 ± 2.85a	1329.68 ± 46.53a	0.6
Glochidone	1.68 ± 0.20	382.34 ± 35.95a	2210.21 ± 56.45a	0.95
Lupeol	1.28 ± 0.21	133.03 ± 26.98a	1340.41 ± 40.19a	0.39
Chrysoeriol	1.12 ± 0.46	169.94 ± 7.17a	2278.96 ± 222.12a	4.53
Cinnamic acid	0.81 ± 0.19	852.36 ± 28.46a	32,231.97 ± 7532.75b	0.41
3,4-Dimethoxybenzoic acid	0.61 ± 0.19	1032.87 ± 66.07b	127,826.08 ± 1213.11c	0.22
Vanillin	0.75 ± 0.21	2846.76 ± 131.14b	144,759.28 ± 17,673c	1.56

Means within a column followed by the same letter of each time exposure are not significantly different at the 0.05 level by Tukey's test

Table 2 The LC_{50} and LC_{90} values (ppm) of compounds isolated from *Maerua siamensis* against 3rd instars of *Aedes aegypti* after 48 h of exposure

Compounds	Slope	LC_{50} ± SE (ppm)	LC_{90} ± SE (ppm)	χ^2
Cappariloside A	0.89 ± 0.37	71.14 ± 5.41a	1980.51 ± 182.71a	1.48
Cappariloside B	0.48 ± 0.24	1.56 ± 0.16b	1980.51 ± 32.66a	1.48
Glochidone	1.57 ± 0.46	158.70 ± 45.78c	1045.70 ± 46.05b	3.62
Lupeol	1.46 ± 0.57	58.87 ± 4.02a	443.32 ± 15.54c	1.56
Chrysoeriol	2.17 ± 0.34	77.55 ± 15.72a	302.46 ± 3.52c	0.25
Cinnamic acid	0.70 ± 0.22	22.79 ± 3.98b	1581.11 ± 149.88d	0.48
3,4-Dimethoxybenzoic acid	0.81 ± 0.20	72.54 ± 3.06a	2732.78 ± 216.78e	0.34
Vanillin	0.98 ± 0.20	130.82 ± 3.71c	2690.07 ± 151.79e	0.78

Means within a column followed by the same letter of each time exposure are not significantly different at the 0.05 level by Tukey's test

glochidone were moderately toxic (LC_{50} = 130.82 and 158.70 ppm, respectively).

Insects are well known to use detoxification enzyme to develop resistance to insecticides by increasing the level of enzymes. For example, Fonseca-González et al. [27] describe high levels of both cytochrome P450 monooxygenases and non-specific esterases in some of the fenitrothion and pyrethroid-resistant *Ae. aegypti* populations in Cambodia. Kasai et al. [28] also suggest that cytochrome P450 monooxygenases play an important role in resistance development for *Ae. aegypti* to pyrethroids. Additionally, Dusfour et al. [29] describe the regulation of cytochrome P450 genes and carboxylesterases in all three populations of *Ae. aegypti* from three French overseas territories worldwide. Thus, studies on the level of detoxification enzymes may be necessary to estimate trends in resistance to new compounds. However, in the present study, all compounds showed no significant effects on carboxylesterase enzyme (CEs) activity compared with controls. Compounds significantly inhibited

Table 3 Carboxylesterase activity (nM para phenyl acetate/mg protein/min) and glutathione-*s*-transferase activity (CDNB product/mg protein/min) of surviving 3rd instar *Aedes aegypti* after treatment with pure compounds from *Maerua siamensis* for 24 h

Compounds	Carboxylesterase activity[a]	CF[b]	Glutathione-*s*-transferase activity[a]	CF[b]
Control	0.030 ± 0.002a	–	0.321 ± 0.121f	–
Cappariloside A	0.031 ± 0.001a	0.97	0.204 ± 0.014b	1.57
Cappariloside B	0.029 ± 0.001a	1.03	0.156 ± 0.005a	2.06
Glochidone	0.031 ± 0.001a	0.97	0.282 ± 0.006d	1.14
Lupeol	0.032 ± 0.001b	0.94	0.343 ± 0.023 g	0.94
Chrysoeriol	0.031 ± 0.001a	0.97	0.280 ± 0.009d	1.15
Cinnamic acid	0.029 ± 0.001a	1.03	0.207 ± 0.002c	1.55
3,4-Dimethoxybenzoic acid	0.031 ± 0.001a	0.97	0.291 ± 0.076e	1.10
Vanillin	0.032 ± 0.001b	0.94	0.493 ± 0.289 h	0.65

[a] Means within a column followed by the same letter are not significantly different at the 0.05 level by Tukey's test

[b] CF is a correction factor = (enzyme activity of control)/(enzyme activity of treatment)

glutathione-*s*-transferase activities between 1.10- and 2.60-fold, which could explain the mortality in mosquito larvae to the compounds.

Conclusions

In conclusion, cappariloside A and cappariloside B are good potential larvicide agents. They show larvicidal activity against *Ae aegypti* larvae with $LC_{50} = 71.14$ and 99.79 ppm at 24 and 48 h, respectively.

Authors' contributions

SN collected samples from Nakhonsawan Campus, Mahidol University, Nakhonsawan, Thailand. SN, VB, and NK designed the experiment. SN, NK, NW, and DR performed the experiments, analysed data and wrote the paper. SN, VB, NK, and PM reviewed and checked all the details. All authors read and approved the final manuscript.

Author details

[1] Nakhonsawan Campus, Mahidol University, Nakhonsawan 60130, Thailand. [2] Animal Toxicology and Physiology Speciality Research Unit, Zoology Department, Faculty of Science, Kasetsart University, Bangkok 10900, Thailand. [3] Biology Department, Faculty of Science and Technology, Rajamangala University of Technology Thanyaburi, Pathumthani, Thailand.

Acknowledgements

This research project was supported by Mahidol University. We thanking the Nakhonsawan Campus Mahidol University. We thank Professor Patoomratana Tuchinda and Mr. Wisuwat Thongpichai for the advice and valuable discussion during this research.

Competing interests

The authors declare that they have no competing interests.

Funding

The Talent Management Fund for Mahidol University financially supported this work.

References

1. Evenhuis NL, Samuel GM. 22 Famlily Culicidae. In: Evenhuis NL, editor. Catalog of the Diptera of the Australasian and Oceanian Regions. Hawaii: Bishop Museum; 2007. p. 191–218 (Retrieved 4 Feb 2012).
2. Halstead SB. Global perspective on dengue research. Dengue Bull. 2000;24:77–82.
3. Seume FW, Casida JE, O'Brien RD. Effects of parathion and malathion separately and jointly upon rat esterases in vivo. Insectic Tox. 1960;8:43–7.
4. Sharma KA, Tiwari U, Gaur MS, Tiwari KW. Assessment of malathion and its effects on leukocytes in human blood samples. J Biomed Res. 2015;29:1–9.
5. Lenoir JS, Cahill TM, Sieber JN, Mcconnell LL, Fellers GM. Summertime transport of current use pesticides from the California Central Valley to Sierra Nevada Mountain Range, USA. Environ Tox Chem. 1999;18(12):2715–22.
6. Newhart K. Environmental fate of malathion. California Environmental Protection Agency Department of Pesticide Regulation Environmental Monitoring Branch. 2006;11:1–20.
7. Pluempanupat S, Kumrungsee N, Pluempanupat W, Ngamkitpinyo K, Chavasiri W, Bullangpoti V, et al. Laboratory evaluation of *Dalbergia oliveri* (Fabaceae: Fabales) extracts and isolated isoflavonoids on *Aedes aegypti* (Diptera: Culicidae) mosquitoes. Ind Crops Prod. 2013;44:653–8.
8. Pavunraj M, Ramesh V, Sakthivelkumar S, Veeramani V, Janarthanan S. Larvicidal and enzyme inhibitory effects of *Acalypha fruticosa* (F.) and *Catharanthus roseus* L (G) DON leaf extracts against Culex quinquefasciatus(say) (Diptera: Culicidae). Asian J Pharm Clin Res. 2017;10(3):213–20.
9. Mohamed AA. Larvicidal activity of *Ocimum sanctum* Linn. (Labiatae) against *Aedes aegypti* (L.) *and Culex quinquefasciatus* (Say). Parasitol Res. 2008;103:1451–3.
10. Abiy Y, Solomon D, Jacob OM, Matthias H, Martin GP. Effect of rotenoids from the seeds of *Millettia dura* on larvae of *Aedes aegypti*. Pest Manag Sci. 2003;59:1159–61.
11. Sivagnaname N, Kalyanasundaram M. Laboratory evaluation of methanolic extract of *Atlantia monophylla* (Family: Rutaceae) against immature stages of mosquitoes and non-target organisms. Mem Inst Oswaldo Cruz Rio de Janeiro. 2004;99(1):115–8.
12. Brummitt RK, Culham A, Seberg O, Heywood VH, editors. Flowering plant families of the world. Firefly Book: Ontario; 2007.
13. Chayamarit K, Capparaceae. Flora of Thailand. Vol. 5. No. 3. In: Larsen K, Smitinand T. editors. Bangkok: Chutima Press., 1991. p. 266–268.
14. Arulanandraj CN, Punithavani T, Indumathy S. Effect of murva (*Maerua oblongifolia*) on alloxan induced diabetes in rats. IJPSR. 2011;2(10):2754–6.
15. Ayo RG, Audu OT, Amupitan JO, Uwaiya E. Phytochemical screening and antimicrobial activity of three plants used in traditional medicine in Northern Nigeria. J Med Plants Res. 2013;7(5):191–7.
16. Manal AI, Bushra EEN. Cytotoxicity study on *Maerua pseudopetalosa* (Glig and Bened.) De Wolf tuber fractions. Afr J Plant Sci. 2015;9(12):490–7.
17. Ihsan C, Ayse K, Peter R. 1H-Indole-3 acetonitrile glycosides from *Capparis spinosa* fruits. Phytochemistry. 1999;50:1205–8.
18. Ayer WA, Flanagan RJ, Reffstrup T. Metabolites of bird's nest fungi-19, new triterpenoid carboxylic acids from *Cyathus striatus* and *Cyathus pygmaeus*. Tetrahedron. 1984;40:2069–82.
19. Keawsa-ard S, Natakankitkul S, Liawruangrath S, Teerawutgulrag A, Trisuwan K, Charoenying P, et al. Anticancer and antibacterial activities of the isolated compounds from *Solanum spirale* Roxb. leaves. Chiang Mai. J Sci. 2012;39(3):445–54.
20. Amami SA, Maitland DJ, Soliman GA. Hepatoprotective activities of *Shouwia thebaica* webb. Bioorg Med Chem Lett. 2006;16:4624–8.
21. Hsieh TJ, Su CC, Chen CY, Liou CH, Lu LH. Using experimental studies and theoretical calculations to analyze the molecular mechanism of coumarin, *p*-hydroxybenzoic acid, and cinnamic acid. J Mol Struct. 2005;741:193–9.
22. Mrinmoy G, Prasenjit M, Tanaya G, Sumanta G, Basudeb B, Prasanta KM. 3,4-Dihydroxybenzoic acid isolated from the leaves of *Ageratum conyzoides* L. Eur J Biotechnol Biosci. 2013;1(4):25–8.
23. Abdel-Mogib M, Ezmirly M, Basaif SA. Phytochemistry of *Dipterygium glaucum* and *Capparis decidua*. J Saudi Chem Soc. 2000;4:103–8.
24. Kumrungsee N, Pluempanupat W, Koul O, Bullangpoti V. Toxicity of essential oil compounds against diamondback moth, *Plutella xylostella*, and their impact on detoxification enzyme activities. J Pest Sci. 2014;87:721–9.
25. Oppenoorth FJ. Localisation of the acetylcholinesterase gene in the housefly, Musca domestica. Entomologia Experimentalis et Applicata. 1979;25:115–7.
26. Gloria NSDS, Frances TTT, Francine DS, Grace G, Alexandre AS, Simone CBG. Larvicidal activity of natural and modified triterpenoids against *Aedes aegypti* (Diptera: Culicidae). Pest Manag Sci. 2016;72:1183–7.
27. Fonseca-González I, Quiñones MA, Lenhart A, Brogdon WG. Insecticide resistance status of *Aedes aegypti* (L.) from Colombia. Pest Mang Sci. 2011;67(4):430–7.
28. Kasai S, Komagata O, Itokawa K, Shono T, Ng LC, Kobayashi M, Tomita T. Mechanisms of pyrethroid resistance in the dengue mosquito vector, *Aedes aegypti*: target site insensitivity, penetration, and metabolism. PLoS Negl Trop Dis. 2014;8(6):1–23.

12

Foliar application of citric and malic acid to stock plants of rose alters the rooting of stem cuttings

Noushin Ghazijahani[1], Ebrahim Hadavi[2*], Moon S. Son[3] and Byoung R. Jeong[3]

Abstract

Background: As the world's best-selling cut flower, there is a high demand for propagated rose plants. Auxins are the dominant growth regulator used to improve the rooting of cutting; however, their application remains a labor-intensive process, and on the other hand, side effects like poor shoot growth exist. Organic acid sprays have enhanced many physiological traits of plants which justified surveying their possible effect on the rooting process in the current study.

Methods: Different concentrations of citric or malic acid were sprayed to stock plants of rose (*Rosa x hybrida* cv. 'Sherbet') prior to taking cuttings. Malic or citric acid (3, 6, or 9 mM) was applied five times on a 14-day interval in a commercial cut flower production greenhouse. Single-node cuttings were prepared and were rooted under mist. Data collected included rooting percentage, root count, root and shoot length and dry weight, and root-to-shoot ratio (R:S).

Results and discussion: The rooting percentage was enhanced by all treatments. Citric acid enhanced root number at 6 mM level followed by 3 mM citric acid. Average root length was increased by all treatments except 3 and 6 mM malic acid. All CA treatments increased shoot length, while it remained unaffected by MA treatments. CA at 3 and 6 mM level and 6 mM MA increased shoot dry weight, but only 9 mM CA increased the root dry weight.

Conclusion: Both organic acids had positive effects on rooting response of 'Sherbet' roses. However, the mechanism of observed responses to organic acids remains to be studied. While they enhance the physiological performance of stock plants, they can also simultaneously both enhance the rooting as well as the shoot growth of the new plantlets.

Keywords: Adventitious rooting, Foliar spray, Organic acids, Propagation, *Rosa* × *hybrida*, Root count

Background

The global import of cut flowers into the US totaled over 1 billion dollars in 2016 [1]. Cut roses are a significant component of imported flowers. Annual growth in the value of rose cut flowers between 2010 and 2014 was 6%, which was the highest among cut flowers [2]. To meet the growing demand, roses are commonly propagated by stem cuttings which are usually treated with a synthetic auxin to promote the formation of adventitious roots. However, applied auxins usually inhibit subsequent bud development of cuttings in apple [3] and rose [4, 5]. Wiesman et al. [6, 7] showed that applied indole-3-butyric acid (IBA) at the cutting base was transported to the upper part of the cutting. Sun et al. [5] concluded that after the applied auxin is transported to the upper part of the cutting, it causes increased ethylene production. Therefore, despite of adequate rooting, high concentrations of auxins may prevent shoot growth or cause the whole young shoots to abscise after the bud-break [8, 9].

The alternate foliar application of auxin on cutting has been tested in many studies and is reviewed by Blythe [10]. In studies on woody plants and some herbaceous species, it is concluded as ineffective [11, 12], while others usually working on herbaceous species could find

*Correspondence: hadavi@kiau.ac.ir
[2] Department of Horticultural Sciences, Islamic Azad University, Karaj Branch, Karaj, Iran

results that were comparable with basal application method [13–17]. However, Blythe reported that a single application of auxin on single-node cuttings of miniature rose did not affect on rooting [18]. The spray of the stock plant with an auxin, prior to cutting collection, has been tried before by Stoutemyer and O'Rourke [19] who sprayed the stock plants of several woody ornamentals with selected concentrations of the auxins and prepared cuttings from the treated plants after several days or weeks. Their results suggested that cuttings taken from the plants treated with auxins show essentially the same rooting responses as cuttings treated with generally accepted methods of application. Whatley et al. [20] sprayed stock plants of rose once 10 days before cutting collection using a solution containing NAA or soaked the cuttings in IBA for 15 min. They observed that cuttings receiving soak treatment rooted better than cuttings receiving the foliar spray, while both treatments rooted better than the non-treated cuttings. Alternative approaches to the problem are tested by Scagel [21], who showed that addition of vesicular–arbuscular mycorrhizal fungi (VAMF) to the rooting medium of two-node cuttings from five cultivars of miniature roses increased the rooting percentage and root count per cutting.

Endogenous organic acids are both a source of carbon skeletons for other biochemical pathways as well as an energy source for cells in the respiratory cycle [22]. In plants, malic acid (MA) is metabolized in mitochondria, acts as a common reserve anion in the plant vacuole, and is a counter ion for K and Ca cations, especially in nitrate-dependent plants [23–25]. Citrate and malate extruded by roots of calcicole plants (plants growing in alkaline soils) facilitate the extraction of phosphorus and iron from such soils [26]. Citric acid (CA) alone or in combination with iron sources has been sprayed on the foliage of plants for successful recovery from iron chlorosis [27–30]. Use of CA alone or in combinations with salicylic acid (SA) and MA increased the essential oil production of *Ocimum basilicum* L. and *Anethum graveolens* L. [31, 32]. Foliar application of CA extended the vase life of lily (*Lilium* spp.) and tuberose (*Polianthes tuberosa* L.) while improving drought tolerance in bean (*Phaseolus vulgaris* L.) [30, 33, 34]. Combination of 1 mM SA with 7 mM CA was superior to others for improving flowering and essential oil production in *Ocimum basilicum* L. [35]. The level of CA inside plant tissue is increased by exogenous application of CA and beneficial effects were observed, which were attributed to observed protective effects on peroxidation-linked membrane deterioration, free radical scavenging, maintenance of membrane stability, an increase in root activity, and activation of an antioxidant response [36]. The increase in peroxidase activity from CA application, which is responsible for H_2O_2 metabolism, has been documented [37]. IBA is responsible for an H_2O_2 increase, which promotes the formation of adventitious roots, possibly via a change in antioxidant enzymes activity; the same rooting response is reported with the addition of either ascorbic acid or H_2O_2 [38].

Considering the role of antioxidative enzyme systems in the adventitious rooting process and the reported effect of CA on it, we hypothesized a possible role for certain organic acids in the rooting process. The 'Sherbet' variety is among the important local varieties with an appealing white flower which was selected for the current study. The objective of this research was to evaluate the applications of CA and MA to stock plants on the rooting of rose cuttings.

Methods

Rose stock plants (*Rosa x hybrid*a cv. 'Sherbet') were grown in a commercial hydroponic cut rose production greenhouse near Gimhae (35°15′47.0″N 128°45′41.9″E), Korea. During the day, heating was applied when the air temperature in the greenhouse was lower than 16 °C (289 K) and ventilation was provided when higher than 30 °C (303). The plants depended on a natural light regime as no supplemental lighting was used. Starting on September 17, the plants were sprayed every 14 days (total of five times) with one of seven treatments. Treatments consisted of CA or MA (Sigma-Aldrich, St. Louis, MO, USA) each at 3, 6, or 9 mM and a distilled water (DW) control. A few drops of surfactant (Tween 20, Fisher Scientific, Hampton, NH) were added to all sprays. The sprays were applied until run-off. The experiment was conducted in a completely randomized design and each treatment was replicated three times. In each replication, there were ten stock plants per treatment (210 total stock plants). Shoot bending (training part of shoots to grow downward) was practiced and cut flower harvest was continued as usual. At 14 days after the fifth spray, the available unopened flower stems were harvested, placed in buckets containing distilled water, and were immediately transferred to a greenhouse of the Horticulture Department at Gyeongsang National University, Jinju, South Korea. Single-node cuttings were cut from the middle part of the shoots; each cutting had a single leaf from which the terminal leaflet was removed. Twelve cuttings (subsamples) were taken from each of the three replications (36 total cuttings per each treatment) and were put immediately in rock-wool cubes (5 cm by 5 cm by 5 cm; Grodan, Denmark) and placed in a fog propagation tunnel for 3 weeks and then under intermittent mist for another 3 weeks inside a research greenhouse at the same institution. The environmental condition maintained during rooting was $98 \pm 2\%$ relative humidity and 18–25 °C (298 K) with 40% shading of the natural sunlight. After a 6-week rooting period, the data were collected. Percentage of cuttings with evident roots (rooting percentage) and the number of roots per cutting were counted (root count) in each plantlet. Then, the roots were detached from each plantlet, and the mean and longest root length were measured in each. The shoots were detached as well and their length was determined. Finally, the detached shoots and roots were pooled for each replication and were subsequently dried in a dry oven at 70 °C (343) for 72 h, and the dry weight of each was measured. The collected data were analyzed for mean separation in 'multivariate' analysis module of SPSS (IBM Corp. Released 2012. IBM SPSS Statistics for Windows, Version 21.0. Armonk, NY), using Duncan's multiple range test option at $p < .05$ level and the correlations were determined by two-tailed Pearson correlation option of the same software package.

Results and discussion

Rooting percentage

The rooting percentage was 94.6% in the control treatment (DW), 97% in 3 mM CA, and 100% in the remaining treatments (Table 1). While much of organic acid treatments statistically show an improved rooting percentage, but the difference is not as much to be considered substantial. Our results support the conclusion by Pivetta et al. who reported that there is no need for IBA application for rooting of rose cuttings [39]. However, earlier reports had suggested the 'Sherbet' variety to be an easy-to-root type [40], though the rooting percentage is only the initial requirement of a successful propagation and could not predict the subsequent performance of plants (detailed experimental data for all traits is provided in Additional file 1).

Root count

This is a key index to judge on the rooting quality. CA at 3 and 6 mM had the greatest mean root count per cutting which was significantly higher than the control but on par with other organic acid treatments (Table 1). Root count was correlated significantly and positively with the mean root length and the longest root length, which suggests that increased root count did not compromise the root length (Table 2).

Longitudinal root and shoot growth parameters

Both mean root length and the longest root length were improved with treatments of CA 3 and 6 mM and MA 9 mM (Table 1). Interestingly, with the exception of 9 mM MA, a similar pattern was noted in shoot growth. As shown in Table 1, MA at 9 mM level caused retarded shoot growth that was even significantly lower than the control treatment. Only 9 mM CA increased root dry weight (Table 1).

Root/shoot ratio

There was a little difference in root/shoot ratio (R:S) between treatments. However, a significant positive correlation between R:S and root dry weights persists, which suggests that an increase in the root growth could have been the cause of increased R:S (Table 2), even though a significant negative correlation between shoot dry weight and R:S suggests that at least part of its increase could be

Table 1 Effect of citric acid (CA) and malic acid (MA) spray treatments on rooting of herbaceous cuttings in 'Sherbet' roses

Treatment	Rooting percentage (%)		Root count		Mean root length (mm)		Longest root length (mm)		Shoot length (mm)		Shoot dry weight (g)		Root dry weight (g)		Root/shoot ratio (DW)	
DW	94.6 (±4.62)	b	4.6 (±.6)	b	25.2 (±1.6)	b	39.2 (±6.6)	c	54.9 (±5.1)	bc	.090 (±.016)	c	.012 (±.002)	b	.14 (±.01)	ab
CA 3 mM	97 (±5.2)	ab	7.1 (±.4)	a	33.1 (±3)	a	51.9 (±7.5)	a	77 (±5.5)	a	.138 (±.022)	ab	.018 (±.002)	ab	.13 (±.03)	ab
CA 6 mM	100 (±0)	a	7.7 (±.8)	a	34.5 (±1.4)	a	51.1 (±3)	a	79.8 (±5.2)	a	.148 (±.033)	a	.017 (±.002)	ab	.12 (±.02)	ab
CA 9 mM	100 (±0)	a	6.5 (±.6)	ab	35.2 (±4.6)	a	48.7 (±3.2)	ab	73.7 (±2.8)	a	.124 (±.003)	abc	.023 (±.007)	a	.19 (±.06)	a
MA 3 mM	100 (0)	a	6.1 (±1.9)	ab	29.7 (±5.5)	ab	41.1 (±7.5)	bc	63 (±6.7)	ab	.113 (±.014)	bc	.017 (±.006)	ab	.15 (±.07)	ab
MA 6 mM	100 (0)	a	6.7 (±1.3)	ab	29.4 (±.5)	ab	47 (±.1)	bc	39.1 (±3.7)	cd	.151 (±.011)	a	.015 (±.005)	ab	.10 (±.03)	b
MA 9 mM	100 (0)	a	6.0 (±1.6)	ab	34.3 (±1.49)	a	49.3 (±1.6)	ab	34 (±23)	d	.096 (±.012)	c	.017 (±.003)	ab	.18 (±.03)	ab

Means with similar letters are not statistically different based on Duncan's test ($p < .05$)

The value in parentheses is the standard deviation

Table 2 Cross correlations between observed traits

	Root count	Mean root length (mm)	Longest root length (mm)	Shoot length (mm)	Shoot dry weight (g)	Root dry weight (g)	Root/shoot ratio (DW)	Rooting percentage (%)
Whole cutting fresh weight (g)	.612**	.599**	.352	.290	.668**	.387	−.085	.501*
Root count		.503*	.578**	.176	.604**	.594**	.149	.315
Mean root length (mm)			.798**	.304	.183	.441*	.268	.442*
Longest root length (mm)				.188	.274	.388	.183	.482*
Shoot length (mm)					.308	.210	−.058	.035
Shoot dry weight (g)						.147	−.498*	.286
Root dry weight (g)							.772**	.272
Root/shoot ratio (DW)								.081

*Correlation is significant at $p<.05$; **correlation is significant at $p<.01$

due to more partitioning of carbohydrates to root system rather than the shoot.

A significant positive correlation between root count and mean root length, longest root length, and root and shoot dry weight was observed (Table 2). This suggests that root count could be a key qualitative feature of rooting, which is correlated to other important rooting parameters.

Six and 9 mM MA created a distinct pattern of growth characterized by less longitudinal shoot growth than the longest root growth, which was opposite to what was observed in the rest of the treatments including the control. However, If we consider the effect of above-mentioned MA levels to be due to increased growth of root as compared to shoot we may expect to note an increase in weight of root or R:S ratio which is not the case as is evident in Table 1. Therefore, we may assume that the mentioned treatments reduced the shoots length, which was possibly compromised somehow to keep the weight unchanged.

In our experiment, all the treatments, except for 3 mM CA, caused improved rooting percentage, even though it was a slight improvement as compared to control. We conclude that MA treatments did not cause a statistically significant improvement in rooting response compared to the DW treatment. However, the mechanism of observed responses to organic acids remains unclear. Induction of citrate production by exogenous citrate is shown in microorganisms and later in plants which could describe the changes in internal homeostasis of organic acids as the possible key to further changes in plant metabolism pattern. Changes in the pattern of carbohydrate distribution in the plant [30, 35, 41] have been reported which could be effective in rooting potential of cuttings. Hu et al. [36] reported that endogenous citric acid levels in the leaf tissue of tall fescue (*Lolium arundinaceum* Schreb.) increased significantly following the external application of citric acid. This coincided with an increase in tolerance of plants to heat stress. They reported that protective effects on peroxidation-linked membrane deterioration, free radicals scavenging, maintenance of membrane stability, an increase in root activity, and activation of antioxidant response and HSP genes coincided with induced heat stress tolerance by citric acid. A similar response may be present in our study; however, more data is needed to understand the mode of action of organic acids on rooting process. Enzymatic, Proteomic and/or genomic approaches may help to better comprehend the pattern of changes exerted by applied foliar organic acid treatments.

Conclusion

By use of organic acids, we could avoid the reported negative effects of auxins on the subsequent shoot growth of treated plants [4, 5, 8, 9]. Organic acid sprays might serve as an alternative to hormonal treatments of rose stock plants. We have to bear in mind that, in addition to the reported effect on rooting in the current study, unlike most PGRs, these organic acids improve the growth and adaptation indices of plants as well which is mentioned earlier. Therefore, these organic acids are considered potential low-cost and environmentally friendly plant biostimulants, which can improve plant growth, adaptability, and rooting parameters as established here.

However, further work is needed to conclude on the commercial aspects of these agents for the nursery industry.

Authors' contributions
NG and MSS and BRJ have participated in study design and field operations, and the data were analyzed and the manuscript was prepared by EH. All authors read and approved the final manuscript.

Author details
[1] Department of Microbiology, Islamic Azad University, Karaj Branch, Karaj, Iran. [2] Department of Horticultural Sciences, Islamic Azad University, Karaj Branch, Karaj, Iran. [3] Department of Horticulture, Gyeongsang National University, Jinju 660-701, Republic of Korea.

Acknowledgements
The authors wish to thank C. H. Hwang and C. H. Ko for their kind help during the experiment.
Chemicals used in this study: citric acid and malic acid.
Species used in this study: Rose (Rosa hybrida L. 'Sherbet').

Competing interests
The authors declare that they have no competing interests.

Funding
The research was cofounded by IAU and GNU.

References
1. Imported: Fresh-cut flowers. 2017. https://www.ustradenumbers.com. Accessed 7 Jul 2017.
2. ITC. Market Analysis and Research, International Trade Centre (ITC). 2014.
3. Christensen MV, Eriksen EN, Andersen AS. Interaction of stock plant irradiance and auxin in the propagation of apple rootstocks by cuttings. Sci Hortic. 1980;12(1):11–7.
4. De Vries D, Dubois LA. The effect of BAP and IBA on sprouting and adventitious root formation of 'Amanda' rose single-node softwood cuttings. Sci Hortic. 1988;34(1):115–21.
5. Sun W-Q, Bassuk NL. Auxin-induced ethylene synthesis during rooting and inhibition of bud break of 'Royalty' rose cuttings. J Am Soc Hortic Sci. 1993;118(5):638–43.
6. Wiesman Z, Riov J, Epstein E. Comparison of movement and metabolism of indole-3-acetic acid and indole-3-butyric acid in mung bean cuttings. Physiol Plant. 1988;74(3):556–60.
7. Wiesman Z, Riov J, Epstein E. Characterization and rooting ability of indole-3-butyric acid conjugates formed during rooting of mung bean cuttings. Plant Physiol. 1989;91(3):1080–4.
8. Hartmann HT, Kester DE, Davis FT. Plant propagation: principles and practices. Englewood Cliffs: Prentice-Hall; 1990.
9. Bassil NV, et al. Propagation of hazelnut stem cuttings using *Agrobacterium rhizogenes*. HortScience. 1991;26(8):1058–60.
10. Blythe EK, et al. Methods of auxin application in cutting propagation: a review of 70 years of scientific discovery and commercial practice. J Environ Hortic. 2007;25(3):166–85.
11. Blythe EK, et al. Cutting propagation of foliage crops using a foliar application of auxin. Sci Hortic. 2004;103(1):31–7.
12. Tincker M. Further experiments with growth substances and the rooting of cuttings. J. Royal Hortic Soc. 1938;63:210–29.
13. Blythe EK, et al. Evaluation of an alternative method of rooting hormone application in cutting propagation. Comb Proc Intl Plant Prop Soc. 2002;52:393–9.
14. Hildreth A, Mitchell J. Spraying is a new method of applying root-promoting substances. Florists' Rev. 1939;84:13.
15. Mitchell JW, Marth PC. Growth regulators for garden, field, and orchard. Chicago: The University of Chicago Press; 1947.
16. Kroin J. Advances using indole-3-butyric acid (IBA) dissolved in water for rooting cuttings, transplanting, and grafting. In: Combined proceedings-international plant propagators society. Univ Washington-Int Plant Propagation Soc; 1992.
17. Blythe EK, et al. Foliar application of auxin for rooting stem cuttings of selected ornamental crops. J Environ Hortic. 2003;21(3):131–6.
18. Blythe EK, et al. Rooting of rose cuttings in response to foliar applications of auxin and surfactant. HortTechnology. 2004;14(4):479–83.
19. Stoutemyer V, Orourke F. Rooting of cuttings from plants sprayed with growth regulating substances. In: Proceedings of the American society for horticultural science. Amer Soc Horticultural Science 701 North Saint Asaph Street, Alexandria; 1945.
20. Whatley B, Thompson S, Williams G Jr. The effects of nutrient solution, foliar spray, 3-indolebutyric acid and dimethyl sulfoxide (DMSO) on rooting of hibiscus cuttings. Comb Proc Intl Plant Prop Soc. 1966;16:287–90.
21. Scagel C. Cultivar specific effects of mycorrhizal fungi on the rooting of miniature rose cuttings. J Environ Hortic. 2001;19(1):15–20.
22. da Silva JAT. The cut flower: postharvest considerations. J Biol Sci. 2003;3:406–42.
23. Day DA, Hanson JB. Pyruvate and malate transport and oxidation in corn mitochondria. Plant Physiol. 1977;59(4):630–5.
24. Osmond C, Laties GG. Compartmentation of malate in relation to ion absorption in beet. Plant Physiol. 1969;44(1):7.
25. Ting I. Towards a model for malate accumulation in plant tissues. Plant Sci Lett. 1981;21(3):215–21.
26. Lopez-Bucio J, Nieto-Jacobo MI. Organic acid metabolism in plants: from adaptive physiology to transgenic varieties for cultivation in extreme soils. Plant Sci. 2000;160(1):1–13.
27. Tagliavini M, et al. Agronomic means for the control of iron deficiency chlorosis in deciduous fruit trees. J Plant Nutr. 2000;23(11–12):2007–22.
28. Abadía J, et al. Organic acids and Fe deficiency: a review. Plant Soil. 2002;241(1):75–86.
29. Álvarez-Fernández A, et al. Foliar fertilization to control iron chlorosis in pear (*Pyrus communis* L.) trees. Plant Soil. 2004;263(1):5–15.
30. Eidyan B, Hadavi E, Moalemi N. Pre-harvest foliar application of iron sulfate and citric acid combined with urea fertigation affects growth and vase life of tuberose (*Polianthes tuberosa* L.)'Por-Par'. Hortic Environ Biotechnol. 2014;55(1):9–13.
31. Jafari N, Hadavi E. Growth and essential oil yield of dill (*Anethum graveolens*) as affected by foliar sprays of citric acid and malic acid. Acta Hortic. 2012;955:287–90.
32. Jaafari N, Hadavi E. Growth and essential oil yield of basil (*Ocimum basilicum* L.) as affected by foliar spray of citric acid and salicylic acid. Z Arznei Gewurzpfl. 2012;17(2):80–3.
33. Darandeh N, Hadavi E, Shoor M. Effect of $FeSO_4$, methanol and H_2SO_4 sprays on chlorophyll content of lilium'Brunello'. Acta Hortic. 2010;900:217–21.
34. El-Tohamy W, et al. Drought tolerance and water status of bean plants (*Phaseolus vulgaris* L.) as affected by citric acid application. J Appl Bot Food Qual. 2013;86(1):212–6.
35. Mirzajani Z, Hadavi E, Kashi A. Changes in the essential oil content and selected traits of sweet basil (*Ocimum basilicum* L.) as induced by foliar sprays of citric acid and salicylic acid. Ind Crops Prod. 2015;76:269–74.
36. Hu L, et al. Exogenous application of citric acid ameliorates the adverse effect of heat stress in tall fescue (*Lolium arundinaceum*). Front Plant Sci. 2016;7:179.
37. Cheniany M, et al. Effect of endogenous phenols and some antioxidant enzyme activities on rooting of Persian walnut (*Juglans regia* L.). Afr J Plant Sci. 2010;4(12):479–87.

38. Li S-W, et al. IBA-induced changes in antioxidant enzymes during adventitious rooting in mung bean seedlings: the role of H_2O_2. Environ Exp Bot. 2009;66(3):442–50.
39. Pivetta K, et al. Effect of type of cuttings and indolbutyric acid on the rooting of rose (*Rosa* sp.' Red success') leafy cuttings during two seasons. Acta Hortic. 1997;482:333–8.
40. Hwang CH, et al. Rooting and growth affected by physiological condition of the stock plant for cutting propagation of roses. Propag Ornam Plants. 2015;15(3):95–100.
41. Talebi M, Hadavi E, Jaafari N. Foliar sprays of citric acid and malic acid modify growth, flowering, and root to shoot ratio of gazania (Gazania rigens L): a comparative analysis by ANOVA and structural equations modeling. Adv Agricul. 2014;2014:147278. https://doi.org/10.1155/2014/147278.

13

Exclosures backed up with community-based soil and water conservation practices increased soil organic carbon stock and microbial biomass carbon distribution, in the northern highlands of Ethiopia

Mengistu Welemariam[1*], Fassil Kebede[2], Bobe Bedadi[1] and Emiru Birhane[3,4]

Abstract

Background: Land degradation is a treat for natural resources in Tigray high lands of northern Ethiopia, where 30–50 percent of the soil productive capacity has been lost in the past 500 years. Restoration and management of degraded lands improve soil health through enhancing soil organic carbon (SOC) stock and microbial biomass carbon (MBC). The knowledge on SOC and MBC concentration and distribution is essential to refine soil management, thereby restoring the ecosystem. This paper quantified the effect of decades old community-based soil and water conservation (SWC) measures, mainly stone terraces, exclosure with and without stone terraces, and non-conserved communal grazing lands on the distribution of MBC and SOC stock.

Methods: Soil sample collection was carried out using systematic sampling design. Transects parallel to each other and to the slope of the landscape were established. In each transect, three landscape positions (i.e., upper, middle, and foot slope) were formed. Composite soil samples were taken from four corners and center of 10 m × 10 m plot of each slope positions under the different SWC measures. Analysis of variance was used to determine the difference in SOC and MBC using SAS 9.2.

Results: Total soil organic carbon concentration was significantly higher in exclosures as compared to terraces and non-conserved grazing lands. The highest mean value of SOC stock (29 Mg C ha^{-1}) was recorded in exclosures with terraces followed by exclosures without terraces (24 Mg C ha^{-1}) and terraces (21 Mg C ha^{-1}), while the lowest (16 Mg C ha^{-1}) was recorded in non-conserved communal grazing lands. Exclosures with terraces improved SOC stock by 64%, followed by exclosures without terraces by 37%, while terraces improved the SOC stock by 25% compared to non-conserved open communal grazing lands in the last 20 years. The upper (0–15 cm) soil depth had significantly ($P < 0.05$) higher (24 Mg C ha^{-1}) SOC stock than the lower (15–30 cm) soil depth (20 Mg C ha^{-1}). Microbial biomass carbon was the highest (640 mg kg^{-1} soil) in exclosures without terraces, followed by exclosures with terraces, (570 mg kg^{-1} soil), terraces (440 mg kg^{-1} soil), and non-conserved communal grazing lands (370 mg kg^{-1} soil).

Conclusion: Exclosures supported with terraces improved and restored the SOC stock and microbial biomass carbon of degraded free grazing lands in the highlands.

Keywords: Conservation, Ethiopia, Land degradation, Microbial biomass carbon, Soil carbon stock

*Correspondence: mengistuwel@gmail.com
[1] College of Agriculture and Environmental Sciences, Haramaya University, P.O.Box 138, Dire Dawa, Ethiopia
Full list of author information is available at the end of the article

Background

Ethiopian highlands have been degraded due to cultivation and deforestation that last for more than 2500 years [1] and around 50 percent of rural highland areas are classified as degraded [2, 3]. The process of land degradation was accelerated due to high human and livestock population pressures; and steep topography [4] in combination with the highly variable (20–40%) and unreliable rainfall, where 70–90% of the rainfall occurs during July and August [1]. The trend is mainly common in Tigray highlands in northern Ethiopia [5]. Overexploitation and mismanagement of these highlands caused serious soil erosion, loss of soil quality, and soil organic carbon [2, 4].

Stone-faced terraces, enforcement of grazing restrictions, and plantation development efforts were implemented as rehabilitation measures since 1970s [3, 6] to tackle land degradation. Majority of the farmers (75%) practice terrace construction on their land [7]. Exclosures have also been important rehabilitation measures to increase biomass production [8].

Area closures (exclosures) were established in the region since 1991 to improve biomass production [9]. They are implemented on biophysically degraded communal grazing lands [8]. Exclosures reduce soil disturbance by decreasing grazing pressure and restricting people's access of uncontrolled cutting of trees and grass for fuel and fodder [10].

Soil management and conservation practices influence soil health through their effect on soil microbiota [11, 12]. Organic carbon and microbial biomass are among the biological indicators describing soil health [13].

Soil organic carbon contains two times as much carbon as the atmosphere and 2.5 times as that of soil biota [14]. It plays a significant role in soil functioning and the global carbon cycle [15] mainly by reducing atmospheric carbon dioxide, and helps to alleviate the problem of global warming and climate change [16]. It also supports biological activity and dictates the physical properties of soil that determine its resistance to erosion [12, 17].

The decline of SOC pool leads to soil quality degradation and reduces biomass production. The loss is enhanced by erosion and other degradation processes [18]. A practical option to increase SOC stock is through SOC sequestration and decrease soil degradation [2] by implementing hillside terraces combined with enrichment plantation in exclosures [9].

Soils and the microbes that live in them regulate the carbon storage, nutrient cycling, and climate change mitigation [19]. Soil microbial biomass (SMB) is responsive to environmental changes [20]. It indicates the soil's capacity to perform ecosystem processes [21]. It also acts as biological early indicator of changes [22, 23]. It makes up 1–5% of total SOC [24] and acts as the most active component in the biochemical process of SOC turnover [25]. A little change in SMB affects directly ecosystem stability [26].

Soil microbial biomass is more sensitive to changes in soil management compared to soil organic matter [27]. Determination of SMB can help to quantify the extent of land degradation and can provide methods for the restoration of degraded ecosystems [28]. The MBC to SOC ratios are also useful measures to manage soil organic carbon content [29].

Restoration and management of degraded lands significantly contribute to enhanced SOC [30] and MBC [31] which are considered as soil health indicators during restoration process [32]. They vary spatially under different management and plant cover [21], though studies that examine the distribution of these soil properties are limited in the area [33].

Quantifying and determining the distribution of SOC stock and MBC are important for site-specific sustainable soil management and to provide a valuable basis for subsequent measurements [15, 34]. Effective carbon sequestration methods can also be implemented [35]. Thus, the objective of this study was to determine the effect of community-based SWC measures mainly terraces and exclosures on distribution of SOC stock and microbial biomass carbon.

The research questions considered include (1) Did the development of terraces on free grazing lands increase concentration of SOC, SOC stock, and microbial biomass carbon? (2) Did the protection of free grazing lands through exclosures significantly increase concentration of SOC, SOC stock, and microbial biomass carbon? (3) Could the support of exclosures with physical SWC measures bring a significant increase in concentration of SOC, MBC, and SOC stock?

Methods

Description of the study area

The study was conducted in Degua Temben district, which is located 50 km west of Mekelle, regional capital of Tigray region, northern Ethiopia. Geographically, it is located at 13°16′23″ to 13°47′44″ Latitude and 39°3′17″ to 39°24′48″ Longitude (see Fig. 1).

The lithology of the study area comprises Mesozoic sedimentary rocks and Tertiary basalt [36]. Soils of the study sites are developed from calcium carbonate-rich parent material of the Agula shale formation, which consists mainly of marble and limestone [37]. According to world reference base [38] soil classification system, *Calcaric Cambisols, Vertic Leptosols, Vertic Cambisols,* and *Lithic Leptosols* are the dominant soil types.

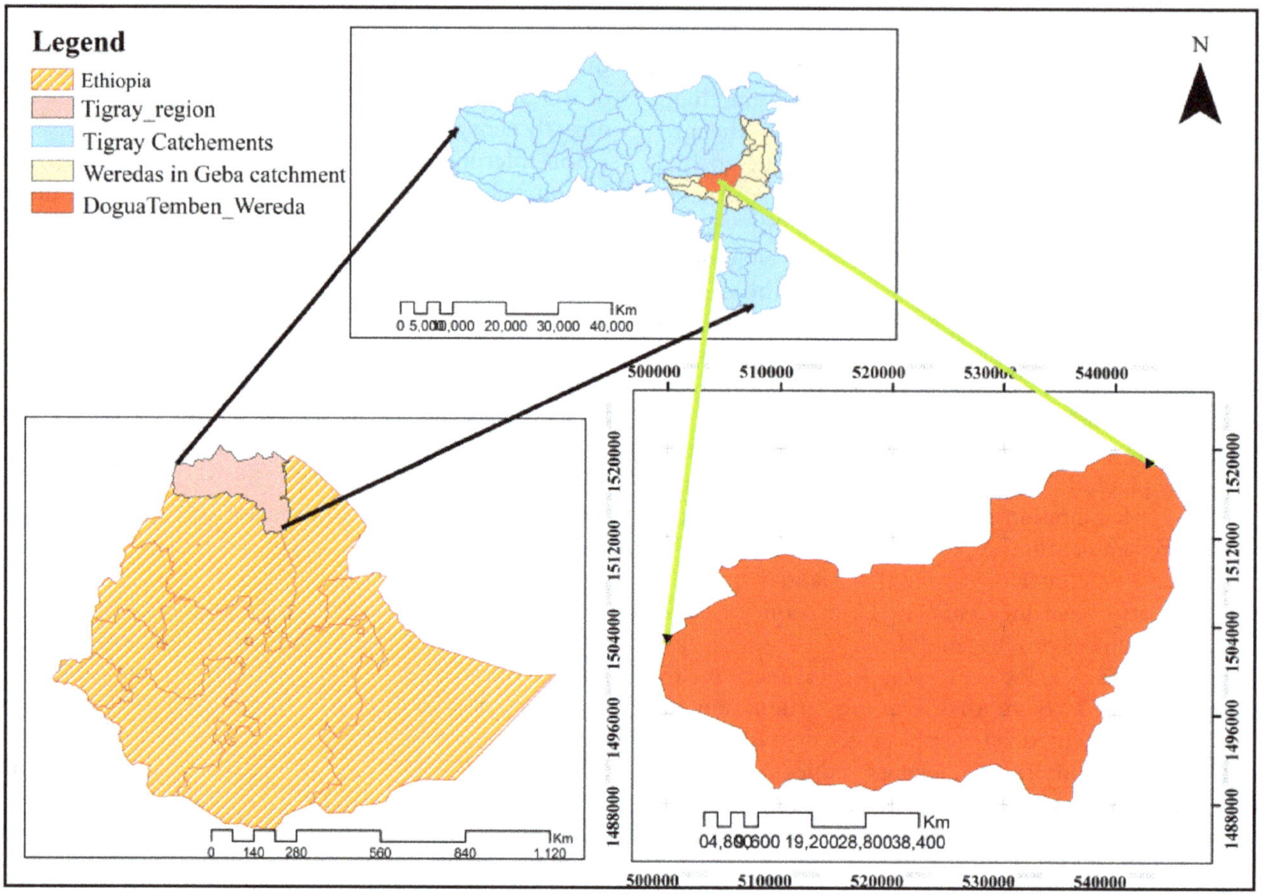

Fig. 1 Location map of the study area

The area receives annual rainfall of 290–900 mm year^{-1} with an average value of 615 mm year^{-1}. The main rainy season is June–August and stops in September. All the study sites are classified as mid-altitude (1800–2200 m above sea level) according to the traditional agro-climate classification system of the country.

Acacia etbaica, Carissa edulis, Dodonaea angustifolia, Stereospermum kunthianum, Rhus Vulgaris, and *Euclea racemosa* are commonly found woody vegetation species in exclosures and in communal grazing lands. The understory vegetation is also dominated by a diverse assemblage of grass and herbs which are palatable to livestock.

Mixed farming system (crop and livestock) is the main of livelihood in the study area. Cultivated lands, forest land, exclosure, and communal grazing lands are the major land uses. Barley (*Hordeum vulgare*), tef (*Eragrostis tef*), wheat (*Triticum aestivum*), and Maize (*Zea mays*) were the major crops in the area.

The site was selected, because it has an extensive exclosures system and terraces [9] with comparable edaphic and topographic features. The common community-based soil and water conservation measures (CBSWC) were terraces and exclosures. These soil and water conservation measures were accepted and practiced by the community for more than two decades.

Characteristics of the main community-based SWC measures

The most commonly accepted and practiced SWC measures (i.e., terraces and exclosures with and without terraces) were established since 1997 by the community. Many of the SWC structures constructed are fully owned by the communities. This has contributed towards ensuring their sustainability [39]. Before their establishment, the selected conserved with CBSWC grazing lands had similar history in terms of grazing with the non-conserved communal grazing lands.

Each of the three selected sites (Kerano, Tesemat, and Alasa) was categorized into four management units described as terraced grazing land, exclosure with terrace, exclosure alone, and non-conserved open communal grazing land. Exclosure with terrace is restricted from the interference of animals, biological and physical SWC measures, mainly stone terraces were commonly implemented in exclosures. Accumulation of sediment, grasses, and litter fall was common on terraces. More woody species are observed compared to the other SWC measures.

In the case of exclosures without SWC, there was no interference of human and livestock. Besides, no other management practices such as physical structures were observed. Trees regenerate naturally and hence have a better vegetation cover than terraced open grazing lands. Erosion types such as sheet, rill, and gully formation are relatively less common compared to terraces and non-conserved communal grazing lands.

The stone terraces in grazing lands were selected as the third management unit, because they are relatively more stable and durable measures than other physical SWC measures. Terraced grazing lands had more accumulated sediments, organic matter, and vegetation cover. Sheet, rill, and gully erosion are less common compared to the open communal grazing lands.

The open, non-conserved communal grazing land was characterized by the low vegetation cover and higher proportions of bare soil with high stone cover. Sheet, rill, and gully erosion were common. It is assumed that the terraces, exclosures, and non-conserved grazing lands had comparable initial conditions at the time of terrace construction and exclosure establishment. Changes in soil SOC and MBC are assumed to be as a consequence of terraces and exclosures establishment.

Sampling technique and sample size

The study was conducted in three nearby sites (Kerano, Tesemat, and Alasa with in the district) and having all the SWC measures. In each SWC measures, three transects separated at a minimum distance of 75 m were established. Transects were parallel to each other and to the topography of the landscape. In each transect, three landscape positions (i.e., upper, middle, and foot slope) were established. The upper slope (US) position is the uppermost portion of each study site, and it can receive little or no overland flow but may contribute runoff to down slope areas. The middle slope (MS) position receives overland flow from the upper slope and contributes runoff to the foot slope (FS). The FS represents the lowest part of each study site, and receives overland flow from both mid and upper slopes [40].

Soil samples were collected from the top 0–15 and 15–30 cm at four corners and center of a 10 m × 10 m size plot using "*X*" sampling design from terraces, exclosure with terrace, exclosure without terraces, and from non-conserved communal grazing lands from three transects spaced at a minimum distance of 75 m [41].

A total of 108 soil samples (i.e., four conservation measures*three slope positions * three samples * three replications) from surface (0–15 cm) for microbial biomass and 216 soil samples (i.e., four conservation measures * three slope positions*three samples * two depths*three replications/sites) from surface (0–15 cm) and subsurface (15–30 cm) were collected for soil organic carbon stock determination.

Determination of soil microbial biomass carbon

The soil samples were sieved through a 2 mm mesh to remove stones, roots, and large organic residues, and then sealed in plastic bags and stored at 4 °C. Soil microbial biomass-C (Cmic) was determined using the substrate-induced respiration (SIR) method [42]. The samples were immediately taken to the laboratory for MBC determination. In the laboratory, 100 g of the field moist soil was thoroughly mixed with 400 mg glucose, and 20 g of the mixture was weighed to four nylon bags. Each bag was mounted into a 500 ml laboratory bottle filled with 20 ml 0.1 M NaOH solution. The bottles were immediately closed with gas-tight caps, and incubated 4 h at 22 °C together with two blanks (without soil samples) for each sample group.

After incubation, the soil samples were immediately removed from the bottles and the absorbed CO_2 was precipitated as barium carbonate by adding 2 ml of 0.5 M barium chloride solution. The remaining sodium hydroxide was titrated against 0.1 M HCl using 3–4 drops of phenolphthalein solution as indicator. The result was calculated according to Schuman et al. [43] (Eq. 1):

$$\mathrm{mg\,CO_2\,100\,g^{-1}h^{-1}} = \frac{(B-S)*2.2*100}{4*\mathrm{SW}*\mathrm{dm}}, \tag{1}$$

where B = mean volume of HCl consumed by blanks (ml), S = mean volume of HCl consumed by samples (ml), 4 = Incubation time (h), 100 = conversion factor (100 g dm), 2.2 = conversion factor (1 ml 0.1 M HCl corresponds to 2.2 tolerant mg CO_2), SW = initial soil weight (g), dm = soil dry matter (%). Assuming a respiratory quotient of 1.1 mg CO_2, 0.100 g^{-1} dm. h^{-1} corresponds to 20.6 mg biomass-C 0.100 g^{-1} dm. Thus, this factor was used to convert the CO_2 into MBC.

Determination of total organic carbon, bulk density, and coarse fragment

Total organic carbon was determined by the wet digestion using potassium dichromate [44]. The microbial quotient was determined as the ratio of MBC/SOC ratio after converting the percentage of SOC carbon into mg 100 gm^{-1} dry soil. Bulk density was determined by core method. Coarse fragment was determined as percentage weight of soil greater than 2 mm.

Determination of soil organic carbon stock

The carbon held in the upper profile is often the most chemically decomposable, and the most directly exposed to natural and anthropogenic disturbances [45]. Therefore, soil organic carbon pool was estimated up to the depth of 30 cm in this study.The data for SOC pool were calculated using the following equation:

$$\text{SOC stock} = \text{SOCi} * \text{BD} * \text{di} * (1 - \text{CFi})/100, \quad (2)$$

where SOCi = soil organic carbon of a given soil depth, Mg C ha^{-1}; BD (Bulk density) = soil mass per sample volume, kilogram soil m^{-3} (equivalent to kg m^{-3}); di = horizon depth or thickness of soil layer, m; CFi = % volume of coarse fragments/100, dimensionless. The rate of change of total organic carbon and SOC stock over time was computed by subtracting the stocks of free grazing land from that of terraces and exclosures and dividing it by the number of years, since the terraces and exclosures were established.

Data analysis

The mean, minimum, maximum, standard error, and coefficient of variation (CV) were determined using descriptive analysis of the original data variables. The significance of the difference between means of each SOC, MBC, and MBC/SOC ratio variable for different SWC measures was carried out by the analysis of variance (ANOVA) using SAS 9.2 [46]. A pairwise mean comparison was done by Duncan Least Significant Difference (LSD). Pearson correlation coefficient was also used for observing any associations between the parameters.

Results

Soil bulk density and organic carbon concentration among the community-based soil and water conservation measures

There was a significant ($P < 0.05$) variation in SOC among the soil and water conservation measures specifically between exclosures, and grazing lands with and without terraces. The highest mean value of SOC (2.9%) was found on exclosures supported with terraces followed by exclosures without terraces (2.8%) and terraces (2.5%), respectively (Table 1). The lowest mean value of SOC (1.9%) was found in non-conserved communal grazing lands. The SOC content followed the order exclosures > stone terraces > non-conserved open communal grazing lands.

Free grazing lands and terraced land had relatively higher BD as compared to both types of exclosures (Table 1). The bulk density of the area was in the order of exclosures without terraces < exclosures with terraces < terraces alone < non-conserved communal grazing lands. The coefficient of variation (CV) for SOC and BD along the different SWC measures was between 17 and 27% for organic carbon and 10–18% for BD, respectively (Table 2). Non-conserved grazing lands had relatively higher percentage of coarse fragments, while exclosures with terraces had significantly lower percentage of coarse fragments as compared to the other soil and water conservation measures (Table 1).

Effect of soil depth and slope position on total soil organic carbon and bulk density

Soil depth resulted in a significant variation in SOC content (Table 3). The upper soil layer (0–15 cm depth) had significantly ($P < 0.05$) higher SOC than the lower one (15–30 cm depth). Soil BD was relatively lower in the upper than lower soil depth. On the other hand, coarse fragment was relatively higher on the upper than lower soil depth. The total SOC and BD in response to soil depth had a coefficient of variation between 20 and 26% for organic carbon and 14 for BD (Table 4), which is a moderate variation. In most of the community-based SWC measures, BD and SOC showed inconsistent variation among slope positions with some increasing trend at lower slope position (Table 5).

Distribution of soil organic carbon stock along the community-based soil and water conservation measures

The SOC stock (Mg ha^{-1}) in exclosures was significantly higher ($P < 0.05$) as compared with the other SWC measures (Table 1). Exclosures with terraces had the highest

Table 1 Effect of community-based soil and water conservation measures on soil properties (Mean ± SEM)

Soil properties	Non-conserved grazing lands	Terraces	Exclosures + terraces	Exclosures alone
SOC (%)	1.9 ± 0.1^c	2.5 ± 0.1^b	2.9 ± 0.1^a	2.8 ± 0.1^a
Bulk density (kg m^{-3})	1280 ± 32^a	1260 ± 28^a	1240 ± 21^a	1220 ± 48^a
Coarse fragment (%)	57.6 ± 0.2^b	53.2 ± 0.2^b	47.4 ± 0.2^a	54.4 ± 0.2^b
SOC stock (Mg ha^{-1})	16.0 ± 1.0^c	21.0 ± 1.3^b	29.0 ± 1.4^a	24 ± 1.3^b

Means followed by the same letter along each column do not differ significantly at $P \leq 0.05$

Table 2 Summary of statistics for SOC, bulk density, and SOC stock in response to SWC measures

SWC	Variable	Mean	Median	Minimum	Maximum	Std Error	CV	Skewness	Kurtosis
Non-conserved grazing land	Organic carbon (%)	1.9	2.1	0.7	3.2	0.1	27	−0.4	0.2
	Bulk density (kg m^{-3})	1284	1362	770	1497	25	14	−0.9	0.0
	SOC stock (Mg ha^{-1})	16	16	3	31	1.0	45	0.2	−0.7
Terracing	Organic carbon (%)	2.5	2.5	1.3	3.6	0.1	17	0.0	0.5
	Bulk density (kg m^{-3})	1260	1288	850	1485	21	12	−0.9	0.4
	SOC stock (Mg ha^{-1})	21	21	6	40	1.0	37	0.1	0.0
Exclosure + terracing	Organic carbon (%)	2.9	2.9	1.3	3.8	0.1	20	−0.4	0.4
	Bulk density (kg m^{-3})	1244	1250	1011	1579	18	10	0.3	−0.4
	SOC stock (Mg ha^{-1})	29	29	9	43	1.0	27	−0.3	−0.5
Exclosure	Organic carbon (%)	2.8	2.8	1.5	3.8	0.1	18	−0.4	0.1
	Bulk density (kg m^{-3})	1220	1308	690	1491	30	18	−0.9	−0.1
	SOC stock (Mg ha^{-1})	24	23	8	52	1.0	39	0.7	0.6

SOC stock (29 Mg ha^{-1}) followed by exclosures without terraces (24 Mg ha^{-1}).

Stone terraces had significantly ($P < 0.05$) higher SOC stock (21 Mg ha^{-1}) than non-conserved communal grazing lands (16 Mg ha^{-1}). Comparison of the effect of soil depth on soil organic carbon stock (Table 3) also showed significantly higher (24 Mg ha^{-1}) in the upper (0–15 cm) and was lower (20 Mg ha^{-1}) in the lower (15–30 cm) soil depth.

Soil organic carbon stock in the different soil and water conservation measures (Table 2) had a coefficient of variation ranging from 27 to 45% which lies under the moderate variability category. The CV for soil organic carbon stock ranged between 38 and 43% in the upper and lower depths, respectively (Table 4), which is also considered to be moderate variation in response to soil depth.

Distribution of soil microbial biomass carbon along the community-based soil and water conservation measures

Exclosures had significantly higher MBC than terraces and non-conserved communal grazing lands (Table 6). The highest value of MBC (640 mg kg^{-1} dry soil) was found in exclosures without terraces followed by exclosures with terraces (570 mg kg^{-1} dry soil) and terraces (440 mg kg^{-1} dry soil), respectively (Table 6). The lowest MBC (370 mg kg^{-1} dry soil) was recorded in non-conserved communal grazing lands. The microbial biomass was in the order of exclosures without terraces > exclosures with terraces > terraces > non-conserved communal grazing lands.

Table 3 Effect of soil depth on total SOC, bulk density, coarse fragment, and SOC stock (Mean ± SEM)

Soil parameters	Depths (cm)	
	0–15 cm	15–30 cm
Organic carbon (%)	2.8 ± 0.1^a	2.2 ± 0.1^b
Bulk density (kg m^{-3})	1244 ± 23^a	1258 ± 25^a
Coarse fragment (%)	55 ± 1.0^a	52 ± 2.0^a
Soil carbon stock (Mg ha^{-1})	24 ± 1.0^a	20 ± 1.0^b

Means of each soil parameter followed by the same letter along each column do not differ significantly at $P \leq 0.05$

The microbial quotient was higher in exclosures than terraces and non-terraced communal grazing land (Table 6). It was higher in exclosures without terraces than exclosures with terraces.

Microbial biomass contents were inconsistent and not significant along the three slope positions of the SWC measures. However, the highest (720 mg kg^{-1}) MBC was found at foot slope in exclosures without terraces, while the lowest (370 mg kg^{-1}) was found at upper slope in non-conserved communal grazing lands (Table 7). As indicated in (Table 8), the CV value for MBC in response to SWC ranged 15–43% which is a moderate variability

Discussion

Exclosures with terraces had significantly higher SOC as compared to the other SWC measures, which could be due to low disturbance, better vegetation cover, and deposition of soil materials. Stone terraces when supported with vegetation enhance soil organic matter accumulation as compared to non-supported ones [47]. Returns of biomass and reduced erosion in exclosures enhanced SOC concentration in arid environments [43, 48].

The presence of significantly lower SOC concentration in non-conserved grazing lands could be due to the fact that grazing reduces vegetation cover and primary production, which in turn exposes the soil to increased erosion losses and reduced carbon inputs [41]. Grazing lands

Table 4 Summary of statistics for total SOC, bulk density, and SOC stock in response to soil depth

Depth (cm)	Variable	Mean	Median	Minimum	Maximum	Std error	CV	Skewness	Kurtosis
0–15	Organic carbon (%)	2.8	2.8	1.3	3.8	0.1	20	0.0	−0.5
	Bulk density (kg m^{-3})	1258	1275	740	1579	17	14	−1.0	1.0
	SOC stock (Mg ha^{-1})	24.2	23.2	5.8	52	0.1	38	0.3	-0.3
15–30	Organic carbon (%)	2.3	2.3	0.7	3.7	0.1	26.2	0.4	0.0
	Bulk density (kg m^{-3})	1247	1260	690	1483	17	14	−0.7	0.1
	SOC stock (Mg ha^{-1})	20	21	3	42	1.0	43	0.2	−0.4

Table 5 Effect of slope position on SOC concentration, SOC stock, and soil bulk density (Mean ± SEM)

SWC measures	Soil parameter	Slope positions		
		Foot slope	Middle slope	Upper slope
Non-conserved grazing lands	SOC (%)	1.9 ± 0.1^{c}	2.1 ± 0.1^{ce}	1.7 ± 0.1^{c}
	Bulk density (kg m^{-3})	1199 ± 40^{bcd}	1310 ± 40^{ad}	1315 ± 40^{a}
	SOC stock (Mg ha^{-1})	18 ± 2^{ac}	15 ± 2^{c}	15 ± 2^{c}
Terraces	SOC (%)	2.6 ± 0.1^{abd}	2.4 ± 0.1^{bd}	2.4 ± 0.1^{de}
	Bulk density(kg m^{-3})	1262.9 ± 40^{abcd}	1276.6 ± 40^{acd}	1240.8 ± 40^{abcd}
	SOC stock (Mg ha^{-1})	22 ± 2^{ab}	20 ± 2^{ac}	22 ± 2^{ab}
Exclosures + terraces	SOC (%)	2.9 ± 0.1^{a}	2.8 ± 0.1^{a}	2.9 ± 0.1^{a}
	Bulk density(kg m^{-3})	1218 ± 40^{abcd}	1233 ± 40^{abcd}	1231 ± 40^{abcd}
	SOC stock (Mg ha^{-1})	31 ± 2^{de}	26 ± 2^{bde}	30 ± 2^{e}
Exclosures alone	SOC (%)	2.7 ± 0.1^{a}	2.7 ± 0.1^{ab}	2.8 ± 0.1^{ab}
	Bulk density (kg m^{-3})	1315 ± 40^{a}	1162 ± 40^{b}	1184 ± 40^{bc}
	SOC stock (Mg ha^{-1})	25 ± 2^{bd}	22 ± 2^{ab}	23 ± 2^{ab}

Means of each soil parameter followed by the same letter do not differ significantly at $P \leq 0.05$

Table 6 Effect of community-based SWC measures on total values of MBC and MBC/SOC ratio (Mean ± SEM)

SWC measures	MBC (mg kg^{-1} soil)	MBC/SOC (%)
Non-conserved grazing lands	370 ± 10^{b}	1.9 ± 0.1^{a}
Terraces	440 ± 30^{b}	1.8 ± 0.2^{a}
Exclosures with terraces	570 ± 40^{a}	2.1 ± 0.1^{a}
Exclosures without terraces	640 ± 50^{a}	2.3 ± 0.2^{a}

Means followed by the same letter across each row do not differ significantly at $P \leq 0.05$

Table 7 Effect of slope position on microbial biomass carbon (mg kg^{-1} soil) (Mean ± SEM)

SWC measures	Slope positions		
	Foot slope	Middle slope	Upper slope
Non-conserved grazing lands	380 ± 70^{bc}	350 ± 70^{b}	370 ± 70^{bc}
Terraces	460 ± 70^{abc}	450 ± 70^{abc}	400 ± 70^{abc}
Exclosures + terraces	550 ± 70^{acd}	560 ± 70^{ad}	580 ± 70^{ad}
Exclosures without terraces	720 ± 70^{d}	560 ± 70^{ad}	640 ± 70^{d}

Means followed by the same letter do not differ significantly at $P \leq 0.05$

with terraces had higher SOC than non-conserved communal grazing lands. This could show that terraces store more organic carbon from the materials that they trap during deposition. Amdemariam et al. [49] reported high SOC in conserved fields compared to non-conserved fields in north western part of Ethiopia. Similarly, Hishe et al. [50] found higher SOC in conserved land scape than non-conserved landscape in Suluh valley of northern highlands Ethiopia. These results indicated practicing soil conservation measures improved soil organic carbon [49].

Non- conserved grazing lands had relatively higher bulk density as compared to exclosures and terraces. This is in line with the result reported by Mekuria and Veldkamp [49], where non-conserved fields exhibit higher mean BD than the conserved fields. This could be due to low soil organic carbon and high soil compaction by livestock grazing in non-conserved open communal grazing lands. The results of other studies also clearly showed the important role of organic matter for rehabilitating degraded soils by improving other soil properties [51, 52]. The BD decreased 1.28–1.26 g cm^{-3} due to terrace

Table 8 Summary statistics on analysis of MBC (mg kg^{-1} soil) in response to SWC measures

SWC	Mean	Median	Minimum	Maximum	Std error	CV	Skewness	Kurtosis
Non-conserved grazing lands	370	370	210	460	10	15	−1	1
Terraces	440	450	170	790	30	35	0.6	0.8
Exclosures + terraces	560	520	250	1250	40	38	1	2.5
Exclosures without terraces	640	630	210	1250	50	43	0.3	−0.5

construction and 1.28–1.22 g cm^{-3} due to exclosure establishment. Mureithi et al. [32] found that, on average, exclosures lowered the BD of top soil from 1.48 to 1.19 g cm^{-3} in range lands of Kenya. Mean value for BD varied from 1.12 g cm^{-3} (in exclosure) to 1.34 g cm^{-3} (in grazing land) in nearby site [4].

Soil organic carbon and BD had CV values 17–27% and 10–18%, respectively. This indicated that they have moderate variation in response to the SWC measures. Because CV of less than 10% has weak variability, CV between 10 and 100% is considered as moderate variability, while CV with higher than 100% is considered as strong variability [53].

The upper soil depth (0–15 cm) had significantly higher SOC than lower (15–30 cm) soil depth. This could indicate the presence of easily decomposable biomass in the upper soil depth. It was found that high SOC at surface soil is due to the input of organic matter from above-ground biomass [34]. Other studies reported that soil organic carbon is mainly formed by the decomposition of plant materials in the soil surface. Soils of Bale Mountains in southeastern Ethiopia showed similar results [54].

A rough and non-consistent increase in SOC was observed at foot slope which could be due to removal of materials from upper slope position and deposition in lower slope positions. Similarly, Mekuria and Veldkamp [48] in northern highlands of Ethiopia and Amare et al. [51] in Anjeni Watershed in central highlands of Ethiopia found higher SOC in lower and depositional zone than upper and loss zones.

Though not consistent in all the SWC measures, a lower bulk density was observed at the foot slope position. This is probably due to increase in SOC down slope. Increase in SOC is one cause for decrease in bulk density [50].

Soil organic carbon stocks are a function of the concentration SOC, soil thickness considered, soil BD, and coarse fragment. The coarse fragment was included as a correction factor to exclude materials that are assumed to have no soil organic carbon. Soil organic carbon stock decreased with increasing depth due to decrease in organic carbon concentration and gravel content. Gravel content and SOC affect SOC stock [34, 55]. The total SOC for the whole 30 cm soil depth was 44 Mg C ha^{-1}. Similarly, a total of 74 Mg C ha^{-1} SOC stock was reported in the top 40 cm in exclosure in a nearby site, which indicated the importance of such restoration measures in addressing SOC depletion in the area [4].

Conversion of free grazing lands to exclosures and supporting with terraces increased SOC stock by 64%, exclosures without terraces by 37% and terraces by 25%. Exclosures increased SOC stock by more than 50% [41]. Protected grasslands contributed to significant carbon sequestration [56]. Increased SOC stock is related to the restoration of natural vegetation, which enhances above-ground and below-ground litter inputs [9, 57]. It was observed in both exclosures types, the canopy of shrubs and under-story vegetation has been restored, and the soil surface is, therefore, protected from erosion [58]. Vegetation restoration and litter accumulation in exclosures significantly increased plant–soil system carbon storage and sequestration [57].

The presence of significantly higher MBC in exclosures could be due to fewer disturbances, which encouraged enrichment of the microbial biomass carbon. The existence of better vegetation cover in exclosures protects the soil from erosion to maintain favorable conditions for microbial growth [27]. High MBC indicated high efficiency of carbon utilization and increase in ecosystem maturity and vice versa [42]. High MBC can also be used as bio-indicator for the recovery of degraded soils [27]. Thus, high MBC in exclosures indicate that they had the potential for the restoration of soil quality [32].

The low MBC in non-conserved communal grazing lands indicates that there is inefficient utilization of carbon by microbes, which could be due to stress such as grazing and soil compaction by livestock. It was also found that erosion had a more severe disturbance to soil microbial biomass [59]. This caused for the loss of carbon in the form of carbon dioxide [60]. Similarly, Gelaw et al. [61] in a nearby site found that environmental disturbance was the main cause for the decrease in microbial biomass carbon.

The MBC/SOC ratio represents the contribution of microbial biomass to organic carbon in soil and is a more useful assessment index for soil health than either MBC or SOC [62]. The MBC/SOC was low in exclosures and terraces than non-conserved communal grazing lands. However, this is in contrast to the report of Anderson and Domsch [42] who found high MBC/SOC in fields with high total organic carbon. This could be due to woody biomass and litter in exclosures with terraces and terraced grazing lands, which is relatively resistant microbial decomposition. In a similar study, the MBC/SOC ratio at 0–15 cm decreased with increasing woody plant stand from 6% in grasslands to 4% in older woodlands, suggesting that woody litter may be less suitable as a microbial substrate compared with grassland litter [63]. Decline of MBC/SOC meant the decrease of available organic matter in soils [64]. Thus, the contribution of MBC to soil organic carbon was relatively lower in exclosures with terraces compared to exclosures without terraces; and in terraced grazing lands than non-conserved communal grazing lands.

Relationship among soil properties

The strong and significant ($P<0.01$) correlation between SOC stock, SOC concentration, and coarse fragment indicated SOC concentration and coarse fragment (Table 9) are important factors that affect SOC stock. Strong and significant ($R^2=0.87$, $P<0.01$) correlation was found between MBC and MBC/SOC, which indicated that MBC has good contribution to SOC concentration and SOC stock. However, SOC was negatively correlated with bulk density. This is in line with similar study done in south Gonder, northwestern highlands of Ethiopia, [47] and in Middle Sulluh Valley, northern Ethiopia [50]. Soil organic carbon was also negatively and significantly ($R^2=-0.1$, $P<0.01$) correlated with MBC to SOC ratio. This could be either the soil organic matter source is resistant or poor quality or the microbes are weak to decompose the organic matter source.

Table 9 Correlation coefficient between the soil properties (n = 108)

	MBC	SOC	MBC/SOC	BD	Coarse fragment
MBC	1				
SOC	0.39**	1			
MBC/SOC	0.87**	-0.1^{ns}	1		
BD	-0.11^{ns}	−0.19*	−0.02	1	
Coarse fragment	0.21^{ns}	0.11^{ns}	0.19	-0.03^{ns}	1
SOC stock	0.32**	0.56**	0.07^{ns}	0.26**	0.76**

ns correlation is not significant

* Correlation is significant at $P\leq0.05$

** Correlation is significant at $P\leq0.01$ confidence interval

Conclusion

The different SWC measures had brought a significant impact on the ability of the soil to sequester SOC and microbial biomass carbon. Exclosures had higher soil organic carbon stock and MBC compared to terraces and non-conserved open communal grazing lands. Soil depth is an important factor that affects SOC concentrations, because it was observed that upper soil depth had higher SOC than lower soil depth, which could be due to the fact that biotic processes like biomass production, decomposition, and above-ground litter are higher in upper soil depth. Soil organic carbon concentration, bulk density, and percentage of coarse fragment had also brought a significant variation in the SOC stock. Establishment of exclosures and construction of terraces in the open communal grazing lands had a positive impact in terms of SOC stock and microbial biomass carbon storage.

Abbreviations

SOC: soil organic carbon; MBC: microbial biomass carbon; CBSWC: community-based soil and water conservation; BD: bulk density; Mg C ha^{-1}: mega gram carbon per hectare; CF: coarse fragment.

Authors' contributions

MW was a Ph.D. student in soil science and participated in proposing the study, ran and managed all experiments, conducted all laboratory analyses performed statistical analyses and interpretation results, and drafted the manuscript. FK and BB as supervisors participated in guiding and reviewing the manuscript. EB was supervisor and participated in designing the field experiment and reviewed the manuscript. All authors read and approved the final manuscript.

Author details

[1] College of Agriculture and Environmental Sciences, Haramaya University, P.O.Box 138, Dire Dawa, Ethiopia. [2] Ethiopian Agricultural Transformation Agency, P.O. Box 708, Addis Ababa, Ethiopia. [3] Department of Land Resources Management and Environmental Protection, Mekelle University, P.O. Box 231, Mekelle, Ethiopia. [4] Faculty of Environmental Sciences and Natural Resource Management, Norwegian University of Life Sciences, PO Box 5003, 1432 Ås, Norway.

Acknowledgements

The authors would like to thank Ministry of Education for providing scholarship to the principal investigator to study his Ph.D. at Haramaya University, Ethiopia. We are also grateful to the anonymous referees for their constructive comments on an earlier version of this manuscript.

Competing interests

The authors declare that they have no competing interests.

Funding

This study was funded by the Ministry of Education of Ethiopia through Haramaya and Wollega University. We would also like to thank the project Steps towards Sustainable Forest Management with the Local Community of Tigray, northern Ethiopia NORAD/NORHED (ETH 13/0018) for the financial support.

References

1. UNDP/ECA/FAO. Sustainable agriculture and environmental rehabilitation in Tigray (SAERT), working document. Mekelle: UNDP/ECA/FAO; 1994. p. 58–123.
2. Shiferaw A, Hurni H, Zeleke G. A review on soil carbon sequestration in Ethiopia to mitigate land degradation and climate change. J Environ Earth Sci. 2013;3(12):187–200.
3. World Food Program (WFP). Scaling up an integrated watershed management approach through social protection programs in Ethiopia: the MERET and PSNP schemes. In: *Hunger Nutrition climate* justice. Case studies: policy responses, local to national. Dublin; 2013.
4. Girmay G, Singh BR. Changes in soil organic carbon stocks and soil quality land-use system effects in northern Ethiopia. Acta Agric Scand Section B Soil Plant Sci. 2012;62:519–30.
5. Gebreegziabher T, Nyssen J, Govaerts B, Getne F, Behailu M, Haile M, Deckers J. Contour furrows for in situ soil and water conservation, Tigray, Northern Ethiopia. Soil Tillage Res J. 2009;103:257–64.
6. Birhanu A. Environmental degradation and management in Ethiopian highlands: review of lessons learned. Int J Environ Prot Policy. 2014;2:24.
7. Nyssen J, Munro N, Haile M, Poesen J, Descheemaeker K, Haregeweyn N, Decker J. Understanding the environmental changes in Tigray: a photographic record over 30 years. Tigray Livelihood Papers. 2007;3:82.
8. Descheemaeker K, Muys B, Nyssen J, Poesen J, Raes D, Haile M, Deckers J. Litter production and organic matter accumulation in exclosures of the Tigray highlands, Ethiopia. Forest Ecol Manag. 2006;233:21–35.
9. Mekuria W, Veldkamp E, Haile M, Nyssen J, Muys B, Gebrehiwot K. Effectiveness of exclosures to restore degraded soils as a result of overgrazing in Tigray Ethiopia. J Arid Environ. 2007;69(2):270–84.
10. Yayneshet T, Eik LO, Moe SR. Seasonal variations in the chemical composition and dry matter degradability of exclosure forages in the semi-arid region of northern Ethiopia. Anim Feed Sci Technol. 2009;148:12–33.
11. Njira KO, Nabwami J. Soil management practices that improve soil health: elucidating their implications on biological indicators. Review paper. J Anim Plant Sci. 2013;18(2):2750–60.
12. Corsi S, Friedrich T, Kassam A, Pisante M, Moraes SJD. Soil organic carbon accumulation and greenhouse gas emission reductions from conservation agriculture. A literature review. Integr Crop Manag. 2012;16:13–9.
13. Weil RR. Significance of soil organic matter to soil quality and health. In: Weil RR, Magdoff F, editors. Soil organic matter in sustainable agriculture. Florida: CRC press; 2004. p. 1–13.
14. Anikwe M. Carbon storage in soils of Southeastern Nigeria under different management practices. Carbon Balance Manag. 2010;5(1):5.
15. Liu ZP, Shao MB, Wang YQB. Large-scale spatial variability and distribution of soil organic carbon across the entire Loess Plateau. China. Soil Res. 2012;50:114–24.
16. Chan Y. Increasing soil organic carbon of agricultural land. Iran: Iran publishing; 2008.
17. Alexandra L, De Bruyn L. The status of soil macrofauna as indicators of soil health to monitor the sustainability of Australian agricultural soils. Ecol Econ. 1997;23:167–78.
18. Lal R, Follett RF, Stewart BA, Kimble JM. Soil carbon sequestration to mitigate climate change and advance food security. Soil Sci. 2007;81:113–27.
19. Rousk J, Smith AR, Jones DL. Does warming and drought have lasting effects on soil ecosystems? Glob Change Biol. 2013. https://doi.org/10.1111/gcb.12338.
20. Haynes RJ. Soil organic matter quality and the size and activity of the microbial biomass: their significance to the quality of agricultural soils. In: Huang Q, Huang PM, Violante A, editors. Soil mineral-microbe-organic interactions: theories and applications. Berlin: Springer; 2008.
21. Loureiro DC, De-polli H, Ceddia MB, Aquino AE. Spatial variability of microbial biomass and organic matter labile pools in a haplic planosol soil, Bragantia. Campinas. 2010;69(1):85–95.
22. Xiu-Mei L, Qi L, Wen-Ju L, Yong J. Distribution of soil enzyme activities and microbial biomass along a latitudinal gradient in farmlands of Songliao plain, Northeast China. Pedosphere. 2008;18(4):409–544.
23. Moussa AS, Van Rensburg L, Kellner K, Bationo A. Soil microbial biomass in semi-arid-communal sandy rangelands in the Western Bophirima district, South Africa. Appl Ecol Environ Res. 2007;5(1):43–56.
24. Jenkinson DS, Ladd JN. Microbial biomass in soil, measurement and turn over. In: Paul EA, Ladd JN, editors. Soil Biochemistry. New York: Marcel Dekker press; 1981. p. 415–71.
25. Marina S, Roberto M, Campiglia E, Grego S. Chemical and biological indicators of soil quality in organic and conventional farming systems in Central Italy. Ecol Indic. 2006;6(4):701–11.
26. Moscatelli MC, Lagomarsino A, Marinari S, Angelis PD, Grego S. Soil microbial indices as bio indicators of environmental changes in a poplar plantation. Ecol Indic. 2005;5(5):171–9.
27. Yan T, Yang L, Campbell CD. Microbial biomass and metabolic quotient of soils under different land use in the three Gorges reservoir area. Geoderma. 2003;115:129–38.
28. Dwivedi V, Soni P. A review on the role of soil microbial biomass in ecorestoration of degraded ecosystem with special reference to mining areas. J Appl Nat Sci. 2011;3(1):151–8.
29. Bhatt M, Banmeru S. Estimates of soil microbial biomass carbon of forest soil types of Gujarat, India. Int J Curr Microbiol Appl Sci. 2014;3(11):817–25.
30. Lal R. The depth distribution of soil organic carbon in relation to land use and management and the potential of carbon. For Ecol Manag. 2005;220:242–58.
31. Haripal K, Sahoo S. Original research article microbial biomass carbon, nitrogen and phosphorus dynamics along a chronosequence of abandoned tropical agroecosystems. Int J Curr Microbiol Appl Sci. 2014;3(9):956–70.
32. Mureithi SM, Verdoodt A, Gachene CK, Njoka JT, Wasonga VO, Neve SD, Ranst E. Impact of enclosure management on soil properties and microbial biomass in a restored semi-arid rangeland, Kenya. J Arid Land. 2014;6(5):561–70.
33. Tamene Gebreyesus, Vlek L, Paul LG. Catchment-scale spatial variability of soil properties and implications on site-specific soil management in northern Ethiopia. Soil Tillage Res. 2011;117:124–39.
34. Xu GC, Li Z, Bin Li, Lu KX, Wang Y. Spatial variability of soil organic carbon in a typical watershed in the source area of the middle Dan River, China. Soil Res. 2013;51(1):41–9.
35. Li M, Zhang X, Zhen Q, Han F. Spatial analysis of soil organic carbon in Zhifanggou catchment of the Loess Plateau. PLOS ONE. 2013;8:e83061.
36. Nyssen JJ, Poesen Moeyersons J, Deckers J, Haile M, Lang A. Human impact on the environment in the Ethiopian and Eritrean highlands, a state of the art. Earth Sci Rev. 2004;64:273–320.
37. Beyth M. Sedimentary basin of Makalle outlier. Am Asso Petrol Geol Bull. 1972;56:2426–39.
38. World reference base (WRB). A framework for international classification, correlation and communication. Italy: FAO; 2006.
39. Woldearegay K. Cost and benefit of catchment management and regreening in Tigray, Ethiopia. Unpublished report. 2012.
40. Mekuria W, Aynekulu E. Exclosure land management for restoration of the soils in degraded communal grazing lands in northern Ethiopia. Land Degrad Dev. 2013;24(6):528–38.
41. Mekuria W. Effectiveness of exclosures to restore ecosystem carbon stock and vegetation in the highlands of Tigray, northern Ethiopia. J Arid Environ. 2010;69:270–84.
42. Anderson JPE, Domsch KH. Measurement of bacterial and fungal contributions to respiration of selected agricultural and forest soils. Can J Microbiol. 1975;21:314–22.
43. Schuman GE, Janzen HH, Herrick JE. Soil carbon dynamics and potential carbon sequestration by rangelands. Environ Pollut. 2002;116(3):391–6.
44. Walkeley A, Black I. An examination of Degtjareff methods for determining soil organic matter and a proposed modification of the chromic acid titration method. Soil Sci. 1934;79:459–65.
45. IPCC. Good practice guidance for land use, land use change and forestry. Hayama: Institute for Global Environmental Strategies (IGES); 2003.
46. SAS (Statistical Analysis System) Institute. SAS user's guide. Proprietary software version 9.2. Cary: SAS Institute, Inc.; 2004.
47. Demelash M, Stahr K. Assessment of integrated soil and water conservation measures on key soil properties in South Gonder, North-Western highlands of Ethiopia. J Soil Sci Environ Manag. 2010;1(7):164–76.
48. Mekuria W, Veldkamp E. Impacts of Land Use Changes on Soil Nutrients and Erosion in Tigray, Ethiopia. Deutscher Tropentag. In: Conference on international agricultural research for development; 2005.

49. Amdemariam T, Selassie YG, Haile M, Yamoh C. Effect of soil and water conservation measures on selected soil physical and chemical properties and barley (*Hordeum* spp.) yield. J Environ Sci Eng. 2011;5:1483–95.
50. Hishe S, James L, Woldeamlak B. Soil and water conservation effects on soil properties in the Middle Silluh Valley, northern Ethiopia. Int Soil Water Conserv Res. 2017;5:231–40.
51. Amare T, Terefe A, Selassie YG, Yitaferu B, Wolfgramm B, Hurni H. Soil properties and crop yields along the terraces and toposequece of Anjeni watershed, Central highlands of Ethiopia. J Agric Sci. 2013;5(2):134–44.
52. Welemariam M, Kebede F, Bedadi B, Birhane E. Effect of community-based soil and water conservation practices on arbuscular mycorrhizal fungi types, spore densities, root colonization, and soil nutrients in the northern highlands of Ethiopia. Chem Biol Technol Agric. 2018;5:9.
53. Nielsen DR, Bouma J. Soil spatial variability. In: Proceedings of a workshop of the ISSS and the SSSA. Las Vegas; 1985.
54. Del E, Del US, Sobre S, Contenido EL, Orgánico C, Del YN, En S. Effects of land use on soil organic carbon and nitrogen in soils of bale, southeastern Ethiopia. Trop Subtrop Agroecosyst. 2011;14:229–35.
55. Alcántara PL, García LB, Espejo AG. Soil organic carbon along an altitudinal gradient in the Despeñaperros Natural Park, southern Spain. Solid Earth. 2015;6:125–34.
56. Su YZ, Zhao HL, Zhang TH. Influences of grazing and exclosure on carbon sequestration in degraded sandy grassland. N Z J Agric Res. 2010;46:321–8.
57. Bird SB, Herrick JE, Wander MM, Wright SF. Spatial heterogeneity of aggregate stability and soil carbon in semi-arid rangeland. Environ Pollut. 2002;116(3):445–55.
58. Welemariam M, Kebede F, Bedadi B, Birhane E. The effect of community-based soil and water conservation practices on abundance and diversity of soil macroinvertebrates in the northern highlands of Ethiopia. Agronomy. 2018;8:56.
59. Xiaojun N, Jianhui Z, Zhengan S. Dynamics of soil organic carbon and microbial biomass carbon in relation to water erosion and tillage erosion. PLoS ONE. 2013;8(5):1–7.
60. Gonzalez-Quiñones V, Stockdale E, Banning NC, Hoyle FC, Sawada Y, Wherrett D, Murphy DV. Soil microbial biomass—interpretation and consideration for soil monitoring. Soil Res. 2011;49(4):287–304.
61. Gelaw AM, Singh BR, Lal R. Soil quality indices for evaluation of tree-based agricultural land uses in a semi-arid watershed in Tigray, Northern Ethiopia. Sustainability. 2015;7:2322–37.
62. Santos VB, Araújo ASF, Leite FC, Nunes AL, Melo WJ. Soil microbial biomass and organic matter fractions during transition from conventional to organic farming systems. Geoderma. 2012;170:227–31.
63. Liao JD, Boutton TW. Soil microbial biomass response to woody plant invasion of grassland. Soil Biol Biochem. 2008;40(5):1207–16.
64. Zhu J, Qiaoling Y, Changjie J. Comparison of soil microbial biomass C, N and P between natural secondary forests and Larix olgensis plantations under temperate climate. In: World congress of soil science, soil solutions for a changing world. Brisbane, Australia; 2010.

14

Investigation of antioxidant and antimicrobial activities of *Euclea schimperi* leaf extracts

Alemayehu Mekonnen*, Minaleshewa Atlabachew and Banchalem Kassie

Abstract

Background: Bioactive components from medicinal plants are said to be responsible for the antioxidant and antimicrobial effects of plant extracts. It was the main purpose of this paper to evaluate the total phenolic content (TPC), total flavonoid content (TFC), and antioxidant and antimicrobial activities of leaf extracts of *Euclea schimperi*.

Results: Strong evidence was found when a positive correlation was found between TPC or TFC and antioxidant power. To put it differently, TPC of the *E. schimperi* leaf extracts varied between 748.78 ± 3.11 mgGAE/100 g of dry extract (methanol) and 40.45 ± 0.66 mgGAE/100 g of dry extract (chloroform) while TFC varied from 3.306 ± 0.142 mgQE/100 g of dry extract (methanol) to 0.334 ± 0.002 mgQE/100 g dry extract (chloroform). Similarly, all extracts exhibited significant amount of antioxidant and antimicrobial activities. Ferric reducing power was ranged between 312.145 ± 1.542 mgAAE/100 g extract (methanol) and 48.177 ± 0.286 mgAAE/100 g (chloroform). DPPH radical scavenging activity (%) lied between 83.4% (methanol) and 58.5% (chloroform). The voltammetric analysis of the crude extract gave three different sharp peaks at the oxidation potential side. Finally, the reducing agent capacity of the extracts was further confirmed by synthesizing silver nanoparticles from $AgNO_3$ solution and the plant crude extracts. Crude extracts of *E. schimperi* leaves showed various degree of antimicrobial activity towards different standard strains with mean zone of inhibition ranging from 0 to 20.67 mm. Both solvent leaf extracts of *E. schimperi* exhibited MIC ranging from 62.5 to 125 mg/mL towards most bacterial species.

Conclusions: *Euclea schimperi* leaf extracts were found to contain high amount of total phenolic and small amount of flavonoid compounds which play a major role in controlling oxidation. In general, therefore, the results prove the effectiveness of the plant for its excellent antioxidant and antimicrobial activities. Accordingly, the positive values of the plant in traditional medicines have been confirmed.

Keywords: *Euclea schimperi*, Free radicals, Antioxidants, Total phenolic content, Total flavonoid content, Antibacterial activity, Minimum inhibitory concentration

Introduction

Traditional medicine is regarded as a combination of knowledge, skill and practice originated from theories, experiences and attitudes native to various cultures that is utilized to treat and diagnose enormous types of physical and mental complications and maintain health [1]. Many people in the world rely on traditional medicine for their primarily health care needs. It is well understood that several medical techniques and procedures have been discovered from the use and knowledge of traditional medicines. This is because the major part of traditional therapy involves the use of plant extracts or their active principles. As a result, natural products have recently become of immense interest owing to their diverse application [2].

Since in the early stage, plants have been used as a source of medicinal compounds and still continued to play a major role in the human health [3]. They are the

*Correspondence: negaalex@yahoo.com; alemayehum@bdu.edu.et
Chemistry Department, Science College, Bahir Dar University, P.O. Box: 79, Bahir Dar, Ethiopia

source of drugs in the traditional and modern medicine, nutraceuticals, food supplements, folk medicines and precursors for synthetic drugs [4]. The discoveries of potential new drugs have been led by screening bioactive compounds from plants having efficient activity against various diseases [5].

Phytochemicals are bioactive and naturally occurring compounds originating in plants, which act against pathogens and provide health benefits for humans [6]. Among the prevalent group of phytochemicals, alkaloids, terpenes, flavonoids and phenolic compounds are widely found in most plants and plant parts [7]. Extracts of plant parts (fruits, seeds, leaves, etc.) are used for medicinal effects, among which some of the classes of compounds have antioxidant activity against reactive species in the body [8].

The most recognized antioxidants derived from medicinal plants are phenolic compounds and have high antioxidant activity against free radicals. However, the extent of the activity depends on the structure of the molecules [8, 9]. These antioxidant constituents are closely associated to the prevention of degenerative diseases, such as cardiovascular and neurological illnesses, oxidative stress malfunctions and cancer [9, 10].

Euclea schimperi, from the family of *Ebenaceae*, is traditionally used in the treatment of wound, teeth infections, eye disorders, headache, pain and spasm. In Ethiopia, the leaf of this plant is used to treat illness such as gonorrhea, eczema and constipation. In addition, they treat the pots in which milk is kept with the smoke of *E. schimperi* branches to prevent the milk from curdling [11, 12]. The leaves is also used for the treatment of skin disorders, snake biting, scabies, leprosy, Tinea capitis, acne, warts, rheumatic pain and elephantiasis [13].

Despite the wide therapeutic potential presented by *E. schimperi* leaves in Ethiopia, studies on the phytochemical profile, antioxidant and antimicrobial activities of the leaves of this plant are still lacking. Only one other MSc thesis abstract posted on Khartoum University website, to our knowledge, has come up with the antioxidant and antibacterial activity of Sudanese *E. schimperi* leaf extract (http://khartoumspace.uofk.edu/handle/123456789/16116). Consequently, in our study, the focus of attention was on the determination of the content of some selected phytochemicals, evaluation of antioxidant and antimicrobial activities of the leaf extracts of Ethiopian *E. schimperi.*

Experimental section

Sample collection

The leaves of *E. schimperi* were collected from Bahir Dar University main campus. The plant was identified by its local name "Dedeho" (Amharic) and registered in Ethiopian folk medicine. A herbarium voucher (Banchalem-001) was identified by Dr. Ali Seid and deposited in the mini-herbarium of a Bahir Dar University (Bahir Dar, Ethiopia). The leaves of the plant were detached from the parent plant and were washed with tape water and placed in shade at room temperature without exposing to sun light and heat for 2 weeks. The air-dried leaves of *E. schimperi were* chopped into small pieces and milled into a uniform powder with a coffee grinder. The powder was packed in polyethylene bags and stored in refrigerator for further use.

Extraction procedure

The powdered leaf of *E. schimperi* was extracted by two different solvents (methanol and chloroform). In each extraction process, 60 g of dry powder was socked in 600 mL of each solvent with ratio of 1:10 (W/V) of the plant material to solvents as per standard method for extraction of medicinal plant [14]. The mixtures were extracted up on shaking with mechanical shaker for 48 h. Then, the extracts were filtered and concentrated using a rotary evaporator under reduced pressure at a maximum of 25 °C. Crude extracts obtained using methanol and chloroform solvents were labeled as ESL_1, and ESL_2, respectively. The yields were calculated and preserved using air tight bottles at 4 °C refrigerator. In all cases, the extracts were tested in triplicates and the mean values were tabulated.

Statistical analysis

Data were presented as mean ± SD from triplicate measurements and significance differences between groups were determined by one-way ANOVA followed by Tukey post hoc test and paired *t* test. Differences were considered significant when $p < 0.05$.

Experimental procedures

Preliminary phytochemical screening

The leaves of *E. schimperi* were subjected to preliminary phytochemical tests following standard procedures described for the screening of phytochemical constituents (alkaloids, polyphenols, flavonoids, saponins, tannins, glycosides, and terpenoids). Methanol and chloroform extracts were screened for the presence of steroids, terpenoids, flavonoids, phenolics, tannins and saponins following standard method [15]. All tests were performed in triplicates.

Determination of total polyphenols

Total phenolic content of *E. schimperi* leaf extracts was determined according to the Folin–Ciocalteu method with slight modification [16]. Gallic acid was used as reference compound for positive control. Exactly 5 mL of plant crude extracts (methanol and chloroform) was diluted with 45 mL of distilled water. Folin–Ciocalteu reagent (0.25 mL) was added to 5 mL of the diluted extract solution followed by 0.5 mL of 7% sodium carbonate (Na_2CO_3) solution. The contents were mixed and the flask made up to volume with distilled water then thoroughly mixed. The mixtures were left for 30 min in the dark and the absorbance was measured at 750 nm.

Determination of total flavonoids

The total flavonoid content of *E. schimperi* leaf methanol and chloroform extracts was determined by aluminum chloride assay with slight modifications [17]. Quercetin was used as standard compound. A sample extract of 2.00 mL was mixed with 4.00 mL of distilled water, followed by an immediate addition of 0.30 mL of 5% $NaNO_2$. After 5 min, 0.30 mL of 10% $AlCl_3$ solution was added. After 6 min, 2.00 mL of 1.00 M NaOH solution was added and then 10 min later, absorbance of the resulting solution was taken at 510 nm.

Measure of antioxidant activities

a. Reducing power assay

The reducing power of the extract was determined according to the method reported by Mitic et al. [18] with slight modification. Ascorbic acid was used as reference. Different concentrations (10, 20, 30, 40%, v/v) of 5.00 mg/mL of stock solutions of *E. schimperi* leaf extracts were prepared for each solvent (methanol, and chloroform). From each solution, 2.50 mL was taken and mixed with 2.5 mL of 0.20 M sodium phosphate buffer (pH 6.6) and 2.5 mL of potassium ferricyanide solution (1%). The mixtures were incubated in water bath at 50 °C for 20 min then, 2.5 mL of trichloroacetic acid (TCA) solution (10% w/v) was added and the resulting mixture was centrifuged at 3000 rpm for 10 min. The supernatant (2.5 mL) was taken, mixed with 2.5 mL of distilled water and 0.5 mL of a ferric chloride solution (0.1% w/v) and the absorbance of the resulting solution was measured at 700 nm.

b. DPPH radical scavenging activity determination

The antioxidant activity of *E. schimperi* leaf extracts was also assessed on the basis of the free radical scavenging effect of DPPH (2, 2-diphenyl-1-picrylhydrazyl) using the method described by [19]. Ascorbic acid was used as positive control. The methanol and chloroform extracts from *E. schimperi* leaves with different concentrations [40, 30, 20, 10% (v/v)] of 5.0 mg/mL stock solutions were prepared. 5.0 mL of 0.004% of DPPH solution was mixed with 1 mL of each extract solution. The solutions were kept in the dark for 30 min to complete the reaction and then the absorbance of the mixture was measured at 517 nm.

c. Voltammetric analysis

The 80 mg/mL concentration of leaf extracts of methanol from *E. schimperi* leaves was subjected to voltammetric analysis to measure antioxidant capacity of the sample. It has to be noted that methanol and chloroform extracts were found to contain high and low amount of antioxidants, respectively. Thus, only methanol extract was considered for further comparative analysis using voltammetric method. Exactly 15.00 mL of each sample extract was diluted to 50.00 mL with acetate buffer solution (0.10 M) of pH 5. Then, cyclic voltammogram of each sample was obtained by scanning the potential from − 1000 to 1400 mV at a scan rate of 100 m Vs^{-1}. Acetate buffer was used as a blank.

All the voltammetric measurements were carried out in a 25-mL cell at room temperature, in a three-electrode configuration. Graphite carbon electrode was used as a working electrode. The reference electrode was an Ag|AgCl salt and the counter electrode was a platinum wire. Cyclic voltammetric measurements were performed in model 700E series electrochemical analyzer connected to a desktop computer. For further relaxation of the peak, the sample was scanned with square wave voltammetry.

d. Synthesis of silver nanoparticles

Antioxidant capacity of *E. schimperi* was measured in terms of reduction ability of silver to synthesize silver nanoparticles (AgNPs) based on the method described by Shankar et al. [20] with slight modification. Two different solvent extracts (methanol and chloroform) were prepared with the same concentration. Methanol extract was prepared with different concentrations (10, 20, 30, 40% v/v) of 5 mg/mL stock solution to measure the reduction potential. 10 mL of 5 mM silver nitrate solution was mixed with 1 mL of each sample in the dark by stirring magnetically for 15 min.

The reduction of metallic silver ions was monitored by measuring the absorbance after different time for different concentration of the reaction. The absorbance of the mixture was scanned between 300 and 600 nm using UV–Vis spectrometer after 2, 6, 9 and 24 h of the reaction begins.

Antibacterial activity determination

Disk diffusion method

The present study was conducted using standard disk diffusion method [21]. Different extract (methanol and chloroform) of *E. schimperi* leaves were used by agar disk diffusion method on Mueller–Hinton Agar (MHA) medium to measure the antimicrobial activity. Standard

Table 1 Result of phytochemical screening of different solvent extracts of *Euclea schimperi* leaves

Phytochemicals	Types of extracts		Color observed	Reagents/methods used
	Methanol	Chloroform		
Saponins	+++	+	Stable froth	Froth test
Tannins	+++	−	Yellow ppt	Lead acetate test
Phenolics	+++	+	Blue black color	Ferric chloride test
Flavonoids	++	−	Yellow ppt	Lead acetate test
Terpenoids	++	−	Reddish brown ring	$CHCl_3 + H_2SO_4$
Steroids	+++	−	Reddish brown color	Sulfuric acid test

+++: present in highest concentration, ++: present in high concentration, +: present in moderate concentration, −: absent [15, 22, 23]

Table 2 Total phenolic content (TPC) expressed as mg GAE/100 g dry extract and total flavonoids content expressed as mg QE/100 g dry extract of different solvent extracts of *Euclea schimperi* leaves

Type of extract	Total phenolics in mg GAE/100 g dry ext.	Total flavonoids mg QE/100 g dry ext.	Antioxidant activity a FRAP (mg AAE/100 g dry extract)	Antioxidant activity a DPPH (mgAAE/100 g dry ext.
Methanol extract	748.78 ± 3.11^a	3.306 ± 0.142^a	312.145 ± 1.542^a	19.707 ± 0.186^a
Chloroform extract	40.45 ± 0.66^b	0.334 ± 0.002^b	48.177 ± 0.286^b	2.717 ± 0.030^b

Sample for three independent measurements (triplicate; $n = 3$, mean ± SD). Values in the same column that are followed by a different letters (a, b) are significantly different $p < 0.05$ by paired *t* test

size Whatman No. 1 filter paper disks, 5.0 mm in diameter, sterilized by moist heat at 121 lb in an autoclave for 15 min was used to determine antibacterial activity.

MHA medium for *S. aureus*, *p. aeruginosa*, *K. pneumoniae*, and *S. pneumoniae* was prepared for disk diffusion test. After sterilization, it was poured into sterilized Petri dish plates and allowed to solidify. Then, 1-day-old fresh culture of bacteria was used for inoculum preparation. A suspension that will be just turbid (~0.5 McFarland standard) by visual inspection was prepared by suspending bacteria in 0.9% NaCl solution and the homogeneous suspension was used for inoculation. Using a sterile cotton swab, bacterial cultures were swabbed on the surface of sterile agar plates. Sterile 5-mm disks was impregnated with 50 μL of extract and placed on the surface of agar plates inoculated with a microbial culture. After overnight incubation, the plates were observed for the zone of inhibition and the diameter of the inhibition zone was measured in millimeter.

Standard antibiotics, erythromycin, chloramphenicol, gentamycin and ciprofloxacin (10 μg/mL) served as positive control and sterile distilled water and chloroform used to dilute plant extract were used as negative control.

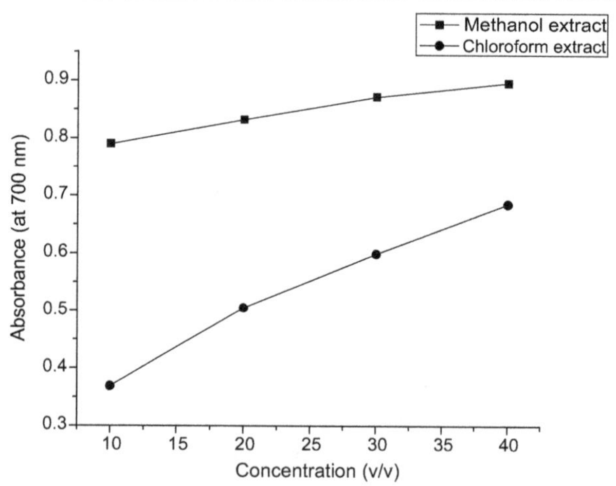

Fig. 1 Reducing power of *Euclea schimperi* leave extracts, expressed as absorbance at 700 nm. Different concentrations (10, 20, 30, 40%, v/v) of 5.00 mg/mL of stock solutions of *Euclea schimperi* leaf extracts were prepared for each solvent (methanol, and chloroform). The graph shows that the reducing power increases with increasing concentration. The antioxidant capacity of methanol extract was significantly higher than that of the chloroform extract

Determination of minimum inhibitory concentration

The broth dilution technique [22] was employed to determine the MIC of the potent extracts. Standardized inoculums of the test organisms was inoculated into series of sterile tubes of 2 mL nutrient broth containing 500, 250, 125, 62.5 and 31.25 mg/mL of each extracts and incubated at 37 °C for 24 h. The lowest concentration of the extract that produced indication of growth (Turbidity) was recorded as the MIC.

Results and discussion

Phytochemical screening tests

The preliminary phytochemical tests are helpful in finding information about chemical constituents present in the plant material [22]. This information could be used to facilitate quantitative estimation and bioassay-guided separation of pharmacologically active compounds from the plant. The preliminary phytochemical screening of the two solvent extracts from *E. schimperi* revealed the presence of several secondary metabolites such as saponins, terpenoids, tannins, steroids, polyphenol and flavonoids in moderate to high yield. As indicated in Table 1 below, significant amount of polyphenols, tannins and saponins were observed in methanol leaf extracts compared to chloroform extract, which was detected from the formation of strongly colored solutions.

Determination of total phenolic content

The total polyphenol content of the leaf extracts was reported in terms of milligram gallic acid equivalent (mg GAE). As follows from Table 2 shown below, gallic acid standard solutions ranging from 50 to 125 µg/mL were used to construct a calibration curve for the determination of the total phenolic content of each sample. It is apparent from Fig. 1 that a straight line with an equation $y = 0.00264x + 0.0699$ and a linear regression coefficient (R^2) of 0.997 was obtained.

As depicted in Table 2, high total phenolic content was observed from *E. schimperi* leaf extracts; namely, methanol extract exhibited the highest level of phenolic content (748.78 ± 3.11) mgGAE/100 g dry extract while chloroform extracts exhibited the least (40.45 ± 0.66) mgGAE/100 g dry extract. This finding, while preliminary, suggests that extraction of phenolic compounds from this plant is influenced by the polarity of the solvent. Several of the recent literature has reported that aqueous mixture of acetone, methanol and ethanol could be better choice for extraction of low- and high-molecular weight phenolic compounds [6]. In this study, extraction of the phenolic compounds using aqueous mixture of methanol (90% methanol in water) was done but it was noticed that a gummy-type extract resulted and it was not possible to completely redissolve the extract using methanol for further studies. This indicates that in the presence of water as an extracting solvent, the crude extract undergoes in situ polymerization. So this result shows that aqueous solvents are not good solvents particularly for the extraction of *E. schimperi* leaves.

Table 3 Absorbance of different solvent extract of *Euclea schimperi* leaves at different concentration for ferric reducing power measurement at 700 nm

Concentration (v/v)	Absorbance at 700 nm of the reduced FRAP product for different extracts	
	Methanol	Chloroform
10%	0.790 ± 0.001[a]	0.369 ± 0.003[b]
20%	0.832 ± 0.002[a]	0.505 ± 0.002[a]
30%	0.872 ± 0.004[c]	0.599 ± 0.003[b]
40%	0.896 ± 0.002[b]	0.685 ± 0.001[a]

Values in the same column that are followed by a different letters (a–e) are significantly different $p < 0.05$ by Tukey's HSD multiple comparison test

Determination of total flavonoid content

Calibration curve was constructed to determine the total flavonoid content of *E. schimperi* leaf extracts. As summarized in Table 2, total flavonoid (TF) contents in the extracts are expressed as mg quercetin equivalent per 100 g of dry extract (QE/100 g of dried extract). Different concentrations of quercetin (2, 4, 6 and 8 µg/mL) were used to construct the calibration curve ($y = 0.0154x + 0.128$, $R^2 = 0.999$). In fact, the total flavonoid content was very small compared to the total phenolics content of leaf extracts of *E. schimperi*. On the other hand, the TFC in methanol extract (3.306 ± 0.142) was relatively higher than that of chloroform extract (0.334 ± 0.002).

Antioxidant potential determination of extracts of *E. schimperi* leaves

Ferric reducing power (FRAP) assay

The reducing power of the extracts from *E. schimperi* leaves, which may serve as a significant reflection of the antioxidant activity was evaluated using potassium ferricyanide reduction method. The antioxidant capacity of the *E. schimperi* leaf extracts with two solvents was evaluated in terms of ascorbic acid equivalent (mg AAE per g dry weight of sample) and this is tabulated in Table 2. For the antioxidant activity determination, aqueous solutions of ascorbic acid standard at different concentrations (0.2, 0.4, 0.6, 0.8 mg/mL) were used to construct the calibration curve and a straight line with an equation $y = 0.3525x + 0.491$ and a linear regression coefficient (R^2) of 0.997 was obtained.

The reducing power of the extracts as a function of different concentrations of extracts is summarized in Table 3 and illustrated in Fig. 1. The graph and the

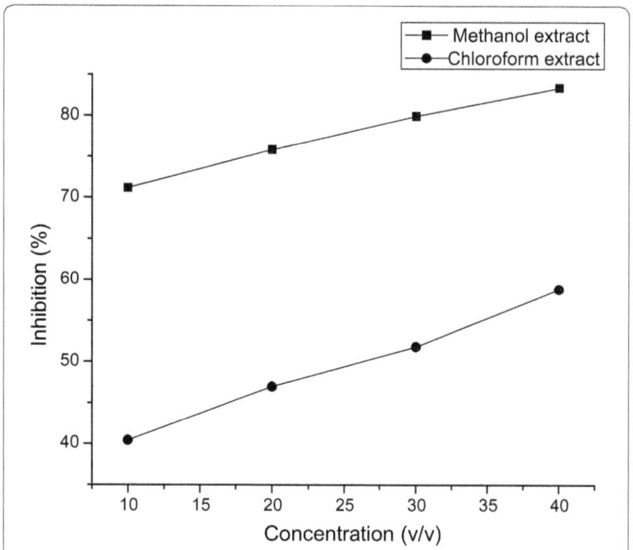

Fig. 2 DPPH radical scavenging activity of *Euclea schimperi* leaf extracts at different concentration. The methanol and chloroform extracts from *Euclea schimperi* leaves with different concentrations (40, 30, 20, 10% v/v) of 5.0 mg/mL stock solutions were prepared. The percent radical scavenging activities of the methanol and chloroform extracts expressed as ascorbic acid equivalents were 19.71 and 2.717 mg AAE/100 g, respectively

analyzed data confirmed that all the extracts exhibited some degree of reducing power. It is apparent from the table that the reducing power increases with increasing concentration. The antioxidant capacity of methanol extract was significantly higher than that of the chloroform extract. This indicates that polar compounds are more responsible for antioxidant activities and a positive correlation was found between TPC and antioxidant activity.

In this assay, the yellow color of the test solution changes to various shades of green when extracts were added which indicates the reducing power of each extracts. The presence of antioxidant causes the conversion of the Fe^{3+}/ferricyanide complex to the ferrous/Fe^{2+} form. Substances which have reduction potential react with potassium ferricyanide (Fe^{3+}) to form potassium ferrocyanide (Fe^{2+}), then reacts with ferric chloride to form ferrous complex that has an absorption maximum at 700 nm.

DPPH radical scavenging activity determination

The DPPH radical scavenging activities of the leaf extracts from *E. schimperi* were estimated by comparing the percentage scavenging activity of the DPPH with a standard, ascorbic acid (Table 2). Figure 2 exemplifies the steady increase in scavenging activity of the methanol and chloroform extracts up to 40% v/v of the 5 mg/mL stock solution of extracts. The methanol and chloroform extracts from *E. schimperi* were able to scavenge the DPPH radical with percentage scavenging activity of 85.4 and 58.5% at the highest concentration, respectively. In other words, the percent radical scavenging activities of the methanol and chloroform extracts expressed as ascorbic acid equivalents were 19.71 and 2.717 mg AAE/100 g, respectively. A comparison of the results with the standard reveals that the DPPH radical scavenging capacity of the two

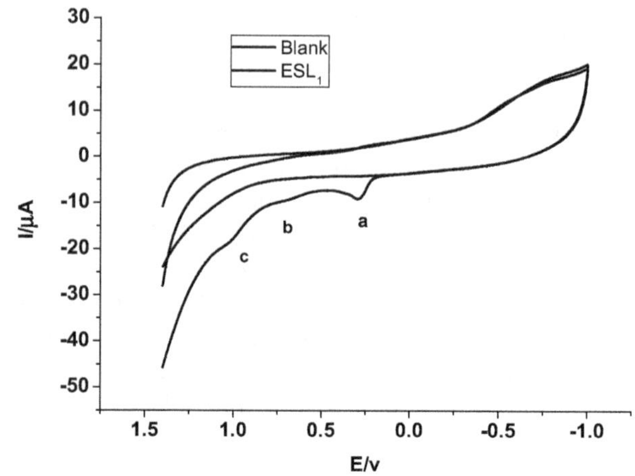

Fig. 3 Cyclic voltammetry analysis of methanol extract of *Euclea schimperi* leaves. Cyclic voltammogram of each sample was obtained by scanning the potential from − 1000 to 1400 mV at a scan rate of 100 m Vs^{-1}. Acetate buffer was used as a blank. Both CV and SWV gave three different peaks and the only different SWV can scatter the peaks clearly

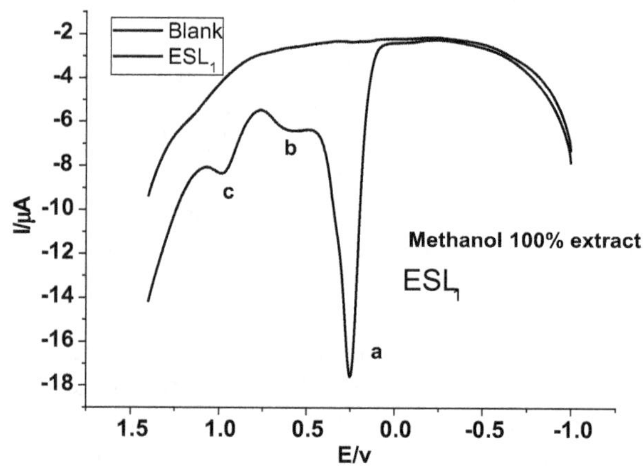

Fig. 4 Square-wave voltammetry analysis of methanol extract of *Euclea schimperi* leaves. Cyclic voltammetric measurements were performed in model 700E series electrochemical analyzer connected to a desktop computer. I_{pa}(µA) = electric current in µA of peak 'a' and E_{Pa}(mV) = oxidation potential in mV of peak 'a'. *Euclea schimperi* leaves extract can act as antioxidant species because there were no any peaks at the reduction side. All three peaks were obtained at the oxidation side

Table 4 Peak parameters of methanol extract of *Euclea schimperi* leaves obtained from voltammetric analysis

Type of extracts	Oxidation peak characters					
	I_{pa}(μA)	E_{Pa}(mV)	I_{pb}(μA)	E_{pb}(mV)	I_{pc}(μA)	E_{pc}(mV)
Methanol	− 17.59	248.35	− 6.41	622.25	− 8.37	989.02

I_{pa}(μA), electric current in μA of peak 'a'; I_{pb}(μA), electric current in μA of peak 'b'; I_{pc}(μA) electric current in μA of peak 'c'; E_{Pa}(mV) oxidation potential in mV of peak 'a'; E_{pb}(mV), oxidation potential in mV of peak 'b'; E_{pc}(mV), oxidation potential in mV of peak 'c'

extracts was weaker than ascorbic acid (89.95 ± 0.116% at the concentration of 125 μg/mL). Another point to consider is methanol extract again retained its superiority in the antioxidant activity. This pattern of DPPH inhibition is commonly observed with plant extracts [18].

Voltammetric analysis, CV

The voltammetric method can be used for the determination of the antioxidant capability in the same manner as the DPPH radical scavenging because of the correlation found between oxidation potentials and antioxidant power. CV method is frequently preferred for various types of physicochemical analysis in redox system due to its high sensitivity, rapidity, simplicity, the possibility of performing analysis in colored or turbid solutions [24]. It can provide information about the number of redox states as well as qualitative information about the stability of these oxidation states and the electron transfer kinetics [25–27].

Square-wave voltammetry (SWV) is another superior electrochemical technique for both mechanistic and electrokinetic studies of various electrode processes. SWV has several advantages. Among these are its excellent sensitivity, the rejection of background currents and its speed. The current produced is proportional to the combined concentration of the antioxidants [28, 29].

The antioxidant activity of the samples is determined by the ease of phenolic compound to be oxidized. It can

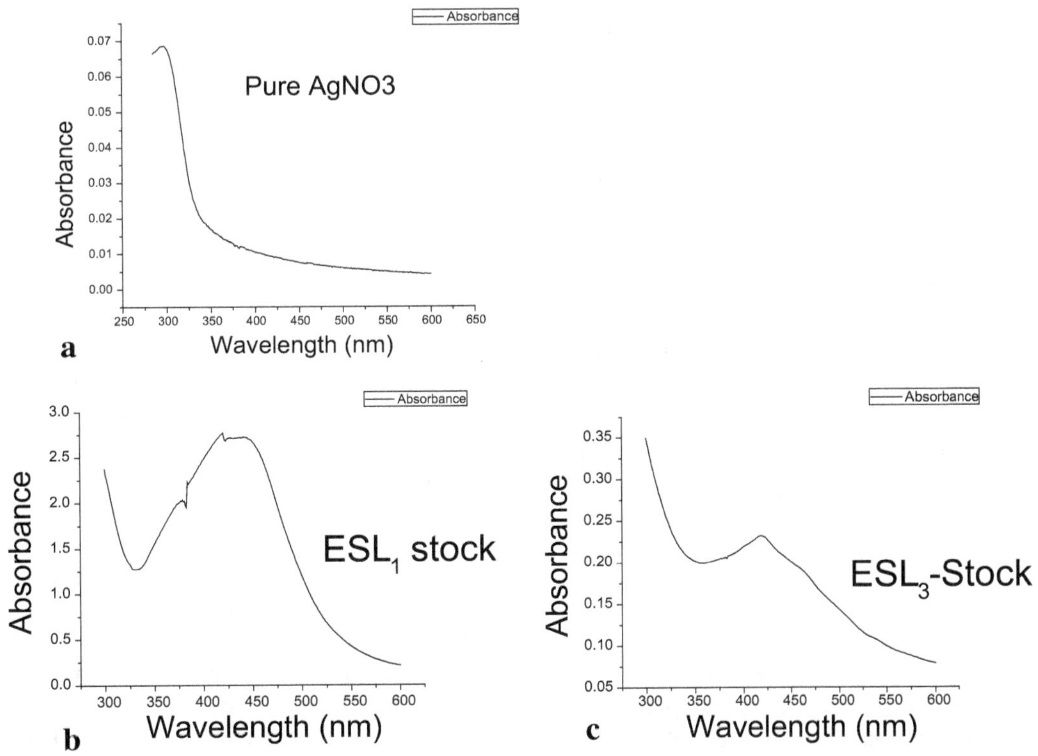

Fig. 5 UV–Vis absorbance of pure silver nitrate solution: **a** methanol extract **b** and chloroform extract **c** of *Euclea schimperi* leaves. The absorbance of the mixture was scanned between 300 and 600 nm using UV–Vis spectrometer after 2.0, 6.0, 9.0 and 24.0 h of the reaction begins. When it is stock, the absorption peak of methanol extract gave two non-smooth or trucks. From the UV–Vis absorbance of methanol and chloroform extracts, it was observed that the reduction of silver ions with chloroform extract was extremely smaller than methanol extracts

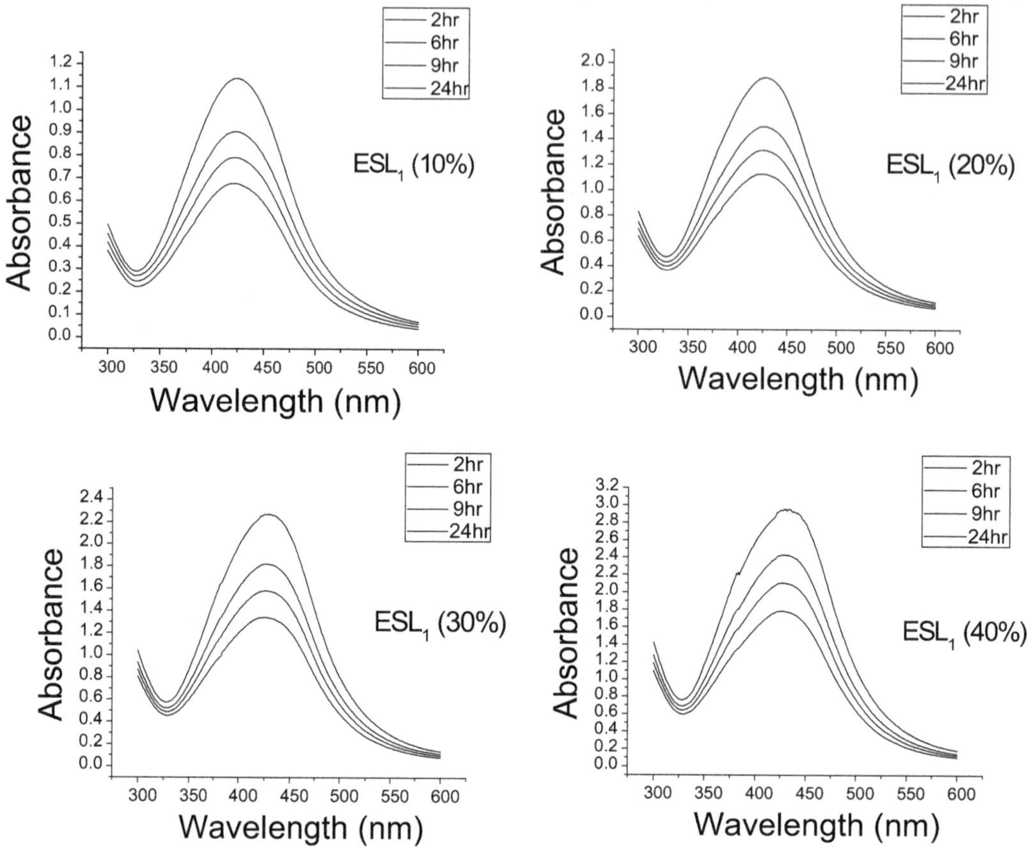

Fig. 6 UV–Vis absorbance of methanol extract of *Euclea schimperi* leaves of different concentrations (10, 20, 30 and 40% v/v) of 5 mg/mL stock solution separately at different time. As the concentration of the extract increases the absorbance of the mixture also increases

be described as easily oxidizable if its positive oxidation potential is as small as possible (Figs. 3 and 4). Both CV and SWV gave three different peaks; the only different was SWV that can scatter the peaks clearly.

As illustrated in Figs. 3 and 4 above, the voltammetric analysis of *E. schimperi* leaves extract with methanol solvent gives three different sharp peaks at the oxidation side. This indicates that the extract contains three reductant or antioxidant species that can be either three different group of reductant or a reductant that can be reduced to another substance in step-wise process. Out of these three different peaks, one was very long and sharp indicating that this represents the major component of the three antioxidant components.

As a result, *E. schimperi* leaves extract can act as antioxidant species because there were no any peaks at the reduction side (all three peaks were obtained at the oxidation side). The peak parameters obtained from the cyclic voltammograms of each sample are tabulated in Table 4.

Synthesis of silver nanoparticles

On account of noble metal nanoparticles are widely applied to areas of human health, there is a growing interest to develop environmentally friendly processes for nanoparticles synthesis [30]. Biological methods of synthesis have paved way for the "greener synthesis" of nanoparticles and these have proven to be better methods due to slower kinetics, offer better manipulation and control over crystal growth and their stabilization [31, 32].

Phytochemicals are involved directly in the reduction of the ions and formation of silver nanoparticles using biological methods [30, 33]. *E. schimperi* leaf extracts contain a variety of phytochemicals such as polyphenols, flavonoids and terpenoids which help in the reduction of metal ion to form nanoparticles. Preliminary identification of nanoparticle formation was carried out by observing the color change of the reaction mixture. It was observed that when $AgNO_3$ solution was mixed with the plant extracts, brown coloration was observed due to the formation of silver nanoparticles. This is a clear indication of the formation of silver nanoparticles [34].

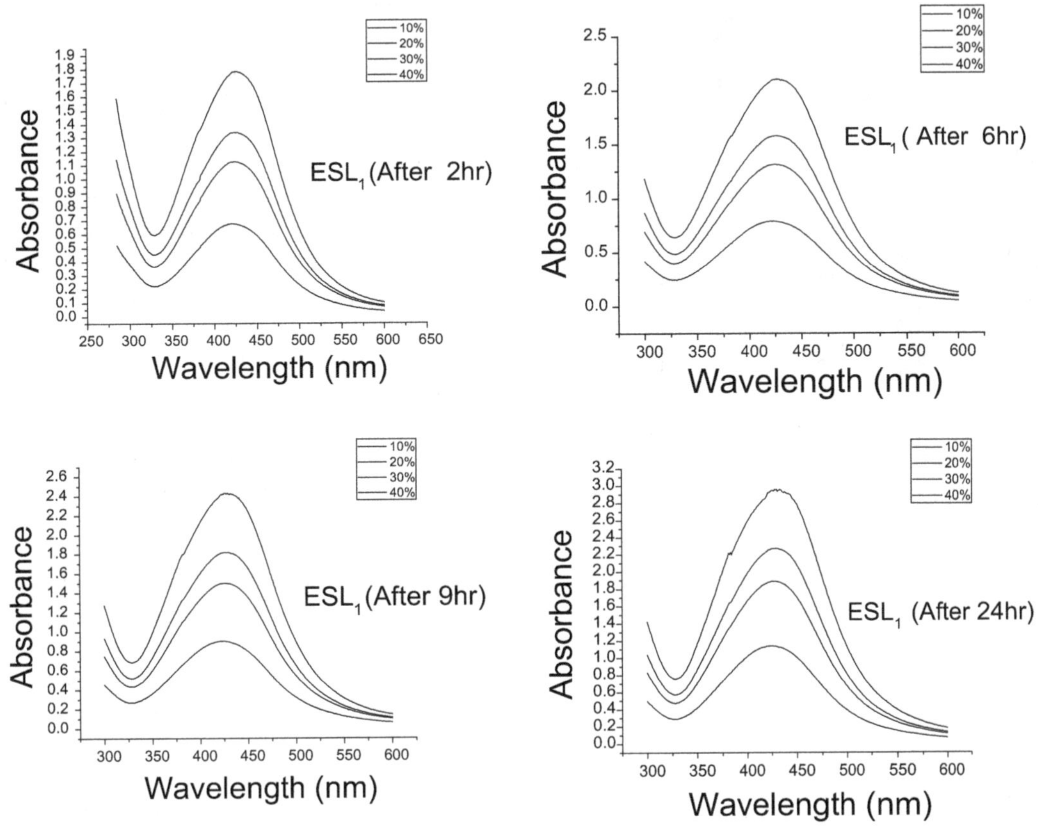

Fig. 7 UV–Vis absorbance of different concentration of methanol extract of *Euclea schimperi* leaves after different reaction time. As the contact or reaction time increases, the formation of silver nanoparticles increases which leads to an increases in absorbance of the mixture. The peak observed in the methanol extract is clearly visible around 428 nm which indicates the reduction of silver ions

Table 5 Comparison of inhibition zone in mm among different crude extracts of *Euclea schimperi* leaves and positive controls against standard drug resistance Gram-positive and Gram-negative bacteria

Types of extracts or positive control	Types of bacteria strain			
	Gram positive		Gram negative	
	S. aurous	*S. pneumoniae*	*P. aeruginosa*	*K. pneumoniae*
Methanol extract	20.67 ± 1.53^{a}	0.00^{a}	17.00 ± 1.00^{a}	0.00^{a}
Chloroform	15.00 ± 1.00^{b}	0.00^{a}	14.33 ± 0.58^{b}	0.00^{a}
Erythromycin	23.00 ± 0.05	22.00 ± 0.20	–	–
Chloramphenicol	24.20 ± 0.30	23.00 ± 0.07	–	–
Gentamycin	–	–	14.00 ± 0.05	18.00 ± 0.35
Ciprofloxacin	–	–	30.30 ± 1.23	20.30 ± 0.89

S. pneumoniae, Streptococcus pneumoniae; S. aureus, Staphylococcus aureus; P. aeruginosa, Pseudomonas aeuroginosa; K. pneumonia, Klebsiella pneumonia

Values in the same column that are followed by a different letters (a, b) are significantly different $p < 0.05$ by paired *t* test

Therefore, the present investigation demonstrates the formation of silver nanoparticles by the reduction of aqueous silver ions from *E. schimperi* leaf extracts using different solvents. The properties of the nanoparticle and its reduction time depend on various characteristics of plant extract, namely concentration of the metal salt and the extracts, contact time and reaction temperature. Nevertheless, this study considered two major factors, contact time and concentration of the plant extracts. As a result, the formation of silver nanoparticles by reduction

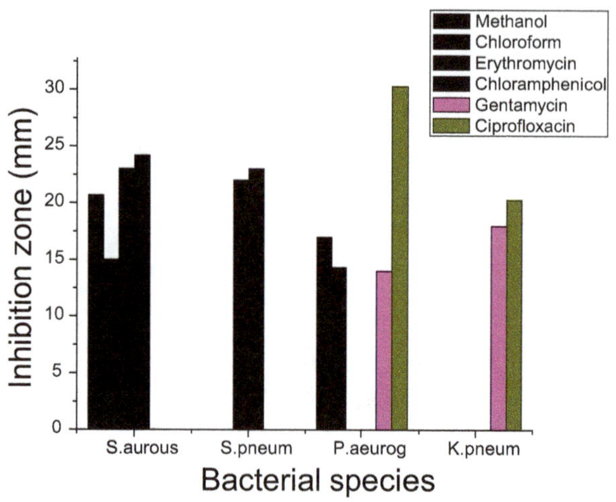

Fig. 8 Comparison of inhibition zone among crude extracts of *Euclea schimperi* leaves using different solvent (methanol and chloroform) against standard Gram-positive and Gram-negative bacteria. Standardized inoculums of the test organisms were inoculated into series of sterile tubes of 2 mL nutrient broth containing 500, 250, 125, 62.5 and 31.25 mg/mL of each extracts and incubated at 37 °C for 24 h. The methanol extract showed maximum antibacterial activity against *S. aureus* and *P. aeuroginosa* with diameter of 20.67 ± 1.53 and 17 ± 1 mm at concentrations of 500 mg/mL, respectively. Chloroform extract also exhibited antibacterial activity against *S. aures* and *P. aeuroginosa* with diameter of 15 ± 1 and 14.33 ± 0.58 mm at concentrations of 500 mg/mL, respectively

of silver ions was measured with different concentrations (10, 20, 30, 40% v/v) of *E. schimperi* leaf extracts at different reaction time (2, 6, 9 and 24 h).

When it is stock, the absorption peak of methanol extract gave two non-smooth or trucks. This may be due to higher deposition of silver nanoparticles. From the UV–Vis absorbance of methanol and chloroform extracts of *E. schimperi* leaves, it was observed that the reduction of silver ions with chloroform extract was extremely smaller than methanol extracts (Fig. 5).

Initially, the reacting mixture was a slightly yellowish liquid; as the reaction proceeded, the solutions became dark brown. Different concentrations (10, 20, 30 and 40%) of methanol extract of *E. schimperi* leaves result in the formation of different absorption peaks. As the concentration of the extract increases the absorbance of the mixture also increases (Fig. 6). The visual inspection of the color change of the mixture from yellow to intense brown was demonstrated with increase in concentration of the leave extracts. This also ascertains the dependence of the formation of silver nanoparticles on the concentration of the extracts. The absorbance of silver nanoparticles in the presence of different concentrations of the extract at different time can be found in Fig. 6.

Another point to consider is as the contact or reaction time increases, the formation of silver nanoparticles increases which leads to an increases in absorbance of the mixture. This result is shown in Fig. 7.

Additionally, as both the concentration of the plant extract and the reaction time increase, non-smooth part of the peak around the same wave length region was increased. This is due to the increase in the concentration of silver nanoparticles formed in the reaction.

As can be seen from the Figs. 6 and 7 above, the peak observed in the methanol extract of *E. schimperi* leaves is clearly visible around 428 nm which indicate the reduction of silver ions.

Determination of antibacterial activity of *E. schimperi* leaf extracts

Agar disk diffusion method

Methanol extract of *E. schimperi* leaves exhibited prominent antibacterial activity against both one Gram-positive and one Gram-negative bacterial species. The chloroform extract of the plant sample also exhibited inhibition zone on some bacterial species but it was not as strong as methanol extract. The inhibition zones obtained by each solvent extract of leaves of *E. schimperi* and the control against each tested microorganism are summarized in Table 5. The results were expressed as mean ± standard deviation of triplicate measurements.

As shown in Fig. 8 and Table 5, zone of inhibition of the tested extract against different bacterial species of the methanol extract showed maximum antibacterial activity against *S. aureus* and *P. aeuroginosa* with diameter of

Table 6 MIC determination of *Euclea schimperi* leave extracts against Gram-positive and Gram-negative bacteria

Bacterial species	Types of extract	Concentration (mg/mL)					
		500.0	250.0	125.0	62.5	31.25	MIC
S. aureus	Methanol	−	−	−	−	+	62.5
	Chloroform	−	−	−	+	+	125
P. aeuroginosa	Methanol	−	−	−	−	+	62.5
	Chloroform	−	−	−	+	+	125

−, inhibition of bacterial growth; +, turbidity (slight bacterial growth)

20.67 ± 1.53 and 17 ± 1 mm at concentrations of 500 mg/mL, respectively. Chloroform extract also exhibited antibacterial activity against *S. aures* and *P. aeuroginosa* with diameter of 15 ± 1 and 14.33 ± 0.58 mm at concentrations of 500 mg/mL, respectively. The antimicrobial activity of methanol leaf extracts of *E. schimperi* was stronger than chloroform extracts towards the tested pathogens. This substantiates that more polar bioactive compounds are extracted more by methanol than chloroform. In other words, alcoholic extracts of tested plants provided better antibacterial effect than other solvents [35]. Both methanol and chloroform extracts showed no activity against *S. pneumoniae* and *K. pneumoniae* (multidrug resistance).

The presence of zone of inhibition around the well was considered positive. The extracts were found to cause marked inhibition of bacteria in a dose-dependent manner. Among bacteria, Gram-positive bacteria have shown higher sensitivity to the extracts compared to Gram-negative bacteria. The inhibition caused by methanol extract was lesser than that of standard antibiotic since the standard antibiotic are found in pure form [36].

Alkaloids, tannins, saponins, glycosides and steroid derived from plants have been shown to have antimicrobial effect and pharmacological activities [37, 38]. Flavonoids, phenols, saponins, steroids, tannins, and terpenes detected from the extracts are responsible to observed antibacterial action of *E. schimperi* leaf extracts. In summary, this antibacterial action may serve as justification to the traditional use of the *E. schimperi* plant against various human disorders.

Determination of minimum inhibitory concentration (MIC)

The MIC was taken as the lowest concentration that prevented bacterial growth and it was determined using a broth dilution test.

As shown in Table 6, the MIC values were evaluated relatively at higher concentrations of the methanol (62.5 mg/mL) and chloroform (125 mg/mL) extracts. Correspondingly, methanol extract afforded notable MIC value at 62.5 mg/mL against both *S. aureus* and *P. aeuroginosa*. An implication of these findings is that *E. schimperi* leaf extracts can inhibit some of the bacterial species at relatively high concentration.

Conclusion

The outcome of our investigation clearly indicated that *E. schimperi* leaf extracts were found to contain high amount of total phenolic and small amount of flavonoids content. Likewise, the extracts revealed significant antioxidant activities against various tests; FRAP, DPPH, voltammetric analysis and silver nanoparticle reduction. Based on the results, the methanol extract showed higher reducing capacity, DPPH radical scavenging activity, anodic peak current and smaller positive oxidation potential than the chloroform extract. The methanol extract can also reduce silver ion and able to form significant amount of silver nanoparticles. Last, the result can also prove the effectiveness of *E. schimperi* leaf extracts for antimicrobial activity.

In general, the methanol extracts demonstrated superior antioxidant and antimicrobial effects than chloroform extracts, which is related to the quantity of phytochemicals present in the crude extracts. To this end, the findings are quite convincing to scientifically confirm the traditional use of this plant for medicinal purpose by the population.

Authors' contributions

AM and MA conceived and designed the experiments: AM analyzed data and wrote the paper; BK performed the experiments. All authors participated in the revisions of the manuscript. All authors read and approved the final manuscript.

Acknowledgements

The authors are thankful to Bahir Dar University for the provision of lab facility and Dr. Maereg Amare for obtaining voltammetric data.

Competing interests

The authors declare that they have no competing interests.

Funding

Not applicable.

References

1. Ignacimuthu S, Ayyanar M, Sankara Sivaraman K. Ethnobotanical investigations among tribes in Madurai district of Tamil Nadu (India). J Ethnobiol Ethnomed. 2006;11(25):1–7.
2. Hammer KA, Carson CF, Riley TV. Antimicrobial activity of essential oils and other plant extracts. J Appl Microbiol. 1999;86:985–90.
3. Giday M, Asfaw Z, Elmqvist T, Woldu Z. An ethnobotanical study of medicinal plants used by the Zay people in Ethiopia. J Ethnopharmacol. 2003;85:43–52.
4. Sukhdev S, Suman PSK, Gennaro L, Dev DR. Extraction technologies for medicinal and aromatic plants. Trieste: International Centre for Science and High Technology; 2008.
5. Hasler CM, Blumberg JB. Phytochemicals: biochemistry and physiology. Introduction. J nutr. 1999;129(3):756S–7S.
6. Evans WC. General methods associated with the phytochemical investigation of herbal products. Trease and Evans pharmacognosy. 15th ed. Philadelphia: Saunders/Elsevier; 2008. p. 137–48.
7. Hamilton AC. Medicinal plants, conservation and livelihoods. Biodivers Conserv. 2004;13:1477–517.
8. Wang H, Liu YM, Qi ZM, Wang SY, Liu SX, Li X, Wang HJ, Xia XC. An overview on natural polysaccharides with antioxidant properties. Curr Med Chem. 2013;20:2899–913.
9. Liu Y, Sun Y, Huang G. Preparation and antioxidant activities of important traditional plant polysaccharides. Int J Biol Macromol. 2018;111:780–6.
10. Huang G, Mei X, Hu J. The antioxidant activities of natural polysaccharides. Curr Drug Targets. 2017;18(11):1296–300.
11. Nwachukwu CU, Umeh CN, Kalu IG, Okere S, Nwoko M. Identification and traditional uses of some common medicinal plants in Ezinihitte Mbaise LGA, of Imo State, Nigeria. Rep Opin. 2010;2:1–10.

12. De Moura KC, Emery FS, Neves-Pinto C, Pinto MD, Dantas AP, Salomão K, Castro SL, Pinto AV. Trypanocidal activity of isolated naphthoquinones from Tabebuia and some heterocyclic derivatives: a review from an interdisciplinary study. J Br Chem Soc. 2001;12:325–38.
13. Gebre-Mariam T, Neubert R, Schmidt PC, Wutzler P, Schmidtke M. Antiviral activities of some Ethiopian medicinal plants used for the treatment of dermatological disorders. J Ethnopharmacol. 2006;104:182–7.
14. Mikayel G, Margarit P, Armen T. Antimicrobial activity of some plant materials used in Armenian traditional medicine. BMC Complement Altern Med. 2017;17:50–8.
15. Gohari AR, Hajimehdipoor H, Saeidnia S, Ajani Y, Hadjiakhoondi A. Antioxidant activity of some medicinal species using FRAP assay. J Med Plants. 2011;10(37):54–60.
16. Amin I, Norazaidah Y, Hainida KE. Antioxidant activity and phenolic content of raw and blanched *Amaranthus* species. Food Chem. 2006;94:47–52.
17. Chang CC, Yang MH, Wen HM, Chern JC. Estimation of total flavonoid content in propolis by two complementary colorimetric methods. J Food Drug Anal. 2002;10:178–82.
18. Milan C, Hana C, Petko D, Maria K, Anton S, Antonin L. Different methods for control and comparison of the antioxidant properties of vegetables. Food Control. 2010;21:518–23.
19. Louli V, Ragoussis N, Magoulas K. Recovery of phenolic antioxidants from wine industry by-products. Bioresour Technol. 2004;92:201–8.
20. Shankar SS, Rai A, Ahmad A, Sastry M. Rapid synthesis of Au, Ag, and bimetallic Au core–Ag shell nanoparticles using Neem (*Azadirachta indica*) leaf broth. J Colloid Interface Sci. 2004;275:496–502.
21. Enzo AP. Traditional medicinal plant extracts and natural products with activity against oral bacteria: Potential application in the prevention and treatment of oral diseases. Evid Based Complement Altern Med. 2011;2011:1–15.
22. Akinyemi KO, Oladapo O, Okwara CE, Ibe CC, Fasure KA. Screening of crude extracts of six medicinal plants used in South-West Nigerian unorthodox medicine for anti-methicillin resistant *Staphylococcus aureus* activity. BMC Complement Altern Med. 2005;5(6):1–7.
23. Tahiya HAA, Amira HSAM, Mohammad AH, Afaf MW, Qasim AR. Comparative study of phytochemical screening, antioxidant and antimicrobial capacities of fresh and dry leaves crude plant extracts of *Datura metel* L. J King Saud Univ. 2014;26(3):237–43.
24. David IG, Bizgan AM, Popa DE, Buleandra M, Moldovan Z, Badea IA, Tekiner TA, Basaga H, Ciucu AA. Rapid determination of total polyphenolic content in tea samples based on caffeic acid voltammetric behaviour on a disposable graphite electrode. Food Chem. 2015;173:1059–65.
25. Jelen F, Kourilova A, Hason S, Kizek R, Trnkova L. Voltammetric study of adenine complex with copper on mercury electrode. Electroanalysis. 2009;21(3–5):439–44.
26. Adam V, Fabrik I, Kizek R, Eckschlager T, Stiborova M, Trnkova L. Vertebrate metallothioneins as target molecules for analytical techniques. TrAC Trends Anal Chem. 2010;29(5):409–18.
27. Hynek D, Krejcova L, Sochor J, Cernei N, Kynicky J, Adam V, Trnkova L, Hubalek J, Vrba R, Kizek R. Study of interactions between cysteine and cadmium (II) ions using automatic pipetting system off-line coupled with electrochemical analyser. Int J Electrochem Sci. 2012;7(3):1802–19.
28. Kilmartin PA, Zou H, Waterhouse AL. Correlation of wine phenolic composition versus cyclic voltammetry response. Am J Enol Vitic. 2002;53(4):294–302.
29. Huang T, Gao P, Hageman MJ. Rapid screening of antioxidants in pharmaceutical formulation development using cyclic voltammetry-potential and limitations. Curr Drug Discov Technol. 2004;1(2):173–9.
30. Song JY, Kim BS. Rapid biological synthesis of silver nanoparticles using plant leaf extracts. Bioprocess Biosyst Eng. 2009;32(1):79–84.
31. Sonvico F, Dubernet C, Colombo P, Couvreur P. Metallic colloid nanotechnology, applications in diagnosis and therapeutics. Curr Pharm Des. 2005;11(16):2091–105.
32. Shipway AN, Katz E, Willner I. Nanoparticle arrays on surfaces for electronic, optical, and sensor applications. Chemphyschem. 2000;1(1):18–52.
33. Bar H, Bhui DK, Sahoo GP, Sarkar P, De SP, Misra A. Green synthesis of silver nanoparticles using latex of *Jatropha curcas*. Colloids Surf A Physicochem Eng Asp. 2009;339:134–9.
34. Jain D, Daima HK, Kachhwaha S, Kothari SL. Synthesis of plant-mediated silver nanoparticles using papaya fruit extract and evaluation of their anti microbial activities. Dig J Nanomat Biostruct. 2009;4:557–63.
35. Bakht N, Humaira F, Madiha A, Ihsan-ul-Haq. Recent trends and methods in antimicrobial drug discovery from plant sources. Austin J Microbiol. 2015;1(1):1–12.
36. Prescott LM, Harley JP, Klein OA. Microbial nutrition, types of media. New York: McGraw Hill Publishers; 2005. p. 95–105.
37. Ieven M, Vanden BD, Mertens F, Vlietinck A, Lammens E. Screening of higher plants for biological activities I. Antimicrobial activity. Planta Medica. 1979;36:311–21.
38. Frei B, Heinrich M, Bork PM, Herrmann D, Jaki B, Kato T, Kuhnt M, Schmitt J, Schühly W, Volken C, Sticher O. Multiple screening of medicinal plants from Oaxaca, Mexico: ethnobotany and bioassays as a basis for phytochemical investigation. Phytomedicine. 1998;5:177–86.

15

Antifungal effect of aqueous and ethanolic extracts of neem leaves, stem bark and seeds on fungal rot diseases of yam and cocoyam

Chukwuma S. Ezeonu[1*], Chinedu Imo[1], Dawn I. Agwaranze[2], Assumpta Iruka[1] and Abraham Joseph[1]

Abstract

Background: Fungal rot diseases of white yam (*Discorea rotundata*) tubers and cocoyam (*Colocasia esculenta*) are the major post-harvest challenges in the availability of these root tubers. The use of biological extract of plants will prevent post-harvest losses without any negative health challenge.

Methods: Microwave-assisted method of extraction was used for both aqueous and ethanolic extracts of neem samples. Fungal-rotted yam and fungal-rotted cocoyam samples were collected, and fungi from them were cultured on potato dextrose agar (PDA). The same medium was used for subculturing until pure fungal isolates were obtained. Pathogenicity tests were carried out on the yam and cocoyam tubers by inoculating with isolated fungi. The neem extracts were integrated into the PDA and their antifungal effects examined.

Results and discussion: From the yam, *Aspergillus ochareus*, *Sclerotium rolfsii* and *Lasidiodiplodia theobromae* were isolated, while *Aspergillus niger*, *Aspergillus oryzae* and *Rhizopus stolonifer* were isolated from cocoyam samples. After 14 days of inoculation, isolated fungi were shown to be highly pathogenic except *R. stolonifer* and *A. ochareus*, which showed moderate pathogenicity. All the plant extracts showed antifungal activities at varied percentage mean zones of inhibition. The inhibition against fungal growth from the isolates of the rotted cocoyam gave mean inhibition between 39.53 ± 8.39% and 96.76 ± 1.60%. Except for ethanolic stem bark extract which gave 39.53 ± 8.39% limited zone of inhibition against *A. niger* growth, the rest of the extract exhibited good zones of inhibition against cocoyam-isolated fungi ranging between 72.87 ± 3.55 and 96.79 ± 1.60%. On fungal isolates from yam, the mean percentage zones of inhibition of all extracts of the neem plant parts showed more effectiveness than the positive control (Ketoconazole) against *L. theobromae* growth. Aqueous stem bark extract with 93.80 ± 1.78% zone of inhibition showed comparable effectiveness to the control (100% inhibition) against *A. ochareus* growth, while against *S. rolfii* growth, there was good inhibition by all the extracts though not to the extent as that of the control.

Conclusion: The fungitoxic potential of these plant extracts on rot-inducing fungi of stored yam and cocoyam suggests that the above extracts used in this study can be put to use by farmers as alternative to commercial/synthetic fungicides.

Keywords: Antifungal, Fungi, Inhibition, Neem, Pathogenicity

*Correspondence: chuksmaristos@yahoo.com
[1] Department of Biochemistry, Federal University Wukari, Wukari, Taraba State, Nigeria
Full list of author information is available at the end of the article

Background

Neem plant (*Azadirachta indica*) is well known for its medicinal properties. Known with different names in different parts of the world (*Indian Lilac, Margosa, Dogonyaro*), neem has been well established for its medicinal properties such as in treatment of various ailments. In Africa, it is popular in the treatment of malaria and for its effectiveness as an antibacterial agent. Recently, *A. indica* like other similar botanicals is gaining prominence in its application in treatment of plant-related diseases. In a research conducted by Mahmoud et al. [1], 5% aqueous leaf extract of neem was shown to cause inhibition in growth of six tested fungal pathogens (*Aspergillus flavus, Aspergillus fumigatus, Aspergillus niger, Aspergillus terreus, Candida albicans* and *Microsporum gypseum*). In another research, aqueous neem extracts inhibited *A. niger* more than *C. albicans*, while alcohol neem extract inhibited *C. albicans* better than *A. niger* [2]. Neem plant is effective against fungal growth, a quality which may be used in increasing the shelf life of plant tubers after harvest.

Yam (*Discorea* spp) belonging to the genus "Dioscorea" is a monocotyledon which is annually produced worldwide. According to IITA [3, 4], Nigeria accounting for 71.9% of yam production worldwide remains the principal producer. Yam-production zones show dominant stretches from Cote d'ivoire through Ghana, Togo, Benin, Nigeria, Cameroon, Gabon, Central African Republic and the Western part of the democratic Republic of Congo, while in East Africa, Sudan and Ethiopia are the major yam producers [5]. According to Opara [6], soft rots in yam are caused mainly by fungi especially *Penicillium* spp, *Fusarium oxysporum* and *Botrydiplodia theobromae*, while dry rots among others is principally caused by *Rosselinia* and *Sphaerostilbe*. Other fungi also include *Rhizopus nodosus* and *Fusarium solani*. Infection in the field can persist and lead to rotting during storage, and control strategy involves treating tubers with systemic fungicide or alkaline material such as Bordeaux mixture and providing adequate aeration and inspecting stored tuber regularly [6]. However, the ideal treatment would be the use of organic fungicide such as neem plant (leave, root and stem bark) extracts; since neem itself is commonly consumed as herb and useful as traditional medicine. Therefore, it should be less toxic and biologically healthy in food preservation. Cocoyam belongs to the *Araceae* family [7]. *Colocasisa esculenta*, otherwise known as taro, is more popular in Nigeria than *Xanthosoma sagittaifolium* variety, also known as tannia [8]. Some of the popular uses of cocoyam in Nigeria include use as soup thickener, consumption as porridge with beans, cassava chips and vegetables as well as formation into chips. It is used as antipoisoning and wound-healing agents by applying the fresh mash in the opening of sores. Thus, it possesses antibiotic properties. It is also formed into flour. This important tuber is consumed both by human and domestic animals. Boiling and fermenting before consumption help to reduce the oxalic acid content in it, which causes itching in some local varieties. Cocoyam is a perennial crop grown mainly for its edible roots [7]. NRCRI [9] report stated that after cassava and yam, cocoyam is the third most important crop among the root and tuber crop cultivated and consumed in Nigeria, and it is nutritionally superior to cassava and yam. Cocoyam is grown for its edible starchy corm and leaves, and is a major source of cheap vegetable carbohydrates, protein, fats, crude fibre, ash, carotene, thiamine, riboflavin, niacin, vitamin C and are more digestible compared to other root crops [10]. Cocoyam is a tuber crop consumed in all states in the south-east [11] and most parts of north-central Nigeria, being a popular root tuber. FAOSTAT [12] stated that in 2008, Nigeria was estimated to produce 5.39 million metric tonnes out of a total of 11.77 million metric tonnes of cocoyam world production. The major setback in cocoyam production according to Igbozulike [7] is deterioration of the corms, and the factors responsible for the loss in quality have been identified to include physiological, mechanical, rodents, birds, insect infestation as well as pathogenic, but most importantly fungal rot.

Methods

Plant collection

Seeds of neem were collected from tree plantation at Mission quarters, Wukari, Taraba State, Nigeria. The mature healthy seeds, leaves and stem bark were also collected randomly from different neem trees. The leaves, stem bark and seeds were dried after being washed under running tap water to eliminate dust and other foreign particles.

Sterilization of materials

All glass wares used in this study were washed with detergent, rinsed and sterilized in a dry, ventilated oven at 160 °C for 2 h. All media were sterilized by autoclaving at a temperature of 121 °C and 15 psi for 20 min. The scalpel, cork borer and inoculating needle were sterilized by dipping them into 70% ethanol and passing them over a Bunsen burner flame until red hot.

Ethanolic and aqueous extraction (microwave-assisted extraction—MAE)

The neem leaves, stem bark and seeds were grounded manually to form the smallest possible size after drying and sieved. The sample (20 g) was measured and added into the extraction container. Ethanol or distilled water (separately for ethanol and aqueous extraction) (200 ml) was then added, tightly covered and placed in the microwave oven. The oven door was closed and the power set between 400 and 680 Watts and allowed to run for a short burst of 3 min and stopped. The set-up was carefully removed and allowed to cool before re-introducing it for another burst of 3 min. This process was repeated until achieving a cumulative extraction period of 45 min. The set-up was allowed to cool before it was opened and filtered. The filtrate was then concentrated and used for antifungal assays.

Preparation of culture medium

Throughout the study, the assayed culture medium employed was LAB M Potato Dextrose Agar (PDA). This medium was used for the growth and maintenance of the fungal isolates. The preparation of Potato Dextrose Agar (PDA) was done according to the manufacturer recipe (39 g in 1 l of water). The medium was sterilized by autoclaving at 121 °C and 15 psi for 20 min for complete dissolution and homogeneity. Thereafter, it was allowed to cool to temperature of between 42 and 45 °C. One capsule of chloramphenicol was added to every 500 ml of sterile cooled PDA so as to prevent bacteria growth [13]. Approximately 15 ml of the cooled amended PDA was poured into each sterile petri dish of 8.6 cm (86 mm) diameter to solidify. The petri dishes that contained the medium were incubated for 24 h at ambient temperature (28 °C) to check for sterility before use as described by Cheesebrough [14].

Isolation of fungi

The isolation technique used by Onyike and Maduewesi [15] was employed in this study. A small section of the yam or cocoyam tissue showing advancing margin of rot and adjoining healthy tissue were cut using sterilized scalpel and cork borer whose surface was sterilized with 70% ethanol. The peeled portions of the yam or cocoyam obtained with the cork borer were placed on the solidified agar. Three peeled portions were placed per plate with equal distance between them. Three replicate plates for each of the rotten portion of site were made for each yam or cocoyam variety. The plates were incubated at 27 ± 2 °C for 7 days. Fungi associated with the yam or cocoyam rot affected tissues were observed and the frequency of isolation determined using method of Okigbo and Ikediugwu [16]. Subculturing was done to obtain pure cultures of the isolates.

Identification of fungal isolates

Subculturing of the isolates was made to obtain pure culture. The colonies growing on the plates were identified macroscopically and microscopically. Some examples are shown in Fig. 1a–f below. Colony colour, type (compact, loose, aerial hyphae), texture (velvety, cottony, coarse) shape and growth pattern were observed. Direct observation of culture under the light microscope (low power) by careful preparation of slides, staining with cotton blue-in-lactophenol was done. Detailed drawings of the diagnostic features and identification manual and guides according to Alexopoulos [17]; Nelson et al. [18]; Rippon [19]; Samson et al. [20]; and Snowdon [21] were used.

Pathogenicity test

The pathogenicity test was carried out to establish which of the fungal isolates caused the rot and to determine whether they could induce similar symptoms on inoculation and be re-isolated, thus fulfilling Koch's postulates. The method of Okafor [22], Okigbo and Ikediugwu [16] were adopted for the pathogenicity study. The pure fungal isolates obtained from infected yam tubers of *Discorea rotundata* and Cocoyam corm of *Colocasia esculenta* variety, respectively, for the period of the collection were used for inoculation. Healthy yam tubers (*D. rotundata* variety) and Cocoyam corm (*C. esculenta* variety) were surface-sterilized with 70% ethanol. Each healthy tuber was bored into about 1 cm deep, with a sterile 6 mm diameter cork borer at three different points on the yam and cocoyam surface (proximal, middle and distal regions). Another 6 mm sterile cork borer was used to cut about 5 mm of mycelia disc from edge of a 5 days old culture of each fungus isolate. The mycelia discs were used to inoculate the holes created by scooping out the plant tissue. The scooped out tissue of the yam and cocoyam were replaced after 5 mm pieces had been cut off to compensate for the thickness of the fungal culture. Two whole tubers of yam and two whole corm of cocoyam were inoculated per fungus. The control set-up consists of tubers that were similarly bored into and inoculated with sterilized PDA agar discs.

The wounds were sealed with petroleum jelly and inoculated tubers were placed in transparent polythene bags whose inside has been moistened with cotton wool soaked in sterile distilled water to maintain a high humidity. The inoculated tubers and corms were kept in the laboratory at room temperature for about 14 days and assessed for rot development by cutting through the points of inoculation where rots developed. The pathogens were re-isolated as previously described and their cultural and morphological characteristics were compared with those of the original isolates.

Antifungal activity of the extracts

Effect of plant extract on mycelia growth of the test fungi was studied using the food poisoning techniques [23]. One millilitre (1 ml) of each plant extract was dispensed per petri dish and 9 ml of the media (molten PDA) was added to each of the petri dish containing extract and carefully spread evenly over the plate, this gave rise to PDA–extract mixture with 10% extract concentration. This was used for examining the inhibition of mycelia growth. The plates were gently rotated to ensure even dispersion of the extracts. The agar extract mixture was allowed to solidify and then inoculated at

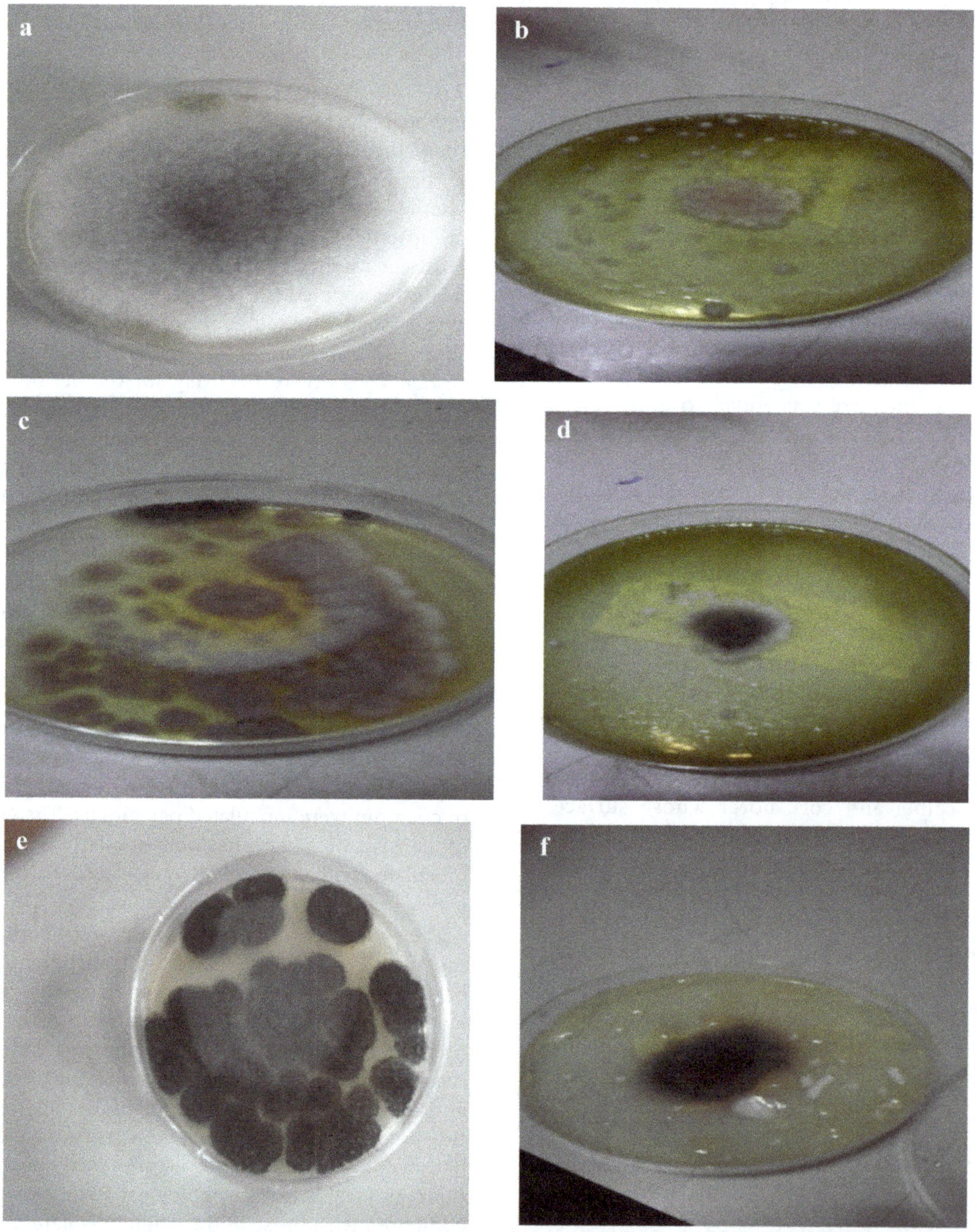

Fig. 1 Images of the pure isolates and inhibited isolates from cocoyam. **a** Rhizopus stolonifera. **b** Plate with inhibitions against Rhizopus stolonifera. **c** Aspergillus oryzae. **d** Plates with inhibitions against Aspergillus oryzae. **e** Aspergillus niger. **f** Plate with inhibitions against Aspergillus niger

the centre with a 4 mm diameter mycelia dish obtained from the colony edge of 7-day old pure cultures of each of the test fungi. Each treatment consists of three replicates. The negative control set-up consist of blank agar plate (no extract) inoculated with the test fungi as described above. Petri-dishes dispensed with molten PDA and one ml of ketoconazole dissolved in distilled water inoculated with each test fungus served as the commercial fungicides. All the plates were incubated at 28 °C for 5 days and examined daily for growth and presence of inhibition. Colony diameter was taken as the mean growth along two directions on two pre-drawn perpendicular lines on the reverse side of the plates. The effectiveness of the extract was recorded in terms of percentage inhibition, which was calculated according to the method described by Whipps [24].

$$\text{Percentage inhibition} = R1 - R2 \times \frac{100}{R1},$$

where $R1$ is the farthest radial distance of pathogen in control plate, while $R2$ is the farthest radial distance of pathogen in extract-incorporated agar plates.

Results and discussion

Pathogenicity test

From the yam, *A. ochareus, S. rolfsii* and *L. theobromae* were isolated, while *A. niger, A. oryzae* and *R. stolonifer* were isolated from cocoyam samples (Tables 1 and 2). Pathogenicity tests in cocoyam revealed that *A. niger, A. oryzae* and *R. stolonifer* caused rot. The nature of rot varies between the inoculated cocoyam tubers with various selected fungi. *A. niger* shows dry and soft rotting, and *A. oryzae* caused dry rot (Table 3). Among isolated fungi (Fig. 2), *A. niger* and *A. oryzae* are the most pathogenic, while *R. oryzae* is the least pathogenic as evident from the weight loss. There was obvious reduction in weight observed in corms of cocoyam exposed to fungal pathogenicity. Order of weight reduction from high to low in the cocoyam exposed to the fungi was 23.43 g (*A. niger*), 16.29 g (*R. stolonifer*) and 14.84 g (*A. oryzae*). The reduction in weight of the control (uninfected cocoyam) was only 1.73 g. Considering the moderate sizes of the cocoyam in Fig. 1, there was indeed high loss in weight and thus great pathogenicity of fungi examined.

Pathogenicity tests in yam tubers (Table 4) revealed rots in them as a result of infestation with *L. theobromae, A. ochareus* and *S. rolfsii.* The nature of rot varies between the inoculated yam tubers with various selected fungi. *L. theobromae* shows wet rotting, *S. rolfsii* caused rot with dirty white colour. Among isolated fungi, *A. ochareus* is the least pathogenic as evident from the weight loss (Fig. 3). Weight loss recorded due to fungal pathogenicity test in yam is shown in the order of weight loss per fungi from the most to the least as 326.25 g (*S. rolfsii*), 199.47 g (*L. theobromae*) and 164.54 g (*A. ochareus*). The control which was not infected showed only 6 g decrease in weight after 14 days. These results established the fact that these fungi has high pathogenicity as depicted by the drastic reduction in weight as well as the tissue rot sighted thereafter.

These results demonstrated the fact that all the fungi investigated for pathogenicity of both yam and cocoyam are indeed pathogenic.

Antifungal activity of the extracts on fungi isolated from cocoyam

Analysis of variance at 95% showed that all the parts (leaf, bark and seed) of the test plant (neem) used in this study significantly ($P<0.05$) inhibited the growth of the fungal organisms against cocoyam rot (Table 5). Ethanolic seed extract (ESE) had the highest inhibitory effect against *A. oryzae* (91.47 ± 2.93% inhibition), while it was also effective against *R. stolonifer* (88.37 ± 1.17% inhibition), and *A. niger* (84.88 ± 3.07% inhibition). The antifungal effects of aqueous seed extract (ASE) on those fungi showed inhibitions of 85.66 ± 2.92% (*A. oryzae*), 81.78 ± 2.92% (*A. niger*) and 79.07 ± 2.33% (*R. stolonifer*), which though without any significant difference, ($P>0.05$) compared to that of ethanolic seed extract (figures are shown above), showed lower inhibition. Thus, both ESE and ASE showed similar antifungal activities.

Ethanolic leave extract (ELE) showed great growth inhibitory effect of 90.70 ± 2.33% against *A. oryzae*, which was significantly higher ($P<0.05$) than the growth

Table 1 Macro- and micro-features of fungi isolates from Yam (*Discorea rotundata*)

S/no	Isolates	Colony characteristics	Microscopy
1.	*Aspergillus ochareus*	Growth on PDA was rapid and fast. Powdery yellow colour almost covering the plates after 72 h	Non-septate conidiophores arising from thick-walled foot cells. Each conidiophore ends in a terminal-enlarged spherical swellings. Conidia borne by phialides arising from a terminal swelling on the conidiophores
2.	*Lasiodiplodia theobromae*	Dirty white with black underneath	Mycelium are septate; chlamydospores are intercalary and terminal
3.	*Sclerotium rolfsii*	Colony is brown to black, globose and compact	Mycelium lack conidia. Mycelium usually light and straight

Table 2 Macro- and micro-features of fungi isolates from cocoyam corm (*Colocasia esculenta*)

S/no	Isolates	Colony characteristics	Microscopy
1.	*Aspergillus niger*	Growth on PDA is rapid and fast. Powdery black colour almost covering the plates after 72 h	Non-septate conidiophores arising from thick-walled foot cells. Each conidiophores ends in a terminal enlarged spherical swellings. Conidia borne by phialides arising from a terminal swelling on the conidiophores
2.	*Aspergillus oryzae*	Growth on PDA is rapid and fast. Powdery brown colour almost covering the plates after 72 h	Non-septate conidiophores arising from thick-walled foot cells. Each conidiophores ends in a terminal enlarged spherical swellings. Conidia borne by phialides arising from a terminal swelling on the conidiophores
3.	*Rhizopus stolonifera*	Rapid and fast growth of about 36 mm after 24 h which soon spread and cover the plates. Growth appear as black pin-head, similar to cotton wool (white in colour)	Mycelium of non-septate hyphae of large diameter. The sporangiophore bears a terminal black spherical sporangium. The sporangiophores are erect and may be branched or unbranched each bearing a single globose sporangium

Table 3 Pathogenicity test result (cocoyam)

S/no	Isolate	Symptoms of infection after 14 days	Pathogenicity
1.	*Aspergillus niger*	Dry rot, soft rot	+++
2.	*Aspergillus oryzae*	Dry rot	+++
3.	*Rhizopus stolonifera*	soft rot	++

+++ = highly pathogenic

++ = moderately pathogenic

Table 4 Pathogenicity test result (yam)

S/no	Isolate	Symptoms of infection after 14 days	Pathogenicity
1.	*Aspergillus ochareus*	Dry rot	++
2.	*Sclerotium rolfsii*	Dry rot	+++
3.	*Lasiodiplodia theobromae*	Wet rot	+++

+++ = highly pathogenic

++ = moderately pathogenic

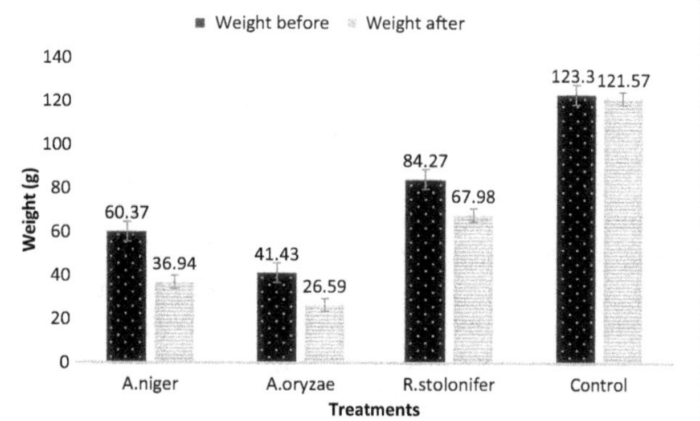

Fig. 2 The mean weight loss of cocoyam after pathogenicity test

inhibition on *R. stolonifer* (76.74 ± 2.33%) and *A. niger* (77.52 ± 1.35%), even though the extract showed significantly high growth inhibition against the three fungi examined. Aqueous leave extract (ALE) showed inhibitory effects of 85.66 ± 2.92% against *A. oryzae*, 75.97 ± 3.55% against *R. stolonifer* and 72.87 ± 3.55% against *A. niger*. There was no significant difference among growth inhibitions ($P > 0.05$) of the aqueous leave extract (ALE) against the fungal rot in cocoyam in this research, even though it had significant antifungal effect on all the fungi.

Ethanolic stem bark extract (EBE) also showed very high growth inhibition against *R. stolonifer*, (91.47 ± 5.85%) and *A. oryzae* (84.51 ± 1.76%) which were significantly ($P < 0.05$) different from that of *A. niger* (39.53 ± 8.39%). It can be inferred that ethanolic stem bark extract (EBE) was not very effective against *A. niger* as shown by the result above.

Aqueous stem bark extract (ABE) showed significant ($P < 0.05$) high growth inhibition against *A. oryzae* (96.79 ± 1.60%), *R. stolonifer* (94.96 ± 1.34%) and *A. niger* (87.21 ± 1.16%).

Meanwhile, ketoconazole which was used as the positive control showed 100% inhibition against the growth of the fungi *A. oryzae* and 89.92% on *A. niger*, but aqueous stem bark extract (ABE) gave inhibitory effect of 94.96 ± 1.34% on the growth of *R. stolonifer.*

Most effective against the growth of *A. niger* is ethanolic seed extract (ESE) (88.37 ± 1.17% inhibition) which does not differ significantly with the control (Ketoconazole) (89.92 ± 1.35% inhibition). Against the growths of *A. oryzae* and *R. stolonifer*, aqueous stem bark extract (ABE) with growth inhibitions of 96.79 ± 1.60% and 94.96 ± 1.34% ,respectively, was most effective. In general, it is reasonable to conclude that all the various typed of extracts of neem plant were effective against the

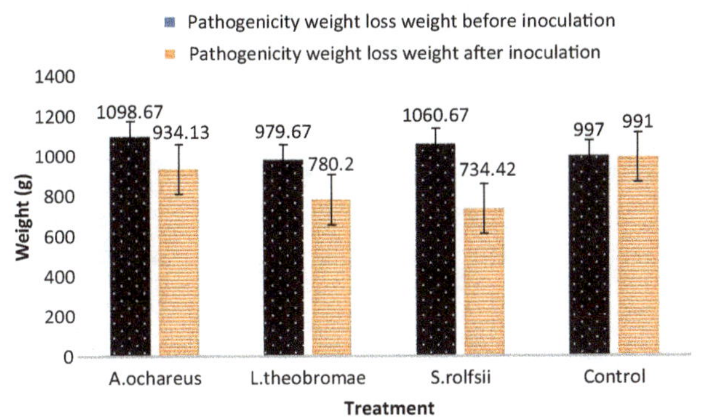

Fig. 3 The mean weight loss of yam after pathogenicity test

growth of the fungai-induced rot in cocoyam examined (except EBE which was not effective against *A. niger*).

Antifungal activity of the extracts on fungi isolated from yam

The effect of extracts on the test organisms of yam rot was significant ($P<0.05$). Aqueous extracts of the stem bark (ABE) gave the highest inhibitory effect of 93.80 ± 1.78% against *A. ochareus* followed by ethanolic extract of leaf (ELE) with 88.37 ± 2.33% inhibition, while ethanolic seed extract (ESE), aqueous seed extract (ASE), aqueous leaf extract (ALE) and ethanolic stem bark extract (EBE) showed inhibition rates of 87.21 ± 1.16%, 82.27 ± 0.50%, 81.78 ± 3.74% and 87.60 ± 1.34%, respectively (Table 6). The inhibitory effect of ABE was significantly ($P<0.05$) higher than that of other extracts, while there was no significant difference in the inhibitory effects of ESE, ELE and EBE on *A. ochareus.* ALE and ASE did not show any significant difference ($P>0.05$), meanwhile ketoconazole which was used as the positive control showed 100% inhibition on the growth of the fungus. The ethanolic extract of leave (ELE) had the highest inhibitory effect on *L. theobromae* by 100%, followed by leave aqueous extract (ALE) and aqueous extract of the stem (ASE) with approximately 92.64% inhibition, respectively. The least inhibitory effect against *L. theobromae* was by aqueous extract of the seed (ASE) with 86.43 ± 4.08% (Table 6). The inhibitory effect of ethanolic extract of the leave (ELE) against *L. theobromae* (100%) was significantly ($P<0.05$) greater than that of all the other extracts including ketoconazole. However, there was no significant difference ($P>0.05$) between inhibitory effects of ethanolic seed extract, aqueous leave extract, ethanolic stem extract and aqueous stem extract. The control showed an effective inhibition of 75.19 ± 1.77% but not significantly ($P<0.05$) higher than all the other extracts (Table 6).

Ethanolic extract of the neem plant seed (ESE) gave the highest inhibitory effect on *Sclerotium rolfsii* (87.99 ± 2.42%), followed by ethanolic leave extract (ELE) (85.66 ± 2.92%), while aqueous extract of stem bark (ABE) showed the least inhibition (70.93 ± 5.81%). The inhibitory effect of ethanolic extract of the leave (ELE) was not significantly different ($P>0.05$) from that of aqueous seed extract (ASE) and ethanolic leave extract (ELE). Also there was no significant difference ($P>0.05$) between the inhibitory effect of aqueous leave extract, aqueous stem extract, ethanolic stem extract and aqueous seed extract.

Ketoconazole showed an effective inhibitory effect (100%) on the growth of the fungus. The values obtained in this study are similar to what was reported by Anukworji et al. [25]. They treated *S. rolfsii* and *L. theobromae* isolated from cocoyam with aqueous and ethanolic extract of neem leaf.

These results agreed with the finding of many researchers. Amadioha [26] discovered that aqueous neem seed and neem extract reduced the growth of the fungi particularly *Pyricularia oryzae* on rice. Bennett and Wallsgrove [27]; Grayer and Harbourne [28], observed that neem phytochemicals (glycosides and saponins) have antifungal activity. Osbourn [29], observed that many saponins exhibit potent antifungal activity and are often present in relatively high levels in healthy plants and as a

Table 5 Percentage mean zones of inhibition of plant extracts on fungal rot of cocoyam

Extract	*A. niger* (%)	*A. oryzae* (%)	*R. stolonifer* (%)
Ethanolic seed extract (ESE)	88.37 ± 1.17^{a}	91.47 ± 2.93^{a}	84.88 ± 3.07^{a}
Aqueous seed extract (ASE)	81.78 ± 2.92^{ab}	85.66 ± 2.92^{ac}	79.07 ± 2.33^{ab}
Ethanolic leave extract (ELE)	77.52 ± 1.35^{b}	90.70 ± 2.33^{ac}	76.74 ± 2.33^{ab}
Aqueous leave extract (ALE)	72.87 ± 3.55^{b}	85.66 ± 2.92^{ac}	75.97 ± 3.55^{b}
Ethanolic stem bark extract (EBE)	39.53 ± 8.39^{c}	84.51 ± 1.76^{c}	91.47 ± 5.85^{cd}
Aqueous stem bark extract (ABE)	87.21 ± 1.16^{ab}	96.79 ± 1.60^{d}	94.96 ± 1.34^{d}
Ketoconazole	89.92 ± 1.35^{a}	100.00 ± 0.00^{d}	88.76 ± 0.68^{c}

Values are mean ± standard deviation (n = 3)

Means in the same column with the same superscript letter are not significantly different at $P<0.05$

Table 6 Percentage mean zones of inhibition of plant extracts on fungal rot of yam

Extract	*A. ochareus* (%)	*L. theobromae* (%)	*S. rolfsii* (%)
Ethanolic seed extract (ESE)	87.21 ± 1.16^{a}	89.53 ± 1.17^{ab}	87.99 ± 2.42^{a}
Aqueous seed extract (ASE)	82.27 ± 0.50^{b}	86.43 ± 4.08^{a}	79.46 ± 2.93^{ab}
Ethanolic leave extract (ELE)	88.37 ± 2.33^{a}	100.00 ± 0.00^{c}	85.66 ± 2.92^{a}
Aqueous leave extract (ALE)	81.78 ± 3.74^{b}	92.64 ± 1.78^{b}	74.80 ± 2.42^{b}
Ethanolic stem bark extract (EBE)	87.60 ± 1.34^{a}	86.82 ± 1.78^{ab}	77.91 ± 4.19^{b}
Aqueous stem bark extract (ABE)	93.80 ± 1.78^{c}	92.64 ± 2.42^{b}	70.93 ± 5.81^{b}
Ketoconazole	100.00 ± 0.00^{d}	75.19 ± 1.77^{d}	100.00 ± 0.00^{c}

Values are mean ± standard deviation (n = 3)

Means in the same column with the same superscript letter are not significantly different at $P < 0.05$

result have been implicated as determinants of a plant's resistance to fungal attack.

This suggests that these phytochemical components might have been responsible for the observed antifungal properties of the test plant used in the present study.

On the contrary, Oluma and Elaigwe [30] observed that extracts of *A. indica* had no inhibitory effect on the mycelial growth and sclerotial formation of *Macrophomina phaseolina*. Premlatha [31] also gave similar report on the ineffectiveness of neem extract against *Rhizoctonia solani*. But Nwachukwu and Umechuruba [32] stated that leaf extracts of neem were shown to be efficacious on major seed-borne diseases of African cocoyam bean seeds and on seed germination and seedling emergent diseases. The active principles present in plants are influenced by many factors, which include the age of plant, extracting solvent, method of extraction and time of harvesting plant materials [33]. According to Okigbo and Ogbonnaya [33], the use of plants extracts for plant disease management is economically viable and poses little environmental risk, and the plants are available to farmers in Nigeria, the tropics and many parts of the world that do not have ready access to other synthetic fungicides. Umar et al. [34] and Makein et al. [35] pointed out that Azadirachtin extracts from the seeds, leaves and bark of the neem tree has been reported to have strong biological activities against insect pest, but with very low toxicity to mammals and the environment. According to Kurucheve et al. [36], the presence of inhibitors to the fungal toxic principle may contribute to differences observed in the antifungal activity of some plant extracts.

Conclusion

All the plant extracts showed antifungal activities at varied percentage mean zones of inhibition. Various neem plant parts extracts showed inhibition against fungal growth from isolates of rotted cocoyam with mean inhibition between 39.53 ± 8.39% and 96.76 ± 1.60%. Apart from ethanolic stem bark extract which gave 39.53 ± 8.39% limited zone of inhibition against *A. niger* growth, the rest of the extract exhibited good zones of inhibition ranging between 72.87 ± 3.55 and 96.79 ± 1.60%. The extracts were effective against fungal isolates from the rotten yam samples. The mean percentage zones of inhibition of all extracts of the neem plant parts showed more effectiveness than the positive control (Ketoconazole) against *L. theobromae* growth. Aqueous stem bark extract with 93.80 ± 1.78% mean zone of inhibition showed comparable effectiveness to the control (100% inhibition) against *A. ochareus* growth, while against *S. rolfii* growth there was good inhibition by all the extracts though not as that of the control. The fungi toxic potential of these plant extracts on rot-inducing fungi of stored yam and cocoyam recommends their use to farmers as alternative to commercial/synthetic fungicides.

Abbreviations

ABE: aqueous stem bark extract; ASE: aqueous seed extract; EBE: ethanolic stem bark extract; ELE: ethanolic leave extract; ESE: ethanolic seed extract.

Authors' contributions

CSE, AI and AJ carried out the field/laboratory operations. CSE designed the study. Data analysis were by CSE, AI, AJ and CI. Supporting literature was by DIA. Preparation of manuscript was carried out by CSE, DIA and CI. All authors read and approved the final manuscript.

Author details

[1] Department of Biochemistry, Federal University Wukari, Wukari, Taraba State, Nigeria. [2] Department of Microbiology, Federal University Wukari, Wukari, Taraba State, Nigeria.

Acknowledgements

The authors acknowledge the immense contributions of Laboratory Technologists of Biology, Biochemistry and Microbiology Departments of the Federal University Wukari, Taraba State, Nigeria.

Competing interests

The authors declare that they have no competing interests.

Funding
This work was not funded by any research grant, it was the efforts of the authors.

References

1. Mahmoud DA, Hassanein NM, Youssef KA, Abou Zeid MA. Antifungal activity of different neem leaf extracts and the nimonol against some important human pathogens. Braz J Microbiol. 2011;42:1007–16.
2. Arumugam PA, Mohamad I, Salim R, Mohamed Z. Antifungal effect of Malaysian neem leaf extract on selected fungal species causing Otomycosis in in-vitro culture medium. Malays J Med Health Sci. 2015;11(2):69–84.
3. IITA. Production cost in the yam based cropping system of South-eastern Nigeria, RCMP Monograph Ibadan, Nigeria. 1991;6:2–26.
4. IITA. Production of yams, present and future perspective. IITA Res Guide. 1993;46:5.
5. Musa YH, Onu JI, Vosanka IP, Anonguku I. Production efficiency of yam in Zing Local Government Area of Taraba State, Nigeria. J Hortic For. 2011;3(12):372–8.
6. Opara LU. YAMS: post-harvest operation. AGST/FAO. In: Danilo Mejía Ed. FAO Technical. INPhO Post harvest compendium. New Zealand; 2003.
7. Igbozulike AO. Evaluation of cocoyam corms processing and storage in Nigeria. Int J Sci Eng Res. 2015;6(7):259–63.
8. Ijioma JC, Effiong JB, Ogbonna MO, Onwuamaoka EA. Determinants of adoption of selected NRCRI cocoyam technologies among farmers in Umuahia South Local Government Area of Abia State, Nigeria. Am Int J Contemp Res. 2014;4:6.
9. National Root Crops Research Institute (NRCRI) Umudike, Cocoyam Program. http://www.nrcri.org. Accessed 13 Feb 2015.
10. Chukwu GOC, Nwosu KI, Madu TU, Chinaka C, Okoye BC. Development of gocing storage method for cocoyam. Munich Personnel RePEc Archives (MPRA) Paper No 17444:25; 2008.
11. Opata PI, Ogbonna PE. Storage profitability and effectiveness of storage methods in yield loss reduction in cocoyam in southeast Nigeria. Afr J Agric Res. 2015;10(49):4496–504. https://doi.org/10.5897/AJAR2015.9756.
12. FAOSTAT. Food and Agricultural Organization: Agricultural Statistics. Rome: FAO; 2010.
13. Green KR. Studies on the epidemiology and control of yam anthracnose. Ph.D. Thesis, University of Reading, Reading, UK. 1994.
14. Cheesbrough M. District laboratory practice in tropical countries. Part 2. London: Cambridge University Press; 1984. p. 134–242.
15. Oniyike RCI, Maduewesi JNC. Variability in pathogenicity of isolates and sensitivity of dithane M-45 and PCNB on cocoyam rot Fungi. Niger J Plant Protect. 1985;9:74–81.
16. Okigbo RN, Ikediugwu FEO. Studies on biological control of post-harvest rot of yam (*Dioscorea* sp.) with *Trichoderma viride*. J Phytopathol. 2000;148:351–5.
17. Alexopoulos CJ. Introductory mycology. 2nd ed. Sydney: Wiley; 1962. p. 119–59.
18. Nelson PE, Tousson TA, Marasssas WFO. Fusarium species, an illustrated manual for identification. University Park: The Pennsylvania state press; 1983. p. 193.
19. Rippon JN. Medical mycology: the pathogenic fungi and the pathogenic actinomycetes. 3rd ed. Philadelphia: W.B. Saunders Co.; 1988. p. 163.
20. Samson RA, Hockstra ES, Van-Orschot AN. Introductory to food borne fungi. Amsterdam: The Netherlands academy of arts and science; 1984. p. 1–25.
21. Snowdon AL. A Colour atlas of postharvest diseases and disorders of fruits and vegetables UK, vol. 2. London: Wolfe scientific books; 1991. p. 416.
22. Okafor N. Microbial rotting of stored yam (*Dioscorea* spp) in Nigeria. Exp Agric. 1966;2:179–82.
23. Sangoyomi TE, Ekpo EJA, Asiedu R. Fungitoxic effects and attributes of *Allium sativum* and *Occimum gratissimum* extracts on *Rhizoctonia solani*, the causal organism of yam *Dioscorea rotundata* (Poir) rot disease. Niger J Mycol. 2009;2(1):166–7.
24. Whipps JM. Effects of media on growth and interactions between a range of soil borne glass pathogens and antagonistic fungi. New Pathol. 1987;107:127–42.
25. Anukworji CA, Ramesh RP, Okigbo RN. Isolation of fungi causing rot of cocoyam (*Colocasia esculenta* (L.) Schott) and control with plant extracts: (*Allium sativum*, L., *Garcinia kola*, heckel., *Azadirachta indica*, L. and *Carica papaya*, L.). Glob Adv Res J Agric Sci. 2012;1(1):202–17.
26. Amadioha AC. Controlling rice blast in vitro and in vivo with extracts of *Azadirachta indica*. Crop Prot. 1999;19:287–90.
27. Bennett RN, Wallsgrove RM. Secondary metabolites in plant defence mechanisms. New Phytol. 1994;127:617–33.
28. Grayer RJ, Harbourn JJ. A survey of antifungal compounds from higher plants. Phytochemistry. 1994;37:19–42.
29. Osbourn AC. Saponins and plant defence—a soap story. Trends Plant Sci. 1996;1:4–9.
30. Oluma HOA, Elaigwe M. Antifungal activity of extracts of some medicinal plants against *Macrophomina phaseolina*. J Bot. 2006;19(1):121–8.
31. Premlatha D. Effect of some soil amendment with some green manure on the survival of sclerotia of *Corticum Sesaki*. Indian Phytopathol. 1982;35:523–5.
32. Nwachukwu EO, Umechuruba CI. Antifungal activities of some leave extracts on seed-borne fungi of African yam bean seeds. J Appl Nat Sci. 2001;1:29–31.
33. Okigbo RN, Ogbonnaya UO. Antifgunal effects of two tropical plant extracts (*Ociumu gratissiumum* and *Afromomum melequeta*) on post-harvest yam (*Dioscorea* spp.) rot. Afr J Biotechnol. 2006;5:5727–31.
34. Umar A, Abdulraham HT, Kokori M. Preliminary studies of the efficiencies of the aqueous extracts of leaf and seed kernel of neem for the control of cowpea bruchid *Callobruchus maculates*. Res J Sci. 2002;8(1 and 2):25–30.
35. Makein HK, Maikai VA, Nok JA. Effect of topical application of neem seed extract on sheep infested with *Amblyomma variegatum*. Afr J Biotechnol. 2007;6(20):2324–7.
36. Kurucheve V, Padmavathi R. Fungitoxicity of selected plant products against *Pythium aphanidermatum*. Indian phytopathol. 1997;50:529–35.

16

Statistical optimization of chitinase production by *Streptomyces rubiginosus* SP24 and efficacy of purified chitinase to control root-knot nematode infection in *Vigna radiata* under controlled conditions

Sneha Jha* and H. A. Modi

Abstract

Background: *Actinomycetes* are known to produce various lytic enzymes such as chitinase which have different biotechnological aspects. Plackett–Burman design (PBD) was applied to evaluate significant components of medium to improve chitinase enzyme production. Response surface methodology was implemented to select those variables which are highly fitted in the model and helped in increasing enzyme production. A second experiment was performed as one of the major applications of statistically optimized and purified chitinase (method and result of purification not shown) for reducing root-knot nematode infection in roots known for causing diseases in plants. Among them, *Meloidogyne* sp. is one which is responsible for decreasing plant growth and fruit yield. For this, *Meloidogyne* sp. (1000 J_2 stage) was used to infect *Vigna radiata* (mung) plant which was previously coated with formulations of purified chitinase and organic manure ratio—1:1 (w/w).

Results: From ANOVA results of PBD data analysis, it was confirmed that colloidal chitin, xylose, peptone and dihydrogen potassium phosphate were found to be significant components for chitinase enzyme production. The best combination of factors (gm) to accomplish the optimum response was found to be colloidal chitin: 1.0, xylose: 2.0, peptone: 0.6 and dihydrogen potassium phosphate: 0.25 for predicted response of 3.81. Seeds coated with purified chitinase (5 ml) isolated from *Streptomyces rubiginosus SP24* helped in enhancing plant growth promotion and decreasing root-knot nematode infection.

Conclusion: Statistically optimized seeds of mung coated with purified chitinase was used for suppressing plant-parasitic nematodes which cause severe harm to the production of crops in terms of plant growth and fruit yield.

Keywords: Chitinase, Plackett–Burman design, Response surface model, Statistical approach, Optimization, *Streptomyces rubiginosus* SP24, *Meloidogyne* sp.

Background

Chitinase (EC 3.2.1.14) belongs to family of glycosyl hydrolase that catalyzes the degradation of chitin, an insoluble linear β-1,4-linked polymer of *N*-acetylglucosamine [10, 52]. They are found in a wide range of microorganisms including bacteria [10, 12], fungi [10, 12], and *Actinomycetes* [10, 36, 20, 54], but their role is diverse. Nowadays, the focus is mainly on *Actinomycetes* which have ample capacity for various types of metabolite production, such as those of antibiotics and extracellular enzymes [22]. The molecular weight of chitinases vary widely in their size, ranging from as low as 20 kDa to about 90 kDa [4, 26]. Chitinase activity was found to be present over a wide range of temperature (psychrophilic

*Correspondence: jhasneha2690@gmail.com
Department of Life Sciences, University School of Sciences, Gujarat University, Ahmedabad, Gujarat 380009, India

to thermophilic) and pH (generally, neutral to basic), depending on the microbe isolated. Chitinolytic enzymes can be divided broadly into two categories: endochitinases (EC 3.2.1.14), which cleave the internal β-1,4-glycosidic bond, and exochitinases, which are further sub-divided into two categories, i.e., chitobiosidases (EC 3.2.1.29) which acts only at the non-reducing end of the chitin chain, and β-*N*-acetylglucosaminidases (EC 3.2.1.30), which cleaves *N*-acetylglucosamine units sequentially from the non-reducing end of the substrate [25, 47].

Optimization process is significant in improving the ultimate production of enzymes and to understand the influence of each variable on the response [38]. In the conventional method for the optimization of enzyme production, the "one factor at a time" approach is used, which involves changing one factor at a time while keeping all the other factors constant [7]. Statistical-based design considers the interaction between independent variables, on the basis of which response is generated. An alternative and more efficient approach is the use of statistical methods. The Plackett–Burman design has been used for screening the main factors and this in turn can be preserved for further optimization [17]. RSM is a collection of statistical techniques that are useful for designing experiments, building models, evaluating the effects of different factors and searching for optimal conditions of the studied factors for desirable responses [17]. In our previous study, optimization using OFAT was done, which was further taken to statistically optimize the medium components for enhanced chitinase production [26] in the presence of all the required nutrients.

Root, stem and leaves of many plants are infected by nematodes causing various morphological as well as physiological changes in plants and in turn leading to disease conditions such as wilting, chlorosis, growth reduction and depletion of the entire plant, ultimately leading to huge economic loss [1, 29, 33, 64]. To control nematodes, anti-nematicidal agents are used worldwide. These anti-nematicidal agents are very much toxic to human health and environment. Among these, root-knot nematode (*Meloidogyne* spp.) is one of the major pathogens that is widespread, most dangerous and difficult to control [13, 28, 53]. Mung (*Vigna radiata*) plants have been reported to be infected by different species of nematodes causing serious damage to plant yield worldwide [62]. Different measures for control of nematode infection in plant have been applied by researchers such as soil amendment with oil cakes of neem, castor, mustard and other plant products for control of plant-parasitic nematodes [2, 3, 9, 32, 35, 50, 57, 61, 57]. The life cycle of nematodes consists of five stages, starting from eggs which develop into root-infecting J_2 juveniles that further invades host by attaching to its root and develop permanent feeding sites [51]. There was considerable decrease in root length, shoot length, fresh and dry weight of plants with increase in inoculum level of nematodes (Haseeb and Butool [19]; Pandey [42]). As a biological agent, chitinase can be used to degrade the nematode cell wall, which is made of chitin [6, 11]. With the use of actinomycetal chitinase as a potential weapon, it is comparatively easy to decrease root-knot nematode disease, which affects total yield, production and quality of crops.

The aim of this research was to apply the Plackett–Burman design as well as Box–Behnken design for screening the media components to enhance the chitinase enzyme production from *Streptomyces rubiginosus SP24* under submerged fermentation. The work was novel as no data on chitinase production from *Streptomyces rubiginosus* was available till date. The efficacy of the optimized as well as purified chitinase thus produced must be utilized for its application studies. Hence, seeds (*Vigna radiate*) coated with purified chitinase in formulation with organic manure was assessed for various plant growth parameters and the number of galls in the root system of the plant to decrease root-knot nematode disease.

Results

Chitinase production by submerged fermentation using Plackett–Burman design

Plackett–Burman design [43] was employed to determine the effect of individual parameters influencing chitinase production by *Streptomyces rubiginosus* SP24 under submerged fermentation (SmF). Factors with $p < 0.05$ were considered to have significant effects on the production of chitinase enzyme and were therefore selected for further optimization studies. Table 1 gives the actual and coded levels of independent variables tested in the design matrix. The design matrix for the screening of significant variables for maximum enzyme production and the corresponding responses (actual and predicted chitinase activity) are shown in Table 2. Optimizing response helps in identification of various significant factors and their desirability.

From Table 2, it was observed that the variation in actual chitinase activity was 0.78–2.36 U/ml. The two values of each variable [maximum (+) and minimum (−)] were chosen such that the difference between the two values (+ and −) was large enough to ensure that it included the peak area for the maximum enzyme production. The effect of the eight nutrient components of submerged fermentation for chitinase production by *Streptomyces rubiginosus* SP24 was studied by the Plackett–Burman design. Among them, colloidal chitin, xylose, peptone and dihydrogen potassium phosphate showed positive effect for chitinase production in the

Table 1 Actual and coded levels of independent variables used in the Plakett-Burman design

Factor	Name	Units (g/100 ml)	Low actual	High actual	Low coded	High coded
A	Colloidal chitin	%	0.1	1	−1	1
B	Peptone	g	0.1	2	−1	1
C	Beef extract	g	0.1	2	−1	1
D	Xylose	g	0.2	2	−1	1
E	Sucrose	g	0.2	2	−1	1
F	Magnesium sulfate	g	0.1	0.5	−1	1
G	Dihydrogen potassium phosphate	g	0.1	0.4	−1	1
H	Calcium carbonate	g	0.1	0.5	−1	1

Table 2 Optimization of variables using Plakett-Burman design for production of chitinase showing the actual and predicted values in terms of U/ml

Run	Colloidal chitin	Peptone	Beef extract	Xylose	Sucrose	Magnesium sulfate	Dihydrogen potassium phosphate	Calcium carbonate	Actual chitinase activity	Predicted chitinase activity
1	1	−1	1	−1	1	1	−1	−1	1.78	1.57
2	1	−1	−1	1	−1	1	−1	1	2.02	2.06
3	−1	1	1	−1	−1	1	−1	1	1.28	1.38
4	1	1	−1	1	1	−1	−1	−1	2.36	2.29
5	−1	1	−1	1	−1	1	1	−1	1.36	1.38
6	−1	1	1	1	1	−1	−1	1	1.96	1.88
7	1	−1	1	1	−1	−1	1	1	1.36	1.56
8	−1	−1	−1	−1	−1	−1	−1	−1	0.96	1.16
9	−1	−1	−1	−1	1	−1	1	1	0.78	0.66
10	1	1	1	−1	−1	−1	1	−1	1.42	1.3
11	−1	−1	1	1	1	1	1	−1	1.28	1.15
12	1	1	−1	−1	1	1	1	1	1.16	1.3

Level (−1 and 1) defines low and high coded values corresponding to the variable (colloidal chitin—0.1 and 1 gm, peptone—0.1 and 2, beef extract—0.1 and 2.0, xylose—0.2 and 2, sucrose—0.2 and 2, magnesium sulfate—0.1 and 0.5, dihydrogen potassium phosphate—0.1 and 0.4, calcium carbonate—0.1 and 0.5)

tested range of concentration. The Pareto chart gives idea regarding the order of significance of the components. After the initial selection, effects below the *T* value limit are likely to be less significant (Fig. 1). The Model *F* value of 17.50 implies that the model is significant as mentioned in Table 3. There is only a 0.09% chance that a "Model *F* value" this large could occur due to noise. Values of "Prob > *F*" < 0.0500 indicate that the model terms are significant.

Optimization of significant variables using the Box–Behnken design

Depending on the Box–Behnken design, four independent variables including colloidal chitin, xylose, peptone, and dihydrogen potassium phosphate were selected for the production of chitinase enzyme (Table 4). The predicted and observed responses are mentioned in Table 5. The results obtained from the Box–Behnken design were then analyzed by standard analysis of variance (ANOVA), and the quadratic regression equation was applied for the prediction of chitinase enzyme production (Table 6). Based on the model, it appeared that colloidal chitin ($p \leq 0.0001$), xylose ($p = 0.0009$), and peptone ($p = 0.0015$) were most significant and dihydrogen potassium phosphate ($p = 0.1670$) was less significant.

The ANOVA result (Table 6) discloses that the quadratic model was highly significant, from "*F*" value of 9.35 and a probability "*p*" value of < 0.0001. There is only a 0.01% chance that a "Model *F* value" this large could occur due to noise. The large value of *F* shows that most of the difference in the response is given by the regression equation. Values of "Prob > *F*" < 0.0500 indicate that model terms are significant. The "lack of fit *F* value" of 0.24 implies that the lack of fit is not significant relative to the pure error. There is a 98.10% chance that a "lack of fit *F* value" this large could occur due to noise.

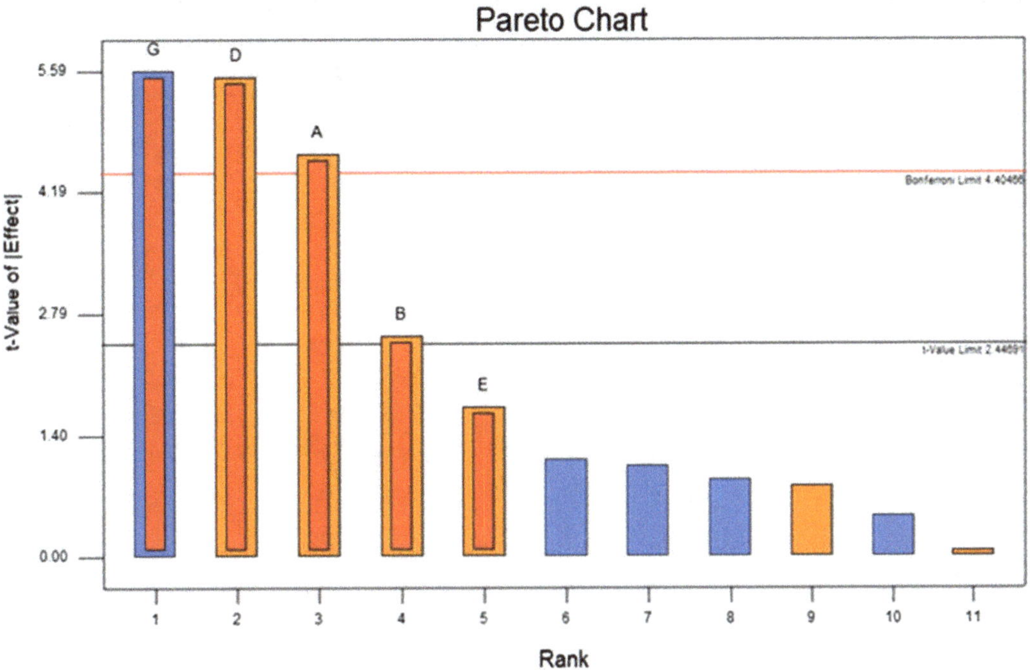

Fig. 1 Pareto chart of independent variables having significant effects. Pareto chart indicates the order of significance of components. After the initial selection, the effects below the T value limit are likely to be less significant. Alphabetical numbering on Pareto chart indicates the significance of variables coded with actual variables: A colloidal chitin; B peptone; D xylose; E sucrose; G dihydrogen potassium phosphate

Table 3 Results of ANOVA for selected factorial model

Source	Sum of squares	*df*	Mean square	*F* value	*p* value prob > *F*
Model	2.1468	4	0.5367	17.50109	0.0009
A colloidal chitin	0.512533	1	0.512533	16.71304	0.0046
B peptone	0.154133	1	0.154133	5.026087	0.0599
D xylose	0.730133	1	0.730133	23.8087	0.0018
G dihydrogen potassium phosphate	0.75	1	0.75	24.45652	0.0017
Residual	0.214667	7	0.030667		
Cor total	2.361467	11			

R^2 0.9091, Adj R^2 0.8572, Pred R^2 0.7329, Adeq Precision 14.449, *df*-degree of freedom

Table 4 Independent variables and their levels for Box–Behnken design

Factor	Name	Low actual	High actual	Low coded	High coded
A	Colloidal chitin	0.1	1	−1	1
B	Xylose	0.2	2	−1	1
C	Peptone	0.2	1	−1	1
D	Dihydrogen potassium phosphate	0.1	0.4	−1	1

Table 5 Chitinase activity from the experimental design for the response surface quadratic model

Std. order	Colloidal chitin	Xylose	Peptone	Dihydrogen potassium phosphate	C.A[a]	C.A[b]
1	0.1	1.1	1	0.25	2.32	2.13
2	0.55	0.2	1	0.25	2.76	2.49
3	0.55	2	0.6	0.1	2.92	2.82
4	0.55	1.1	0.6	0.25	2.12	2.52
5	0.55	0.2	0.6	0.1	1.56	1.88
6	0.55	1.1	0.6	0.25	2.18	2.52
7	0.55	1.1	0.2	0.4	1.96	2.25
8	1	1.1	1	0.25	3.96	3.78
9	0.1	1.1	0.6	0.1	1.68	1.52
10	1	0.2	0.6	0.25	2.38	2.87
11	0.55	0.2	0.2	0.25	1.28	1.61
12	0.1	1.1	0.6	0.4	1.76	1.86
13	0.1	1.1	0.2	0.25	1.62	1.25
14	0.55	2	1	0.25	3.52	3.43
15	0.55	2	0.6	0.4	2.98	3.16
16	0.55	1.1	0.6	0.25	1.92	2.52
17	0.1	0.2	0.6	0.25	1.38	1.22
18	1	2	0.6	0.25	4.28	3.81
19	0.55	2	0.2	0.25	2.78	2.55
20	1	1.1	0.2	0.25	2.56	2.9
21	0.1	2	0.6	0.25	1.24	2.16
22	1	1.1	0.6	0.1	2.72	3.17
23	0.55	1.1	1	0.4	2.82	3.13
24	0.55	0.2	0.6	0.4	2.72	2.22
25	0.55	1.1	0.2	0.1	2.62	1.91
26	0.55	1.1	0.6	0.25	3.48	2.52
27	0.55	1.1	1	0.1	2.72	2.79
28	0.55	1.1	0.6	0.25	2.82	2.52
29	1	1.1	0.6	0.4	4.02	3.51

C.A[a] = actual chitinase activity (U/ml) C.A[b] = predicted chitinase activity (U/ml)

Non-significant lack of fit is required to fit the model which is obtained if $p > 0.1$.

Final equation in terms of actual factors:

$$\text{Chitinase activity} = +2.52 + 0.83 \times A + 0.47 \times B + 0.44 \times C + 0.17 \times D + 0.51 \times A \times B + 0.18 \times A \times C + 0.30 \times A \times D - 0.18 \times B \times C - 0.28 \times B \times D + 0.19 \times C \times D,$$

where A is colloidal chitin, B is xylose, C is peptone and D is dihydrogen potassium phosphate.

The three-dimensional (3D) plots give an idea of the combined effect of two independent variables for chitinase enzyme production, when other variables were kept constant (Fig. 2a–d). The perturbation plot showed the outcome of all factors at a point in the design space (Fig. 3). Among the variables used for RSM, xylose had a significant effect on chitinase enzyme production compared with other variables. The perturbation graph shows that the factor colloidal chitin had a significant role in enzyme production. Before optimization, the enzyme production predicted was 0.66 U/ml, and threefold increased enzyme production was achieved after optimizing the medium by RSM. The predicted maximum chitinase enzyme activity was estimated to be 3.81 U/ml, while that of the actual chitinase enzyme activity obtained was 4.28 U/ml.

Table 6 Results of ANOVA for response surface quadratic model

Source	Sum of squares	*df*	Mean square	*F*-value	*p* value Prob > *F*
Model	15.64013333	10	1.56401333	9.352075121	< 0.0001
A colloidal chitin	8.200533333	1	8.20053333	49.03539	< 0.0001
B xylose	2.6508	1	2.6508	15.85055588	0.0009
C peptone	2.3232	1	2.6508	13.89165965	0.0015
D Dihydrogen potassium phosphate	0.3468	1	0.3468	2.073703326	0.1670
AB	1.0404	1	1.0404	6.221109979	0.0226
AC	0.1225	1	0.1225	0.732493245	0.4033
AD	0.3721	1	0.3721	2.224985605	0.1531
BC	0.1369	1	0.1369	0.818598574	0.3775
BD	0.3025	1	0.3025	1.808809851	0.1954
CD	0.1444	1	0.1444	0.863445099	0.3651
Residual	3.010266667	18	0.16723704		
Lack of fit	1.364346667	14	0.09745333	0.236836136	0.9810
Pure error	1.64592	4	0.41148		
Cor total	18.6504	28			

R-Squared 0.838595, Adj R-Squared 0.748926, Pred R-Squared 0.662642, Adeq Precision 11.52083

Biocontrol of root-knot nematode disease (*Meloidogyne* sp.) using purified chitinase in formulation

Under pot study, it was noted that there was significant increase in the number of leaves in set-1 which was inoculated with purified chitinase: organic manure (1:1) [22], and set-2, which was inoculated with purified chitinase and nematodes (10 ml) at the beginning of experiment [22]. But, it was found on comparison that the number of leaves was less and equal to that found in the control, which received only water [16], and set-3 [16], which was inoculated with commercially available nematicidal agent (fenamiphos) and nematodes (10 ml) at the beginning of the experiment (pot—9, 10 and 11) as described in Fig. 4.

There was an increase found in the overall plant growth parameters in case of set-1 compared to the other three sets. Plant shoot length was found to be higher (14.83) in set-1 and lowest in the control (9.36), while plant root length was found to be higher in set-1 (6.83) and lowest in set-2 (5.13). Fresh shoot weight was found to be higher in set-1 (0.444) and lowest in set-3 (0.224), while fresh root weight was found to be higher in the control (0.04) and lowest in set-3 (0.01). Dry shoot weight as well as dry root weight was found to be higher in set-1 (0.093, 0.018) and lowest in set-3 (0.018, 0.0053). Chlorophyll a was found to increase in set-2 (0.3) and chlorophyll b was found to increase in set-1 (0.495), while carotenoid content was found to be slightly higher in set-1 (0.138) compared to that in set-3 (0.137). There was statistical difference found between them according to Duncan's test, within the column in Table 7.

The inoculum level (10 ml) of *Meloidogyne* sp. was found to affect plant growth directly by affecting plant growth parameters such as plant height, fresh and dry weight of shoot and root, chlorophyll *a* and *b* as well as carotenoid content and the number of galls/root system. As the nematode density (1000) increased, the corresponding root-knot index was found to increase (set-2 and set-3) as mentioned in Table 8. There was comparatively less number of galls/plant in set-2 as compared to set-3 which showed the ability of purified chitinase to decrease infection.

Discussion

Optimizing response helps in identification of various significant factors and its desirability. Chitin is an important factor influencing chitinase production by *Streptomyces lividans* and *Streptomyces* sp. Da11 [18, 60]. Supplementation of glucose in colloidal chitin medium was found to decrease chitinase production in *Streptomyces* sp. [37]. Excluding chitin, other carbon sources such as sucrose, glucose, maltose, lactose and arabinose were not found to be significant in increasing chitinase production [16, 46, 58]. According to Gangwar et al. [14], additional carbon sources such as maltose, xylose, fructose, and lactose enhanced chitinase yield in *Streptomyces violascens* NRRL B2700 by 1.6 to sixfold and our results were similar to the above statement, as it was found that xylose as a carbon source was found to enhance chitinase production. Comparative optimization between organic and inorganic nitrogen sources revealed that organic nitrogen sources such as yeast extract and peptone were significant factors influencing chitinase activity as compared to inorganic sources, according to Vaidya et al. [59]. This is in accordance with

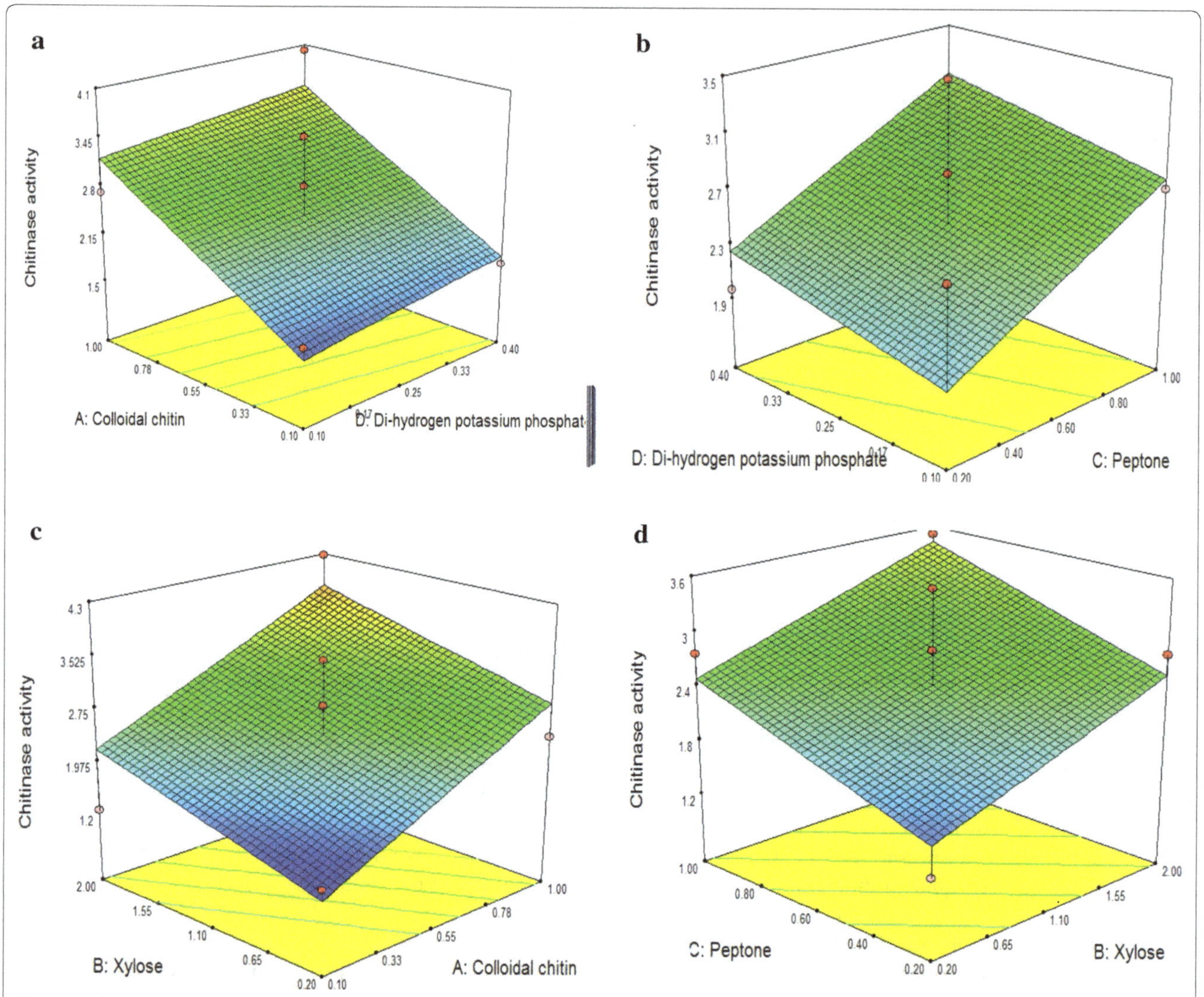

Fig. 2 a–d 3D plot showing the effect of two variables against chitinase activity while all other factors are kept constant

our result showing that the addition of yeast extract and peptone to the medium enhanced chitinase activity, leading to increased production from *Streptomyces rubiginosus* SP24. Enhanced production of chitinase was obtained in strains of *Streptomyces* sp. NK1057, NK528 and NK951 by 29, 9.3 and 28%, respectively, as described by Sarrafzadeh and Hussein [49] under optimized conditions. As reported by Sandhya et al. [48], optimum chitin concentration for chitinase induction is in the range of 10–20 g/1000 ml in case of microbes and our results are in agreement with this. During the course of the present study, the optimum colloidal chitin concentration for higher chitinase production by *Streptomyces rubiginosus* SP-24 is 1.0 g/100 ml, which is considerably lower than the range studied by Sandhya et al. [48]. Tasharrofi et al. [56] reported that the optimum chitin concentration for chitinase production by *B. pumilus* was 4.76 g/1000 ml, which was considerably higher than the present investigation. The minimum use of substrate with higher production of enzyme may be favorable, as this may be an added advantage for industrial use. Lower concentrations of chitin sometimes fail to increase chitinase production in microbes, and as an alternative some other nitrogen sources such as peptone and urea are reported to increase chitinase production [59], indicating that other mechanisms may be involved. Production of chitinase is increased on addition of minerals such as KBr and $MgSO_4$ to the production medium as described by Gohel et al. [15]. From the model, it is clear that dihydrogen potassium phosphate exerts positive effect by increasing chitinase activity, in turn leading to increase in the production of chitinase.

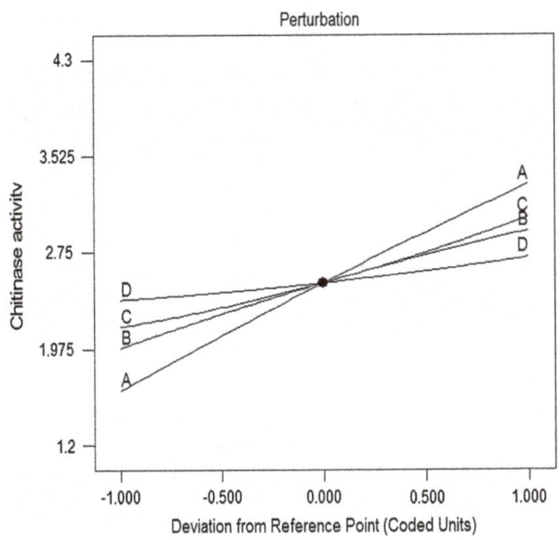

Fig. 3 Perturbation curve showing the maximum chitinase production by factor A (colloidal chitin)

Biocontrol of root-knot nematode disease *(Meloidogyne* sp.) using purified chitinase in formulation

A characteristic study of chitinase treatment for nematicidal activity was studied for *Paenibacillus illinoisensis KJA-424* by Jung et al. [27]. The present study was, therefore, conducted to determine if purified chitinase produced by *Streptomyces rubiginosus SP24* was associated with the inhibition of egg hatching of root-knot nematode with a hypothesis that there was an association between microbes and plant-parasitic nematodes. From the results of plant growth parameters, it can be concluded that purified chitinase indirectly showed inhibition of the egg hatching of *Meloidogyne* sp. by showing significant increase in plant growth parameters (Table 7). As reported by Cronin et al. [8], in vitro studies resulted in 70% inhibition of egg hatching by *Globodera rostochiensis* (Ro1), when purified chitinase was used. Production of chitinase by *Pseudomonas aureofaciens* was able to attack the egg shell rather than the cuticle of *Criconemella xenoplax* [63]. Another significant factor that helped in increased plant growth is the number of leaves which significantly increased when chitinase in formulation with organic manure was coated on mung seeds compared to the control (Fig. 4). This might have occurred due to the direct damage of chitinase to the juvenile nematodes (J_2) caused by chitinase activity. *Actinomycetes* are known for the production of various enzymes having lytic property such as lipase, protease, chitinase, amylase, xylanase, cellulase, and mannanase [5] and might be capable of affecting eggshells of root-knot nematode [33–41]. As reported by Mercer et al. [34], chitinase interferes with the egg

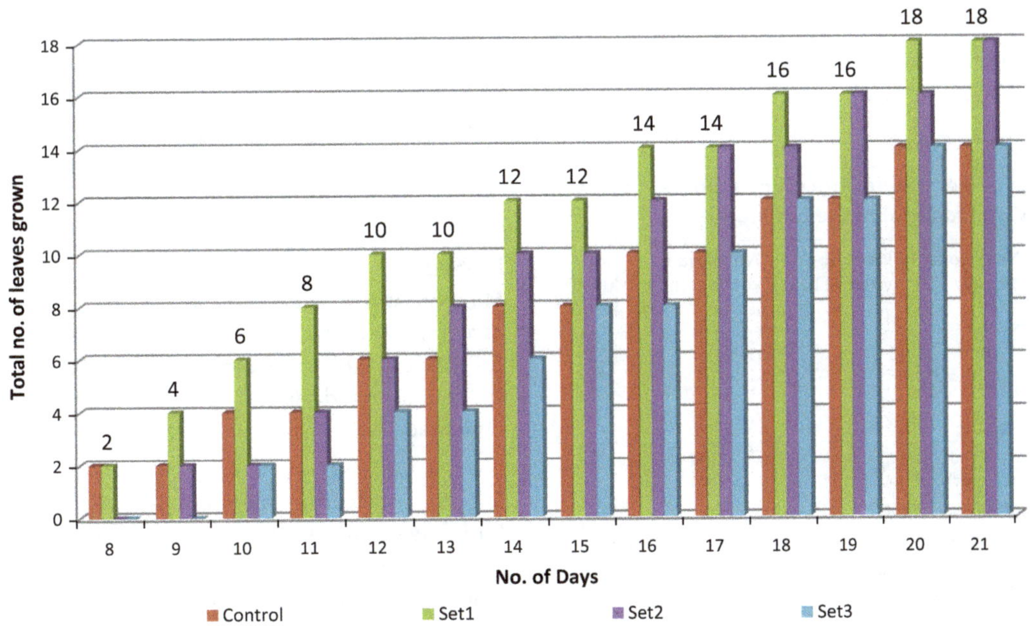

Fig. 4 Growth of leaves in mung plant (*Vigna radiata*) during the experimental period (day-1 to day 21). Results shown here in bar graph are the mean replicates of three. There was no growth of leaves from day 1 to day 7. Hence, it is not shown in the graph. Control—addition of only water; set-1 contained only chitinase:organic manure formulation); set-2 contained chitinase:organic manure formulation as well as 10 ml of inoculum volume of nematode suspension added near the roots of the plant; set-3 contained fenamiphos:organic manure formulation as well as 10 ml of inoculum volume of nematode suspension added near the roots of plant at the beginning of the experiment

Table 7 Influence of purified chitinase and commercially available anti-nematicidal agent on plant growth promotion (*Vigna radiata*) under controlled conditions

Plant growth parameters	Control	Set-1	Set-2	Set-3
Shoot length	9.36^{b}	14.83^{b}	13.2^{b}	12.46^{b}
Root length	6.6^{c}	6.83^{c}	5.13^{c}	6.32^{c}
Total plant height	15.96^{a}	21.66^{a}	18.33^{a}	18.83^{a}
Shoot weight	0.408^{de}	0.444^{d}	0.308^{de}	0.224^{de}
Root weight	0.04^{f}	0.029^{f}	0.011^{g}	0.01^{f}
Total plant fresh weight	0.448^{d}	0.474^{d}	0.319^{de}	0.234^{de}
Shoot weight	0.038^{f}	0.093^{f}	0.028^{g}	0.018^{f}
Root weight	0.009^{f}	0.018^{f}	0.007^{g}	0.0053^{f}
Total plant dry weight	0.048^{f}	0.111^{f}	0.035^{g}	0.023^{f}
Chlorophyll *a*	0.119^{f}	0.27^{e}	0.3d^{e}	0.29^{de}
Chlorophyll *b*	0.225^{ef}	0.495^{d}	0.482^{d}	0.477^{d}
Carotenoid	0.085^{f}	0.138^{f}	0.132^{ef}	0.137^{f}

Control (2 pots), set-1, 2 and 3 are replicates of three pots, and each set is the mean average of three seedlings/pot. Control—addition of only water (pots 1 and 2)

Set-1 contained only chitinase:organic manure formulation (pots 3, 4 and 5)

Set-2 contained chitinase:organic manure formulation as well as 10 ml of inoculum volume of nematodal suspension added near roots of plant (pots 6, 7, and 8)

Set-3 contained fenamiphos: organic manure formulation as well as 10 ml of inoculum volume of nematode suspension added near the roots of the plant at the beginning of the experiment (pots 9, 10, and 11)

Statistical analysis results showed values with the same letters (a, b, c, d, e, f, g) within the column were not significantly different and values with different letters within the column were significantly different from each other according to Duncan's test

hatching of *M. hapla*. Endo-chitinase production by isolate KJA-424 led to a change in the shape of egg as well as its rupturing [27]. The results of pot study showed that chitinase caused lysis of the eggshell of *Meloidogyne* sp. that resulted in egg hatching inhibition. The lysis of the eggshell by chitinolytic activity plays an important role in the control of root-knot nematode (*Meloidogyne* sp.), as seen with increased chlorophyll a and b content as well as carotenoid content of the plant (Table 7). The inoculum level (10 ml) used in formulation with purified chitinase significantly decreased the infection to the plant (Table 8) as compared with a formulation containing anti-nematicidal agent.

Table 8 Different nematode inoculum volumes affecting mung plant (*Vigna radiata*) in the pot study

Sr. no.	Nematode density	Inoculum volume (ml)	No. of galls/ root system	Root-knot index
Control/set-1	0	0	0	0
Set-2	1000	10	5	2
Set-3	1000	20	9	2

Control (2 pots), set-1, 2 and 3 are replicates of three pots, and each set is the mean average of three seedlings/pot

Control—addition of only water (pots 1 and 2)

Set-1 contained only chitinase:organic manure formulation (pots 3, 4 and 5)

Set-2 contained chitinase:organic manure formulation as well as 10 ml of inoculum volume of nematode suspension added near the roots of the plant (pots 6, 7, and 8)

Set-3 contained fenamiphos:organic manure formulation as well as 10 ml of inoculum volume of nematode suspension added near the roots of the plant at the beginning of the experiment (pots 9, 10, and 11)

The number of galls per root system was counted as described by Taylor and Sasser [55], where gall index 1 was on a scale of 0–5: 0 = 0 galls; 1 = 1–2 galls; 2 = 3–10 galls; 3 = 11–30 galls; 4 = 31–100 galls; 5 > 100 galls

Conclusion

In the present study, through Plakett–Burman design, it was found that higher chitinase activity resulted due to four major variables (colloidal chitin, xylose, peptone and dihydrogen potassium phosphate). The statistical model as designed by the Box–Behnken tool of the Design-Expert software has showed significant elevation in chitinase activity in the presence of selected variables of the defined media. It was confirmed that supplementation (g/100 ml) with colloidal chitin 1.0, xylose 2.0 and peptone 0.6 helped in enhancing chitinase production. Similar optimization process may be used to improve the chitinase activity, in turn leading to its improved production from novel isolate without any extra expenditure. Plant-parasitic nematodes are found in all agricultural sections of the world. They severely influence and cause harm to the productivity of crops, where environmental factors support their survival and dispersal particularly in tropical and sub-tropical regions of the world. The damage caused by nematodes to plants is directly proportional to the number of nematodes in soil and their reproduction rate on the plant. Results showed that purified chitinase used in formulation had good property to decrease the disease and increase plant growth-promoting traits.

Methods

Microorganism and culture conditions

Streptomyces rubiginosus SP24 (accession number KT198721) isolated from the rhizosphere of *Gossypium* sp. was selected, as it was able to produce extracellular chitinase [26].

Experimental culture medium

50 ml seed culture medium used for growth of *Streptomyces rubiginosus SP24* contained the following composition (gm/100 ml): colloidal chitin, 1%; sodium nitrate, 0.2; magnesium sulfate, 0.1; dihydrogen potassium phosphate, 0.1; potassium chloride, 0.05; calcium carbonate, 0.1; iron sulfate heptahydrate, 0.001; pH 7.0 at 30 °C under shaking conditions (100 rpm) for 72 h in orbital shaker (REMI Model no. 396LAG).

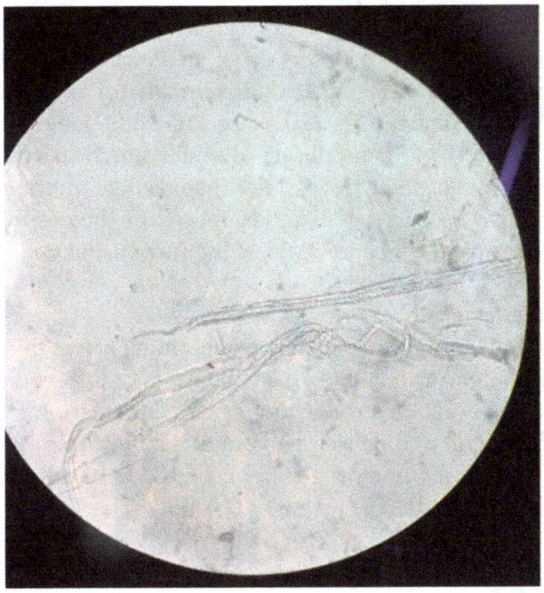

Fig. 5 Stereomicroscopic view. Inoculum J_2 stage (*Meloidogyne* sp.) selected for biocontrol of root-knot nematode disease

Chitinase assay

The reaction mixture contained 1 ml of 1% colloidal chitin in phosphate buffer (0.05 M, pH 7.0) and 1 ml culture filtrate was incubated at 37 °C for 1 h. Suitable substrate and enzyme blanks were included. Chitinase activity was assayed by following the spectrophotometric method of Reissig et al. [45]. The reaction was terminated by adding 0.1 ml of 0.08 M potassium tetraborate (pH 9.2) to 0.5 ml of the reaction mixture. It was then boiled in a water bath for 3 min. Then 3 ml of diluted *p*-dimethylaminobenzaldehyde [10 g of DMAB dissolved in 100 ml of glacial acetic acid containing 12.5% 10 N (v/v) hydrochloric acid] reagent was added and again incubated at 37 °C for 15 min. The released product in the reaction mixture was read at 585 nm in a visible spectrophotometer (Systronics). Chitinase activity was determined using *N*-acetylglucosamine (Hi-media) as the standard. One unit of chitinase activity is defined as the amount of enzyme, which produces 1 μ mol of *N*-acetylglucosamine in 1 ml of reaction mixture under standard assay condition [31].

Optimization of process parameters

Plackett–Burman design: An efficient method for selection of medium components was used to determine the factors that have significant influence on chitinase production (Plackett and Burman [43]).

Experimental Design: Carbon sources, nitrogen sources, colloidal chitin, and inorganic ions were selected for increasing the production of chitinase. The design was formulated according to Plackett–Burman design as well as Box–Behnken design using Design-Expert 7.0.0 (Stat-Ease). Plackett–Burman screening method was designed with a two-level factorial design (−1) for low level and (+1) for high level for the identification of important medium components with consideration of their main effects.

The design was developed to evaluate the interactions of various process variables with one another. The variables that had the most significant effect, required for maximal chitinase production, were investigated by screening k variables in $k+1$ experiments. Eight independent variables were screened in 12 runs in one block according to the design. The most important effect of each variable was calculated as the difference between the average of measurements made at the high level (+1) and the average of measurements observed at low level (−1) of that factor The significance of the variables was determined by calculating the p value (Prob > F) through standard regression analysis which indicates whether the model is significant or not. The dependent (response) variable studied was chitinase activity. After completion of the experiment, the results entered were compared to that of the predicted value and data obtained were subjected to the analysis of variance (ANOVA).

Box–Behnken design

After estimating the comparative significance of independent variables, four most significant variables were selected for further determination of their optimal level with respect to chitinase activity as a response. For this reason, the Box–Behnken design was applied. The effects of colloidal chitin, xylose, peptone and dihydrogen potassium phosphate on chitinase activity were evaluated. The levels of these factors were optimized for maximum chitinase production (the response). A 29-trial experimental design was used, where each variable was tested at two different levels: low (−1) and high (+1). The significance of variables was determined by calculating the p value (Prob > F) through standard regression analysis, which indicates whether the model is significant or not.

Software and data analysis

Design-Expert version 7.0 trial version (Stat-Ease Inc., Minneapolis, MN, USA) as a statistical software was used for analysis and interpretation of experimental design.

Purification of optimized chitinase enzyme

Four methods of purification (ammonium sulfate precipitation, dialysis, concentrators and Sephadex G-50 gel permeation chromatography) were used for isolation of chitinase from *Streptomyces rubiginosus SP24* and has already been published in our previous work (Jha and Modi [23]). Hence, the methods and results are not described here.

Biocontrol of root-knot nematode disease *(Meloidogyne* sp.) in unsterilized soil by *Streptomyces rubiginosus* SP24

Experimental design for pot study

For the screening experiment as well as inoculation and termination of experiment, modified method of Rajeswari and Ramakrishnan [44] was used. Pot study was carried out in triplicates under controlled conditions with 1:4 soil–sand ratio (now referred to as soil). Seeds of *Vigna radiata* were surface sterilized by gently shaking with 70% ethanol (5 min) and 20% sodium hypochlorite solution, followed by rinsing with sterile distilled water. Seeds were allowed to soak in a Petri dish with sterile distilled water overnight [24]. 5 gm of organic manure was mixed with 5 ml of purified chitinase, and 4.28 U/ml (chitinase thus obtained after applying RSM was further purified with molecular techniques—ammonium sulfate precipitation, dialysis, using concentrators and with Sephadex G-50 column chromatography) enzyme was used for coating seeds of *Vigna radiate* using 1% CMC as an adhesive. Three seeds were planted in each pot and watered on a regular basis. Control sets were maintained, with no treatment as well as no nematodes added. Set-1 contained only chitinase:organic manure formulation, set-2 contained chitinase:organic manure formulation as well as 10 ml of inoculum volume of nematode suspension added near the roots of the plant and set-3 contained fenamiphos:organic manure formulation as well as 10 ml of inoculum volume of nematode suspension added near the roots of the plant. Each treatment was replicated three times and arranged in a completely randomized block design. Plants were watered appropriately and maintained. The number of leaves was counted each day till termination of the experiment. The plants were examined 21 days after sowing and then uprooted for determining fresh as well as dry weights of shoot, root, chlorophyll a and b content, carotenoid content, nematode population in roots and number of galls/root system.

Preparation of nematode inoculum

Second-stage juveniles (J_2) of the root-knot nematode were used as inoculum in the present investigation. 100 gm of infected soil sample was strained on a kitchen strainer, washed with water and collected in a 2000 ml beaker. The strained water was further collected and sieved through a 400-mesh size strainer and the collected debris containing the soil sample on the strainer was transferred to a centrifuge tube. A small amount of water was added and the solution centrifuged at 3500 rpm for 3 min. The supernatant was discarded and the pellets obtained in the centrifuge tube were collected and added to sucrose solution (45.4 gm/100 ml). The resulting suspension was centrifuged at 3500 rpm for 3 min. Egg masses were placed in a coarse sieve fitted with double-layered tissue paper in Petri plates. These Petri plates were then placed in an incubator at 23 ± 2 °C. The hatched juveniles (J_2) were collected in a beaker every 24 h (Fig. 5) and observed under stereomicroscope to identify J_2-stage juveniles. The average number of juveniles was then calculated to represent the number of juveniles per 10 ml and 20 ml of the suspension.

Inoculation of Meloidogyne *sp. in pot*

For inoculation, the suspension containing J_2 was taken in a micropipette and added near the roots of the seedlings. The holes were covered with soil after inoculation. Nematode inoculation density consisted of a uniform quantity of suspension containing either 0 or 1000 (10 ml) freshly hatched second-stage juveniles (J_2)/pot.

Termination of the experiment

The plants were uprooted 21 days after inoculations and the following parameters studied. The following growth-related parameters were determined for each treatment of the different experiments.

(1) Plant fresh weight (root and shoot), (2) plant dry weight (root and shoot), (3) number of leaves/plant, (4) chlorophyll a and b content, and (5) number of root galls/root system.

Plant fresh and dry weight

Fresh as well as dry weights of shoots and roots were recorded for each treatment. Plants of each treatment were taken out from the pots and soil particles adhering to roots were removed by washing in tap water and labeled appropriately. In the laboratory, fresh weights of shoot and root were determined with the help of a weighing balance. For determining the dry weights, plants from each treatment were wrapped in blotting paper sheets, labeled and then dried in a hot air oven at 60 °C for 24 h before weighing.

Estimation of chlorophyll and carotenoids

The leaves of the mung bean plants were homogenized with 80% acetone. The homogenate was filtered and centrifuged at 3000 rpm for 15 min at 4 °C. The supernatant was collected and the volume was made up to 25 ml with 80% acetone. Chlorophylls and carotenoid contents were estimated in accordance with Maclachlan and Zalik [30]. The absorbance of the supernatant was recorded at 645 and 663 nm for chlorophyll contents, and at 480 and 510 nm for carotenoid contents, on a spectrophotometer. The amount of chlorophylls and carotenoids was expressed as mg/gm fresh weight.

Number of root galls per root system

Roots of uprooted plants were washed under the tap water and examined for the presence of root galls. The number of galls per root system was counted as described by Taylor and Sasser [55], where Gall index 1 was on a scale of 0–5: 0 = 0 galls; 1 = 1–2 galls; 2 = 3–10 galls; 3 = 11–30 galls; 4 = 31–100 galls; 5 > 100 galls.

Statistical analysis

The experiment was designed in such a way that three replicates were obtained for each plant growth parameter and the mean results were analyzed by performing analysis of variance test (ANOVA), followed by Duncan's post hoc test using software IBM SPSS 23.0 trial version [21] (IBM Corp. Released 2015).

Abbreviations

PBD: Plackett–Burman design; RSM: response surface methodology; Mung: *Vigna radiata*; SmF: submerged fermentation; Coded values: maximum (+) and minimum (−); 3D: three dimensional; ANOVA: analysis of variance; J_2: second-stage juveniles.

Authors' contributions

SJ designed and carried out the experiments and wrote the manuscript. HM supervised and directed the research outcomes as well as overall improved and finalized the manuscript. Both authors read and approved the final manuscript.

Acknowledgements

We acknowledge the financial assistance in terms of fellowship provided by the Department of Science and Technology, Ministry of Science and Technology, New Delhi, India.

We are thankful to the Department of Science and Technology-INSPIRE (IF140045) for their financial support in terms of fellowship.

Competing interests

The authors declare that they have no competing interests.

References

1. Abad P, Favery B, Rosso M, Castagnone-Sereno P. Root-knot nematode parasitism and host response: molecular basis of a sophisticated interaction. Mol Plant Pathol. 2003;4:217–24.
2. Akhtar M, Mahmood I. Organic amendments in relation to nematode management with particular reference to India. Integr Pest Manag. 1996;1(4):201–15.
3. Alam MM. Neem in biocontrol. In: Jairajpuri MS, Alam MM, Ahmad I, editors. Nematode biocontrol (Aspects and Prospects). Delhi: CBS Publishers and Distributers; 1990. p. 17–40.
4. Balakrishnan S, Duraisamy G, Manokaran K, Ganesan R, Chinthamani A, Chandrasekar U. Production and purification of chitinase by *Streptomyces* sp. from soil. J Adv Sci Res. 2012;3(3):25–9.
5. Bhaturiwala RA, Jha SC, Jain NK, Modi HA. Enzyme profiling of selected chitinase producing *Actinomycetes*. European Journal of Biotechnology and Bioscience. 2017;5(1):39–43.
6. Challis GL, Hopwood DA. Synergy and contingency as driving forces for the evolution of multiple secondary metabolite production by *Streptomyces species*. Proc Natl Acad Sci. 2003;100:14555–61.
7. Chen HC. Response surface methodology for optimizing citric acid fermentation by *Aspergillus foetidus*. Process Biochem. 1994;29:399–405.
8. Cronin DY, Moenne-Loccoz C. Dunne, Gara F O'. Inhibition of egg hatch of the potato cyst nematode *Globodera rostochiensis* by chitinase-producing bacteria. Eur J Plant Pathol. 1997;103:433–40.
9. Deepa SP, Subramanian S, Ramakrishnan S. Biomanagement of citrus nematode *Tylenchulus semipenetrans* Cobb on lemon, *Citrus limonia* L. J Biopest. 2011;4(2):205–7.
10. Dhanya RP, Anusha R, Citarasu T. Studies on the Bioactive *Actinomycetes* of Estuarine Soils of South West Coast of India. Int J Curr Microbiol Appl Sci. 2015;4(2):158–65.
11. Dicklow MB, Acosta N, Zuckerman B. A novel *Streptomyces* species for controlling plant-parasitic nematode. J Chem Ecol. 1993;19:159–73.
12. Farag AM, Abd-Elnabey HM, Ibrahim HAH, El-Shenawy M. Purification, characterization and antimicrobial activity of chitinase from marine-derived Aspergillus terreus. Egypt J Aquatic Res. 2016;42(2):185–92.
13. Fourie H, McDonald AH. Nematodes. ARCLNR Leaflet. Crop Prot. Ser. 2000;18:4–4.
14. Gangwar M, Singh V, Pandey AK, Tripathi CK, Mishra BN. Purification and characterization of chitinase from *Streptomyces violascens* NRRL B2700. Indian J Exp Biol. 2016;54(1):64–71.
15. Gohel V, Chaudhary T, Vyas P, Chhatpar HS. Statistical screening of medium components for the production of chitinase by the marine isolate *Pantoea dispersa*. Biochem Eng J. 2006;28:50–6.
16. Gupta R, Saxena RK, Chaturvedi P, Virdi JS. Chitinase production by *Streptomyces viridificans*: its potential in fungal cell wall lysis. J Appl Bacteriol. 1995;78:378–83.
17. Haddar A, Fakhfakh-Zouari N, Hmidet N, Frikha F, Nasri M. Kamoun AS Low-cost fermentation medium for alkaline protease production by *Bacillus mojavensis* A21 using hulled grain of wheat and sardinella peptone. J Biosci Bioeng. 2010;10(3):288–94.
18. Han Y, Zhiyong L, Miao X, Zhang F. Statistical optimization of medium components to improve the chitinase activity of *Streptomyces* sp. Da11 associated with the South China Sea sponge *Craniella australiensis*. Process Biochem. 2008;43:1088–93.
19. Haseeb A, Butool F. Pathogenicity and reproduction of Meloidogyne incognita on Ocimum sanctum. Inter Nematology Netw Newsl. 1989;6:27–30.
20. Hsu SC, Lockwood JL. Powdered Chitin agar as a selective medium for Actinomycetes in water and soil. Appl. Microbial. 1975;29(422):426.
21. IBM Corp. Released 2015. IBM SPSS Statistics for Windows, Version 23.0. Armonk, NY: IBM Corp.
22. Inbar E, Green SJ, Hadar Y, Minz D. Competing factors of compost concentration and proximity to root affect the Distribution of *Streptomycetes*. Microbiol Ecol. 2005;50:73–81.
23. Jha SC and Modi HA. Purification and characterization of chitinase from novel isolate Streptomyces rubiginosus SP 24. Proceedings of conference : Synthetic biology approaches in drug discovery and microbial biotechnology. 2018; 107-114. ISSN No. 2321-1520.
24. Jha CK, Patel B, Saraf M. Stimulation of the growth of Jatropha curcas by the plant promoting bacterium *Enterobacter cancerogenus* MSA2. World J Microbiol Biotechnol. 2012;28:891–9.
25. Jha S, Jha CK, Modi HA, Saraf M. Microbial Chitinases: Manifestation and Prospective (Chapter: 6). In: Garg N, Aeron A, editors. Industrial microbiology: microbes in process. Nova Science Publishers: INC. USA; 2013. p. 146–62.
26. Jha S, Modi HA, Jha CK. Characterization of extracellular chitinase produced from *Streptomyces rubiginosus* isolated from rhizosphere of *Gossypium* sp. Cogent Food Agric. 2016;2:1198225.
27. Jung WJ, An KN, Jin YL, Park RD, Kim KY, Shon BK, Kim TH. Effect of chitinase-producing *Paenibacillus illinoisensis* KJA-424 on egg hatching of root-knot nematode (Meloidogyne incognita). J Microbiol Biotechnol. 2002;12(6):865–71.
28. Kaskavalci G. Effect of soil solarization and organic amendments treatment for controlling Meloidogyne incognita in tomato cultivars in Western Anatolia. Turk. J. Agric. For. 2007;31:159–67.
29. Loveys RR, Bird AF. The influence of nematodes on photosynthesis in tomato plants. Physiol Plant Pathol. 1973;1973(3):525–9.

30. Maclachlan C, Zalik S. Plastid structure, chlorophyll concentration and free amino acid composition of a chlorophyll mutant of barley. Can J Bot. 1963;41:1053–62.
31. Mathivanan N, Kabilan V, Murugesan K. Purification, characterization, and antifungal activity of chitinase from *Fusarium chlamydosporum*, a mycoparasite to groundnut rust, *Puccinia arachidis*. Can J Microbiol. 1998;44:646–51.
32. Mc Geehan S. Impact of waste materials and organic amendments on soil properties and vegetative performance. Appl Environ Soil Sci. 2012. https://doi.org/10.1155/2012/907831.
33. Melakeberhan H, Webster JM, Brooke RC. Improved techniques for measuring the CO_2 exchange rate of *Meloidogyne* nematode bean plants. Nematologica. 1984;30:213–21.
34. Mercer CF, Greenwood DR, Grant JL. Effect of plant and microbial chitinases on the eggs and juveniles of *Meloidogyne hapla* Chitwood. Nematologica. 1992;38:227–36.
35. Mohan K. Comparison of inorganic and organic nematicides on the population of soil nematodes in hybrid of *Saccharum* species. J Biopest. 2011;4(2):201–4.
36. Nagpure A, Gupta R. Purification and characterization of an extracellular chitinase from antagonistic *Streptomyces violaceusniger*. J Basic Microbiol. 2013;53(5):429–39.
37. Nawani NN, Kapadnis BP. Optimization of chitinase production using statistics based experimental designs. Process Biochem. 2005;40:651–60.
38. Nelson PR. Exact critical points for the analysis of means. Comm Statistics. 1982;11:699–709.
39. Ohk SH, Yoo YJ, Kim BK, Bai DH. Purification and characterization of *Streptococus mutans* cell wall hydrolase from *Bacillus subtilis* YL-1004. J Microbiol Biotechnol. 2001;11:957–63.
40. Paik HD, Lee NK, Lee KH, Hwang YI, Pan JG. Identification and partial characterization of cerein BS229, a bacteriocin produced by *Bacillus cereus* BS229. J Microbiol Biotechnol. 2000;10:195–200.
41. Parker N, Hayward AC, Stirling GR. Effect of chitinolytic soil bacteria on root-knot nematode eggs, p. 157. In: 5th International congress plant pathology, abstracts of papers, Kyoto, Japan; 1988.
42. Pandey A. Recent process developments in solid-state fermentation. Process Biochem. 1992;27(2):109–17.
43. Plackett RL, Burman JP. The design of optimum multi-factorial experiments. Biometrika. 1946;33(4):305–25.
44. Rajeswari M, Ramakrishnan S. Influence of *Streptomyces fradiae* against Root knot nematode *Meloidogyne incognita* in Tomato. Res J Agric For Sci. 2015;3(1):6–11.
45. Reissig JL, Strominger JL, Leloir LF. A modification colorimetric method for the estimation of *N*-acetylamino sugars. J Biol Chem. 1955;27:959–66.
46. Ridout CJ, Coley-Smith JR, Lynch JR. Fractionation of extracellular enzymes from a mycoparasitic strain of *Trichoderma harzianum*. Enzyme Microb Technol. 1988;10:180–7.
47. Sahai AS, Manocha MS. Chitinases of fungi and plants: their involvement in morphogenesis and host parasitic interaction. FEMS Microbiol Rev. 1993;11:317–38.
48. Sandhya C, Adapa LK, Nampoothiri KM, Binod P, Szakacs G, Pandey A. Extracellular chitinase production by *Trichoderma harzianum* in submerged fermentation. J Basic Microbiol. 2004;44:49–58.
49. Sarrafzadeh Hussein M. Nutritional requirements of *Bacillus thuringiensis* during different phases of Growth, Sporulation and Germination evaluated by Plackett–Burman Method, Iran. J Chem Eng. 2012;31:131–6.
50. Sivakumar M, Gunasekaran K. Management of root-knot nematodes in tomato, chilli and brinjal by neem oil formulations. J Biopest. 2011;4(2):198–200.
51. Subramanian P, Choi I-C, Mani V, Park J, Subramaniyam S, Choi K-H, Sim JS, Lee CM, Koo JC, Hahn B-S. Stage-Wise Identification and Analysis of miRNA from Root-Knot Nematode *Meloidogyne incognita*. Int J Mol Sci. 2016;17(10):1758. https://doi.org/10.3390/ijms17101758.
52. Subramaniam S, Venkatraman R, Narayanan GK. Studies on Production of Enzyme Chitinase from *Streptomyces* sp. and its anti-fungal activity. J Pharm Res. 2012;5(3):1409–13.
53. Sharma P, Schaubel DE, Sima CS, Merion RM, Lok ASF. Re-weighting the Model for End-Stage Liver Disease Score Components. Gastroenterology. 2008;135(5):1575–81.
54. Sowmya B, Gomathi D, Kalaiselvi M, Ravikumar G, Arulraj C and Uma C. Production and Purification of Chitinase by Streptomyces sp. from Soil. J Adv Sci Res 2012;(3):25-29.
55. Taylor AL, Sasser JN, Raleigh NC. Biology, identification and control of root-knot nematodes (*Meloidogyne* species) N.C. State University Graphics; 1978.
56. Tasharrofi N, Adrangi S, Mehdi Fazeli M, Hossein Rastegar H, Khoshayand MR, Faramarzi MA. Optimization of chitinase production by *Bacillus pumilus* using Plackett–Burman design and response surface methodology. Iran J Pharm Res Autumn. 2011;10(4):759–68.
57. Tejada M, Hernandez MT, Garcia C. Soil restoration using composted plant residues: effects on soil properties. Soil Tillage Res. 2009;102:109–17.
58. Ulhoa CJ, Peberdy JF. Regulation of chitinase synthesis in *Trichoderma harzianum*. J Gen Microbiol. 1991;137:2163–9.
59. Vaidya RJ, Shah IM, Vyas PR, Chhatpar HS. Production of chitinase and its optimization from a novel isolate *Alcaligenes xylosoxidans*: potential in antifungal biocontrol. World J Microbiol Biotechnol. 2001;17:691–6.
60. Vionis AP, Niemeyer F, Karagouni AD, Schrempf H. Production and processing of a 59-kilodalton exochitinase during growth of *Streptomyces lividans* carrying p CHIO12 in soil microcosms amended with crab or fungal chitin. Appl Environ Microbiol. 1996;62:1774–80.
61. Wani AH. Management of root-knot nematode, *Meloidogyne incognita*, on okra and mung by soil amendment with oil cakes and leaves of different plants. Nematologia Mediterranea. 2006;34:83–7.
62. Wani AH, Bhat MY. Control of root-knot nematode, *Meloidogyne incognita* by urea coated with Nimin or other natural oils on mung, *Vigna radiata* (L.) R. Wilczek J Biopest. 2012;5:255–8.
63. Westcott SW, Kluepfel DA. Inhibition of *Criconemella xenoplax* egg hatch by *Pseudomonas aureofaciens*. Phytopathology. 1993;83:1245.
64. Williamson VM, Gleason CA. Plant-nematode interactions. Curr Opin Plant Biol. 2003;6:327–33.

17

Composition and structure of natural organic matter through advanced nuclear magnetic resonance techniques

Dainan Zhang[1], Dandan Duan[1], Youda Huang[1,2], Yu Yang[1] and Yong Ran[1*]

Abstract

Natural organic matter (NOM) plays important roles in biological, chemical, and physical processes within the terrestrial and aquatic ecosystem. Despite its importance, a clear and exhaustive knowledge on NOM chemistry still lacks. Aiming to prove that advanced solid-state ^{13}C nuclear magnetic resonance (NMR) techniques may contribute to fill such a gap, in this paper we reported relevant examples of its applicability to NOM components, such as biomass, deposition material, sediments, and kerogen samples. It is found that nonhydrolyzable organic carbons (NHC), chars, and polymethylene carbons are important in the investigated samples. The structure of each of the NHC fractions is similar to that of kerogens, highlighting the importance of selective preservation of NOM to the kerogen origin in the investigated aquatic ecosystems. Moreover, during the artificial maturation experiments of kerogen, the chemical and structural characteristics such as protonated aromatic, nonprotonated carbons, and aromatic cluster size play important roles in the origin and variation of nanoporosity during kerogen maturation.

Keywords: Natural organic matter, Biomass, Algae, Pollen, Atmospheric deposited particles, Sediments, Kerogen, Advanced solid-state ^{13}C NMR

Background

Terrigenous NOM generally includes a large fraction of biologically and chemically resistant (recalcitrant) compounds such as nonhydrolyzable organic matter, kerogen, and residual NOM, resulting from an enhanced degradation carried out by soil microbes [2, 19]. Despite its intrinsic recalcitrance, terrigenous NOM can also be recycled in the terrestrial environment. Ancient terrigenous NOM fractions also comprise kerogen or old soil NOM and humic substances, albeit to a lesser extent (<25%; [14]). Terrigenous NOM supplied by rivers undergoes intensive reworking in estuaries and deltas [9], with some estimates suggesting losses of two-thirds of the global fluvial input of particulate NOM (POM) in these environments [21]. Part of terrigenous NOM may be lost if leached (conversion to dissolved NOM), completely mineralized, exported, or buried. However, the controlling mechanisms for these processes are highly variable and are not yet well known [49].

Biomass residues (such as pollens and algae) are transformed via chemical, biological, and physical processes into more stable forms during humification processes. During humification, the number of aromatic and paraffinic carbons increases, whereas the level of alkyl C–O carbons decreases. Therefore, humification and degradation processes result in the loss of some of the characteristic signals of structurally identifiable materials such as carbohydrates and lead to the enhancement of signals from refractory structures, such as aromatic and alkyl compounds. It is well established that refractory biomacromolecules (such as lignin and sporopollenin biopolymers) are a recalcitrant and important constituent of NOM [3, 11].

*Correspondence: yran@gig.ac.cn
[1] State Key Laboratory of Organic Geochemistry, Guangzhou Institute of Geochemistry, Chinese Academy of Sciences, Guangzhou 510640, China
Full list of author information is available at the end of the article

Atmospheric particulate matter contains organic matter and inorganic materials and its sources can be either anthropogenic or natural. Its composition is closely related to issues such as direct and indirect climate modifications, air pollution, and reduced visibility. In the literature, most of the studies on the characterization of organic matter in atmospheric particulate matter have been conducted on the water soluble fraction using ^{1}H NMR [5]. Note that even 2D ^{1}H-^{13}C HSQC and HMQC can only detect protonated carbons. Thus, these studies completely lose the nonprotonated structural information. In addition, the characterization of insoluble fractions is also examined. To some extent, solid-state NMR is essential for providing relatively complete structural information on organic matter in atmospheric organic matter. To the best of our knowledge, there are only a number of cases using solid-state NMR to study its organic matter. In almost all these studies, simple ^{13}C cross polarization/magic angle spinning (CP/MAS) and spectral editing techniques have been used. The high-spinning speed multi-ramped amplitude cross polarization/magic angle spinning (multiCP/MAS) technique was applied for acquiring quantitative ^{13}C NMR spectra [20]. This multiple-cross polarization (multiCP) technique provides a simple, robust way to obtain quantitative solid-state ^{13}C NMR spectra of organic materials, with good signal-to-noise ratio. The spectral editing techniques (i.e., dipolar dephasing and a ^{13}C chemical-shift-anisotropy filter) are employed to investigate the subspectra of the chemical groups. Therefore, the advanced solid-state techniques to the characterization are yet to be applied.

Various physical and chemical methods have been used, whereas advanced solid-state NMR techniques have not been applied yet, despite they are potentially capable to significantly improve the knowledge on black carbon (BC) and kerogen carbon (KC). Methods for BC quantification in soils and sediments have been extensively evaluated [13, 16, 32, 35, 42]. By contrast, less efforts have been directed at the quantification of KC in soils and sediments, and the current protocols are mainly restricted to the fields of organic–petrographic analyses and chemical treatments [6, 12, 22, 39, 40, 46]. Kerogen is the most abundant organic matter in the Earth's crust [48]. Major input of kerogen to soils and sediments is from the natural weathering processes of sedimentary rocks and from coal since the latter, which is mainly composed of kerogen, is dispersed during mining operations. Kerogen typically has a three-dimensional structure with aromatic nuclei cross-linked by aliphatic chain-like bridges. These nuclei consist of aromatic sheets separated by voids with an average width of 30–40 nm. The aliphatic bridges contain linear or branched chains, and/or S- or O-containing functional groups [10]. Although most kerogens are distributed in deeply buried reservoirs [48], they can be present in significant amounts in soils and sediments. For example, (Ran et al. [40] found that Borden aquifer with a total OC content of 0.21 g kg^{-1} consisted of at least 186 g kg^{-1} of KC. Song et al. [46] reported 240–480 g kg^{-1} of KC out of total OC in one soil and three sediments. However, all existing KC quantification methods may suffer from operational shortcomings such as fine NOM particle loss, and incomplete amorphous organic carbon removal [13, 40, 46].

Kerogen is the sedimentary organic matter, which generates petroleum and natural gas. In this respect, many efforts have been devoted to its isolation and to the characterization of its chemical structure and evolution during sediment burial, aiming at a better prediction of oil and gas pools. With kerogen being a complex mixture of organic materials tightly mixed with minerals in sediments, its characterization represents a challenging goal. This notwithstanding, many progresses have been so far done and are due not only to the development of oil exploration via geological sample availability, but also to the huge increase in analytic capabilities during the last 40 years. However, despite the numerous advances made with respect to kerogen knowledge, important points are still far from being elucidated.

Solid-state NMR spectroscopy is commonly used to characterize complex humic substances [7, 18, 33, 38] including humic and fulvic acids [1, 17, 44, 47] and humin Preston and Newman [36, 37]. The primary technique used in these studies is the solid-state ^{13}C CP-MAS technique. Our understanding of chemical structures of humic substances has markedly improved over decades of extensive investigations using the solid-state ^{13}C CP-MAS spectroscopy [47, 51]. For example, it is well documented that marine humic acids generally contain less aromatics than lignin-derived terrigenous ones [27, 33]. Despite the knowledge gained by these NMR studies using ^{13}C CP-MAS, the full potential of solid-state NMR for characterizing humic substances has yet to be realized [33]. ^{13}C CP-MAS has drawbacks such as loss of structural information and poor quantification [8, 23, 30, 41, 45]. Moreover, due to their complex and heterogeneous nature, ^{13}C CP-MAS or direct polarization magic angle spinning (DP-MAS) NMR spectra of NOM generally reveal broad, poorly resolved peaks pertaining to major functional groups. Therefore, ambiguous assignments can be caused by the overlapping chemical shifts of numerous different structural moieties [33]. Thus, other analytic methods have also been used in sediment NOM characterization as reviewed in the previous review [19, 33].

During the recent years, many advanced solid-state ^{13}C NMR techniques have been developed, improved, and applied for characterizing complexity of NOM in algaes, plants, soils, sediments, and kerogens [23, 43]; Mao and Schmidt-Rohr [24–26], Mao et al. [27–29]. These advanced solid-state NMR techniques include ^{13}C DP/MAS NMR, quantitative DP/MAS plus dipolar dephasing, ^{13}C chemical-shift anisotropy filter (CSA), and CH, CH_2 spectral editing [28]. They overcome the drawbacks of conventional CP/MAS techniques and enable us to obtain more detailed chemical and structural information on various types of NOM. The identification and quantification of specific functional groups such as CH_2, CH_3, aromatic CH, aromatic C–O, protonated aromatic C, nonprotonated aromatic C, carbonyl C, and carboxyl C can be achieved. The structural parameters such as aromaticity and aromatic cluster size can also be estimated by using spectral analysis and long-range ^1H-^{13}C recoupled dipolar dephasing [4, 28]. These advanced techniques provide reliable measurements and have proven to be powerful tools for the characterization of NOM [28]. In the following sections, we will discuss several case studies (algae organic matter, pollen organic matter, sediment organic matter, atmospheric deposition particles, and kerogens) to prove the applicability of advanced 1D solid-state in this research field. Moreover, solid-state NMR experiments were performed on a Bruker spectrometer operating at 400-MHz ^1H and 100-MHz ^{13}C frequencies in all reported cases.

Algal organic matter

The three algae (Spirulina, Seaweed, and Porphyra) samples were fractionated into the lipid (LP), lipid free (LF), alkaline nonhydrolyzable carbon (ANHC), and acid nonhydrolyzable carbon (NHC) to obtain the refractory component. The major steps for isolation of the algal fractions were described elsewhere [54]. Briefly, the bulk samples (OS) were extracted to separate lipids by using Soxhlet extraction with 2:1 CH_2Cl_2/CH_3OH (v/v) for 24 h. The LP fractions were dried at 100 °C. Subsamples of the LF fractions were saponified for 1 h in 1 N KOH in 85:15 methanol/H_2O (v/v). The ANHC fractions were hydrolyzed twice with 2 N trifluoroacetic acid (TFA) at 100 °C for 3 h. Subsequently, samples were hydrolyzed in 4 and 6 N TFA at 100 °C for 18 h. Finally, the residual hydrolyzable organic matter was removed with 6 M HCl at 110 °C for 24 h.

The multiCP/MAS spectra were measured at a spinning speed of 14 kHz, where spinning sidebands are fairly small (<3%) and have little overlap with center bands. The 90° pulse lengths were 4.3 μs for ^1H and 4 μs for ^{13}C. To achieve dead time-free detection, which is indispensable for spectra with broad lines, all spectra were recorded with a Hahn echo generated by an exorcycled 180° pulse applied one rotation period (t_r) after the end of cross polarization. The ^1H decoupling field strength was $|\gamma B_1|/2\pi = 65$ kHz during the period of $2\ t_r = 0.14$ ms duration before the Hahn echo, and ca. 55 kHz during signal detection. The ramp for CP was implemented with 11 steps of 0.1 ms duration and a 1% amplitude increment (90–100%). The duration of the repolarization period (t_z) in multiCP was 0.7 s. Corresponding multiCP spectra of nonprotonated C and mobile segments were selected by recoupled dipolar dephasing (multiCP/DD) with a dipolar dephasing time of 68 μs. The ^{13}C spectra of three commercial algae and their LF and NHC fractions are shown in Fig. 1. Thin lines and thick lines represent quantitative ^{13}C NMR spectra of all C (multiCP/MAS) and of nonprotonated or mobile C (multiCP/MAS with recoupled dipolar dephasing), respectively. The relative percentages of different C functional groups are listed in Table 1 [55]. In general, the three bulk commercial algae samples contain high alkyl C–O (31.7–43.1% of all C) and alkyl C (17.7–24.7%), but low aromatic C (5.91–7.89%). After organic solvent extraction, the $(CH_2)n$ percentages decrease in the LF fractions relative to the corresponding bulk samples for Spirulina and Seaweed (Table 1). But for Porphyra, the $(CH_2)n$ signal varies slightly and is quite similar to that of the OS, likely due to the lower content of lipid in Porphyra. The NHC fractions of three commercial algae generally contain much lower alkyl C–O (8.72–12.6% of all C) than their corresponding OS (31.7–43.1%) and LF samples (28.9–43.4%), indicating that the majority of polysaccharides and carbohydrates were removed during alkaline and acid hydrolysis. In addition, the aromatic C including aromatic C–C, aromatic C–H and aromatic C–O, and ketone/aldehyde C are much more abundant in the NHC fractions than in corresponding OS and LF samples. The signal at 27.5–31.8 ppm is the most intense in the NHC fraction of Porphyra, suggesting that its NHC fraction contains the highest polymethylene carbon among the three commercial algae.

Pollen organic matter

The two pollens (*Nelumbo nucifera* and *Brassica campestris* L.) were sequentially fractionated into the lipid (LP), lipid free (LF), residue obtained after successive TFA hydrolysis (TFAR), residue obtained after saponification (SR), and acid nonhydrolyzable carbon (NHC) to obtain the refractory component, as described elsewhere [53]. In brief, the lyophilized OS samples were extracted to separate lipids via Soxhlet extraction with CH_2Cl_2:CH_3OH (2:1, v/v) for 24 h. The LP fractions were dried at 60 °C. Subsamples of the LF fractions were hydrolyzed twice with 2 N TFA at 100 °C for 3 h. Subsequently, samples were hydrolyzed in 4 and 6 N TFA at 100 °C for 18 h. The

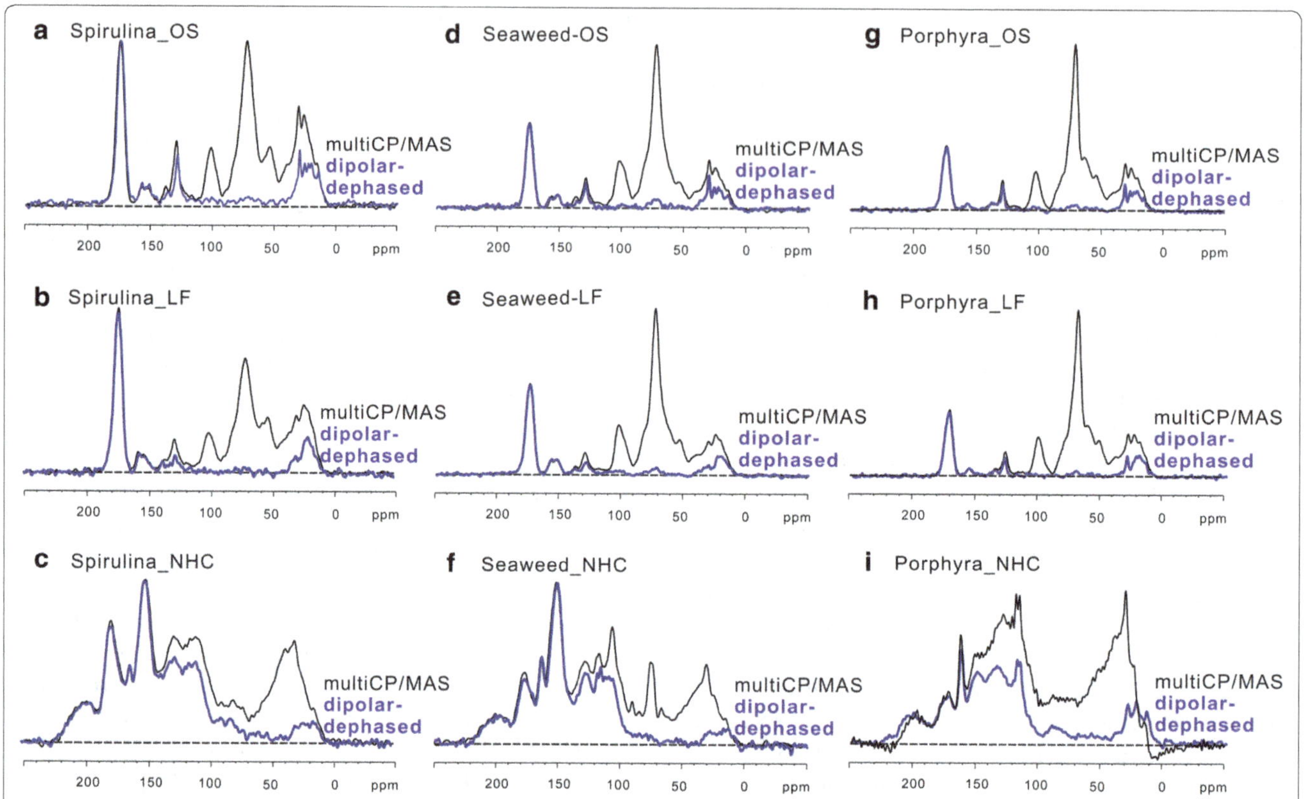

Fig. 1 MultiCP ^{13}C spectra (*thin lines*) and multiCP ^{13}C spectra after recoupled dipolar dephasing (*thick lines*, 2 s recycle delay) of the bulk algal samples (OS) and related fractions (LF and NHC)

residue obtained after successive TFA hydrolysis (TFAR) fractions was saponified by refluxing for 1 h in 5% KOH in 2-methoxy ethanol/H_2O (88/12, v/v). The residue obtained after saponification (SR) fractions was treated by 6 N HCl at 110 °C for 24 h. The remaining residues made up the final NHC fraction. The cross polarization (CP) experiments included ^{13}C cross polarization/total sideband suppression (CP/TOSS), CP/TOSS plus dipolar dephasing (CP/TOSS/DD), and ^{13}C chemical-shift-anisotropy (CSA) filter. The ^{13}C solid NMR spectra of the investigated bulk pollens and their fraction samples represent ^{13}C CP/TOSS NMR spectra of all C (Additional file 1: Figure S1). A pair of spectral editing techniques, dipolar dephasing, and ^{13}C CSA were employed to study the structures of the bulk pollens and their fractions in more detail [53].

The bulk lotus and rape pollen samples contain the largest amounts of alkyl C–O (53.9 and 47.7%, respectively) (Additional file 1: Table S1). The relative proportions of alkyl C–O for TFAR, SR, and NHC fractions (11.6–19.1%) decrease greatly compared to those for LF fractions (44.7–52.4%). After successive TFA and 6 N HCl treatments, alkyl C increases to a greater extent than does aromatic C. The TFAR, SR, and NHC fractions display the highest relative amount of alkyl carbon (44.1–53.2%), which includes polymethylene-based functionalities. The NHC fractions mainly consist of alkyl carbon (44.1–53.2%), aromatic moieties (110–165 ppm) (20.5–25.8%), alkyl C–O (11.6–12.6%), and O–CH_3/NCH (8.82–10.4%). The NHC fractions have extremely high poly(methylene) carbon percent, up to 21.0%. NMR spectroscopy indicates that the structure of the NHC fractions in pollen samples is similar to that of sporopollenin, which principally contains aliphatic, aromatic, ether, and carbonyl and carboxylic groups in varying degrees [1, 15, 31, 50].

Based on the ^{13}C NMR spectra of the classes reported by Nelson and Baldock [34], the constituents of the bulk pollen samples and their fractions are largely aliphatic components (including sporopollenin), carbohydrates, protein, and lignin as estimated by the four listed biochemical classes (Additional file 1: Table S2). Carbohydrate and aliphatic components account for 63.3–68.0 and 12.5–15.4% of the total organic carbon in the bulk pollen samples, respectively, followed by protein and lignin ranging from 9.34 to 14.9% and from 5.83 to 8.67% of the OC content, respectively. The aliphatic components of LF fractions range from 10.8 to 11.7%, which are

Table 1 Quantification (%) of functional groups from the ^{13}C multiCP/MAS NMR spectra for the bulk algal samples (OS) and related fractions (LF and NHC)

Samples	Alkyl 0–45 ppm	OCH_3/NCH 45–60 ppm	Alkyl C–O 60–95 ppm	O–Alkyl–O 95–110 ppm	Arom C 110–145 ppm	Arom C–O 145–165 ppm	COO/NC=O 165–190 ppm	Ketone/aldehyde 190–220 ppm	$(CH_2)n$ 27.5–33.5 ppm	Polar C[a]
OS										
Spirulina	24.7	8.47	31.7	6.88	7.89	2.63	17.6	0.14	6.45	60.5
Seaweed	17.7	6.37	42.8	9.35	6.68	2.73	13.9	0.50	4.67	66.3
Porphyra	20.7	9.84	43.1	7.53	5.91	1.01	11.9	0.00	5.27	65.9
LF										
Spirulina	23.3	9.72	28.9	6.17	6.76	3.53	21.3	0.36	4.79	63.8
Seaweed	17.1	8.18	41.9	9.19	5.31	3.23	14.8	0.20	3.62	68.3
Porphyra	20.6	9.85	43.4	7.58	5.48	1.25	11.9	0.00	4.97	66.4
NHC										
Spirulina	17.7	5.17	8.72	8.60	23.3	16.3	13.8	6.32	4.22	50.3
Seaweed	14.9	4.02	12.6	11.1	22.3	19.5	10.4	5.23	3.91	51.8
Porphyra	19.1	8.79	11.0	6.45	36.4	13.4	4.96	0.00	7.15	38.2

[a] *Polar C* polar aliphatic (45–95 ppm) + polar aromatic and C=O (145–220 ppm)

lower than in their OS samples. In NHC fractions, alkyl C and aromatic C (especially aromatic C–C) were more visible because of the removal of carbohydrates. The NHC fractions are composed of aliphatic components (52.1–58.2%), lignin (20.5–25.9%), carbohydrate (11.5–13.4%), and protein (7.96–10.4%), consistent with the previous studies that sporopollenin is not a uniform macromolecule but rather a series of closely related biopolymers with different chemical groups occurring in varying amounts [15, 50].

Sediment organic matter

Three surface sediment samples (B1, A01, and A04) were sequentially fractionated into the demineralized (DM), lipid (LP), lipid free (LF), and acid nonhydrolyzable carbon (NHC) to obtain the refractory component, as described elsewhere [52]. Briefly, to remove paramagnetic material and enrich the relative OC content in samples, the bulk samples (OS) were tested with 1 N HCl/10% HF to obtain the DM fraction. The free lipids (LP) were extracted repeatedly from an aliquot of DM with a 2:1 CH_2Cl_2/CH_3OH (v/v) using soxhlet extraction for 24 h, and then dried at 50 °C. The solid residue (lipid-free fraction, LF) was dried at 50 °C. The LF fractions were hydrolyzed twice with 2 N TFA at 100 °C for 3 h. Subsequently, samples were hydrolyzed in 4 and 6 N TFA at 100 °C for 18 h. Finally, the residual hydrolyzable organic matter was removed with 6 M HCl at 110 °C for 24 h.

The ^{13}C solid NMR spectra in the demineralized carbon (DM), LF, and NHC fractions for the three sediments collected from the Pearl River Delta represent ^{13}C CP/TOSS NMR spectra of all C. In addition, the spectra resulting from two spectral editing, dipolar dephasing, and CSA are shown in Additional file 1: Figure S2, and described elsewhere [52]. The ^{13}C CP/TOSS data show that the alkyl carbon components account for 25.6–32.4% of the total organic carbon in the DM fractions (Table 2). The LP fractions mainly consist of alkyl carbon. As a result, alkyl carbon contents of the LF fractions decrease to 22.0–27.3% relative to the DM fractions. The LF fractions have higher aromatic C contents (38.8–39.2%) than DM fractions (32.8–35.9%) do.

Based on the ^{13}C NMR spectra, the contribution of four biomacromolecule classes (lignin, protein, carbohydrate, and aliphatic component) in sediments and their isolated fractions is estimated according to Nelson and Baldock (Table 2). As the sediments were reported to contain char or black carbon (BC), the char end member is added. The solid-state ^{13}C CP/MAS NMR spectra of char are cited from one previous investigation [34]. The result shows that aliphatic components and carbohydrate account for 22.7–34.8 and 21.3–25.5% of the total OC contents in the DM fractions, followed by char and lignin ranging from 15.6 to 24.1% and from 13.7 to 16.8% of the total OC contents, respectively (Table 3). After organic solvent extraction, the aliphatic components of LF fractions range from 17.9 to 25.6%, which are lower than those of their DM fractions. The NHC fractions (normalized to OC contents) are composed of char (30.8–40.9%), aliphatic component (25.0–33.9%), lignin (14.0–17.4%), carbohydrate (11.2–11.9%), and protein (6.71–8.27%). The NHC fractions are enriched in aromatic moieties due to the removal of carbohydrates and proteins.

Atmospheric deposition particles

Atmospheric dry and wet deposition particles were seasonally collected with several stainless steel buckets (i.d. 50 ± 0.5 cm, height 50 cm, flat bottom) in the Guangzhou Institute of Geochemistry (DHS) in Guangzhou, an urban area of Pearl River Delta (PRD), South China from Apr. 2010 to Jul. 2011. Bulk dry and wet deposition samples were collected by using 25 L plastic buckets and were carried back to the laboratory. The samples were centrifuged by using a high-speed refrigerated centrifuge (Changsha, China) to collect the suspended particulate materials (SPM). Then the separated atmospheric SPM were analyzed by advanced solid-state ^{13}C NMR techniques combined with elemental analysis and Rock-Eval pyrolysis for their chemical and structural characterization. In order to calculate the deposition fluxes, the rainfall amount records at each sampling site were periodically obtained from the reservoir administration offices or from the local meteorological departments.

The results show that the deposition fluxes of SPM range from 47.58 to 65.07 mg/d/m^2 (data not shown), and are higher in spring and fall than in summer and winter. The total organic carbon (TOC) of SPM ranges from 13.15 to 17.09% in DHS.

The NMR data indicate a similar distribution of dominated nonpolar alkyl C and alkyl C–O in all the deposited SPM, followed by aromatic, and aromatic C–O, COO, NC=O, ketone, quinone, and aldehyde carbon (Duan et al. unpublished data). Relative high contents of COO are observed in the suburb area during summer and fall seasons. Aromaticity is in the range of 11.04–18.4%, whereas polymethylene carbon varies from 5.00 to 5.97%. The percentages of nonprotonated aromatic carbon accounting for aromatic carbon range from 5.5 to 14.81%. The above results are further confirmed by the data of the Rock-Eval pyrolysis and the five-component model, as used in the above sediment section. The Rock-Eval shows that the S2 concentrations range from 43.67 to 61.34 mg HC/g and the hydrogen index (HI) values are in the range of 332–380 mg HC/g TOC, suggesting the presence of high aliphatic components (Table 4). Moreover, the S1

Table 2 Functional groups (%) detected in ^{13}C CP/TOSS NMR Spectra of demineralized sediments (DM) collected at different sites (B1, A01, and A04) from the Pearl river delta and their fractions (LF and NHC)

Samples	Alkyl C 0–48 ppm	OCH_3/NCH 48–65 ppm	Alkyl C–O 65–91 ppm	O-alkyl-O 91–110 ppm	Arom C–C 91–150 ppm	Arom C–H 91–150 ppm	Arom C–O 150–165 ppm	COO/NC=O 150–165 ppm	$(CH_2)n$	Aromatic C[a]	Polar C[b]
DM											
B1	27.4	12.0	15.9	3.27	18.6	13.0	4.36	5.57	8.30	35.9	46.1
A01	32.4	11.8	13.6	3.56	15.6	12.3	4.81	5.84	9.50	32.8	45.6
A04	25.6	11.8	14.6	4.83	23.0	7.38	5.46	7.29	7.70	35.9	46.9
LF											
B1	22.0	12.6	17.2	3.08	19.6	14.9	4.68	5.89	6.50	39.2	46.9
A01	27.3	12.9	13.6	3.04	19.8	11.3	5.30	6.73	8.30	36.4	46.8
A04	23.0	12.2	15.6	3.25	21.0	12.7	5.19	7.21	6.40	38.8	46.6
NHC											
B1	24.6	9.39	8.12	1.05	23.6	22.7	5.62	4.95	7.90	51.9	36.0
A01	29.2	9.43	8.37	1.32	22.1	18.7	5.94	4.95	10.7	46.7	39.4
A04	24.0	8.81	8.15	1.18	26.2	19.1	6.68	5.87	7.70	52.0	37.2

[a] *Aromatic C* aromatic C–C (91–150 ppm) + aromatic C–H (91–150 ppm) + aromatic C–O (150–165 ppm) (%)

[b] *Polar C* polar aliphatic (48–91 ppm) + polar aromatic and C=O (150–190 ppm)

Table 3 Contributions (%) of five model components for the sediments and their fractions

Samples	Carbohydrate	Protein	Lignin	Aliphatic	Char
DM					
B1	25.5	11.4	15.8	26.6	20.8
A01	21.3	11.4	16.8	34.8	15.6
A04	24.7	14.8	13.7	22.7	24.1
LP					
B1	17.8	10.9	19.0	52.3	0.00
A01	10.9	4.31	10.9	73.9	0.00
A04	13.7	14.5	14.3	57.4	0.00
LF					
B1	28.8	12.9	15.8	17.9	24.7
A01	23.1	14.9	16.4	25.6	20.1
A04	26.4	14.5	13.4	18.9	26.7
NHC					
B1	11.4	7.82	15.8	25.1	39.8
A01	11.2	6.71	17.4	33.9	30.8
A04	11.9	8.27	14.0	25.0	40.9

and S2 values are positively related to aliphatic carbon. The carbohydrates accounting for 32.47–47.95% of the total carbon are the main components in the atmospheric bulk deposition, followed by aliphatic components (19.88–27.33%), proteins (12.65–17.87%), lignin (16.68–26.86%), and BC (0.00–1.61%) (spectra not shown). In general, the chemical compounds of alkyl groups are more abundant than those of aromatic groups.

Bulk kerogen and thermally stimulated kerogen

Kerogen was isolated from the Maoming shale and sequentially treated at different temperatures (250, 300, 350, 400, 450, and 500 °C) to understand changes in chemical, structural, and nanoporous characteristics during artificial maturation. Quantitative ^{13}C direct polarization magic angle spinning (DP/MAS) NMR experiments were performed at a spinning speed of 14 kHz. The 90° ^{13}C pulse length was 4 μs. Nonprotonated carbons and mobile carbon fractions were quantified using a combination of the DP/MAS technique with a recoupled dipolar dephasing (DP/MAS/DD).

Advanced solid-state ^{13}C NMR techniques (DP/MAS NMR, DP/MAS-DD, and CSA filter) were employed and combined with Rock–Eval pyrolysis to characterize the kerogen sample, whereas the nanoporosity and surface properties were elucidated using N_2 and CO_2 adsorption techniques. The results show that the H/C and O/C atomic ratios decrease, but vitrinite reflectance (R_o) and maturation (T_{max}) values increase with the increasing temperature, which are consistent with the declining trends for S2, S3, HI, and oxygen index (OI) parameters derived from the Rock-Eval pyrolysis (data not shown), suggesting that the dehydration, decarboxylation, and condensation predominately occurred throughout the artificial thermal process.

The aliphatic and carbonyl carbons of the kerogen samples significantly decrease, but the aromatic nonprotonated and protonated carbons significantly increase when the temperature increases (Fig. 2). Aromaticity (F_{aro}) is an important parameter for the kerogen structure characterization. The aromaticity ranges from 47.4 to 94.7% for the investigated kerogens (Table 5). As the simulation temperature increases, F_{aro} (%) increases slightly from 47.4% at 25 °C to 54.9% at 300 °C, and then sharply increases from 82.7% at 400 °C to 94.7% at 500 °C. Moreover, F_{aro} (%) is highly correlated with R_o ($R^2 = 0.93$, $p < 0.01$) and T_{max} ($R^2 = 0.73$, $p < 0.01$). In addition, the nonpolar nonprotonated aromatic carbon (F_{naro}) is highly correlated with R_o (%) (Fig. 2). The above statistics analysis demonstrates that F_{aro} (%) and F_{naro} (%) could be used as appropriate proxies for the kerogen maturity. Furthermore, the average aromatic cluster size can be obtained by the following equation: $n_C \geq 6/\chi^2_{edge} \geq 6/\chi^2_{edge, max}$, which gives a lower limit to the number of carbons of an

Table 4 Rock-Eval parameters of atmospheric deposited particles in DHS

DHS	S1[a] (mg of HC g^{-1} dw)	S2[b] (mg of HC g^{-1} dw)	S3[c] (mg of CO^2 g^{-1} dw)	HI (hydrogen index) (mg of HC g^{-1} TOC)	OI (oxygen index) (mg of CO_2 g^{-1} TOC)	TOC
Spring (Apr. 2010–May 2010)	9.62	43.67	13.93	332	23	13.15
Fall (Sep. 2010–Nov 2010)	9.74	53.04	18.25	351	20	15.09
Winter (Dec. 2010–Feb. 2011)	10.86	55.03	17.04	380	10	14.50
Summer (May 2011–Jul. 2011)	22.68	61.34	22.35	359	23	17.09

[a] S1 is generated during pyrolysis under 300 °C

[b] S2 is generated during pyrolysis from 300 to 650 °C

[c] S3 is originated from CO and CO_2 detected by infrared (IR) during pyrolysis

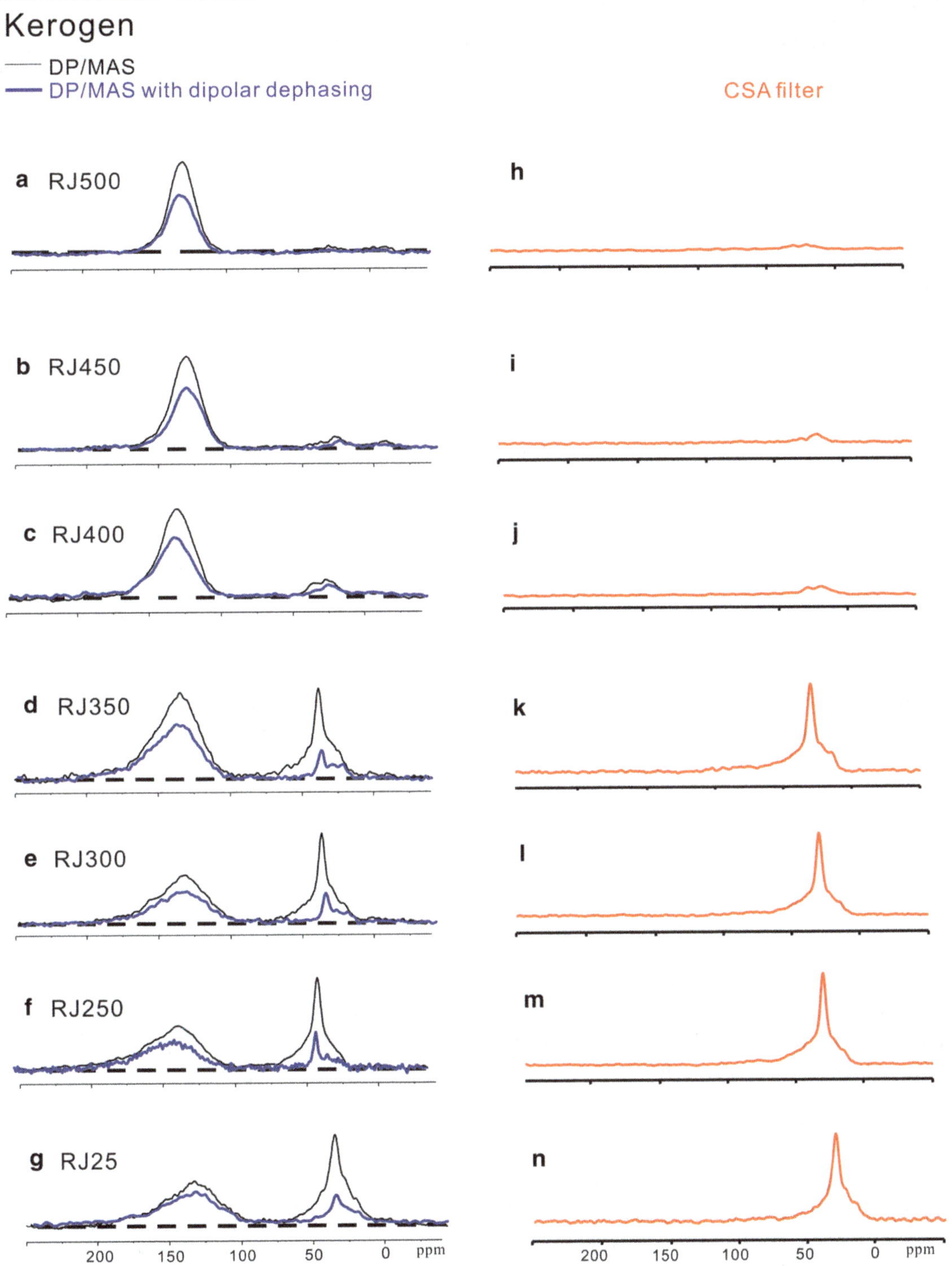

Fig. 2 ^{13}C NMR DP/MAS (*thin lines* **a**–**g**), DP with dipolar dephasing (*solid lines*, **a**–**g**, 20 s recycle delay), and CSA filtered spectra (**h**–**n**, 1 s recycle delay) for thermally treated kerogens

aromatic cluster in the kerogen [28]. The minimum aromatic cluster size ranges from 10 carbons at 350 °C to 38 carbons at 500 °C. In addition, the nanopore volumes ($V_{o,\ d\text{-}co2}$) increase from 28.5 at 25 °C to 80.3 μL/g at 450 °C but significantly decrease to 78.9 μL/g on the kerogen at 500 °C. It is found that aromaticity and aromatic cluster size are tightly related to the nanoporous volumes for the thermally matured kerogen samples. In conclusion, the

Table 5 NMR parameters of thermally stimulated kerogens

Kerogen	R_o (%)[a]	F_{aro} (%)[b]	F_{alkyl} (%)[c]	F_{Oaro} (%)[d]	F_{naro} (%)[e]	F_{Haro} (%)[f]	F_{Haro}/F_{aro}	F_{aro}/F_{alkyl}	$X^{g}_{edge, min}$	$n^{h}_{C, min}$
RJ25	0.36	47.4	46.3	7.57	38.7	1.10	0.02	1.02	0.18	4
R250	0.48	51.4	41.7	7.92	42.6	0.90	0.02	1.23	0.17	5
R300	0.52	54.9	40.1	7.65	44.4	2.80	0.05	1.37	0.19	6
R350	0.76	63.4	30.9	9.87	51.2	2.30	0.04	2.05	0.19	10
R400	1.40	82.7	14.1	8.90	56.6	17.2	0.21	5.89	0.32	22
R450	2.10	90.6	7.62	7.07	61.7	21.8	0.24	11.9	0.32	34
R500	2.69	94.7	4.94	5.46	63.0	26.2	0.28	19.2	0.33	38

[a] R_o is the vitrinite reflectance

[b] F_{aro} is the aromaticity

[c] F_{alkyl} is the fraction of nonpolar alkyl carbon

[d] F_{Oaro} is the fraction of aromatic C–O carbon

[e] F_{naro} is the fraction of nonpolar nonprotonated aromatic carbon

[f] F_{Haro} is the fraction of protonated aromatic carbon

[g] $X_{edge, min}$ is the fraction of minimum aromatic edge along the edges of the aromatic rings

[h] $n_{C, min}$ is the minimum aromatic cluster size

advanced NMR techniques are very useful for investigating the structure and maturation degrees of the thermally simulated kerogen samples.

Conclusions

NMR techniques have facilitated our ability to identify the key NOM composition, structure, transformation, and fates, and their changes as a function of environmental factors. Terrestrial environments are important sites of organic carbon cycling. In this manuscript, several case studies have been reported to prove the applicability of NMR tools to study the structures and compositions of NOM. This study characterized the isolated NHC from biomass, atmospheric deposition, and sediments in the Pearl River Delta by using advanced NMR analytical technique and Rock-Eval pyrolysis technique. It is feasible that biochemical macromolecules can be effectively removed, and condensed NOM (NHC and BC) can be effectively isolated from the investigated samples with the multiple step hydrolysis procedure. The NHC fractions are important or even dominant components of the NOM in the investigated samples. Systematic characterization of NHC shows that it is originated from polymethylene C (algaenan, sporopollenin, cutin, and cuticle), and/or bitumen, kerogen, aged terrigenous NOM (lignin), and black carbon. Moreover, during the artificial maturation experiments of kerogens, as their chemical composition and structure are altered, their nanoporosity is changed to some extent. The chemical and structural characteristics such as aromatic protonated, nonprotonated carbons, and aromatic cluster size play important roles in the development of nanoporosity during kerogen maturation. In summary, the structure of each of the NHC fractions is similar to that of kerogen, highlighting the importance of selective preservation of NOM to the kerogen origin in the investigated aquatic ecosystems. The advanced NMR techniques are very useful to investigate the compositions, constituents, and structures for the biomass, sediment NOM, atmospheric SPM, and thermally matured kerogen samples.

Abbreviations

ANHC: alkaline nonhydrolyzable carbon; BC: black carbon; ^{13}C CSA: ^{13}C chemical shift anisotropy filter; ^{13}C CP/MAS: ^{13}C cross polarization/magic angle spinning; ^{13}C CP/TOSS: ^{13}C cross polarization/total sideband suppression; ^{13}C DP/MAS: ^{13}C direct polarization/magic angle spinning; ^{13}C DP/MAS-DD: ^{13}C cross polarization/magic angle dipolar dephasing; Multi CP: high-spinning speed multi-ramped amplitude cross polarization; DHS: Guangzhou Institute of Geochemistry; DM: demineralized carbon; F_{aro}: aromaticity; HMQC: proton-detected heteronuclear multiple quantum coherence; HSQC: proton-detected heteronuclear single quantum coherence; KC: kerogen carbon; LP: lipid; LF: lipid free; NHC: acid nonhydrolyzable carbon; NMR: nuclear magnetic resonance; NOM: natural organic matter; OS: bulk sample; POM: particulate NOM; PRD: Pearl River Delta; SR: residue obtained after saponification; TFA: trifluoroacetic acid; TFAR: residue obtained after successive TFA hydrolysis; TOC: total organic carbon; V_o: nanoporosity.

Authors' contributions

DNZ and DDD have collected the biomass, sediment, atmospheric samples, and done the fractionated extraction and NMR spectra, and written the manuscript, and YDH and YY have collected the sediment and kerogen samples and the NMR spectra. YR has designed the investigation and written the manuscript. All authors read and approved the final manuscript.

Author details

[1] State Key Laboratory of Organic Geochemistry, Guangzhou Institute of Geochemistry, Chinese Academy of Sciences, Guangzhou 510640, China.
[2] University of Chinese Academy of Sciences, Beijing 100049, China.

Acknowledgements
The authors thank Dr. Jingdong Mao in Old Dominion University, USA for the analytic measurements of the samples by using advanced NMR techniques. This is a contribution from GIGCAS, No. IS-2337.

Competing interests
The authors declare that they have no competing interests.

Funding
This investigation was supported by a key joint project of the National Natural Foundation of China (NNSFC) the Guangdong Province Natural Science Foundation (U1201235), and a general project of the NNSFC (41473103).

References

1. Ahlers F, Bubert H, Steuernage S, Wiermann R. The nature of oxygen in sporopollenin from the pollen of *Typha angustifolia* L. Zeitschrift für Naturforschung C. 2000;55(3–4):129–36.
2. Bergamaschi BA, Tsamakis E, Keil RG, Eglinton TI, Montlucon DB, Hedges JI. The effect of grain size and surface area on organic matter, lignin and carbohydrate concentration, and molecular compositions in Peru Margin sediments. Geochim Cosmochim Acta. 1997;61:1247–60.
3. Blokker P, Schouten S, van den Ende H, de Leeuw JW, Hatcher PG, Sinninghe Damsté JS. Chemical structure of algaenans from the fresh water algae *Tetraedron minimum*, *Scenedesmus communis* and *Pediastrum boryanum*. Org Geochem. 1998;29:1453–68.
4. Brewer CE, Schmidt-Rohr K, Satrio JA, Brown RC. Characterization of biochar from fast pyrolysis and gasification systems. Environ Prog Sustain Energy. 2009;28:386–96.
5. Chalbot MCG, Kavouras IG. Nuclear magnetic resonance spectroscopy for determining the functional content of organic aerosols: a review. Environ Pollut. 2014;191:232–49.
6. Cornelissen G, Gustafsson O, Bucheli TD, Jonker MTO, Koelmans AA, Van Noort PCM. Extensive sorption of organic compounds to black carbon, coal, and kerogen in sediments and soils: mechanisms and consequences for distribution, bioaccumulation, and biodegradation. Environ Sci Technol. 2005;39:6881–95.
7. Conte P, Piccolo A. Conformational arrangement of dissolved humic substances. Influence of solution composition on association of humic molecules. Environ Sci Technol. 1999;33:1682–90.
8. Duer MJ. Introduction to solid-state NMR spectroscopy. Oxford: Blackwell; 2004.
9. Eglinton TI, Repeta DJ. Organic matter in the contemporary ocean. In: Elderfield H, editor. The oceans and marine geochemistry. London: Elsevier Pergamon; 2004.
10. Engel MH, Macko SA. Organic geochemistry: principle and application. New York: Plenum Press; 1993.
11. Gelin F, Volkman JK, Largeau C, Derenne S, Damsté JS, De Leeuw JW. Distribution of aliphatic, nonhydrolyzable biopolymers in marine microalgae. Org Geochem. 1999;30(2):147–59.
12. Ge'linas Y, Prentice KM, Baldock JA, Hedges JI. An improved thermal oxidation method for the quantification of soot/graphitic black carbon in sediments and soils. Environ Sci Technol. 2001;35(17):e3519–25.
13. Ge'linas Y, Baldock JA, Hedges JI. Organic carbon composition of marine sediments: effect of oxygen exposure on oil generation potential. Science. 2001;294:e145–8.
14. Goni MA, Yunker MB, Macdonald RW, Eglinton TI. The supply and preservation of ancient and modem components of organic carbon in the Canadian Beaufort Shelf of the Arctic Ocean. Mar Chem. 2005;93:53–73.
15. Guilford WJ, Schneider DM, Labovitz J, Opella SJ. High resolution solid state ^{13}C NMR spectroscopy of sporopollenins from different plant taxa. Plant Physiol. 1988;86(1):134–6.
16. Gustafsson O, Haghseta K, Chan F, McFarlane A, Gschwend PM. Quantification of the dilute sedimentary soot phase: implications for PAH speciation and bioavailability. Environ Sci Technol. 1997;31(1):e203–9.
17. Hatcher PG, Lerch HE III, Bates AL, Verheyen TV. Solid-state ^{13}C nuclear magnetic resonance studies of coalified gymnosperm xylem tissue from Australian brown coals. Org Geochem. 1989;14:145–55.
18. Hayes CE, Edelstein WA, Schenck JF, Mueller OM, Eash M. An efficient, highly homogeneous radiofrequency coil for whole-body NMR imaging at 1.5 T. J Magn Reson. 1985;63(3):622–8.
19. Hedges JI, Oades JM. Comparative organic geochemistries of soils and marine sediments. Org Geochem. 1997;27:319–61.
20. Johnson RL, Schmidt-Rohr K. Quantitative solid-state ^{13}C NMR with signal enhancement by multiple cross polarization. J Magn Reson. 2014;239:44–9.
21. Keil RG, Mayer LM, Quay PD, Richey JE, Hedges JI. Loss of organic matter from riverine particles in deltas. Geochim Cosmochim Acta. 1997;61:1507–11.
22. Kleineidam S, Rügner H, Ligouis B, Grathwohl P. Organic matter facies and equilibrium sorption of phenanthrene. Environ Sci Technol. 1999;33(10):1637–44.
23. Mao JD, Hu WG, Schmidt-Rohr K, Davies G, Ghabbour EA, Xing BS. Quantitative characterization of humic substances by solid-state carbon-13 nuclear magnetic resonance. Soil Sci Soc Am. 2000;64:873–84.
24. Mao JD, Schmidt-Rohr K. Accurate quantification of aromaticity and nonprotonated aromatic carbon fraction in natural organic matter by 13C solid-state nuclear magnetic resonance. Environ Sci Technol. 2004;38:2680–4.
25. Mao JD, Schmidt-Rohr K. Separation of aromatic-carbon 13 C NMR signals from di-oxygenated alkyl bands by a chemical-shift-anisotropy filter. Solid State Nucl Magn Reson. 2004;26:36–45.
26. Mao JD, Schmidt-Rohr K. Methylene spectral editing in solid-state ^{13}C NMR by three-spin coherence selection. J Magn Reson. 2005;176:1–6.
27. Mao JD, Cory RM, McKnight DM, Schmidt-Rohr K. Characterization of a nitrogen-rich fulvic acid and its precursor algae from solid state NMR. Org Geochem. 2007;38:1277–92.
28. Mao JD, Fang X, Lan Y, Schimmelmann A, Mastalerz M, Xu L, Schmidt-Rohr K. Chemical and nanometer-scale structure of kerogen and its change during thermal maturation investigated by advanced solid-state ^{13}C NMR spectroscopy. Geochim Cosmochim Acta. 2010;74:2110–27.
29. Mao JD, Fang X, Schmidt-Rohr K, Carmo AM, Hundal LS, Thompson ML. Molecular-scale heterogeneity of humic acid in particle-size fractions of two Iowa soils. Geoderma. 2007;140:17–29.
30. Mehring M, Weber H, Müller W, Wegner G. High resolution 13 C NMR in undoped cis-and trans-polyacetylene. Solid State Commun. 1983;45(12):1079–82.
31. Meuter-Gerhards A, Schwerdtfeger C, Steuernagel S, Wilmesmeier S, Wiermann R. Studies on sporopollenin structure during pollen development. Zeitschrift für Naturforschung C. 1995;50(7–8):487–92.
32. Middelburg JJ, Nieuwenhuize J, Breugel P. Black carbon in marine sediments. Mar Chem. 1999;65(4):e245–52.
33. Mopper K, Stubbins A, Ritchie JD, Bialk HM, Hatcher PG. Advanced instrumental approaches for characterization of marine dissolved organic matter: extraction techniques, mass spectrometry, and nuclear magnetic resonance spectroscopy. Chem Rev. 2007;107:419–42.
34. Nelson PN, Baldock JA. Estimating the molecular composition of a diverse range of natural organic materials from solid-state ^{13}C NMR and elemental analyses. Biogeochemistry. 2005;72:1–34.
35. Nguyen TH, Brown R, Ball WP. An evaluation of thermal resistance as a measure of black carbon content in diesel soot, wood char, and sediment. Org Geochem. 2004;35(3):217–34.
36. Preston CM, Newman RH. Demonstration of spatial heterogeneity in the organic matter of de-ashed humin samples by solid-state ^{13}C CPMAS NMR. Can J Soil Sci. 1992;72(1):13–9.
37. Preston C, Newman R. A long-term effect of N fertilization on the ^{13}C CPMAS NMR of de-ashed soil humin in a second-growth Douglas-fir stand of coastal British Columbia. Geoderma. 1995;68:229–41.
38. Preston CM. Applications of NMR to soil organic matter analysis: history and prospects. Soil Sci. 1996;161:144–66.
39. Ran Y, Huang W, Rao PSC, Sheng G, Fu J. The role of condensed organic matter in the nonlinear sorption of hydrophobic organic contaminants by a peat and sediments. J Environ Qual. 2002;31:1953–62.
40. Ran Y, Xiao B, Huang W, Peng P, Liu D, Fu J, Sheng G. Kerogen in aquifer material and its strong sorption for nonionic organic pollutants. J Environ Qual. 2003;32:1701.

41. Schmidt-Rohr K, Spiess HW. Multidimensional solid-state NMR and polymers. Hoboken: Elsevier; 1994.
42. Schmidt MWI, Skjemstad J, Czimczik CI, Glaser B, Prentice KM, Gelinas Y, Kuhlbusch TAJ. Comparative analysis of black carbon in soils. Global Biogeochem Cycles. 2001;15(1):163–7.
43. Schmidt-Rohr K, Mao JD. Efficient CH-group selection and identification in ^{13}C solid-state NMR by dipolar DEPT and 1H chemical-shift filtering. J Am Chem Soc. 2002;124:13938–48.
44. Smernik RJ, Olk D, Mahieu N. Quantitative solid-state 13C NMR spectroscopy of organic matter fractions in lowland rice soils. Eur J Soil Sci. 2004;55:367–79.
45. Smernik RJ, Kookana RS, Skjemstad JO. NMR characterization of ^{13}C-benzene sorbed to natural and prepared charcoals. Environ Sci Technol. 2006;40:1764–9.
46. Song JZ, Peng PA, Huang WL. Black carbon and kerogen in soils and sediments. 1. Quantification and characterization. Environ Sci Technol. 2002;36:3960–7.
47. Thorn KA. Structural characteristics of the IHSS Suwannee River fulvic and humic acids determined by solution state C-13 NMR spectroscopy. Sci Total Environ. 1987;62:175–83.
48. Tissot BP, Welte DH. Petroleum Formation and Occurrence. New York: Springer; 1984.
49. Wassmann P, Bauerfeind E, Fortier M, Fukuchi M, Hargrave B, Moran B, Noji T, Nothig EM, Olli K, Peinert R, Sasaki H, Shevchenko V. Particulate organic carbon flux to the Arctic Ocean sea floor. Organic Carbon Cycle in the Arctic Ocean; 2004. p. 101–138.
50. Wilmesmeier S, Steuernagel S, Wiermann R. Comparative FTIR and ^{13}C CP/MAS NMR spectroscopic investigations on sporopollenin of different systematic origins. Zeitschrift für Naturforschung C. 1993;48(9–10):697–701.
51. Wilson MA. NMR techniques and applications in geochemistry and soil chemistry. Ocford: Pergamon Press; 1987.
52. Zhang D, Duan D, Huang Y, Xiong Y, Yang Y, Ran Y. Role of structure, accessibility and microporosity on sorption of phenanthrene and nonylphenol by sediments and their fractions. Environ Pollut. 2016. doi:10.1016/j.envpol.2016.05.052.
53. Zhang D, Duan D, Huang Y, Yang Y, Ran Y. Novel phenanthrene sorption mechanism by two pollens and their fractions. Environ Sci Technol. 2016;50(14):7305–14.
54. Zhang D, Ran C, Yang Y, Ran Y. Biosorption of phenanthrene by pure algae and field-collected planktons and their fractions. Chemosphere. 2013;93:61–8.
55. Zhang D, Ran Y, Cao X, Mao J, Cui J, Schmidt-Rohr K. Biosorption of nonylphenol by pure algae, field-collected planktons and their fractions. Environ Pollut. 2015;198:61–9.

18

Potential of three microbial bio-effectors to promote maize growth and nutrient acquisition from alternative phosphorous fertilizers in contrasting soils

Cécile Thonar[1*], Jonas Duus Stevens Lekfeldt[2], Vincenza Cozzolino[3], Dominika Kundel[1], Martin Kulhánek[4], Carla Mosimann[1,5], Günter Neumann[6], Alessandro Piccolo[3], Martin Rex[7], Sarah Symanczik[1], Florian Walder[1,8], Markus Weinmann[6], Andreas de Neergaard[2] and Paul Mäder[1]

Abstract

Background: Agricultural production is challenged by the limitation of non-renewable resources. Alternative fertilizers are promoted but they often have a lower availability of key macronutrients, especially phosphorus (P). Biological inoculants, the so-called bio-effectors (BEs), may be combined with these fertilizers to improve the nutrient use efficiency.

Methods: The goal of this study was to assess the potential of three BEs in combination with alternative fertilizers (e.g., composted manure, biogas digestate, green compost) to promote plant growth and nutrient uptake in soils typical for various European regions. Pot experiments were conducted in Czech Republic, Denmark, Germany, Italy, and Switzerland where the same variety of maize was grown in local soils deficient in P in combination with alternative fertilizers and the same set of BEs (*Trichoderma*, *Pseudomonas*, and *Bacillus* strains). Common guidelines for pot experiment implementation and performance were developed to allow data comparison, and soils were analyzed by the same laboratory.

Results: Efficiency of BEs to improve maize growth and nutrient uptake differed strongly according to soil properties and fertilizer combined. Promising results were mostly obtained with BEs in combination with organic fertilizers such as composted animal manures, fresh digestate of organic wastes, and sewage sludge. In only one experiment, the nutrient use efficiency of mineral recycling fertilizers was improved by BE inoculation.

Conclusions: These BE effects are to a large extent due to improved root growth and P mobilization via accelerated mineralization.

Keywords: Biofector, Bio-effector, Bio-inoculants, *Bacillus*, *Pseudomonas*, *Trichoderma*, Maize, Organic fertilizer, PGPR, Phosphorus, Recycling fertilizer

Background

To meet global demands for agricultural crops, farming systems in industrialized countries have undergone profound transformations. On the one hand, high application rates of synthetic fertilizers and manure together with the use of pesticides, irrigation, and short crop rotations have increased yields and have helped to reduce hunger in these countries [1]. On the other hand, modern high-input agriculture has created a multitude of environmental problems. In this context, groundwater pollution and the eutrophication of aquatic systems due to soil erosion and nutrient leaching and run-off from agricultural fields are of great concern [1, 2]. Additionally, the extensive use of synthetic

*Correspondence: cecile.thonar@fibl.org; cecile.thonar@gmail.com
[1] Soil Sciences Department, Research Institute of Organic Agriculture FiBL, Ackerstrasse 113, 5070 Frick, Switzerland
Full list of author information is available at the end of the article

fertilizers is contributing to climate change as fertilizer production and use cause significant greenhouse gas emissions [3]. Moreover, for some plant nutrients such as phosphorous (P), natural fixation processes in soil leads to a decrease in the efficiency of the fertilizers as only a fraction of the applied P fertilizer is taken up by plants. Depending on the soil pH and mineralogy, a considerable amount of P fertilizers gets either adsorbed to iron, aluminum oxides, or calcium and is no longer directly available to the plant [4]. Consequently, surplus fertilizer inputs are often maintained to meet the crop's demands. Finally, commonly used P fertilizers in modern agriculture are regarded as non-renewable resources with a limited supply, and the reserves are concentrated in a few countries around the globe [4, 5]. Hence, judicious use of the resource by effective recycling is required for long-term sustainability.

Today, increasing attention is being paid to the downside of high-input agricultural systems and much research is aimed at developing alternative ways to produce sufficient food in a sustainable and environmentally sound way. A number of different approaches have been investigated with the intention to reduce fertilizer inputs into agroecosystems, including breeding plant varieties with better P acquisition efficiency [6], fertilizer placement strategies [7], or application of soil organisms and natural extracts with plant growth-promoting potential [8–10]. These so-called bio-effectors (BEs) include plant growth-promoting rhizobacteria (PGPR) for instance *Pseudomonas* and *Bacillus* spp., as well plant-associated fungi, such as arbuscular mycorrhizal fungi (AMF) or *Trichoderma spp.* [11]. Moreover, BEs comprise non-living agricultural bio-stimulants (ABs) such as extraction products from macroalgae, compost, and plants [12]. The plant growth-promoting effect of BEs is based on different mechanisms, for example, the control of plant pathogens either by stimulation of the plant's defense mechanisms or by microbial antagonism [11, 13]. Some BEs induce plant growth by stimulating the plant's own hormone production such as auxins or cytokinins [13, 14], or by producing directly hormone-like compounds [15]. Regarding P acquisition, some BEs are associated with increased phosphatase activities in the soil [16–18] and have also shown the ability to liberate P from soil particles [9, 16, 19, 20] and from P-recycling fertilizers [21, 22]. Another important mechanism is based on the so-called "mycorrhiza helper" effect [23] which has been observed for many bacterial BEs [24]. This term refers to a stimulation of AMF which fulfill important functions regarding P uptake by plants.

Although BEs have shown the potential to promote plant growth, their effectiveness is influenced by numerous biotic and abiotic factors. Consequently, depending on the properties of the soil and the structure of the naturally occurring rhizosphere community, their plant growth-promoting effect has often shown limited reproducibility. As reported in Mosimann [25], soil properties such as pH can have impacts on the BEs' ability to exert plant growth-promoting effects. Moreover, the amount of total and plant available P seems to be a decisive factor for the performance of bacterial BEs [26, 27]. Also, competition with naturally occurring rhizosphere microorganisms can limit the survival of introduced BEs [28]. This indicates that more work is needed to unravel the conditions under which BEs can exert plant growth-promoting effects, in order to achieve more homogenous and predictable effects. With this new knowledge, specifically adapted BEs could be developed which might improve fertilizer use efficiency and decrease the dependence of agriculture on large external inputs of synthetic fertilizers.

Maize (*Zea mays*), a major crop globally, is particularly sensitive to low phosphorus availability [29] and has shown responsiveness to BE applications [30–32]. For this reason, maize was selected as the study crop to evaluate the potential of three BEs: the bacterial BEs *Pseudomonas* sp. DSMZ 13134 and *Bacillus amyloliquefaciens* FZB42 and the fungal BE *Trichoderma harzianum* T-22. We applied all BEs individually in combination with different organic and inorganic P fertilizers in order to study the impact of the different P sources on the BEs' ability to promote P acquisition and growth of maize. In order to shed more light on the impact of soil properties on the effectiveness of the applied BEs, we used six soils originating from different sites in Europe. We hypothesized that BE application improves plant P acquisition from soil and from recycling fertilizers, and that soil properties, e.g., pH and P levels (total P and plant available P), influence the effectiveness of the three BEs.

Methods

Overall experimental design

Maize growth (variety Colisée, KWS Saat, Germany) was investigated in pots using soils collected from six different sites (Table 1) and by applying a factorial design including the two following factors: P fertilizer (minimum 3 levels) and microbial BE application (2–4 levels). Table 2 gives an overview of the treatments applied to each soil.

Collection of soils and soil characterization

Topsoil was collected from the six different locations. The soils were characterized by one analytical lab (LA Chemie, University of Hohenheim, Germany) for pH (0.01% $CaCl_2$), texture, organic carbon content [33], and plant available soil P (CAL method for neutral and alkaline soils [33] and DL method for acidic soils [33]). Additional information regarding geographic origin of the

Table 1 Soil characteristics

Soil-ID	Geographic origin	Management	Soil type	Texture			Soil pH	C org	Phosphorus
				Clay	Sand	Silt	($CaCl_2$)		CAL/DL[d]
				(%)	(%)	(%)		(%)	(mg P kg^{-1})
1-Buus[a]	Buus (Switzerland)	Rotation: Lay rotation Fertilization: organic, low input Site type: farmer field site	Loamy soil	29.9	3.90	66.2	6.6	2.64	6.5
2-Castel[a]	Castel Volturno (Italy)	Rotation: Maize-clover Fertilization: mineral, low input Site type: farmer field site	Vertic Xerofluvent	36.5	19.0	44.5	7.3	1.34	19.2
3-Humpolec[a]	Humpolec (Czech Republic)	Rotation: P-WW-SB[c] Fertilization: mineral, low input Site type: experimental station	Cambisol Loamy sand	15.9	52.9	31.2	5.2	1.63	45.0/59.8
4-Lukavec[a]	Lukavec (Czech Republic)	Rotation: long-term protection area around field experiment Fertilization: none Site type: experimental station	Cambisol Loamy sand	14.0	56.0	30.0	5.4	1.5	48.0/49.8
5-Vörden[b]	Vörden (Germany)	Rotation: before 2009 grazing area for wildlife – since 2009 field forage cultivation Fertilization: low input (organic and mineral) Site type: farmer field site	Cambisol	12.0	41.1	46.9	5.0	0.72	11.4/9.4
6-Taastrup[a]	Taastrup (Denmark)	Rotation: Barley-barley Fertilization: no P fertilizer since 1964 Site type: long-term field trial	Sandy loam	13.4	55.4	31.2	5.8	1.09	15.3/17.9

[a] Data were recorded on the pure soil

[b] Data were recorded on the 2:1 soil: sand mixture used in the pot experiments

[c] *P* potato, *WW* winter wheat, *SB* spring barley

[d] P measured after double lactate (DL) extraction is also given for soil with pH < 6 (exception for soil 3)

soils, management of the sites, and soil type is given in Table 1. Most soils had low to very low levels of available P, which makes them ideal for testing the efficiency of BEs in improving P availability and/or P uptake of added fertilizers by maize. The pH was ranging from slightly acidic to medium alkaline.

BE treatments

The three BE strains tested (Table 3) were formulated within three commercial products, namely Trianum-P (Koppert, The Netherlands) containing *Trichoderma harzianum* strain T-22 (BE1), Proradix WP (Sourcon Padena, Germany) containing *Pseudomonas* sp. strain DSMZ 13134 (BE2), and Rhizo Vital 42 fl. (Abitep, Germany) containing *Bacillus amyloliquefaciens* strain FZB42 (BE3). The choice of these three BEs is based on their published ability to act as potential bio-stimulants and biofertilizers. The strain *Trichoderma harzianum* Rifai 1295-22 (T22) has demonstrated abilities to increase the solubilization of sparingly soluble calcium phosphates [34] and to promote plant growth in various plant species [35, 36] including maize [37]. The strain *Pseudomonas* DSMZ 13134 is an efficient root colonizer [38] and has been shown to solubilize inorganic phosphate under in vitro conditions [39]. Moreover, it has recently shown the potential to improve plant growth and phosphorus acquisition when inoculated to maize [10, 25]. The gram-positive strain *Bacillus amyloliquefaciens* FZB42 is mainly studied for its biocontrol potential but other studies also indicate its ability to act as a biofertilizer [40, 41]. The *Trichoderma* strain was inoculated at a concentration of 2.5×10^4 spores per g of substrate and the two bacterial strains at a concentration of 2×10^6 cfu per g of substrate. A non-inoculated control (BE0) was included for every soil and in combination with every P treatment. The exact amount of inoculated cells or spores in the different products was checked in preliminary work where inoculation suspensions were plated on specific media and counted afterward. This has led to a standardized inoculation protocol that was later applied for all pot experiments.

P fertilization treatments

The three common P treatments were negative control without the addition of P fertilizer (P0), Triple Super Phosphate (TSP), and Rock Phosphate (RP). In most

Table 2 Overview of the experimental design and setup at the different locations (1: Switzerland, 2: Italy, 3 and 4: Czech Republic, 5: Germany, 6: Denmark)

	1-Buus	2-Castel	3-Humpolec	4-Lukavec	5-Vörden	6-Taastrup
Factors tested (experimental design)						
Crop						
Maize (variety: Colisée)	x	x	x	x	x	x
Bio-effectors						
No bio-effectors (BE0)	x	x	x	x	x	x
BE1: Trianum-P	x	x	x	x		
BE2: Proradix	x	x	x	x	x	x
BE3: Rhizovital	x	x	x	x	x	
Fertilizers						
No P fertilizer (P0)	x	x	x	x	x	x
Triple Super phosphate	x	x	x	x	x	
Rock phosphate	x	x	x	x	x	
Local sewage sludge ashes					x	x
P-enriched BOF slag					x	
Fresh Digestate	x					
Local composts	x				x	
Composted cattle manure		x				
Composted horse manure		x				
Local sewage sludges					x	x
Experimental setup						
Amount of substrate per pot (kg-DW equivalent)	2.5	2.5	5	5	5	2.5
Number of plants per pot at harvest	1	1	3	3	1	1
Growth period (weeks)	8	8	15	15	8	6.5
Number of replicates	5	5	5	5	4	3
Soil to sand ratio	2:1	2:1	2:1	2:1	2:1	1:1
Growing facilities[a]	GC	GC	GH	GH	GH	GH

[a] GC means growth chamber and GH greenhouse

Table 3 Bio-effector strains applied

Product	Producer	Abbr.	Species	Type of organism	Formulation	Application rate (cfu g^{-1} soil)
Control	n.a.	BE0	n.a.	n.a.	n.a.	n.a.
Trianum-P	Koppert, The Netherlands	BE1	*Trichoderma harzianum*, strain T-22	Fungus	Powder	2.5×10^4
Proradix WP	Sourcon Padena, Germany	BE2	*Pseudomonas sp.* strain DSMZ 13134	Bacteria	Powder	2×10^6
Rhizovital 42 fl.	ABiTEP, Germany	BE3	*Bacillus amyloliquefaciens*, strain FZB42	Bacteria	Liquid	2×10^6

n.a. means not applicable

cases, a number of additional P treatments (Table 4) were investigated and consisted of organic fertilizers and/or recycled inorganic fertilizers. These fertilizers were locally available and previously studied for their properties by the individual groups. As a result, some soils were tested with organic fertilizers and others only with recycled inorganic fertilizers (in addition to TSP and RP). Except for the P0 treatment, all pots received P at a dose of 50 mg P per kg of dry substrate. The TSP (20% P) was provided by Landor in Switzerland and RP (12% P) by Herbert Molitor in Germany. The additional P treatments consisted of recycled inorganic fertilizers (sewage sludge ashes and P-enriched BOF (basic oxygen furnace) slags [42]) and organic fertilizers (fresh biogas digestate, mature compost from green

Table 4 Main characteristics of the fertilizers used

Fertilizers	Abbr.	Inorg/org	Dry matter content (%)	Total P (mg P g^{-1})	Water extractable P (mg P g^{-1})	Total N (mg N g^{-1})	Ammonium ($NH_4^+ - N$) (mg N g^{-1})
Triple super phosphate	TSP	Inorg	100	201.0	201.0	0	0
Rock phosphate	RP	Inorg	100	115.4	0.22	0	0
Sewage sludge ashes (Germany)	SSA-DE	Inorg	100	103.5	0.79	n.d.	n.d.
Sewage sludge ashes (Denmark)	SSA-DK	Inorg	100	89	n.d.	n.d.	n.d.
P-enriched BOF slag (Germany)	LSD/SSA	Inorg	100	18.0	0.0	n.d.	n.d.
Fresh digestate (Switzerland)	FrDi	Org	48.0	4.36	0.130	14.6	0.61
Compost (Switzerland)	Comp-CH	Org	59.6	2.45	0.043	10.4	0.24
Compost (Germany)	Comp-DE	Org	57.2	1.42	0.67	n.d.	n.d.
Composted cattle manure (Italy)	C-MComp	Org	60.0	7.75	n.d.	22.7	n.d.
Composted horse manure (Italy)	H-MComp	Org	67.0	10.0	n.d.	14.2	n.d.
Sewage sludge (Germany)	SS-DE	Org	99	38.1	0.52	n.d.	n.d.
Sewage sludge (Denmark)	SS-DK	Org	23.0	30.4	1.5	n.d.	n.d.

n.d. means not determined

wastes, composted cow and horse manures, and sewage sludge). They were characterized by the local laboratories (Table 4) for their dry matter content, total P content, water extractable P, total potassium (K) and N content, and their ammonium N content.

Pot experiment setup, growing conditions, and harvest

After sieving, the soil was mixed with quartz sand (0.6–1.2 mm) in the ratio 2:1 or 1:1 (soil dry weight (DW)/sand DW; W/W). Each pot contained the equivalent of 2.5 kg or 5 kg DW of the experimental substrate (Table 2). All pots received N (100 mg N kg^{-1} substrate) and K (166 mg K kg^{-1} substrate) in form of calcium nitrate and Kalimagnesia, respectively. Where alternative P treatments were applied, N and K were measured and reduced from the basal dose of N and K (except for the German and the Danish trials where the addition of N and K was not reduced).

The N, K, and P fertilizers were added individually to every pot, and water addition was adjusted to reach 60–70% of the substrates water holding capacity (WHC). Three seeds were sown per pot, and the BE suspension was added in the seeding hole (5 ml per seeding hole = 15 ml per pot). BE suspensions were prepared under sterile conditions by diluting the products with 2.5 mM $CaSO_4$. The surface of the pots was then covered by a fine layer of quartz sand to avoid formation of surface crusts after watering.

The pots were transferred to growth chamber or greenhouse (see Table 2) and watered according to plant's need in order to keep the initial WHC (increased to 70–80% after two weeks). Thinning (including the root systems) was performed one week after sowing, leaving one plant in each pot. During the growth period, plant height was measured every second week and final harvest took place 8 weeks after sowing. Any deviation from this standard protocol (size pot, growing period, number of plant per pot, standard P fertilizers) is described in Table 2. Every treatment was replicated five times (or at minimum three times, see Table 2), and pots were randomly distributed in the growth chambers or greenhouses. The guidelines for establishing the pot experiments were common to all soils, and deviations to experimental setup are due to particular constraints of the different involved groups.

Plant analyses

At harvest, shoots were collected, dried (65 °C), and stored for further analyses. Root systems were washed from the substrate, and a subsample was kept (only in Swiss and Italian soils) to assess percentage root length colonization (RLC %) by arbuscular mycorrhizal fungi (AMF) following the protocol of Phillips and Hayman [43] and Brundrett et al. [44]. The remaining root was weighed and dried for total root DW assessment. Dried shoot material was milled for later determination of total N and P concentration. Shoot nitrogen concentration was measured using CHN analyzers, while shoot P concentration was determined spectrophotometrically or using flow injection analysis (Danish experiment) after acid extraction.

Data analyses and statistics

Residuals of the data (shoot DW, N and P content in shoot and RLC %) were tested for normality using the Shapiro–Wilk test. For every soil, a two-way ANOVA was performed to test the significance of each factor

(P fertilizer and BE strain) and their interaction. Since this interaction was always significant and for the sake of the graphical representation (in Figs. 2, 3, 4, and 5), a one-way ANOVA was performed within each P treatment and for each soil to determine which BE treatment (BE1, BE2, or BE3) was significantly different from the un-inoculated controls (Tukey's test, $P < 0.05$). Due to differences in growing conditions (Table 2, lower part) leading to substantial differences in the aboveground biomass response among single experiments, we normalized the aboveground biomass, P, and N content relative to the negative control treatment (P0, BE0) in the following way (exemplified for biomass):

$$normalized\ biomass_{sample} = \frac{biomass_{sample}}{\overline{biomass_{control(P0,BE0)}}}$$

Furthermore, in order to quantify the relative effect of BE additions across soils and P fertilizer types, absolute values for shoot DW (g plant^{-1}) were converted to shoot biomass improvement (in %) for every BE (BE1, BE2, and BE3) calculated relative to the BE0 control. For this, we applied the following equation:

$$improvement_{\%} = 100 \cdot \left(\frac{biomass_{sample(P_x,BE_y)} - \overline{biomass_{sample(P_x,BE_0)}}}{\overline{biomass_{sample(P_x,BE_0)}}} \right)$$

A value of 0% means that the shoot biomass of maize inoculated with the BE strain in question did not improve shoot biomass as compared to the un-inoculated control (using the same P fertilizer), whereas a value below or above 0% indicates that the specific BE treatment led to a decrease or an increase in aboveground biomass, respectively. In order to investigate the relative effect of BEs across different experiments, we categorized the added P fertilizers as either organic or inorganic. Data from experiments conducted in the Czech Republic were not considered here as they only included inorganic fertilizers (TSP and RP). For each BE, we performed a separate two-way ANOVA on these data with "soil" and "fertilizer type" and their interaction as factors, in order to analyze if the effect of BE was dependent on soil and/or fertilizer type (Table 6).

For each fertilizer, the fertilizer efficiency (FE %) relative to TSP (positive control) was calculated as the mean efficiency measured in n replicated pots. The efficiency in individual pots was calculated as follows:

$$FE_{\%} = 100 \cdot \left(\frac{biomass_{sample} - \overline{biomass_{control(P0,BE0)}}}{\overline{biomass_{TSP}} - \overline{biomass_{control(P0,BE0)}}} \right)$$

Finally, the P uptake efficiency (PUE %) from the different fertilizers was calculated as follows:

$$PUE_{\%} = 100 \cdot \left(\frac{P\ content_{sample} - \overline{P\ content_{control(P0,BE0)}}}{\overline{P\ content_{TSP}} - \overline{P\ content_{control(P0,BE0)}}} \right)$$

Results

The aboveground biomass was highly variable across the experiments (Fig. 1). A two-way ANOVA on raw biomass data (Fig. 1) for the negative (P0) and positive controls (TSP) without addition of BE (BE0) yielded highly significant effects of soil ($P < 0.001$) and P fertilizer ($P < 0.001$), as well as a highly significant interaction between the two factors ($P < 0.001$). In five out of six soils, there was a significant positive effect of TSP addition on the aboveground biomass (Fig. 1, Tukey's test, $P < 0.05$). The Humpolec soil is the only soil not affected by TSP addition with regards to aboveground biomass production.

In absence of BE inoculation (BE0), plants responded differently to the addition of the various P fertilizers. This was true for normalized biomass (Fig. 2), normalized P content (Fig. 3), and normalized N content (Fig. 4). The most effective fertilizers were the P-enriched BOF slag (Vörden trial) and in general the composted material from green wastes (FrDi, Comp-CH, Comp-D (see Table 4 for abbr.)) and sewage sludges from Denmark (incinerated or not).

The following combinations of BE and P fertilizers resulted in a higher normalized shoot biomass (Fig. 2), when compared to the negative control treatment (BE0) receiving the same P fertilizer (treatments

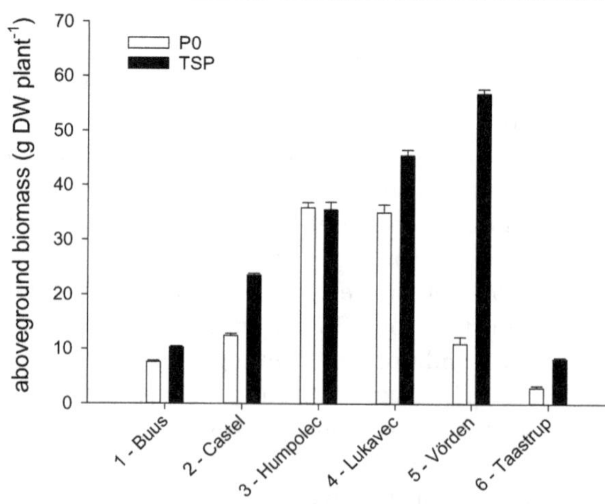

Fig. 1 Aboveground biomass of maize plant at harvest in the different trials: 1: Switzerland, 2: Italy, 3 and 4: Czech Republic, 5: Germany, 6: Denmark [without BE inoculation and either supplemented with Triple Super Phosphate (TSP) or not (P0)]. *Error bars* represent Standard Error of the Mean (SEM)

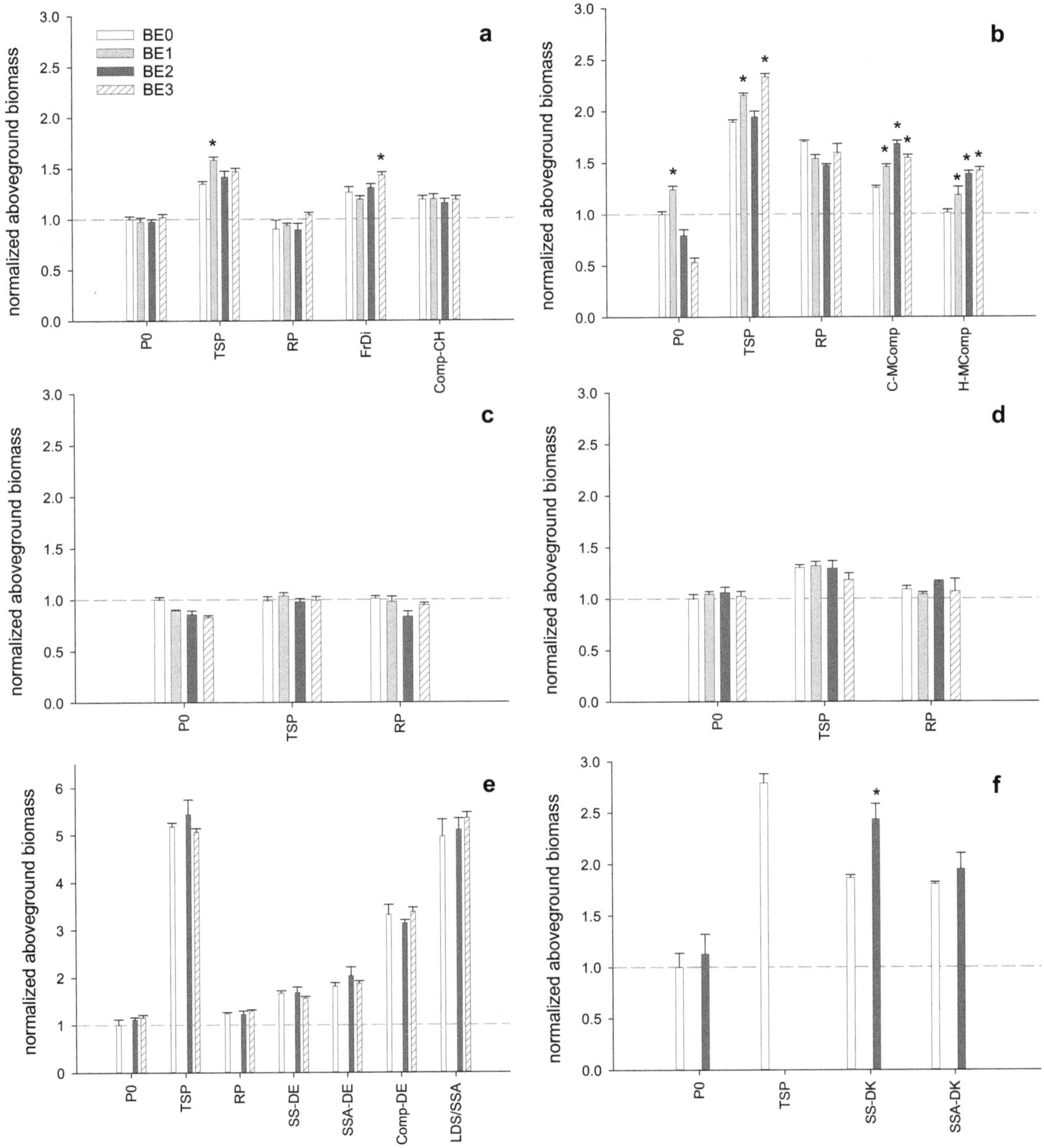

Fig. 2 Normalized aboveground biomass for the different trials: 1-Buus Switzerland (**a**), 2-Castel Italy (**b**), 3-Humpolec Czech R. (**c**), 4-Lukavec Czech R. (**d**), 5-Vörden Germany (**e**), 6-Taastrup Denmark (**f**). *Bars* represent the mean of five replicates ($n = 4$ for **e**, and $n = 3$ for **f**). *Error bars* represent SEM. BE treatments (BE1-3) resulting in a significantly higher normalized biomass compared to the BE0 control within each P fertilizer treatment and within each trial are indicated with an *asterisk* (*Tukey's test with $P < 0.05$). See Tables 3 and 4 for explanation of BE and P treatment abbreviations

highlighted with an asterisk): BE1 combined with TSP in the Buus (Fig. 2a) and Castel soils (Fig. 2b); BE2 combined with sewage sludge in the Taastrup soil (Fig. 2f); BE3 combined with fresh digestate and TSP in the Buus (Fig. 2a) and Castel (Fig. 2b) soils, respectively; and all BEs combined with composted animal manures

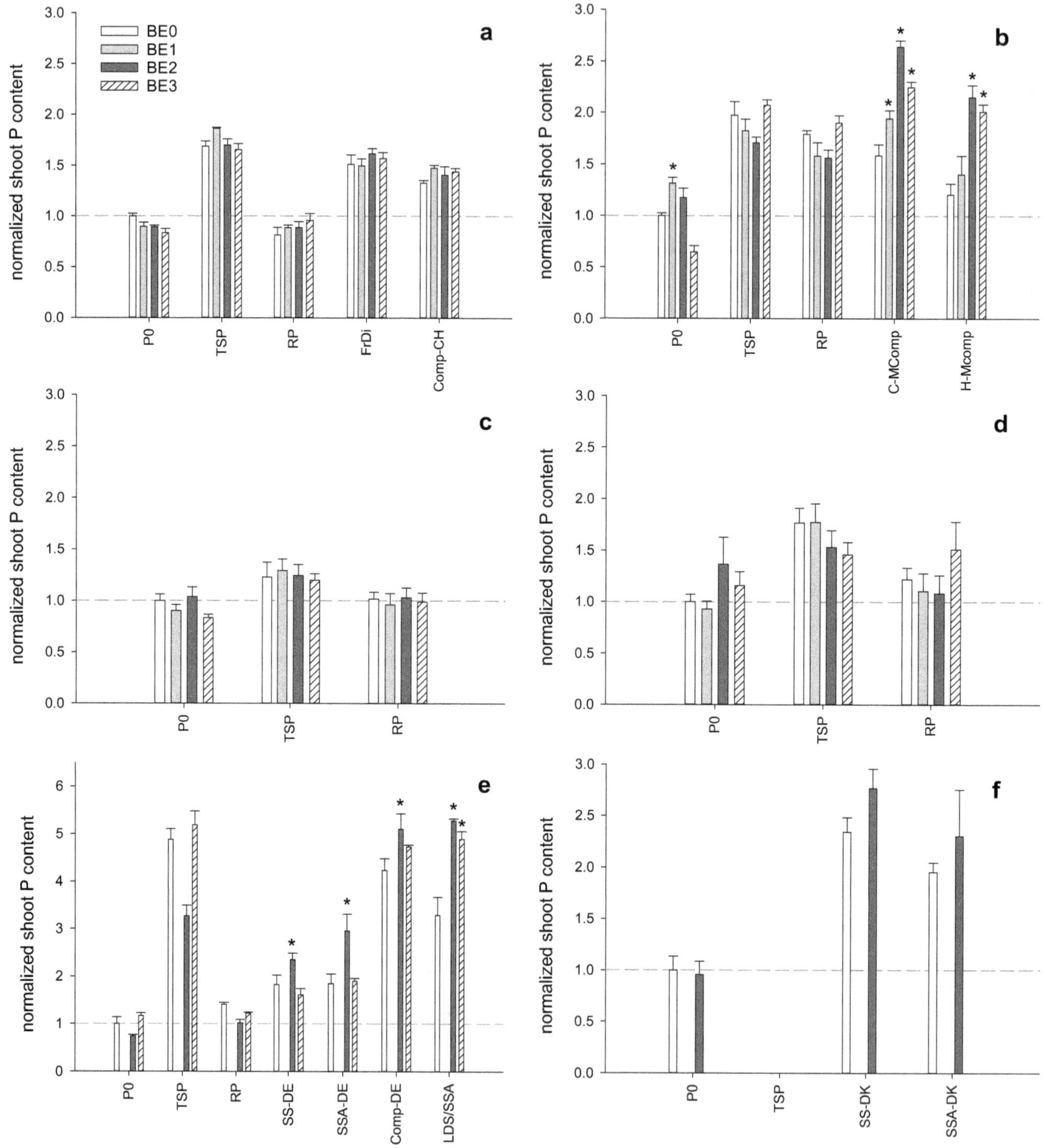

Fig. 3 Normalized aboveground P content for the different trials: 1-Buus Switzerland (**a**), 2-Castel Italy (**b**), 3-Humpolec Czech R. (**c**), 4-Lukavec Czech R. (**d**), 5-Vörden Germany (**e**), 6-Taastrup Denmark (**f**). *Bars* represent the mean of five replicates ($n = 4$ for **e**, and $n = 3$ for **f**). *Error bars* represent SEM. BE treatments (BE1-3) resulting in a significantly higher normalized P content compared to the BE0 control within each P fertilizer treatment and within each trial are indicated with an *asterisk* (*Tukey's test with $P < 0.05$). See Tables 3 and 4 for explanation of BE and P treatment abbreviations

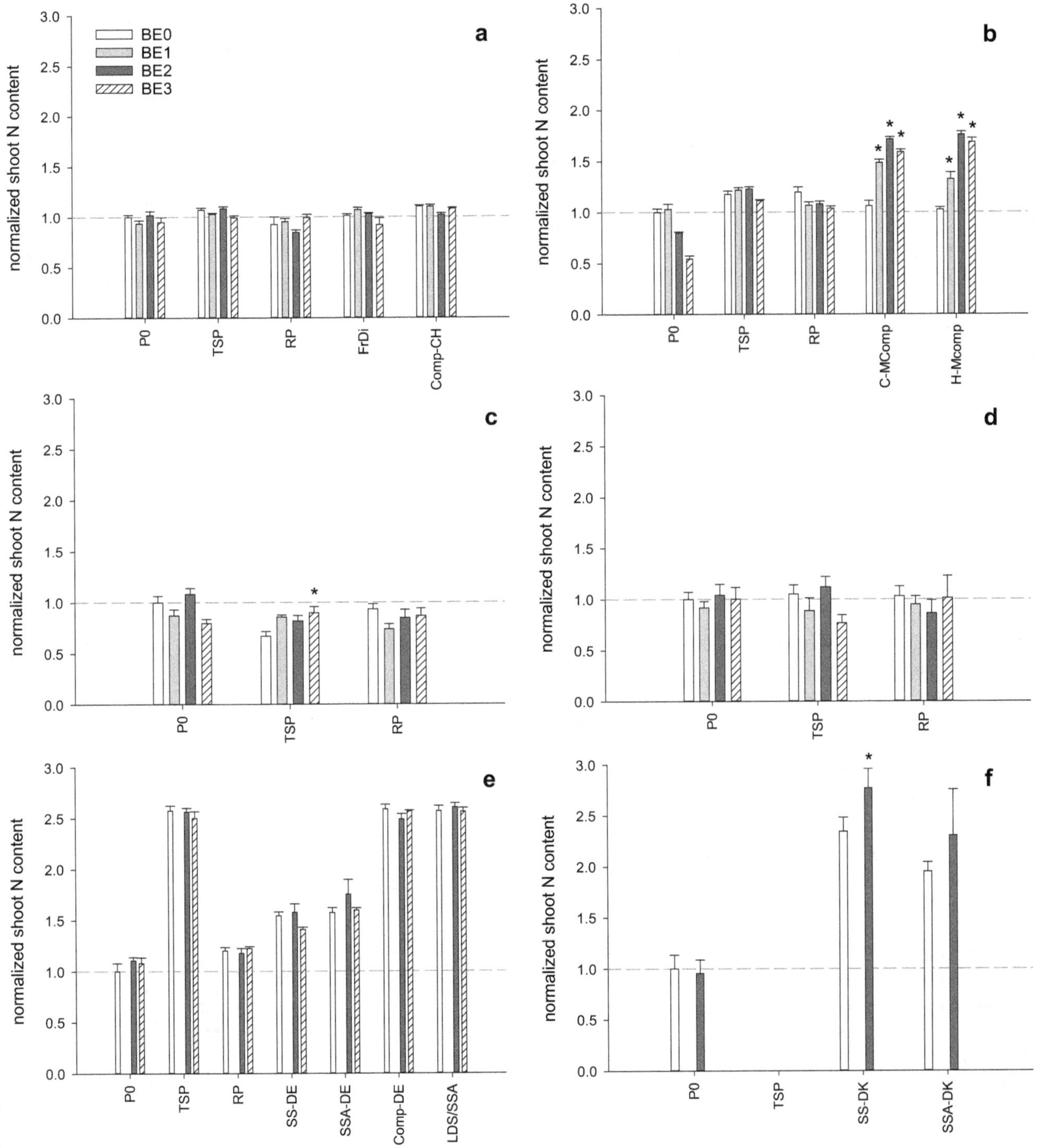

Fig. 4 Normalized aboveground N content for the different trials: 1-Buus Switzerland (**a**), 2-Castel Italy (**b**), 3-Humpolec Czech R. (**c**), 4-Lukavec Czech R. (**d**), 5-Vörden Germany (**e**), 6-Taastrup Denmark (**f**). *Bars* represent the mean of five replicates ($n = 4$ for **e**, and $n = 3$ for (**f**). *Error bars* represent SEM. BE treatments (BE1-3) resulting in a significantly higher normalized N content compared to the BE0 control within each P fertilizer treatment and within each trial are indicated with an *asterisk* (*Tukey's test with $P < 0.05$). See Tables 3 and 4 for explanation of BE and P treatment abbreviations

in the Castel soil (Fig. 2b). In the Humpolec, Lukavec, and Vörden soils, BE inoculation did not result in any increased normalized shoot biomass (Fig. 2 c–e).

Across the four experiments included in Table 6 (all except the two Czech experiments), BE inoculations induced an increase in shoot DW of 5–7.5%, as compared to un-inoculated controls (Table 6). For all three tested BEs, soil influenced the BE-derived increase in shoot biomass ($P < 0.01$). When looking across fertilizer types, the largest BE effects were observed in the Castel soil for BE1 (+12%), in the Taastrup soil for BE2 (+17%), and in the Buus soil for BE3 (8%). The factor "fertilizer type" significantly influenced the improvement in biomass when plants were inoculated with BE2 and BE3 (Table 6, $P < 0.0001$), but no such effect of "fertilizer type" was observed with BE1 (Table 6, $P > 0.05$). Furthermore, there was a significant interaction between "soil" and "fertilizer type" for all BE treatments (Table 6, $P < 0.0001$).

Across all soils, there was a significant positive effect of adding BE2 (+14%) and BE3 (+13%) when organic fertilizers were applied (Table 6, $P < 0.05$). For inorganic fertilizers, there was only a significant effect of BE3 (+7%, $P < 0.05$). When no fertilizer was added, none of the three BEs resulted in an improvement of the shoot biomass (Table 6, $P > 0.05$).

The normalized aboveground P content was significantly different between P fertilizers in all experiments (Fig. 3). In the Castel soil, inoculation with BE1 led to a significantly larger P content in the P0 and cow manure compost treatments, while inoculation with BE2 and BE3 resulted in an increased P content in both animal manure treatments (Fig. 3b). In the Vörden soil, inoculation with BE2 resulted in an increased P content when combined with any of the recycling fertilizers (Fig. 3e). In the remaining soils (Buus, Humpolec, Lukavec, and Taastrup), BE inoculation did not result in increased aboveground P contents. In the Castel soil, the higher P content due to BE2 addition (combined with cow and horse manures) is reflecting the significant increase in P concentration in the aboveground biomass (Table 7). BE3 increased P concentration only in the P0 treatment, while BE1 did not induce any P concentration increase. In the Vörden soil, where plants also accumulated more P in the aboveground biomass (Fig. 3e), P concentration was also higher (as compared to the corresponding BE0 control), but only when sewage sludge ashes were combined with BE2 and when P-enriched BOF slags were combined with BE2 or BE3.

The normalized aboveground N content (Fig. 4) showed less difference between P fertilization treatments than the normalized P content did (Fig. 3). Only in Vörden (Fig. 4e) and Taastrup (Fig. 4f) soils, a clear effect of P fertilizers on aboveground N content was observed. In the Castel soil, the effect of P fertilization depended on BE addition since we only observed a significantly larger normalized aboveground N content when pots had been inoculated with BEs 1–3 (Fig. 4b). In the Taastrup soil, a significantly higher aboveground N content was observed in plants inoculated with BE2 and when sewage sludge was used as fertilizer. In the Humpolec soil, we found a significant increase in aboveground N content due to inoculation with BE3, when TSP was used as P fertilizer (Fig. 4c). However, the recorded N content (Fig. 4c) in this treatment (TSP/BE3) was numerically smaller than in the negative control (P0/BE0). In the remaining soils (Buus, Lukavec, and Vörden), BE inoculation did not result in increased aboveground N contents.

Intensity of RLC % by AMF was only measured in roots of plants growing in the Buus and Castel soils (Fig. 5). For un-inoculated treatments which were not amended with P fertilizer (BE0, P0), the two soils induced different

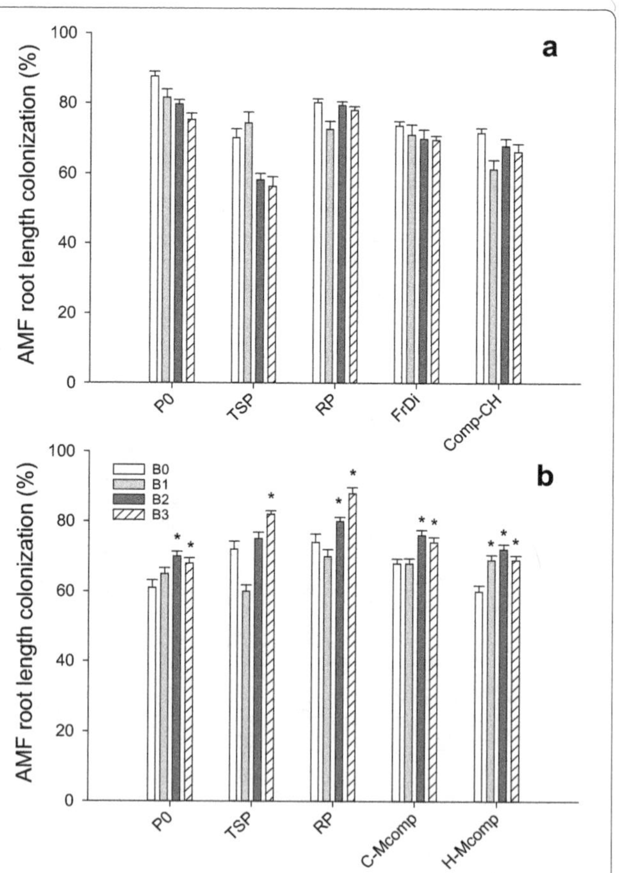

Fig. 5 Root length colonization percentage by AMF (in %) in Buus (**a**) and Castel soil (**b**). *Bars* represent the mean of five replicates. *Error bars* represent SEM. BE treatments (BE1-3) resulting in a significantly higher RLC % compared to the BE0 control within each P fertilizer treatment and within each trial is indicated with an *asterisk* (*Tukey's test with $P < 0.05$). See Tables 3 and 4 for explanation of BE and P treatment abbreviations

levels of RLC % (87% for the Buus soil and 60% for the Castel soil). In the Buus soil, BE inoculation leads to either no change in RLC % or a decrease, especially for plants amended with TSP (Fig. 5a). In contrast, the roots from the Castel soil had significantly higher RLC % in BE-inoculated pots, especially for BE3 (independent of P fertilizer applied) and for the combination of BE2 and animal manures (Fig. 5b).

Discussion

Variable levels of soil fertility and harvested aboveground biomass (P0, BE0 and TSP, BE0)

Due to variation in the experimental setup (e.g., pot size and growth duration), different aboveground biomasses were harvested ranging from 4 to 36 g per plant in the negative control treatment (P0, BE0). The different increases due to TSP addition (TSP, BE0) indicated that soils had varying initial fertility levels (Figs. 1, 2) with the Czech soils (Humpolec and Lukavec) being the less depleted in P. Plants growing in the Vörden (DE), the Taastrup (DK), and, to a lesser extent, in the Castel (IT) soil, were obviously P limited (increase between 2- to fivefold due to TSP addition), while the plants in the Buus (CH) soil showed a moderate P increase (1.6-fold). The same response to TSP addition was observed with spring wheat with the Vörden and Taastrup soils [45], where a similar experimental setup was used. The Taastrup soil originates from a long-term field trial where no nutrients have been added for approximately 50 years [46], thereby explaining the response to application of easily available P sources. The Castel soil was taken from an alkaline site where native P is largely bound in calcium complexes. The Buus soil is from an organic field site where the plant available P fraction is indeed low (P CAL = 6.5 mg P/kg), but with a large organic P fraction (data not shown) which likely contributed to P supply during the experiment causing a moderate response to TSP addition.

Variable P fertilizer effects (P, BE0)

We observed a range of responses due to the application of different P fertilizers. The calculated fertilizer efficiencies (Table 5) indicate that, in the absence of inoculated BEs, recycling fertilizers (organic or inorganic) were superior to the rock phosphate fertilizer (the Castel experiment was an exception). The fertilizer and P use efficiencies for sewage sludge, and sewage sludge ashes were quite similar within the same experiment but quite different between the two experiments where these products were included (16–22% for the Vörden experiment and 45–49% for the Taastrup experiment). This is in line with the data on water extractable P in the sludges where the level in the Danish sludge was three times higher than in the German sludge (Table 4). Furthermore, the availability of P in sewage sludge ashes can be quite different depending on the processing of the sludge in the water treatment plant [47].The very high fertilizer efficiency recorded for the P-enriched BOF slag has been observed earlier in wheat [45] and maize [42].

Moderate effects in the absence of added P (P0, BE1-3)

When no P was added, the three BEs neither improved aboveground biomass nor P uptake. In one case (Castel-IT), there was even a reduced aboveground biomass due to the application of BE2 and BE3. The absence of response after BE application was observed in many other experiments [45, 48, 49] and in particular in soils characterized by extraordinary low levels of available P (20 mg kg^{-1} or lower) or that were limiting in other soil nutrients [10, 50, 51]. In contrast, other studies have shown that for maize and other crops, the application of BEs was more efficient in the absence of added P fertilizers [36–38]. However, in these cases, soil P availability was distinctly higher and P CAL ranged between 35 and 45 mg/kg.

The observed absence of plant responses to BE applications when soils are extremely low in available P could be explained by direct growth limitation of the host plants, which finally limits sufficient production of roots and root exudates to initiate beneficial plant–microbial interactions with indigenous or inoculated soil microbes in the rhizosphere. This phenomenon is well documented, e.g., for mycorrhizal associations frequently absent on extremely nutrient-impoverished soils [52, 53] or for the establishment of the Rhizobium symbiosis in leguminous plants on low P soils which is promoted by starter applications with soluble P [54]. Moreover, in the absence of significant soil P pools that can be mobilized by BEs, their presence will have no beneficial effect on plant P supply, or may even be in competition with plants by immobilizing P in microbial tissue.

BE effects depend on soil and fertilizer type

In the present study, it was observed that for the three tested BEs, the soil strongly modulated the observed BE effect (Table 6). Various studies have shown that soil pH, soil organic carbon content, soil microbial activity, and soil nutrient status are all influencing the success of BE application [16, 25, 55–57]. Our findings indicate that the most responsive soils to BE application (in combination with P fertilizers) were either alkaline (Castel-IT) or depleted in soil nutrients (Taastrup-DK and Vörden-DE for P acquisition), confirming the influence of soil pH and soil nutrient statues on the effectiveness of BEs. The type of fertilizer used along with BEs was also significantly influencing the magnitude of BE effects (Table 6). Since all tested fertilizers were different (with the exception of

Table 5 Relative fertilizer efficiencies (FE) and P use efficiencies (PUE)

P fertilizer	Abbr.	1-Buus (Switzerland)		2-Castel Volturno (Italy)		3-Humpolec (Czech R.)		4-Lukavec (Czech R.)		5-Vörden (Germany)		6-Taastrup (Denmark)	
		FE (% of TSP)	PUE (% of TSP)	FE (% of TSP)	PUE (% of TSP)	FE (% of TSP)	PUE (% of TSP)	FE (% of TSP)	PUE (% of TSP)	FE (% of TSP)	PUE (% of TSP)	FE (% of TSP)	PUE (% of TSP)
Negative control	P0	0bc	0c	0d	0d	0a	0a	0b	0b	0d	0c	0c	n.a.
Triple super phosphate	TSP	100a	100a	100a	100a	100a	100a	100a	100a	100a	100ab	100a	n.a.
Rock phosphate	RP	−28c	−27c	79b	81ab	−70.1a	7a	30b	28b	6cd	11 c	n.a.	n.a.
Fresh digestate	FrDi	76a	74ab	n.a.	n.a.	n.a.	n.a.	n.a.	n.a.	n.a.	n.a.	n.a.	n.a.
Composts	Comp-CH, Comp-DE	54ab	48b	n.a.	n.a.	n.a.	n.a.	n.a.	n.a.	55b	84 ab	n.a.	n.a.
Composted cow manure	C-MComp	n.a.	n.a.	28c	60bc	n.a.	n.a.	n.a.	n.a.	n.a.	n.a.	n.a.	n.a.
Composted horse manure	H-MComp	n.a.	n.a.	2d	21cd	n.a.	n.a.	n.a.	n.a.	n.a.	n.a.	n.a.	n.a.
Sewage sludges	SS-DE, SS-DK	n.a.	n.a.	n.a.	n.a.	n.a.	n.a.	n.a.	n.a.	16cd	21c	49b	n.a.
Sewage sludge ashes	SSA-DE, SSA-DK	n.a.	n.a.	n.a.	n.a.	n.a.	n.a.	n.a.	n.a.	20cd	22c	45b	n.a.
P-enriched BOF slag	LSD/SSA	n.a.	n.a.	n.a.	n.a.	n.a.	n.a.	n.a.	n.a.	95 a	59b	n.a.	n.a.

The efficiencies are calculated for the BE0 treatments only. Efficiencies are calculated relative to TSP as the positive control (see "Methods" section for details). For each column, different letters after mean values represent significantly different means (Tukey's test, $P < 0.05$)

n.a. means not applicable

Table 6 Shoot biomass improvement (%) of the different BE treatments relative to the un-inoculated control (BE0)

	Trianum-P (BE1)	Proradix (BE2)	Rhizovital (BE3)
Overall BE effect (across all treatments)	7.5%	5.2%	6.0%
Number of observations	50	87	78
BE effect in the different soils (across all fertilizers types)			
Buus (Switzerland)	2.7%b[a]	0.4%b[a]	8%a[a]
Castel (Italy)	12.3%a[a]	7.8%b[a]	6.7%a[a]
Vörden (Germany)	Not measured	3.4%ab[a]	3.6%b[a]
Taastrup (Denmark)	Not measured	17.0%a[a]	Not measured
BE effect with the different fertilizer types (across all soils)[b]			
No fertilizer added	10.8%a[a]	−1.2%b[a]	−11.5% b[a]
Organic fertilizer	6.9%a[a]	*14.0%a[a]*	*13.0% a[a]*
Inorganic fertilizer	6.4%a[a]	1.5%b[a]	7.3% a[a]
ANOVA on the effect of soil, fertilizer type and their interaction			
Soil (*p* value)	0.0002 (*df* = 1)	0.0070 (*df* = 3)	0.0054 (*df* = 2)
Fertilizer type (*p*-value)	ns (*df* = 2)	<0.0001 (*df* = 2)	<0.0001 (*df* = 2)
Soil x fertilizer type (*p*-value)	<0.0001 (*df* = 2)	<0.0001 (*df* = 6)	<0.0001 (*df* = 4)

df means degree of freedom

See "Methods" section for description of the calculations

[a] Levels not connected by the same letter are significantly different (Tukey's HSD test $P < 0.05$)

[b] Underlined combinations of BE and P fertilizer type indicate when BE addition significantly improved shoot biomass as compared to the un-inoculated control with the same fertilizer type (*t*-test with $P < 0.05$)

Table 7 Aboveground P concentration (mg/g)

	BE0	BE1	BE2	BE3
1-Buus				
Control P0	1.57a	1.45ab	1.44ab	1.29b
TSP	1.97a	1.85a	1.89a	1.77a
RP	1.42a	1.49a	1.63a	1.44a
FrDi	1.87ab	1.97a	1.94a	1.72b
Comp-CH	1.76a	1.95a	1.91a	1.91a
2-Castel Volturno				
Control P0	0.96c	1.01bc	1.40a	1.15b
TSP	1.00a	0.81b	0.84ab	0.85ab
RP	1.00a	0.98a	1.01a	1.15a
C-MComp	1.20b	1.27b	1.50a	1.39ab
H-MComp	1.13b	1.12b	1.48a	1.35ab
3-Humpolec				
Control P0	1.03a	1.03a	1.26a	1.04a
TSP	1.28a	1.27a	1.29a	1.25a
RP	1.03a	1.01a	1.26a	1.07a
4-Lukavec				
Control P0	0.88a	0.78a	1.10a	0.98a
TSP	1.19a	1.19a	1.04a	1.11a
RP	0.99a	0.92a	0.82a	1.21a
5-Vörden				
Control P0	1.12a	n.a.	0.76b	1.14a
TSP	1.06a	n.a.	0.68b	1.15a
RP	1.28a	n.a.	0.94b	1.05b
SS-DE	1.23ab	n.a.	1.58a	1.16b
SSA-DE	1.15b	n.a.	1.62a	1.14b
Comp-DE	1.45a	n.a.	1.84a	1.58a
LSD/SSA	0.75b	n.a.	1.17a	1.03a
6-Taastrup				
Control P0	1.99a	n.a.	1.71b	n.a.
SS-DK	2.49a	n.a.	2.28a	n.a.
SSA-DK	2.15a	n.a.	2.31a	n.a.

For each row, means which are followed by different letters are significantly different (Tukey's HSD test $P < 0.05$, performed within each P treatment and for each soil)

n.a. means not applicable

TSP and RP that were tested in almost all experiments), we categorized them in two types (organic or inorganic) in order to investigate if there were some more general trends in the interaction between P fertilizer applied and the effect of the three BEs across the experiments. Clearly, BE2 and BE3 (both bacterial) applied with organic fertilizers were the best combinations and in particular when these BEs were combined with composted animal manures (for BE2 and BE3) or with the Danish sewage sludge (for BE2). In the Vörden soil, sewage sludge was also combined with BE2. But here, BE2 failed to exert any effect on the obtained biomass, showing again the complexity of these processes and why the interaction between soil and fertilizer type was significant in the performed two-way ANOVA (Table 6). In two cases (Buus and Castel soils), BE1 (fungal) was able to increase the aboveground biomass when combined with TSP, indicating that the three BEs have likely different modes of action (see below). The study of Lekfeldt et al. [45] indicates that, to some extent, these effects and interactions are also crop specific since they did not observe any positive effects of BE1-3 on plant growth or P uptake when wheat was fertilized with recycling fertilizers.

Possible modes of action

Several studies [27, 39, 58] have shown that, in in vitro conditions, many BEs are able to dissolve insoluble tricalcium phosphates by releasing protons and/or organic acids (e.g., gluconic acids), decreasing the pH of the growth medium and/or act as cation chelators (Al, Fe, and Ca) [59]. Nevertheless, in pot experiments where soil low in available P is used as the growth substrate, this solubilization ability was limited, as almost none of the BEs could increase plant P content in P0 and RP treatments, with the exception of BE1 in the alkaline Castel soil (P0 treatment,). This limited solubilization ability under in situ soil conditions was also observed in other studies [27, 45, 49] and may be explained by the soil buffering capacity that inhibits the BE-induced soil acidification [33] that otherwise would have favored solubilization of calcium phosphates from soil and/or RP. However, we observed that in the Vörden soil, the amendment of sewage sludge ashes or P-enriched BOF slags (both inorganic fertilizers containing P in calcium phosphates complexes) increased plant P uptake when they were inoculated with BE2 or BE3 (Fig. 3e). Since this was not associated with increased N content in shoot (Fig. 4e), we can exclude that the better acquisition of P is the result of improved root growth. Rather, it is possible that the higher P content and concentration (Fig. 3e and Table 7) measured in the shoots are indeed the result of P solubilization (from these fertilizers), induced by BEs and in particular BE2. Yet, it remains to be understood why the biomass of these plants was not larger than the BE0 control and why the same BE could not solubilize P when combined with rock phosphate in the same soil.

Inoculated plants with higher biomass, P, and N contents (in the Castel soil, BE1-3 with composted animal manures; or in the Taastrup soil, BE2 with sewage sludge) could result from BE-induced mineralization of organic P and N contained in the tested organic fertilizers. For P, several studies have shown that phosphatase activities (acid and alkaline) measured in soil can be increased upon inoculation with BEs [16–18, 60]. Although it was not measured in our experiments, we argue that increased mobilization of P from organic fertilizers via higher phosphatase activities is likely one of the major mechanisms behind the successful BE-organic fertilizer combinations observed in this study. Nevertheless, plants can also produce phosphatases, and therefore BEs may also have induced growth of more roots and these may in return have produced more of these enzymes. Indeed, the fact that BEs combined with TSP lead in some cases to higher aboveground biomass suggests that BEs can increase root growth and therefore the volume of soil explored for nutrient uptake. This mechanism has been described in several studies [61–64] and results from the direct or indirect production of root growth regulators such as auxin.

The volume of soil explored for nutrient uptake can also be increased via higher root length colonization percentage (RLC %) by AMF. In this study, the only roots analyzed for RLC % were from the experiments with the Buus and Castel soils. In the latter case, an increase in RLC % after BE inoculation was recorded with or without application of P fertilizers (Fig. 5b). This mycorrhizal boosting effect is known for several microorganisms [65–67] and in particular for BE2 [68]. In the Castel soil, the increased RLC % due to BE inoculation in combination with the P fertilization treatments P0, RP, and TSP did not necessarily translate into higher biomass or acquired P. This indicates that the mycorrhizal boosting effect of BE inoculation in that soil was of minor importance for plant P uptake and growth, although we do not know to which degree it contributed to the improvement in P uptake and growth for the plants inoculated with the three BEs and amended with the animal manures. In the Buus soil (Fig. 5a), RLC % was not increased after BE inoculation. This can be explained in part by the fact that roots were already highly colonized (almost 90%). But it also seems that in some combinations, BE inoculation had a negative impact on RLC % under these conditions, and this could be due to competition. Yet, the clear mechanisms remain to be explained.

Conclusion

We have observed that in our experiments, the most responsive soils for the three tested BEs were either alkaline or highly depleted in nutrients, and that a positive effect of the BEs was dependent on an input of P fertilizer under these conditions. Of all tested fertilizers, the combination of BE with organic fertilizers (in particular composted animal manures) was the most promising. Nevertheless, the combination of BE and organic fertilizers was not always improving biomass production or P uptake (e.g., Buus soil). The next challenge will be to understand what properties of organic fertilizers make them more compatible with a given BE and how this interacts with soil parameters, such as soil pH, soil microbial activities, organic matter content, and soil P and N status. Regarding the mechanisms, the results from this study suggest that the three BEs likely improved root growth (volume of soil explored for nutrients uptake) and P mobilization via accelerated mineralization, whereas chemical P solubilization likely took place in very few cases. Although higher mycorrhizal RLC% were observed upon inoculation with the three BEs in the alkaline soil of Castel, it remains to be determined if this was the principal mechanism contributing to improved P acquisition and aboveground biomass in that experiment.

Clarification of mechanisms together with new experiments with more diverse soils and characterization of BE-compatible fertilizers will give further insights about the window of conditions leading to successful use of BEs in agriculture. Nevertheless, our results show which kind of soils and fertilizers type is more responsive to BE inoculation.

Authors' contributions
The study was elaborated by PM, GN, AP, AdN, MK, MR, VC, MW, and CT. The experiments were performed by SS, CT, CM, FW, MR, MK, JSDL, and VC. The manuscript was written by CT, JSDL, and DK. All authors read and approved the final manuscript.

Author details
[1] Soil Sciences Department, Research Institute of Organic Agriculture FiBL, Ackerstrasse 113, 5070 Frick, Switzerland. [2] Department of Plant and Environmental Sciences, University of Copenhagen, Copenhagen, Denmark. [3] CERMANU-University of Naples Federico II, Naples, Italy. [4] Department of Agro-environmental Chemistry and Plant Nutrition, Czech University of Life Sciences, Prague, Czech Republic. [5] Agrofutura AG, Brugg, Switzerland. [6] Institute of Crop Science 340h, University of Hohenheim, Stuttgart, Germany. [7] Institut für Baustoff Forschung, Duisburg, Germany. [8] Agroscope Reckenholz, Zurich, Switzerland.

Acknowledgements
This study was funded by the European Community's Seventh Framework Program 662 (FP7/2007–2013) under Grant Agreement no 312117 (BIOFECTOR) and by CORE Organic II (FP7 ERA-NET) under the Grant Agreement no 249667 (Improve-P).

Competing interests
All authors confirm that they have no competing interests in any of the products tested in this work, or any results presented in the manuscript.

References

1. Tilman D, Cassman KG, Matson PA, Naylor R, Polasky S. Agricultural sustainability and intensive production practices. Nature. 2002;418(6898):671–7.
2. Tilman D, Fargione J, Wolff B, D'Antonio C, Dobson A, et al. Forecasting agriculturally driven global environmental change. Science. 2001;292(5515):281–4.
3. Vermeulen SJ, Campbell BM, Ingram JSI. Climate change and food systems. Ann Rev Environ Res. 2012;37(1):195–222.
4. Smil V. Phosphorous in the environment: natural Flows and Human Interferences. Annu Rev Energy Env. 2000;25(1):53–88.
5. Cordell D, Drangert JO, White S. The story of phosphorus: global food security and food for thought. Glob Environ Change. 2009;19(2):292–305.
6. Lynch JP, Brown KM. Topsoil foraging—an architectural adaptation of plants to low phosphorus availability. Plant Soil. 2001;237(2):225–37.
7. Dunbabin V, Armstrong R, Officer S, Norton R. Identifying fertiliser management strategies to maximise nitrogen and phosphorus acquisition by wheat in two contrasting soils from Victoria, Australia. Aus J Soil Res. 2009;47:74–90.
8. Adesemoye A, Kloepper J. Plant-microbes interactions in enhanced fertilizer-use efficiency. Appl Microbiol Biotechnol. 2009;85(1):1–12.
9. Fröhlich A, Buddrus-Schiemann K, Durner J, Hartmann A, Von R, et al. Response of barley to root colonization by Pseudomonas sp. DSMZ 13134 under laboratory, greenhouse, and field conditions. J Plant Interact. 2011;7(1):1–9.
10. Nkebiwe PM, Weinmann M, Müller T. Improving fertilizer-depot exploitation and maize growth by inoculation with plant growth-promoting bacteria: from lab to field. Chem Biol Technol Agric. 2016;3(1):15–15.
11. Berg G. Plant-microbe interactions promoting plant growth and health: perspectives for controlled use of microorganisms in agriculture. Appl Microbiol Biotechnol. 2009;84(1):11–8.
12. Sharma HSS, Fleming C, Selby C. Plant biostimulants: a review on the processing of macroalgae and use of extracts for crop management to reduce abiotic and biotic stresses. J Appl Phycol. 2014;26(1):465–90.
13. Lugtenberg B, Kamilova F. Plant-growth-promoting rhizobacteria. Ann Rev Microbiol. 2009;63:541–56.
14. McNear DH Jr. The rhizosphere-roots, soil and everything in between. Nat Educ Knowl. 2013;4(3):1–1.
15. Sharma HSS, Selby C, Carmichael E, McRoberts C, Rao JR, et al. Physicochemical analyses of plant biostimulant formulations and characterisation of commercial products by instrumental techniques. Chem Biol Technol Agric. 2016;3(1):1–17.
16. Mäder P, Kaiser F, Adholeya A, Singh R, Uppal HS, et al. Inoculation of root microorganisms for sustainable wheat–rice and wheat–black gram rotations in India. Soil Biol Biochem. 2011;43(3):609–19.
17. Krey T, Caus M, Baum C, Ruppel S, Eichler-Lobermann B. Interactive effects of plant growth-promoting rhizobacteria and organic fertilization on P nutrition of Zea mays L. and Brassica napus L. J Plant Nutr Soil Sci. 2011;174(4):602–13.
18. Hussain M, Asghar H, Arshad M, Shahbaz M. Screening of multi-traits rhizobacteria to improve maize growth under axenic conditions. J Anim Plant Sci. 2013;23(2):514–20.
19. Cassán F, Vanderleyden J, Spaepen S. Physiological and agronomical aspects of phytohormone production by model plant-growth-promoting rhizobacteria (pgpr) belonging to the genus Azospirillum. J Plant Growth Regul. 2013;33(2):440–59.
20. Rodríguez H, Fraga R. Phosphate solubilizing bacteria and their role in plant growth promotion. Biotechnology Advances. 1999; 17(4): 319-339.
21. Gaind S, Gaur A. Impact of fly ash and phosphate solubilising bacteria on soybean productivity. Bioresour Technol. 2002;85(3):313–5.
22. Jastrzębska M, Kostrzewska M, Treder K, Jastrzębski W, Makowski P. Phosphorus biofertilizers from ash and bones—agronomic evaluation of functional properties. J Agric Sci. 2016;8(6):58–70.
23. Garbaye J. Mycorrhization helper bacteria: a new dimension to the mycorrhizal symbiosis. Acta Bot Gallica. 1994;141(4):517–21.
24. Artursson V, Finlay R, Jansson J. Interactions between arbuscular mycorrhizal fungi and bacteria and their potential for stimulating plant growth. Environ Microbiol. 2006;8(1):1–10.
25. Mosimann C, Oberhänsli T, Ziegler D, Nassal D, Kandeler E, et al. Tracing of two pseudomonas strains in the root and rhizoplane of maize, as related to their plant growth-promoting effect in contrasting soils. Front Microbiol. 2017;7:2150.
26. Ryan MH, van Herwaarden AF, Angus JF, Kirkegaard JA. Reduced growth of autumn-sown wheat in a low-P soil is associated with high colonisation by arbuscular mycorrhizal fungi. Plant Soil. 2005;270(1–2):275–86.
27. Menzies N, Bell M, Dart P. Soil additives to stimulate N fixing and P availability—science, myth and legend. Barton: Dubbo Grains Research Update, GRDC; 2009. p. 172–88.
28. Buddrus-Schiemann K, Schmid M, Schreiner K, Welzl G, Hartmann A. Root colonization by *Pseudomonas* sp. DSMZ 13134 and impact on the indigenous rhizosphere bacterial community of barley. Microb Ecol. 2010;60(2):381–93.
29. Postma JA, Lynch JP. Theoretical evidence for the functional benefit of root cortical aerenchyma in soils with low phosphorus availability. Ann Bot. 2011;107(5):829–41.
30. Walker V, Bertrand C, Bellvert F, Moenne-Loccoz Y, Bally R, et al. Host plant secondary metabolite profiling shows a complex, strain-dependent response of maize to plant growth-promoting rhizobacteria of the genus Azospirillum. New Phytol. 2011;189(2):494–506.
31. Gholami A, Shahsavani S, Nezarat S. The Effect of plant growth promoting Rhizobacteria (PGPR) on germination, seedling growth and yield of maize. Proc World Acad Sci. 2009;3(1):19–24.
32. Rosas SB, Avanzini G, Carlier E, Pasluosta C, Pastor N, et al. Root colonization and growth promotion of wheat and maize by *Pseudomonas aurantiaca* SR1. Soil Biol Biochem. 2009;41(9):1802–6.
33. VDLUFA. VDLUFA-Methodenbuch. Band I: Die Untersuchung von Böden. 4 ed. 1991, Darmstadt: VDLUFA-Verlag.
34. Altomare C, Norvell W, Björkman T, Harman G. Solubilization of phosphates and micronutrients by the plant-growth-promoting and biocontrol fungus *Trichoderma harzianum* Rifai 1295-22. Appl Environ Microbiol. 1999;65(7):2926–33.
35. Adams P, De-Leij F, Lynch J. *Trichoderma harzianum* Rifai 1295-22 mediates growth promotion of crack willow (Salix fragilis) saplings in both clean and metal-contaminated soil. Microb Ecol. 2007;54(2):306–13.
36. Rudresh DL, Shivaprakash MK, Prasad RD. Tricalcium phosphate solubilizing abilities of *Trichoderma* spp. in relation to P uptake and growth and yield parameters of chickpea (Cicer arietinum L.). Can J Microbiol. 2005;51(3):217–22.

37. Akladious S, Abbas S. Application of *Trichoderma harzianum* T22 as a biofertilizer potential in maize growth. J Plant Nutr. 2014;37(1):30–49.
38. Buddrus-Schiemann K, Schmid M, Schreiner K, Welzl G, Hartmann A. Root colonization by *Pseudomonas* sp. DSMZ 13134 and impact on the indigenous rhizosphere bacterial community of barley. Microb Ecol. 2010;60(2):381–93.
39. Fröhlich A, Buddrus-Schiemann K, Durner J, Hartmann A, von Rad U. Response of barley to root colonization by Pseudomonas sp. DSMZ 13134 under laboratory, greenhouse, and field conditions. J Plant Interact. 2012;7(1):1–9.
40. Borriss R. Use of plant-associated Bacillus strains as biofertilizers and biocontrol agents in agriculture, in Bacteria in agrobiology: plant growth responses. Berlin: Springer; 2011. p. 41–76.
41. Bákonyi N, Gajdos E, Lévai L, Veres S, Tóth B, et al. Comparison of effects of different biofertilisers on early development of cucumber and wheat seedlings. in Zbornik Radova 44. Hrvatski i 4 Medunarodni Simpozij Agronoma, Opatija, Hrvatska, 16-20 Veljače 2009. 2009. Poljoprivredni Fakultet Sveučilišta Josipa Jurja Strossmayera u Osijeku.
42. Severin M, Breuer J, Rex M, Stemann J, Adam C, et al. Phosphate fertilizer value of heat treated sewage sludge ash. Plant Soil Environ. 2014;60(12):555–61.
43. Phillips JM, Hayman DS. Improved procedures for clearing roots and staining parasitic and vesicular-arbuscular mycorrhizal fungi for rapid assessment of infection. Trans Br Mycol Soc. 1970;55(1):158.
44. Brundrett M, Piché Y, Peterson R. A new method for observing the morphology of vesicular-arbuscular mycorrhizae. Can J Bot. 1984;62:2128–34.
45. Lekfeldt J, Rex M, Merclc F, Kulhánek M, Tlustoš P, et al. Effect of bioeffectors and recycled P-fertiliser products on the growth of spring wheat. Chem Biol Technol Agric. 2016 **(in press)**.
46. van der Bom F, Magid J, Stoumann Jensen L. Long-term P and K fertilisation strategies and balances affect soil availability indices, crop yield depression risk and N use efficiency. Eur J Agron. **(under revision)**.
47. Nanzer S, Oberson A, Huthwelker T, Eggenberger U, Frossard E. The molecular environment of phosphorus in sewage sludge ash: implications for bioavailability. J Environ Qual. 2014;43(3):1050–60.
48. Kundel D. Bio-effector application: impact on growth performance of maize and effect on associated arbuscular mycorrhizal fungi. University of Constance: Konstanz; 2016.
49. Neumann G. 2nd Periodic report. Resource preservation by application of bioeffectors in European crop production., technical report. 2015.
50. Dutta D, Bandyopadhyay P. Performance of chickpea (Cicer arietinum L.) to application of phosphorus and bio-fertilizer in laterite soil. Arch Agron Soil Sci. 2009;55(2):147–55.
51. Kaur G, Reddy M. Role of phosphate-solubilizing bacteria in improving the soil fertility and crop productivity in organic farming. Arch Agron Soil Sci. 2014;60(4):549–64.
52. Lambers H, Ahmedi I, Berkowitz O, Dunne C, Finnegan PM, et al. Phosphorus nutrition of phosphorus-sensitive Australian native plants: threats to plant communities in a global biodiversity hotspot. Conserv Physiol. 2013;1(1):1–23.
53. Teste FP, Laliberte E, Lambers H, Auer Y, Kramer S, et al. Mycorrhizal fungal biomass and scavenging declines in phosphorus-impoverished soils during ecosystem retrogression. Soil Biol Biochem. 2016;92:119–32.
54. Mmbaga GW, Mtei KM, Ndakidemi PA. Extrapolations on the use of rhizobium inoculants supplemented with phosphorus (P) and potassium (K) on growth and nutrition of legumes. Agric Sci. 2014;5:1207–26.
55. Egamberdiyeva D. The effect of plant growth promoting bacteria on growth and nutrient uptake of maize in two different soils. Appl Soil Ecol. 2007;36(2–3):184–9.
56. Shaharoona B, Arshad M, Zahir ZA, Khalid A. Performance of Pseudomonas spp. containing ACC-deaminase for improving growth and yield of maize (Zea mays L.) in the presence of nitrogenous fertilizer. Soil Biol Biochem. 2006;38(9):2971–5.
57. Fliessbach A, Winkler M, Lutz M, Oberholzer H, Mader P. Soil amendment with Pseudomonas fluorescens CHA0: lasting effects on soil biological properties in soils low in microbial biomass and activity. Microb Ecol. 2009;57(4):611–23.
58. Khan M, Zaidi A, Wani P. Role of phosphate-solubilizing microorganisms in sustainable agriculture—a review. Agron Sustain Dev. 2007;27(1):29–43.
59. Jha A, Jha S, Baidya D. Ecological diversity, mechanism, and biotechnology of phosphate-solubilizing bacteria for enhanced crop production, in phosphate solubilizing microorganisms. Berlin: Springer; 2014. p. 157–74.
60. Hu X, Roberts D, Xie L, Maul J, Yu C, et al. Development of a biologically based fertilizer, incorporating *Bacillus megaterium* A6, for improved phosphorus nutrition of oilseed rape. Can J Microbiol. 2013;59(4):231–6.
61. Vacheron J, Desbrosses G, Bouffaud ML, Touraine B, Moenne-Loccoz Y, et al. Plant growth-promoting rhizobacteria and root system functioning. Front Plant Sci. 2013;4:356–75.
62. Dobbelaere S, Vanderleyden J, Okon Y. Plant growth-promoting effects of diazotrophs in the rhizosphere. Crit Rev Plant Sci. 2003;22(2):107–49.
63. Combes-Meynet E, Pothier J, Moenne-Loccoz Y, Prigent-Combaret C. The pseudomonas secondary metabolite 2,4-diacetylphloroglucinol is a signal inducing rhizoplane expression of Azospirillum genes involved in plant-growth promotion. Molr Plant Microbe Interact. 2011;24(2):271–84.
64. Chamam A, Sanguin H, Bellvert F, Meiffren G, Comte G, et al. Plant secondary metabolite profiling evidences strain-dependent effect in the Azospirillum-Oryza sativa association. Phytochemistry. 2013;87:65–77.
65. Bidondo LF, Colombo R, Bompadre J, Benavides M, Scorza V, et al. Cultivable bacteria associated with infective propagules of arbuscular mycorrhizal fungi. Implications for mycorrhizal activity. Appl Soil Ecol. 2016;105:86–90.
66. Panneerselvam P, Mohandas S, Saritha B, Upreti KK, Poovarasan S, et al. Glomus mosseae associated bacteria and their influence on stimulation of mycorrhizal colonization, sporulation, and growth promotion in guava (Psidium guajava L.) seedlings. Biol Agric Hortic. 2012;28(4):267–79.
67. Srinath J, Bagyaraj DJ, Satyanarayana BN. Enhanced growth and nutrition of micropropagated Ficus benjamina to Glomus mosseae co-inoculated with *Trichoderma harzianum* and Bacillus coagulans. World J Microbiol Biotechnol. 2003;19(1):69–72.
68. Yusran Y, Roemheld V, Mueller T. Effects of *Pseudomonas* sp. "Proradix" and *Bacillus amyloliquefaciens* FZB42 on the establishment of amf infection, nutrient acquisition and growth of tomato affected by *Fusarium oxysporum* Schlecht f.sp. radicis-lycopersici Jarvis and shoemaker. eScholoarship Repository, University of California. 2009; 1106.

19

HRMAS NMR spectroscopy applications in agriculture

Pierluigi Mazzei* and Alessandro Piccolo

Abstract

The relatively recent and advanced high-resolution magic angle spinning (HRMAS) NMR technique enables the direct application of NMR spectroscopy to semi-solid and gel-like samples. It combines the advantages of both solid- and liquid-state NMR by allowing to concomitantly measure intact and non-manipulated samples. Based on both 1D and 2D homo- and heteronuclear NMR spectra, HRMAS evaluates the composition of fresh semi-solid samples with a similar resolution as that of classical liquid-state NMR techniques. The enhanced spectral quality still obtained for semi-solid samples is mainly due to the MAS system, whose rapid spinning and sample orientation minimize the anisotropic processes that prevent the acquisition of meaningful NMR spectra for non-liquid materials. Moreover, HRMAS allows us to use edited pulse sequences which, especially in the case of biological tissues or agrofood products, may provide a simultaneous information on polar and non-polar components without the need of preliminary sample extraction. Additionally, this technique may differentiate molecular species according to their degree of mobility in hydrated matrices. The evident versatile potential of the HRMAS NMR makes this technique particularly useful for life science molecular studies. Despite the focus of HRMAS has been greatly devoted on clinical biomedicine, materials chemistry, and metabolomics, there are already enough studies that show useful applications on agricultural issues. This report reviews the latest representative studies that employ HRMAS NMR on systems related to agricultural chemistry, requiring the characterization and dynamics of soil components, plant tissues, agrofood products, and in vivo organisms.

Keywords: HRMAS NMR, Soil organic matter, Plant tissues, Agrofood products, Metabolomics, Soil fertilization, Environment

Introduction

Advanced and reliable analytical techniques are demanded to face up the ever more challenging issues related to the wide field of agricultural chemistry. The main topics of large interest are the following: (i) the improvement of knowledge on the complex chemical processes occurring in soil and involving its humic and mineral components; (ii) the unraveling of processes related to the biological stimulation of plant growth and relationship between soil components and plant roots; (iii) the bioavailability and fate of pollutants in the agro-environment; (iv) the assessment of quality, geographical origin, and traceability of agrofood products; (v) the development of practices in bio-agriculture aimed to improve crop quality. Although the complexity of these problems calls for a multidisciplinary analytical approach, the most determinant and critical responses commonly reside in chemical and molecular information. Nuclear magnetic resonance (NMR) and liquid or gas chromatographic systems coupled to advanced mass spectrometers (LC-MS or GC-MS, respectively) are the most used analytical methods to obtain accurate and detailed molecular information. Despite the poor sensitivity in respect to mass spectrometry, NMR spectroscopy enables to reach a complementary structural and conformational information of molecular systems in a non-destructive way.

Conceptually, the NMR phenomenon occurs when nuclei of magnetically active (spin number $\neq 0$) atoms are immersed in a static magnetic field and exposed to a second oscillating magnetic field. Following the pulse

*Correspondence: pierluigi.mazzei@unina.it
Centro Interdipartimentale per la Risonanza Magnetica Nucleare per l'Ambiente, l'Agro-Alimentare ed i Nuovi Materiali (CERMANU), Università di Napoli Federico II, Via Università 100, 80055 Portici, Italy

interaction between the applied electromagnetic radiation and the dipolar moments of the nuclei subjected to the static field, the resonance energy is relaxed producing a free induction decay that contains detailed information about the structure, dynamics, reaction state, and chemical environment of the target molecular material [1, 2]. The combination of phase- and length-modulated pulses specific of the selected NMR pulse sequences enables a controlled manipulation of nuclear spins that ultimately provides structural and dynamical information and reveals through-bonding or through-space correlations among nuclei, in either mono- or multidimensional systems [3].

An interesting aspect of NMR spectroscopy is its versatility since it enables to examine samples in either liquid- or solid-state, by selecting the proper NMR probe where the samples are placed. However, both techniques are not devoid of inherent disadvantages. NMR spectroscopy of solid materials suffers for strong dipolar and quadrupolar interactions, chemical shift anisotropy (CSA), and magnetic susceptibility. Despite the adoption of different technological solutions for reducing the effects of these phenomena, they are still responsible for the low resolution of solid-state spectra and the exclusion of any direct analysis of hydrogen nuclei. Conversely, totally dissolved samples for liquid-state NMR spectroscopy contain molecules that tumble isotropically, thus fully minimizing the anisotropicity of solid matter and enabling acquisition of high-resolution NMR spectra. Nevertheless, the use of liquid-state NMR spectroscopy for the molecular identification of agricultural matrices (plant roots, leaves and fruits, animal tissues, soils, etc.) requires preliminary procedures such as extraction, purification, and concentration, which are not only time-consuming, but also entail a sample manipulation possibly conducive of loss and/or degradation of components [4].

In the late 1990, the new high-resolution magic angle spinning (HRMAS) technique was introduced to enable NMR applications directly on heterogeneous semi-solid and gel-like samples [5]. HRMAS is a combination of solid- and liquid-state NMR techniques since it allows us to obtain spectra with a resolution similar to that of liquid-state spectroscopy but on intact and non-manipulated semi-solid materials. In fact, these samples maintain a sufficient degree of molecular mobility to afford an efficient minimization of factors that affect NMR measurements of solid materials. The principles of HRMAS have been extensively discussed previously [6–10]. Briefly, each semi-solid sample is reduced to small pieces, fit into a HRMAS rotor, and added with a small amount of deuterated solvent to ensure both the molecular spin mobility and the deuterium lock. The rotor is then tightly sealed by a screwed-in insert to prevent dehydration or loss of solvent during spectral acquisition. The rotor is finally capped (Fig. 1), loaded into the HRMAS probe, and spun at a relatively moderate rate (ν_R) around an axis inclined of an angle of 54.7° (the so-called "magic angle") to the static magnetic field (B_0). The MAS system permits to suppress the dipolar interactions, the chemical shift anisotropy (CSA), and the magnetic susceptibility arising at the interface between the solid and the liquid phase. These extensive spectral-broadening interactions are described in a Hamiltonian that contains an angular dependence of $(3\cos^2\theta - 1)/2$, where θ is the angle between the static magnetic field and the internuclear vector responsible for the interactions. When the sample rotates around an axis inclined of 54.7° to the magnetic field, all broadening factors are theoretically nulled [7, 8].

The high resolution of HRMAS spectra is due to the reduction of H–H dipolar couplings to few hundred Hz in many semi-solid samples, in respect to the 20–50 kHz usually found for rigid solid materials. In fact, the still significant segmental molecular motions in semi-solid matter allow the MAS system to efficiently minimize dipolar couplings [8]. Moreover, the signal resolution achievable by HRMAS is sufficient to perform most of 1D and 2D homo and heteronuclear NMR experiments, inasmuch as those conducted by liquid-state NMR

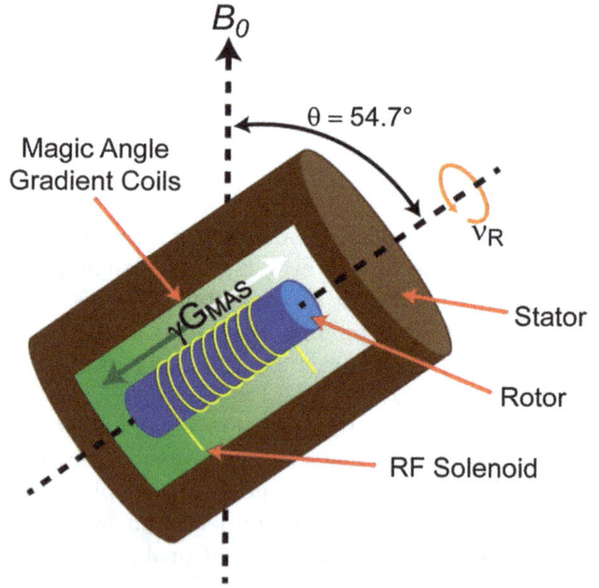

Fig. 1 Scheme of HRMAS stator. The *figure* shows the disposition of both RF solenoid and gradient coils as well as indicates the inclination of *rotation axis* at the magic angle ($\theta = 54.7°$) in respect of static magnetic field (B_0). During HRMAS acquisitions, each sample has to be spun at a moderate rate (ν_R) around an axis inclined at the magic angle. (*Figure* reprinted with permission of authors, © 2012 Alam TM, Jenkins JE. Published in Advanced Aspects of Spectroscopy (Intech Open) under CC BY 3.0 license. Available from: http://dx.doi.org/10.5772/48340)

spectroscopy. Therefore, HRMAS represents the best solution to also study complex molecular systems, such as membrane proteins, large molecular-weight polymers, whose dipole–dipole interactions are too strong to be sufficiently averaged by the simple molecular motion as that expected in liquid-state NMR spectroscopy. It is also critical in HRMAS spectroscopy that the spin rotation has to reach a value at least of the same order of magnitude as the line-broadening interactions, while avoiding an excessively fast rotation that may hasten sample degradation and reduce the measurements reliability. A relatively moderate spin rate ranging between 3 and 6 kHz is thus recommended, in order to reach best compromise between resolution and sample stability. An advantage of the HRMAS technique over solid-state MAS resides in the presence in HRMAS probes of gradients coils (Fig. 1), which may not only further improve spectral resolution, but also select coherence pathways and permit diffusion experiments to differentiate molecular species according to their degree of mobility in hydrated matrices. Moreover, edited (namely "filtered") pulse sequences, such as T_2- and diffusion-based sequences, can be used in HRMAS systems [10–12]. This allows HRMAS spectroscopy to provide information on both polar and non-polar components of biological tissues or agrofood products without preliminary separations though extractions.

All this indicates that HRMAS NMR spectroscopy is a versatile and effective technique for studying complex molecular systems. Despite that most of HRMAS applications had been focusing on biomedical studies and materials chemistry [7, 13, 14], a body of literature has already pointed out that HRMAS may result determinant in investigations related to agricultural issues. Hence, this review aims to report on the most interesting applications of HRMAS NMR in agricultural chemistry.

HRMAS applications in agriculture

Soil humic matter, environmental studies, and agricultural practices

NMR spectroscopy has been repeatedly applied to elucidate the distribution of carbon molecules in soils of different origins [15]. However, HRMAS may be particularly useful in the quest of clarifying the soil complex chemical processes due to better capacity to characterize the molecular structures of soil humic and humo-mineral components than solid- or liquid-state NMR. In fact, HRMAS was proved to enable the identification of molecular structures at the solid-aqueous interface of a whole soil through the acquisition of 1D, selective-1D, and 2D NMR spectra [16]. The same authors observed that the detection of aromatic signals may be significantly enhanced when the soil is swollen in dimethylsulfoxide (DMSO) rather than in water. This was attributed to the capacity of DMSO and other dipolar aprotic solvents to cleave hydrogen bonds and solvate aromatic components [17]. A 3D HRMAS spectroscopy was applied to examine tomato cutin and then compared to solution-state spectra of soil extracts from the A_h horizon under an oak forest [18]. The authors claimed that the molecular structures identified in the soil humic matter may resemble the biomolecules observed in the tomato cutin.

The HRMAS technique was further employed to identify the components of a peat humic acid mostly involved in the organomineral complexes with both kaolinite and montmorillonite clay minerals and how the interactions were influenced by both clay mineralogy and several solution properties, such as pH, ionic strength, and the presence of Na^+ and Ca^{2+} cations [19]. A study on complexes between calcium-exchanged montmorillonite and either fulvic acids or soil alkaline extracts was conducted by ^1H HRMAS NMR and showed that aliphatic and, partially, aromatic components bound preferentially to the clay surface [20]. These results seemed to confirm that a mechanism of organic matter preservation in soil occurs by adsorption on clay minerals of hydrophobic products of biological tissues degradation [21].

The decomposition process of ^{13}C- and ^{15}N-enriched pine and wheatgrass litter was followed by HRMAS NMR spectroscopy [22]. While the 1D ^{13}C spectra of fresh plant tissues well differentiated the composition of fresh pine and grass materials, the differences became negligible after decomposition and suggested a similar accumulation of recalcitrant carbon species, thus again supporting a mechanism of progressive selection of hydrophobic degradation products regardless of the plant origin [21]. Moreover, this study also highlighted the advantage of the use of diffusion-based filter to selectively identify and monitor the signals of fast-diffusing components developed during degradation process [22]. Analogously, ^{13}C and ^{15}N labeled soil microbial biomass and leachate degradation were monitored by HRMAS over time in order to better understand the degradation process [23]. This work also showed that the biomass degradation pathways are similar to those of plant material under similar degrading conditions, thus suggesting that the difference between recalcitrant carbon from different sources is negligible after decomposition. Moreover, ^1H–^{15}N HSQC (hetero single quantum coherence) 2D NMR spectra suggested that degradation either selectively preserves or produces specific peptides incorporated in yet unknown structures [23].

A variety of natural organic matters (NOM) (wood, kerogen, bitumen, and whole sediments) were subjected to both Fourier transform ion cyclotron resonance mass spectrometry (FTICR-MS) and NMR techniques [24]. Interestingly, this work showed that HRMAS applied to the whole NOM samples was superior to the traditional liquid-state NMR measurements of extracts from the same NOM materials [24]. Another study combined HRMAS and DRIFT (Diffuse reflectance infrared Fourier transform spectroscopy) techniques to identify the relationship between the structure of soil humic substances and long-term (40 years) soil management practices, including organic and/or mineral fertilizations [25]. The application of T_2-based (namely CPMG or spin-echo) NMR filters to reduce signals heterogeneity in ^1H spectra of humic substances allowed to develop cluster analyses to differentiate humic matter as a function of soil treatments [25]. In the same line, conventional and diffusion-edited HRMAS proton spectra were applied to evaluate the long-term stabilization mechanisms of organic materials in soils undergone solid cattle manure and crop residues amendments and fractionated into free, intra-macroaggregate, intra-microaggregate, mineral-associated, and dissolved organic matter [26]. The authors compared the peaks detected in 1D HRMAS spectra of HF-treated soil with those of a bovine albumin protein and cultured soil microbes and suggested that the material associated with the clay mostly resembles the molecular components derived from microbial biomass [26]. Such finding was accompanied by a further study in which diffusion-edited HRMAS NMR spectra of HF-treated soil aggregates of different dimensions were related to the role of microbes in conservation tillage systems [27]. A tight relationship between soil hydrophobicity and humic composition was revealed by relating HRMAS spectra of soil humic fractions and physical parameters such as soil wettability, pore size distribution, and soil aggregate stability [28]. These results confirmed previous works that the hydrophobicity of humic fractions can affect aggregate wetting and slaking by modifying the pore size distribution with a shift from micro- and mesopores to ultramicropores. The HRMAS technique was also applied to monitor the modifications of both chemical and biological properties of SOM upon the strong pedoturbation caused by colonies of snow vole (*Chionomys nivalis Martins*) in a mountain environment [29]. The authors reported that snow vole activity was capable to significantly affect the humification process and SOM cycling, with effects on the hormone-like bioactivity of humic fractions.

HRMAS NMR spectroscopy was also applied in environmental chemistry studies, such as the interactions between a pesticide and soil mineral components. The adsorbed or intercalated forms of herbicide 4-chloro-2-methylphenoxyacetic acid on a highly hydrated clay, such as hydrotalcite, were clearly distinguished by ^1H-HRMAS. In fact, the adsorbed herbicide gave sharp signals indicating high mobility, while intercalated herbicide gave very wide unresolved spectra due to its strong interaction with the solid matrix [30]. The chemical shift drift and line broadening of HRMAS signals also served to reveal the interactions occurring between the herbicide trifluralin and a soil sampled from the A_h surface horizon under an oak forest [16]. A HRMAS investigation on organomineral complexes obtained by coating either montmorillonite or kaolinite clays by peat humic acids showed that these humic-clay complexes, especially in acidic conditions, were responsible for the increasing sorption of the non-ionic and hydrophobic phenanthrene contaminant [19]. The advanced NMR pulse sequence STDD (saturation transfer double difference) was used to prove the pH-dependent (D_2O or sodium phosphate D_2O buffer solution at pH7) interactions between a peat soil and several contaminants, including 1-naphthol and the pesticides trifluralin, acifluorfen, and 4-nitro-3-(trifluoromethyl) phenol [31]. The authors created epitope maps (Fig. 2) on the basis of HRMAS STDD results, which revealed the involvement of pesticides functional groups in the contaminant-soil associations and enabled a prediction of predominant sorption mechanisms. Recently, this study was extended by exploiting results from diffusion, relaxation, cross polarization, and saturation transfer difference NMR experiments, which permitted to follow the sequestration of contaminants from the liquid phase into the whole soil [32].

The HRMAS technique has been successfully applied to follow the effects of inorganic and organic contaminants in biological tissues. In a study of the influence of arsenic-contaminated irrigation water on *Lactuca sativa* L. grown in either sandy or clay loam soils, HRMAS was used to verify how fresh lettuce leaves were affected by the arsenic uptake and soil type [33]. This work showed that the metabolomic approach developed by HRMAS enabled the detection of the significant changes in phenolic compounds contained in As-treated plants. The leaves of lettuce plants exposed to the pesticide Mancozeb were monitored by HRMAS [34]. The metabolomic profile of the plant leaves revealed that the pesticide affected the concentration of a number of lettuce metabolites, thus indicating a large oxidative stress and a Krebs cycle up-regulation at a late stage of plant growth. The effect of the exposure of two organochlorine pesticides lindane and chlordecone on the maize root metabolome was recently explored by HRMAS [35]. The metabolomic profiles revealed some exposure-dependent changes in the amount of several maize metabolites, such as carbohydrates, amino acids, tricarboxylic acid cycle intermediates, and fatty acids, which were included in a diagram of biosynthetic pathways [35].

Other works have reported the adoption of HRMAS technique to study the effects of different agricultural

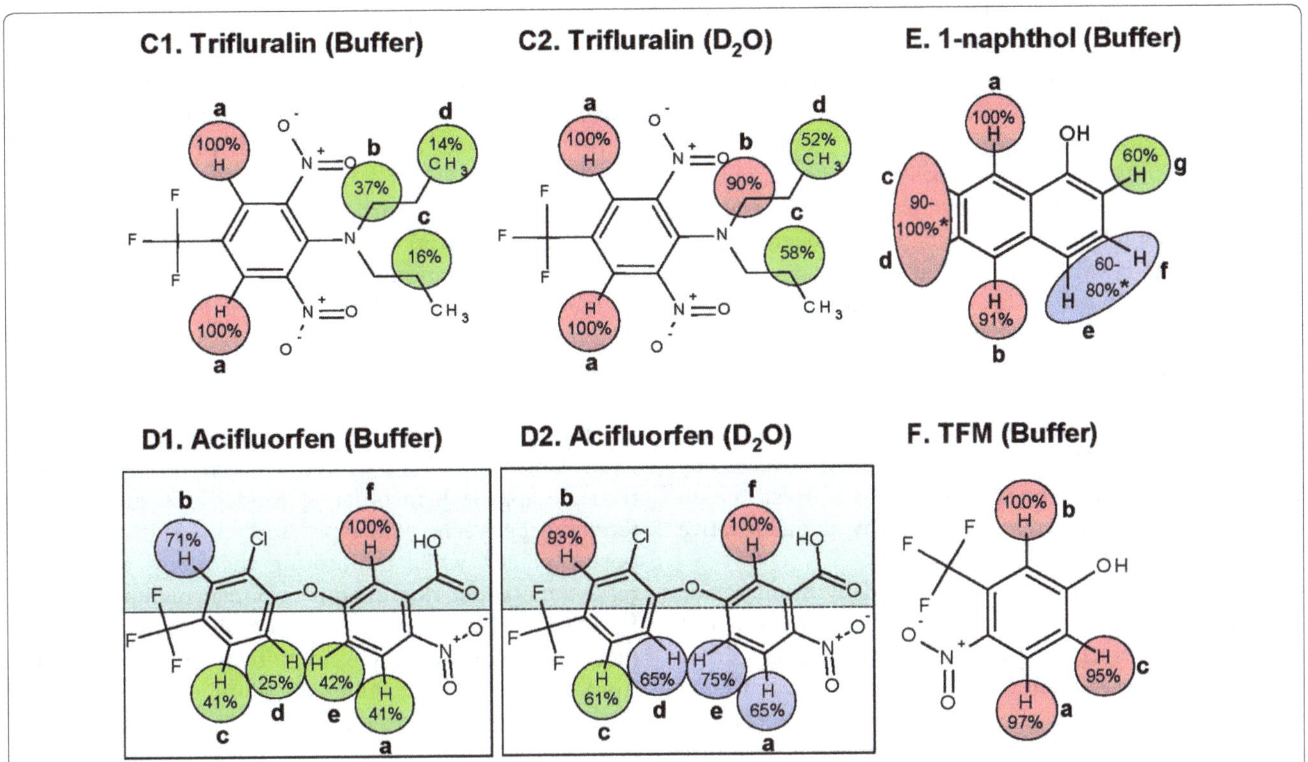

Fig. 2 Epitope maps of the four pesticides dissolved in the pH7 buffer or D_2O. The percentage values for each epitope map are all relative to the strongest binding proton in each pesticide, which is expressed as 100%. Protons which have 80–100% binding are highlighted by *red circles*, protons with 65–79% with *blue*, and 64% and lower with *green*. The *shaded region* in *D1* and *D2* simply highlights the side of the molecule that has the strongest interaction with the soil surface. An *asterisk* (*) indicates that absolute quantification was not possible at these positions due to spectral overlap.

practices on plants metabolic profiles, thus relating the soil management to agrofood products. For example, HRMAS NMR-based metabolomics was applied with the aim of proving the different compositional qualities of three potato cultivars grown under either conventional or organic system [36]. HRMAS results were also determinant to show that conventional and organic management treatments affect differently the composition of fresh olive fruits [37].

Plants

The possibility given by HRMAS to acquire meaningful NMR spectra directly on whole fresh tissues represents an advantage for studying the molecular composition of plant shoots, flowers, fruits, and roots. Several literature works have used this technique to characterize the complex molecular content and quality of whole plant tissues, and better understand the effects on plants of external stimuli or specific treatments. Multidimensional HRMAS together with CPMAS was applied to characterize cutin, an insoluble component of plant leaf and fruit cuticles. In particular, tomato cutin was shown to contain a structural skeleton composed predominantly by an aliphatic polyester with some olefinic and aromatic moieties [38]. Moreover, these authors inferred that α-branched fatty acids/esters in cutin could be cross-linked to esters of mid-chain hydroxyls [38]. Later, the same authors adopted a similar analytical approach to identify the structure of cutan and a cutin/cutan mixture from the *Agave americana* leaf cuticle [39]. The structure of suburban, a protective biopolymer associated with suberin and located in the periderm of above ground parts of plants or in the endoderm of roots, was also investigated by combining HRMAS NMR to other techniques, such as CPMAS NMR, Scanning Electron Microscopy (SEM), flash-pyrolysis Gas Chromatography coupled to Mass Spectrometry (GC–MS), and FT-IR (Fourier Transform-InfraRed spectroscopy) [40]. The integration of results from these different techniques enabled the authors to propose a model structural unit for the specific suburban from *Betula nigra* bark under study. The algae–fungi symbionts called lichens were successfully differentiated on the basis of chemotaxonomy by using a combination of HRMAS (conventional and CPMG-edited proton spectra) and FT-IR techniques [41]. Despite the difficulty in distinguishing lichens morphologically, the authors proposed a non-destructive HRMAS-based method enabling unambiguous differentiation results, while concomitantly avoiding a laborious and time-consuming sample pre-treatment [41].

HRMAS allowed to obtain a metabolic profile of both shoots and roots of four genotypes of *Withania somnifera* (L.), which is a plant with several pharmaceutical and nutraceutical applications [42]. The authors were able to detect through HRMAS the presence of withaferin A and withanone molecules, which are secondary metabolites produced by this plant and important natural medicinal compounds [42]. Proton and carbon 1D and 2D HRMAS NMR spectra enabled the identification and assignment of several bioactive triterpenoids and phenolics in Koroneiki olive leaves [43]. HRMAS has been successfully used to quantify the free sugars content in fresh melon fruits, whose results were validated by a quantitative enzymatic assay [44].

The presumed beneficial role exerted by humic acids in plant regeneration was verified by HRMAS on root and shoot calli explanted by pear and quince plants, after treatments with different concentrations of peat humic acids [45]. The authors showed that the selected humic acids induced a positive and dose-dependent response only in pear plants [45]. Recently, HRMAS revealed the molecular composition of *crocus sativus* petals and detected several bioactive compounds such as kinsenoside, goodyeroside A, and 3-hydroxy-γ-butyrolactone [46]. By comparing HRMAS results with those of ethanol extracts of the same petals analyzed by liquid-state NMR, it was observed that extraction favored hydrolysis reactions, thus decreasing the reliability of NMR results in the liquid-state.

An advanced NMR application for the identification of the composition of woody angiosperm plants reports a detailed characterization of ^{13}C-labeled poplar plants, when these were preliminarily grown in $^{13}CO_2$-enriched atmosphere and supplied with [$^{13}C_6$] glucose to roots [47]. The authors used HRMAS and liquid-state NMR techniques to analyze the whole poplar tissues and their ethanol/water extracts, respectively. Remarkably, they exploited the ^{13}C labeling of plants by making use of 3D experiments, such as ^{1}H-^{13}C HCCH-COSY and ^{1}H-^{13}C HCCH-TOCSY, to unambiguously assign several signals. Recently, Mazzei and coworkers applied ^{1}H HRMAS metabolomics to identify the effects on young *Solanum lycopersicum* (tomato) plants of seeds treatment with two bioactive secondary metabolites (6-pentyl-2H-pyran-2-one and harzianic acid) extracted from different *trichoderma* fungal strains [14]. A metabolomic profiling of tomato leaves was achieved with 1D proton and 2D ^{1}H-^{13}C HRMAS NMR spectra (Figs. 3, 4), whereby the most intense proton and carbon signals were duly assigned. The authors observed a significant enhancement of seedlings germination accompanied by a change in leaf metabolome that was treatment-specific, thereby inferring that the method may help in developing new biofertilizers based on fungi extracts [14].

The HRMAS technique has been also applied to evaluate transgenic plants. ^{1}H HRMAS NMR spectra were successfully combined with the Batch Processing multivariate approach in order to model and interpret differences between wild-type and PttMYB76-modified poplar plants, while monitoring the time- or growth-related metabolic fluctuations in poplar plants [48]. Both HRMAS NMR and IR enabled the differentiation of the metabolome of three different cultivars of beans deriving from either wild or transgenic genotype [49]. Moreover, the authors also reported an overexpression of quercetin and myricetin flavonoids in the transgenic beans. Roots and leaves of *swingle citrumelo* were investigated to prove the capability of a selected transgene to induce an overexpression of proline, an amino acid commonly associated with plant responses to certain environmental stress [50]. In particular, the authors used HRMAS to compare the metabolome of wild-type plants with that of transgenic T-35S plants carrying the VaP5CSF129A transgene encoding a mutated enzyme involved in proline biosynthesis. Spectroscopic results showed that the transgenic plants exhibited a genotype-dependent response that entailed different levels of proline, proline betaine, sucrose, and acetic acid [50]. In another study, the composition of wild-type *Arabidopsis thaliana* was compared to that of cellulose-deficient mutant *ectopic lignification1* through HRMAS-based experiments. Interestingly, the mutant form showed a larger abundance of methanol, fatty acids and/or lipids, glutamine, phenylalanine, starch, and nucleic acids [51].

Agrofood products

The quest from producers and consumers for quality of agrofood products, such as edible fruits, vegetables,

(See figure on next page.)

Fig. 3 ^{1}H HRMAS CPMG NMR spectrum of control tomato leaves acquired at a spin rate of 5 kHz. Three spectral regions are shown: **a** 0.8–3.3 ppm, **b** 3.34–4.27 ppm, and **c** 4.4–9.25 ppm. The labels refer to assignment of the most intense signals detected in all treatments (*F* fructose, *G* glucose, *Gal* galactose, *S* sucrose, *ACho* acetylcholine, *GABA* γ-aminobutyric acid, and *UDP* uridine diphosphate).

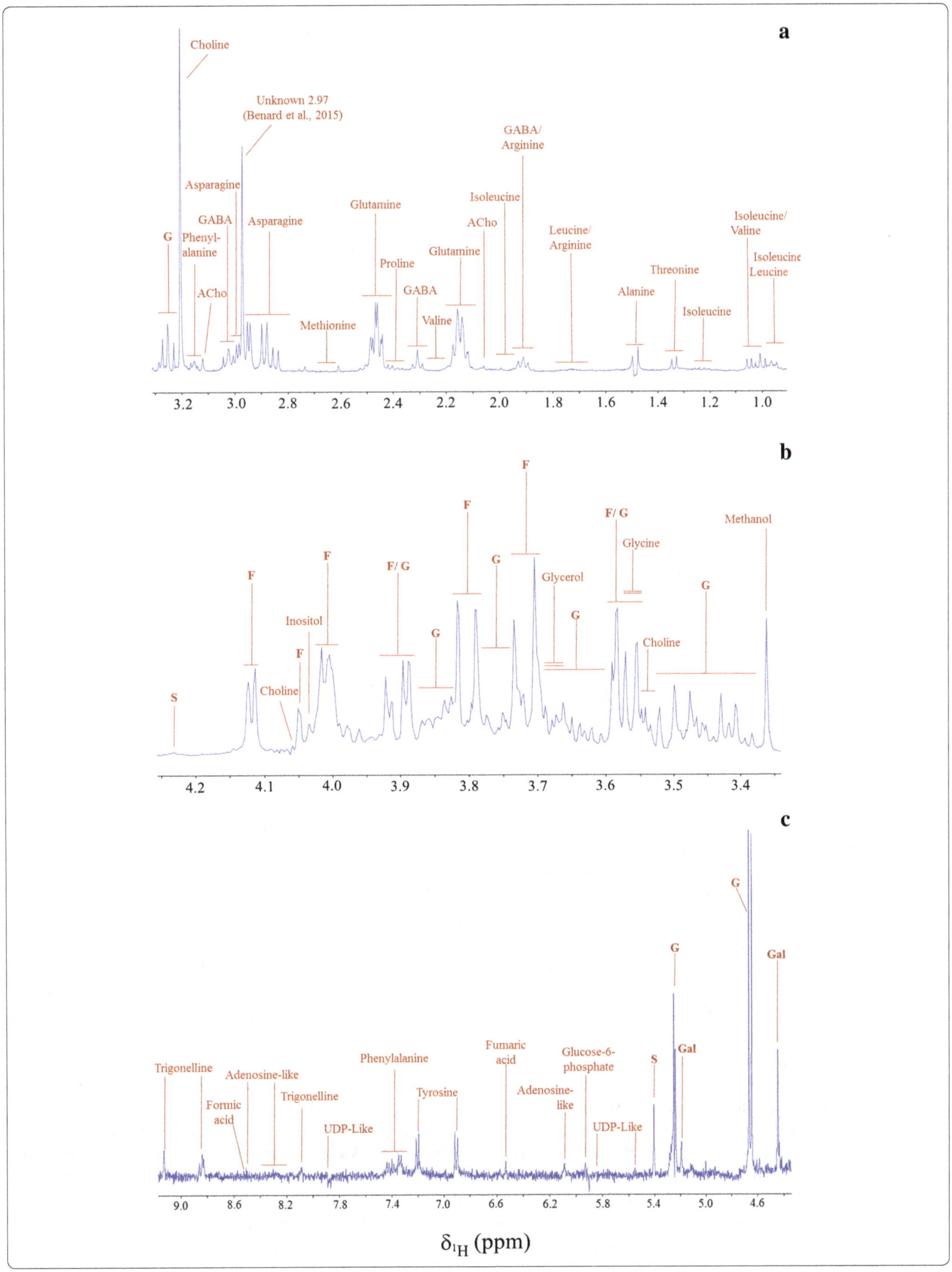
a
Choline
Unknown 2.97
(Benard et al., 2015)
GABA/
Arginine
Asparagine
Isoleucine
Glutamine
GABA
Asparagine
ACho
Isoleucine/
Valine
G
Phenyl-
alanine
Leucine/
Arginine
Proline
Glutamine
Isoleucine
Leucine
Threonine
ACho
GABA
Alanine
Methionine
Valine
Isoleucine
3.2 3.0 2.8 2.6 2.4 2.2 2.0 1.8 1.6 1.4 1.2 1.0
b
F
F
F/ G
Methanol
Glycine
F
F/ G
G
F
Glycerol
G
Inositol
G
G
F
Choline
Choline
S
4.2 4.1 4.0 3.9 3.8 3.7 3.6 3.5 3.4
c
G
G
Gal
Gal
Fumaric
acid
Glucose-6-
phosphate
S
Phenylalanine
Trigonelline
Adenosine-like
Tyrosine
Adenosine-
like
Formic
acid
Trigonelline
UDP-Like
UDP-Like
9.0 8.6 8.2 7.8 7.4 7.0 6.6 6.2 5.8 5.4 5.0 4.6
$\delta_{^1H}$ (ppm)

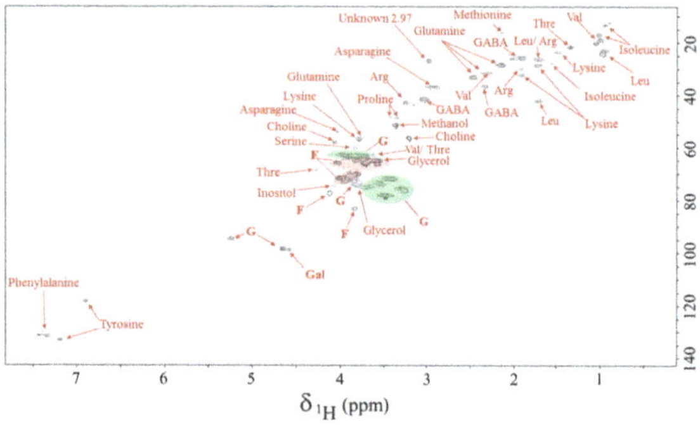

Fig. 4 ^{1}H–^{13}C HSQC NMR spectrum of control tomato leaves optimized for a short-range JCH coupling of 145 Hz. The labels refer to the assignment of the most intense signals detected in all treatments (*F* fructose, *G* glucose, *Gal* galactose, *S* sucrose, *GABA* γ-aminobutyric acid, *Leu* leucine, *Arg* arginine, *Val* valine, and *Thre* threonine). The *ellipsoid regions* circumscribe the dense cluster of F (*red*) or G (*green*) signals.

cheeses, and cereals, calls for the adoption of reliable analytical techniques and methods that warrant the food nutritional value, traceability, and safety. In this context, the HRMAS NMR capacity to precisely detect and quantify specific compound classes in complex biological materials indicates this technique as an excellent candidate to assess the general quality of agrofood products.

In the case of fruit studies, ^{1}H and ^{13}C HRMAS NMR spectra were helpful in identifying the progressive changes in composition of mango pulp during 19 days of ripening [52]. In fact, ripening produced a decrease of citric acid accompanied by an increase of lactic acid, gamma aminobutyric acid, phenylalanine, and niacin. The effects of relatively low pressures on strawberries during storage were estimated by following the physical tissue damage, and the changes in molecular quality by Magnetic Resonance Imaging (MRI) and HRMAS proton spectra, respectively [53]. HRMAS was used to assess the chemical composition of coarsely ground coffee beans of Arabica and Robusta varieties, as well as the variation in content of Maillard reaction-related compounds and coffee aroma constituents, as a function of coffee roasting temperature from 30 to 215 °C [54].

A metabolomic approach by ^{1}H-HRMAS helped to neatly differentiate several cultivars of apples [55] and sugarcanes [56]. A detailed metabolic profiling of different tissues (flavedo, albedo, seeds, and pulp) of both lemon and citrus fruits was conducted through HRMAS as well as they characterized several compounds associated with the development of mold in the citron flavedo [57]. Recently, HRMAS has also been used to efficiently trace the geographical origin of 60 fermented and dried cocoa beans obtained from major growing areas in Africa, Central/South America, Asia/Oceania [58].

A number of interesting works have described the application of HRMAS NMR to edible vegetables. For example, seeds, peel, flesh, and puree of tomatoes were studied by HRMAS to produce a metabolomic profiling of Spanish Almeria tomatoes and identify the molecular markers correlated to ripening process [59]. Later, the same research group used HRMAS to differentiate three varieties of tomatoes and the basis of their composition [60]. In particular, they observed that the content in metabolites relevant for the tomato taste, such as fructose and organic acids, depended on the specific variety [60]. Ritota et al. [61] reported the use of HRMAS to differentiate Italian sweet peppers (*Capsicum annum* L.) as a function of both cultivar and geographical origin [61]. Later, the same research group used a similar ^{1}H- and ^{13}C-HRMAS approach to characterize two Italian cultivars of garlic (*Allium sativum* L.) [62]. They found several organosulfurs, allicin, and some allyl-organosulfurs in garlic samples and confirmed their presence by SPE-GC-MS analysis. Lam et al. [63] suggested that HRMAS-based experiments also enabled to differentiate and describe both structural and metabolic profiles of intact ^{13}C-enriched wheat, broccoli, and maze seeds [63].

Dairy products were also intensively studied by HRMAS NMR. Already in 2004, HRMAS was applied to trace the metabolomic profile of the Italian Parmigiano Reggiano cheese [64] and to correlate its concentration of several free amino acids and low molecular-weight metabolites to the cheese ripening during a 24 months-long monitoring [65]. Furthermore, Shintu and Caldarelli [66] also described, through HRMAS, the metabolic differences of twenty Emmental cheeses from five different countries [66]. A combination of several multivariate statistical analyses of ^{1}H HRMAS results was used to differentiate the Italian "mozzarella di bufala campana" on the basis of the geographical origin of that specific soft-cheese product [67]. Their work relied not only on conventional proton spectra, but also on both T_2- and diffusion-edited ^{1}H spectra, which enabled to isolate and resolve the proton signals of relatively low molecular-weight molecules (long spin–spin relaxation times) and lipidic components (smaller diffusivity), respectively (Fig. 5). Moreover, they found evidence that the relevant abundances of isobutylic alcohol, lactic acid, and acetic acid, were the markers which would help to identify the aged samples of "mozzarella di bufala campana" [67].

The HRMAS has been also employed to examine cereals, cereal-derived, and apicultural products. This technique combined with HPLC analyses successfully discriminated Italian Royal Jellies from Chinese ones [68]. Moreover, an HRMAS approach was recently

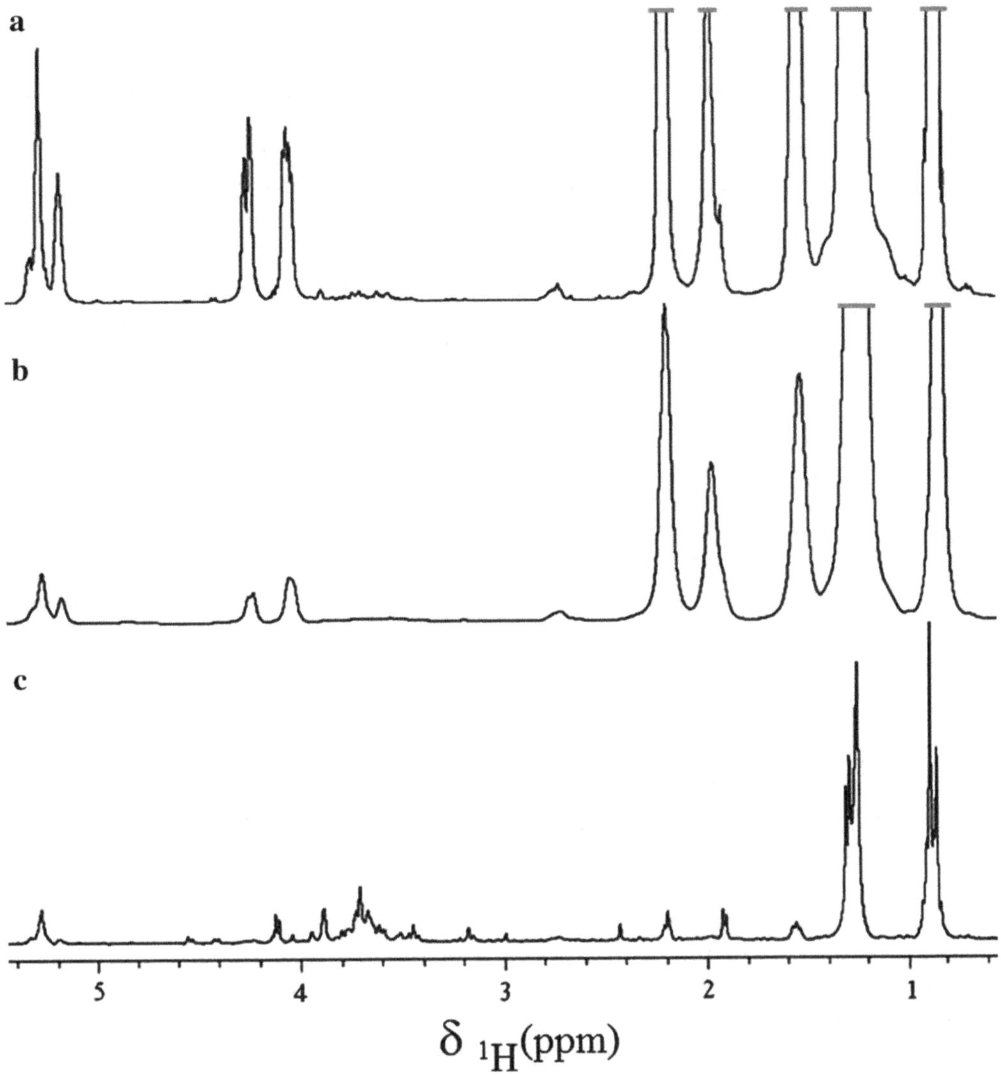

Fig. 5 ^{1}H HRMAS spectra of a sample of buffalo-milk mozzarella: **a** conventional ^{1}H spectrum; **b** diffusion-edited spectrum; **c** CPMG-edited spectrum.

exploited to distinguish waxy from non-waxy Chinese rice grains, deriving from different *Oryza sativa* L. cultivars [69]. Both ^{1}H and ^{31}P HRMAS NMR spectra were used to elucidate the granule hydration in either wheat- or maize-derived starches at different temperatures [70]. Most interestingly, the authors modulated the MAS spin rate to distinguish between immobilized (water-inaccessible starch domains) and mobile components, thus inferring that their distribution depended on the starch granules assembly [70].

Plant response to biotic and abiotic stress

As for other biological materials HRMAS has been usefully applied to rapidly monitor the plant stress in response to both abiotic and biotic factors, such as drought, salinity, nutrients deficiency, or pathogens influence. For instance, a study by HRMAS of both seedling shoots and roots enabled to identify a neat metabolomic response induced in two cultivars of rice by drought and salinity stress [71]. In particular, the abiotic stress favored the accumulation of free amino acids and saccharides. Concomitantly, the same authors presented preliminary results on the metabolomic changes occurring in rice leaves due to the penetration and colonization of three strains of the *Magnaporthe grisea* phytopathogenic fungus [71].

Both liquid-state and HRMAS NMR techniques showed that drought response in wheat kernels consisted in a resistance occurring when wheat was subjected to drought early during grain-filling [72]. Moreover, they found an influence on proteins metabolism, and fumaric acid was identified as a potential marker for drought stress in mature kernels. The HRMAS technique was combined with MRI to investigate the poisonous *Jatropha curcas* L. plant infected by a *Jatropha mosaic begomovirus* [73]. This study associated morphological, anatomical, and metabolic profiling of virus-infected tissue of *J. curcas* plant. A twofold accumulation of intermediates of the Tricarboxylic acid (TCA) cycle in virus-infected plants was found, which suggested a larger rate of respiration [73]. In another work, both leaves and fruit peels of *citrus sinensis* plants were proved by HRMAS to be infected by the bacterium *Xanthomonas axonopodis pv. Citri*, which is responsible for the citrus cancer. The authors revealed that the defensive metabolism activated by *Citrus sinensis* induced relevant changes in amino acids, carbohydrates, organic acids, and terpenoids content in the plant [74].

Due to the concern on iron (Fe) plant deficiency, which is potentially responsible for poor yields and crop quality, any understanding of the metabolomic plant response related to poor Fe uptake is important to design strategies to deal with iron-deficiency. With this aim, Fe-sufficient and Fe-deficient soybean leaf extracts and whole leaves were studied by liquid-state and HRMAS NMR, respectively [75]. The authors observed that Fe-deficient soybeans revealed an enhanced TCA cycle activity, and the activation of mechanisms against the oxidative stress. HRMAS NMR has been recently applied to study the metabolomic changes in the germination of wheat seeds treated with fungicides bavistin and thiram [76]. Both fungicides were found to delay the germination cycle in wheat seeds, while thiram, a non-systemic fungicide, affected the metabolic processes to a larger extent than the systemic fungicide bavistin. Moreover, HRMAS identified the metabolic changes exerted by salinity and silicon (H_4SiO_4-based fertilizers) treatments (either alone or combined) on the quality of four Marmande tomato varieties [77]. In particular, silicon addition appeared to enhance the effect of light intensity on photoassimilates available to fruits, without a clear effect on the organoleptic quality.

In vivo HRMAS NMR spectroscopy

Due to its capacity to produce meaningful spectra of semi-solid and fresh samples, HRMAS represents an eligible technique to perform in vivo analysis of living micro- and macroorganisms. In this respect, Fredlund et al. [78] proposed a pioneering method to in vivo study microorganisms subjected to stress conditions [78]. They focused on the biocontrol of yeast *Pichia anomala J121* and verified how their intracellular metabolites varied as a function of an induced oxygen limitation. Specifically, they tested two conditions: (i) initial aerobic conditions with restricted oxygen access during the yeast's growth period, and, (ii) initial microaerobic conditions followed by anaerobiosis. Biomass production was larger in treatment (i), whereas the specific ethanol production rate upon glucose addition was similar in both treatments, thus implying that oxygen availability affected the respiration, but not the yeast fermentation. Following glucose depletion, ethanol was oxidized to acetate in treatment (i), but continued to be produced in (ii) [78].

HRMAS was later applied to monitor the antituberculosis ethionamide pro-drug activation in a strain of *Mycobacterium smegmatis* [79]. This work showed that both conventional and diffusion-edited ^1H HRMAS NMR spectra enabled the detection of a drug metabolite directly inside intact cells of living bacteria. Their experimental work was based on an optimized procedure of filling the 4 mm HRMAS rotor with microorganisms. They were successful to show the feasibility even at moderate magnetic fields of NMR detection in bacterial cells of an unlabeled metabolite at a pharmacologically relevant (i.e., submillimolar) concentration [79]. ^1H and ^{31}P HRMAS NMR spectroscopy was also applied to in vivo evaluate the composition of *Aporrectodea caliginosa* earthworms [80]. In particular, a metabolic profiling of studied earthworms was achieved in order to monitor the effects of an acute exposure to the herbicide glyphosate. The authors claimed for a low toxicity of the herbicide as well as they argued its possible trapping into the cutaneous mucus which suggested a role of earthworms in the horizontal dispersion and stabilization of glyphosate in the drilosphere. Moreover, ^{31}P-^1H HETCOR (HETeroCORrelation) 2D NMR spectra were helpful in the identification of several phosphorylated metabolites such as phospholombricine and lombricine [80]. *Daphnia magna* crustaceans, which are conventionally used as aquatic model organism in ecotoxicology to evaluate the short- and long-term impact of toxic compounds at physiological and genomic levels, were analyzed by HRMAS NMR [81]. The authors identified the metabolic variations occurring in daphnids as a function of successive life stages and at different physiological statuses. More recently, HRMAS was used to examine in vivo the composition of both a wild type and a mutant (GST2) of the *Drosophila melanogaster* fruit fly [82]. Increased levels of triglycerides observed in the mutant were interpreted as indicative of insulin resistance with implications for mitochondrial disfunction. Finally, also the CMP (Comprehensive Multi Phase), a recent NMR probe enabling the analysis of samples in all phases, including the semi-solid HRMAS-like approach, was used to investigate whole living organisms. For example, it has been recently reported that 1D and 2D CMP-NMR experiments permitted to evaluate the most abundant metabolites in living *Hyalella Azteca* crustaceans [83]. In addition, the authors studied the survival extent of these amphipods as a function of oxygen availability, temperature, and organism orientation into the NMR rotor. It is noteworthy the strategy to feed with ^{13}C-labeled *Chlamydomonas reinhardtii* algae the *H. Azteca* prior their NMR analysis in order to significantly increase the spectroscopic response for carbon signals.

Concluding remarks

HRMAS is a relatively recent NMR technique that was developed to allow direct examination of semi-solid fresh samples without any pre-extraction. Here, we described the most relevant applications of HRMAS in studies related to agricultural issues. Specifically, we showed how HRMAS provides, by both multidimensional and multinuclear spectroscopy, a clear and informative molecular characterization of complex heterogeneous systems (i.e., soil organic matter, plant-derived materials) and enables to unravel the environmental reactivity of inorganic and organic materials. Moreover, when combined with chemometrics, HRMAS represents a reliable tool to study edible agrofood products and assess, directly and indirectly, their nutritional value, geographical origin, authenticity, genotype, and safety. The HRMAS becomes ever more important for the investigation of metabolic processes in plants as related to specific management practices or to responses to either abiotic stress or phytopathogens infections. Moreover, the possibility to acquire edited spectra (T_2- and diffusion-based ones are the most commonly used) by HRMAS, further enhances the potential of this technique for quantitative analyses of complex biological systems.

The wide variety of materials (including in vivo micro- and macroorganisms) successfully studied by HRMAS suggests that it is the best choice when standard solution- and solid-state NMR techniques are not applicable or unsatisfying. Moreover, the rapidity and relative simplicity of the HRMAS technique, well coupled to the inherent reduction of solvents consumption and waste production, represent further elements in making this methodology very attractive for a wide spectrum of scientific investigations.

Authors' contributions

PM and AP contributed equally in the literature selection and discussion. Both authors read and approved the final manuscript.

Acknowledgements

Not applicable.

Competing interests

The authors declare that they have no competing interests.

References

1. Levitt M. Spin dynamics: basics of nuclear magnetic resonance. 2nd ed. Chichester: Wiley; 2008.
2. Mazzei P, Piccolo A. Interactions between natural organic matter and organic pollutants as revealed by NMR spectroscopy. Magn Reson Chem. 2015;53:667–78.
3. Jacobsen NE. NMR spectroscopy explained—simplified theory, applications and examples for organic chemistry and structural biology. New Jersey: Wiley; 2007.
4. Valentini M, Ritota M, Cafiero C, Cozzolino S, Leita L, Sequi P. The HRMAS–NMR tool in foodstuff characterization. Magn Reson Chem. 2011;49:S121–5.
5. Maas WE, Laukien FH, Cory DG. Gradient, high resolution, magic angle sample spinning NMR. J Am Chem Soc. 1996;118(51):13085–6.
6. Doty FD, Entzminger G, Yang AY. Magnetism in high-resolution NMR probe design. II: HR MAS. Concept Magn Res. 1998;10:239–60.
7. Alam TA, Jenkins JE. HR-MAS NMR spectroscopy in material science. In Farrukh MA, Editor. Advanced aspects of spectroscopy. Intech Open; 2012. p. 279–306.
8. Beckonert O, Coen M, Keun HC, Wang Y, Ebbels TMD, Holmes E, et al. High-resolution magic-angle-spinning NMR spectroscopy for metabolic profiling of intact tissues. Nat Protoc. 2010;5(6):1019–32.
9. Renault M, Shintu L, Piotto M, Caldarelli S. Slow-spinning low-sideband HR-MAS NMR spectroscopy: delicate analysis of biological samples. Sci Rep. 2013;3(3349):1–10.
10. Farooq H, CourtierMurias D, Soong R, Bermel W, Kingery WM, Simpson AJ. HRMAS NMR spectroscopy: a practical guide for natural samples. Curr Org Chem. 2013;17(24):3013–31.
11. Liu M, Nicholson JK, Lindon JC. High-resolution diffusion and relaxation edited one- and two-dimensional 1H NMR spectroscopy of biological fluids. Anal Chem. 1996;68:3370–6.
12. Courtier-Murias D, Farooq H, Masoom H, Botana A, Soong R, Longstaffe JG, et al. Comprehensive multiphase NMR spectroscopy: basic experimental approaches to differentiate phases in heterogeneous samples. J Magn Reson. 2012;217:61–76.
13. Mazzei P, Piccolo A, Nugnes L, Mascolo M, De Rosa G, Staibano S. Metabolic profile of uterine leiomyomas intact tissue using high resolution magic-angle spinning ^{1}H NMR spectroscopy. NMR Biomed. 2010;23:1–10.
14. Mazzei P, Vinale F, Woo SL, Pascale A, Lorito M, Piccolo A. Metabolomics by proton high-resolution magic-angle-spinning nuclear magnetic resonance of tomato plants treated with two secondary metabolites isolated from trichoderma. J Agric Food Chem. 2016;64:3538–45.
15. Šmejkalová D, Spaccini R, Piccolo A. Multivariate analysis of CPMAS ^{13}C-NMR spectra of soils and humic matter as a tool to evaluate organic carbon quality in natural systems. Eur J Soil Sci. 2008;59:496–504.
16. Simpson A, Kingery W, Shaw D, Spraul M, Humpfer E, Dvortsak P. The Application of 1H HR-MAS NMR spectroscopy for the study of structures and associations of organic components at the solid-aqueous interface of a whole soil. Environ Sci Technol. 2001;35:3321–5.
17. Piccolo A, Campanella L, Petronio BM. ^{13}C–NMR spectra of humic substances extracted with different mechanisms. Soil Sci Soc Am J. 1990;54:750–5.
18. Simpson A, Kingery W, Hatcher P. The identification of plant derived structures in humic materials using three-dimensional NMR spectroscopy. Environ Sci Technol. 2003;37:337–42.
19. Feng X, Simpson AJ, Simpson MJ. Investigating the role of mineral-bound humic acid in phenanthrene sorption. Environ Sci Technol. 2006;40:3260–6.
20. Simpson AJ, Simpson MJ, Kingery WL, Lefebvre BA, Moser A, Williams AJ, et al. The application of 1H high-resolution magic-angle spinning NMR for the study of clay-organic associations in natural and synthetic complexes. Langmuir. 2006;22:4498–503.
21. Piccolo A. The supramolecular structure of humic substances. A novel understanding of humus chemistry and implications in soil science. Adv Agron. 2002;75:57–134.
22. Kelleher BP, Simpson MJ, Simpson AJ. Assessing the fate and transformation of plant residues in the terrestrial environment using HR-MAS NMR spectroscopy. Geochim Cosmochim Acta. 2006;70:4080–94.

23. Spence A, Simpson AJ, Mcnally DJ, Moran BW, McCaul MV, et al. The degradation characteristics of microbial biomass in soil. Geochim Cosmochim Acta. 2011;75:2571–81.
24. Zhong J, Sleighter RL, Salmon E, McKee GA, Hatcher PG. Combining advanced NMR techniques with ultrahigh resolution mass spectrometry: a new strategy for molecular scale characterization of macromolecular components of soil and sedimentary organic matter. Org Geochem. 2011;42:903–16.
25. Ferrari E, Francioso O, Nardi S, Saladini M, Dal Ferro N, Morari F. DRIFT and HR MAS NMR characterization of humic substances from a soil treated with different organic and mineral fertilizers. J Mol Struct. 2011;998:216–24.
26. Courtier-Murias D, Simpson AJ, Marzadori C, Baldoni G, Ciavatta C, Fernández JM, et al. Unraveling the long-term stabilization mechanisms of organic materials in soils by physical fractionation and NMR spectroscopy. Agric Ecosyst Environ. 2013;171:9–18.
27. Plaza C, Courtier-Murias D, Fernández JM, Polo A, Simpson AJ. Physical, chemical, and biochemical mechanisms of soil organic matter stabilization under conservation tillage systems: a central role for microbes and microbial by-products in C sequestration. Soil Biol Biochem. 2013;57:124–34.
28. Dal Ferro N, Berti A, Francioso O, Ferrari E, Matthews GP, Morari F. Investigating the effects of wettability and pore size distribution on aggregate stability: the role of soil organic matter and the humic fraction. Eur J Soil Sci. 2012;63:152–64.
29. Pizzeghello D, Cocco S, Francioso O, Ferrari E, Cardinali A, Nardi S, et al. Snow vole (Chionomys nivalis Martins) affects the redistribution of soil organic matter and hormone-like activity in the alpine ecosystem: ecological implications. Ecol Evol. 2015;5(20):4542–54.
30. Combourieu B, Inacio J, Delort AM, Forano C. Differentiation of mobile and immobile pesticides on anionic clays by 1H HR MAS NMR spectroscopy. Chem Commun. 2001;21:2214–5.
31. Shirzadi A, Simpson MJ, Kumar R, Baer AR, Xu Y, Simpson AJ. Molecular interactions of pesticides at the soil-water interface. Environ Sci Technol. 2008;42:5514–20.
32. Masoom H, Courtier-Murias D, Soong R, Maas WE, Fey M, Kumar R, et al. From spill to sequestration: the molecular journey of contamination via comprehensive multiphase NMR. Environ Sci Technol. 2015;49:13983–91.
33. Beni C, Marconi S, Boccia P, Ciampa A, Diana G, Aromolo R, et al. Use of arsenic contaminated irrigation water for lettuce cropping: effects on soil, groundwater, and vegetal. Biol Trace Elem Res. 2011;143:518–29.
34. Pereira SI, Figueiredo PI, Barros AS, Dias MC, Santos C, Duarte IF, Gil AM. Changes in the metabolome of lettuce leaves due to exposure to mancozeb pesticide. Food Chem. 2014;154:291–8.
35. Blondel C, Khelalfa F, Reynaud S, Fauvelle F, Raveton M. Effect of organochlorine pesticides exposure on the maize root metabolome assessed using high-resolution magic-angle spinning 1H NMR spectroscopy. Environ Pollut. 2016;214:539–48.
36. Pacifico D, Casciani L, Ritota M, Mandolino G, Onofri C, Moschella A, et al. NMR-based metabolomics for organic farming traceability of early potatoes. J Agric Food Chem. 2013;61:11201–11.
37. Rosati A, Cafiero C, Paoletti A, Alfei B, Caporali S, Casciani L, et al. Effect of agronomical practices on carpology, fruit and oil composition, and oil sensory properties, in olive (*Olea europaea* L.). Food Chem. 2014;159:236–43.
38. Deshmukh AP, Simpson AJ, Hatcher PG. Evidence for cross-linking in tomato cutin using HR-MAS NMR spectroscopy. Phytochemistry. 2003;64:1163–70.
39. Deshmukh AP, Simpson AJ, Hadad CM, Hatcher PG. Insights into the structure of cutin and cutan from Agave americana leaf cuticle using HRMAS NMR spectroscopy. Org Geochem. 2005;36:1072–85.
40. Turner JW, Hartman BE, Hatcher PG. Structural characterization of suberan isolated from river birch (*Betula nigra*) bark. Org Geochem. 2013;57:41–53.
41. Alcantara GB, Honda NK, Ferreira MMC, Ferreira AG. Chemometric analysis applied in 1H HR-MAS NMR and FT-IR data for chemotaxonomic distinction of intact lichen samples. Anal Chim Acta. 2007;595:3–8.
42. Bharti SK, Bhatia A, Tewari SK, Sidhu OP, Roy R. Application of HR-MAS NMR spectroscopy for studying chemotype variations of *Withania somnifera* (L.) Dunal. Magn Reson Chem. 2011;49:659–67.
43. Manolopoulou E, Spinella A, Caponetti E, Dais P, Spyros A. High resolution-magic angle spinning NMR study of olive leaves. In: van Duynhoven J, Belton PS, Webb GA, van As H, editors. Magnetic resonance in food science: food for thought. London: Royal Society of Chemistry; 2013. p. 164–70.
44. Delgado-Goni T, Campo S, Martın-Sitjar J, Cabanas ME, San Segundo B, Arus C. Assessment of a 1H high-resolution magic angle spinning NMR spectroscopy procedure for free sugars quantification in intact plant tissue. Planta. 2013;238:397–413.
45. Marino G, Righi V, Simoni A, Schenetti L, Mucci A, Tugnoli V, et al. Effect of a peat humic acid on morphogenesis in leaf explants of *Pyrus communis* and *Cydonia oblonga*. Metabolomic analysis at an early stage of regeneration. J Agric Food Chem. 2013;61:4979–87.
46. Righi V, Parenti F, Tugnoli V, Schenetti L, Mucci A. Crocus sativus petals: waste or valuable resource? The answer of high-resolution and High-resolution magic angle spinning nuclear magnetic resonance. J Agric Food Chem. 2015;63:8439–44.
47. Mori T, Tsuboi Y, Ishida N, Nishikubo N, Demura T, Kikuchi J. Multidimensional high-resolution magic angle spinning and solution-state NMR characterization of 13C-labeled plant metabolites and lignocellulose. Sci Rep. 2015;5(11848):1–12.
48. Wiklund S, Karlsson M, Antti H, Johnels D, Sjöström M, Wingsle G, et al. A new metabonomic strategy for analysing the growth process of the poplar tree. Plant Biotechnol J. 2005;3:353–62.
49. Choze R, Alcantara GB, Filho EGA, Silva LMA, Faria JC, Liao LM. Distinction between a transgenic and a conventional common bean genotype by 1H HR-MAS NMR. Food Chem. 2013;141:2841–7.
50. Oliveira CS, Carlos EF, Vieira LGE, Liãoc LM, Alcantara GB. HR-MAS NMR metabolomics of 'Swingle' citrumelo rootstock genetically modified to overproduce proline. Magn Reson Chem. 2014;52:422–9.
51. Wheeler HL, Soong R, Courtier-Murias D, Botana A, Fortier-Mcgill B, Maas WE, et al. Comprehensive multiphase NMR: a promising technology to study plants in their native state. Magn Reson Chem. 2015;53:735–44.
52. Gil AM, Duarte IF, Delgadillo I, Colquhoun IJ, Casuscelli F, Humpfer E, Spraul M. Study of the compositional changes of mango during ripening by use of nuclear magnetic resonance spectroscopy. J Agric Food Chem. 2000;48(5):1524–36.
53. Otero L, Préstamo G. Effects of pressure processing on strawberry studied by nuclear magnetic resonance. Innov Food Sci Emerg Technol. 2009;10:434–40.
54. Ciampa A, Renzi G, Taglienti A, Sequi P, Valentini M. Studies on coffee roasting process by means of nuclear magnetic resonance spectroscopy. J Food Quality. 2010;33:199–211.
55. Vermathen M, Marzorati M, Baumgartner D, Good C, Vermathen P. Investigation of different apple cultivars by high resolution magic angle spinning NMR. A feasibility study. J Agric Food Chem. 2011;59:12784–93.
56. Alves Filho EG, Silva LMA, Choze R, Lião LM, Honda NK, Alcantara GB. Discrimination of sugarcane according to cultivar by 1H NMR and chemometric analyses. J Braz Chem Soc. 2012;23(2):273–9.
57. Mucci A, Parenti F, Righi V, Schenetti L. Citron and lemon under the lens of HR-MAS NMR spectroscopy. Food Chem. 2013;141:3167–76.
58. Marseglia A, Acquotti D, Consonni R, Cagliani LR, Palla G, Caligiani C. HR MAS 1H NMR and chemometrics as useful tool to assess the geographical origin of cocoa beans—comparison with HR 1H NMR. Food Res Int. 2016;85:273–81.
59. Pérez EMS, Iglesias MJ, Ortiz FL, Pérez I, Galera MM. Study of the suitability of HRMAS NMR for metabolic profiling of tomatoes: application to tissue differentiation and fruit ripening. Food Chem. 2010;122:877–87.
60. Pérez EMS, López JG, Iglesias MJ, Ortiz FL, Toresano F, Camacho F. HRMAS-nuclear magnetic resonance spectroscopy characterization of tomato "flavor varieties" from Almería (Spain). Food Res Int. 2011;44:3212–21.
61. Ritota M, Comarini F, Sequi P, Valentini M. Metabolomic characterization of Italian sweet pepper (*Capsicum annum* L.) by means of HRMAS-NMR spectroscopy and multivariate analysis. J Agric Food Chem. 2010;58(17):9675–84.
62. Ritota M, Casciani L, Han BZ, Cozzolino S, Leita L, Sequi P, et al. Traceability of Italian garlic (*Allium sativum* L.) by means of HRMAS-NMR spectroscopy and multivariate data analysis. Food Chem. 2012;135:684–93.
63. Lam L, Soong R, Sutrisno A, de Visser R, Simpson MJ, Wheeler HL, et al. Comprehensive multiphase NMR Spectroscopy of Intact 13C Labeled Seeds. J Agric Food Chem. 2014;62:107–15.

64. Shintu L, Ziarelli F, Caldarelli S. Is high-resolution magic angle spinning NMR a practical speciation tool for cheese samples? Parmigiano Reggiano as a case study. Magn Reson Chem. 2004;42:396–401.
65. Shintu L, Caldarelli S. High-resolution MAS NMR and chemometrics: characterization of the ripening of Parmigiano Reggiano Cheese. J Agric Food Chem. 2005;53:4026–31.
66. Shintu L, Caldarelli S. Toward the determination of the geographical origin of emmental(er) Cheese via high resolution MAS NMR: a preliminary investigation. J Agric Food Chem. 2006;54:4148–54.
67. Mazzei P, Piccolo A. ^{1}H HRMAS-NMR metabolomic to assess quality and traceability of mozzarella cheese from Campania buffalo milk. Food Chem. 2012;132:1620–7.
68. Messia MC, Sobolev AP, Gómez-Caravaca AM, Lamanna R, D'Amico I, Caboni MF, Marconi E, et al. HR-MAS NMR metabolic profiling, furosine and (E)-10-Hydroxy-2-decenoic acid for qualitative and geographical discrimination of royal jelly. J Apic Res. 2013;52(3):141–8.
69. Song EH, Kim HJ, Jeong J, Chung HJ, Kim HY, Bang E, et al. A ^{1}H HR-MAS NMR-based metabolomic study for metabolic characterization of rice grain from various *Oryza sativa* L. cultivars. J Agric Food Chem. 2016;64:3009–16.
70. Larsen FH, Blennow A, Engelsen SB. Starch granule hydration—a MAS NMR Investigation. Food Biophys. 2008;3:25–32.
71. Fumagalli E, Baldoni E, Abbruscato P, Piffanelli P, Genga A, Lamanna R, et al. NMR techniques coupled with multivariate statistical analysis: tools to analyse *Oryza sativa* metabolic content under stress conditions. J Agron Crop Sci. 2009;195:77–88.
72. Winning H, Viereck N, Wollenweber B, Larsen FH, Jacobsen S, Søndergaard I, et al. Exploring abiotic stress on asynchronous protein metabolism in single kernels of wheat studied by NMR spectroscopy and chemometrics. J Exp Bot. 2009;60(1):291–300.
73. Sidhu OP, Annarao S, Pathre U, Snehi SK, Raj SK, Roy R, et al. Metabolic and histopathological alterations of Jatropha mosaic begomovirus-infected *Jatropha curcas* L. by HR-MAS NMR spectroscopy and magnetic resonance imaging. Planta. 2010;232:85–93.
74. Silva LMA, Alves Filho EG, Choze R, Lião LM, Alcantara GB. 1H HRMAS NMR spectroscopy and chemometrics for evaluation of metabolic changes in *Citrus sinensis* caused by Xanthomonas axonopodis pv. Citri. J Braz Chem Soc. 2012;23(6):1054–61.
75. Lima MRM, Diaz SO, Lamego I, Grusak MA, Vasconcelos MW, Gil AM. Nuclear magnetic resonance metabolomics of iron deficiency in soybean leaves. J Proteome Res. 2014;13:3075–87.
76. Gogna N, Dorai K. HR-MAS NMR-based metabolomic approach to study the effect of fungicidal stress on wheat seed germination. Curr Sci. 2015;108(9):1694–701.
77. Iglesias MJ, García-López J, Collados-Luján JF, López-Ortiz F, Díaz M, Toresano F, et al. Differential response to environmental and nutritional factors of high-quality tomato varieties. Food Chem. 2015;176:278–87.
78. Fredlund E, Broberg A, Boysen ME, Kenne L, Schnürer J. Metabolite profiles of the biocontrol yeast *Pichia anomala* J121 grown under oxygen limitation. Appl Microbiol Biotechnol. 2004;64:403–9.
79. Hanoulle X, Wieruszeski JM, Rousselot-Pailley P, Landrieu I, Baulard AR, Lippens G. Monitoring of the ethionamide pro-drug activation in mycobacteria by 1H high resolution magic angle spinning NMR. Biochem Biophys Res Commun. 2005;331:452–8.
80. Bon D, Gilard V, Massou S, Pérès G, Malet-Martino M, Martino R, et al. In vivo 31P and 1H HR-MAS NMR spectroscopy analysis of the unstarved *Aporrectodea caliginosa* (Lumbricidae). Biol Fertil Soils. 2006;43:191–8.
81. Bunescu A, Garric J, Vollat B, Canet-Soulas E, Graveron-Demilly D, Fauvelle F. In vivo proton HRMAS NMR metabolic profile of the freshwater cladoceran Daphnia magna. Mol Biosyst. 2010;6:121–5.
82. Righi V, Apidianakis Y, Psychogios N, Rahme LG, Tompkins RG, Tzika A. In vivo high-resolution magic angle spinning proton NMR spectroscopy of *Drosophila melanogaster* flies as a model system to investigate mitochondrial dysfunction in *Drosophila* GST2 mutants. Int J Mol Med. 2014;34:327–33.
83. Liaghati Mobarhan Y, FortierMcGill B, Soong R, Maas WE, Fey M, Monette M, et al. Comprehensive multiphase NMR applied to a living organism. Chem Sci. 2016;7:4856–66.

20

Water-mediated NOE: a promising tool for interrogating interfacial clay–xenobiotic interactions

Ronald Soong, Adolfo Botana, Jasmine Wang, Hashim Farooq, Denis Courtier-Murias and Andre Simpson*

Abstract

Background: The sorption of anthropogenic compounds on clay minerals is a complex molecular process with important implications for the fate of agrochemicals and organic pollutants in the environment.

Results: The present study illustrates the use of a water-mediated NOE approach to study clay binding interactions. This method exploits the interfacial water layer on clay surfaces as a hydrogen reservoir for magnetization transfer. The interactions of four different xenobiotics with clay suspension were investigated through this method to demonstrate its capability to screen for the clay–xenobiotic molecular affinity. Further, based onthe NOE build-up rates, epitope map of clay–xenobiotic interactions can be generated, explaining theorientation and mechanism of the interactions.

Conclusions: The water-mediated NOE approach has the potential to reveal key insights into the role that interfacial water plays in the binding process, providing a better understanding of the partitioning of anthropogenic compounds from bulk water into aqueous clay suspensions.

Background

The enormous number of different xenobiotic substances in our environment is becoming a significant health concern [1–3]. These xenobiotic substances, including hormones, antibiotics and numerous pesticides, are commonly released to the environment through human activities [2–4]. As these substances accumulate in our environment, the question of their bioavailability becomes increasingly important. For instance, the occurrence of hormones, particularly estrogens, has recently gained attention due to their toxicological significance and their influence on biological activities in aquatic organisms [2, 4]. Although the acute health effects of these xenobiotics are well documented, the effect of chronic exposure remains an open question [1–3, 5, 6]. The intrinsic interactions between these xenobiotics and soils hold the key to understanding both their environmental accumulation and their bioavailability [7–11]. Therefore, a better understanding as to how xenobiotics are sequestered into complex rock, soil, and sediments is required [12–14].

Soil is represented by a complex mixture consisting of mineral and organic constituents that are in solid, gaseous, and aqueous states [15]. Amongst the inorganic constituents in soil, clays constitute the major component (often around 30% w/w) and are important in the preservation of labile organic compounds. Clays are inorganic materials composed prevalently of layers of SiO_4 and/or AlO_4 [16]. Because of their anionic nature, the surfaces of clays are covered with small cations. For instance, a host of cations, such as Na^+, Ca^{2+}, Mg^{2+}, K^+, and NH_4^+ are capable of binding to clay surfaces, giving them different colloidal property [16]. In addition to sequestering cations from solution, clays are capable of adsorbing water molecules on their surfaces through hydrogen bonding, allowing them to expand and swell several times their dry mass [17, 18]. Due to this excellent colloidal property, clay is often been employed in a range of applications, from mining to agriculture [19].

So far, most soil–organic interaction studies mainly concerned the association of xenobiotic with humic substances [13, 20, 21]. These studies reveal that the

*Correspondence: andre.simpson@utoronto.ca
Environmental NMR Center, University of Toronto Scarborough, 1265 Military Trails, Toronto, ON M1C 1A4, Canada

xenobiotics are sequestered into soils through weak interactions driven synergistically by both hydrophobic and hydrophilic interactions [21]. Despite the inorganic nature of clay, it is able to interact with a number of organic molecules such as fatty acids and aromatic compounds [22, 23]. In addition, recent studies have demonstrated that soluble organic matters (SOM) from terrestrial source are able to bind to clay particles in solution [12, 13]. Although the binding between clay and organic matter has already been demonstrated, the exact mechanism of interactions at the interface remains an open question due to the lack of suitable high-resolution structure determination techniques. Since these organic compounds exhibit a range of binding affinity to clay particles, traditional high-resolution diffraction techniques, such as SANS and powder XRD, are ill-suited to tackle such problems [19, 24]. In fact, this requires a technique that can interrogate molecular interactions over a wide range of timescales.

Nuclear magnetic resonance (NMR) spectroscopy has evolved into an indispensable technique capable of providing insights into different intermolecular interactions over a range of time scales (ms to ns) as shown in Fig. 1 [14, 20, 25–28]. For a complete review of NMR techniques and applications to understand interactions in environmental samples, readers should refer to an excellent review by Mazzei and Piccolo [28]. In environmental research, NMR is becoming an important tool used to highlight specific interactions between different constituents in soils [21, 29–32]. Over the years, numerous detailed NMR studies have provided valuable information on the underlying factors regarding the different binding properties of a number of anthropogenic contaminants [31–35]. Recently, a novel NMR technology was introduced, termed as comprehensive multiphase (CMP) NMR [38]. This approach combines hardware from solution, gel, and solid-state NMR into a single probe and permits environmental samples to be investigated in situ, in their fully hydrated state, and in all phases, allowing for all phases to be simultaneously investigated [28, 35–38]. A number of NMR techniques can be used to investigate interactions between organic soil constituents [31, 36, 39]. A brief summary illustrating the different NMR techniques and the length/time scales of interaction is shown in Fig. 1. Most of these techniques are based on the transfer of magnetization through physical interactions and result in a detectable NMR signal. The magnitude of this magnetization transfer is proportional to the strength of the interactions involved. Therefore, NMR becomes the tool of choice to unravel the complexity behind various intermolecular interactions. Amongst these experiments, magnetization transfer through nuclear overhauser effects (NOE) and saturation transfer difference (STD) are commonly used to determine intermolecular interactions [31, 39–41]. For most NMR experiments, magnetization transfer requires the ability to saturate an observable NMR signal. Commonly this involves interactions between one organic material (for example soil organic matter) and an organic ligand (for example a contaminant). However, in the case of minerals and clays, which do not contain an organic phase, this can be more challenging.

NMR has been successfully applied in investigating clay–organic exposure in recent studies [21, 42]. Most of these studies often involved solid-state NMR (SSNMR) techniques with compounds irreversibly bound on clay surfaces [21, 43, 44]. For instance, ^{29}Si NMR combined with cross polarization (^{1}H–^{29}Si) under magic angle spinning (CP-MAS) was used to investigate the binding of

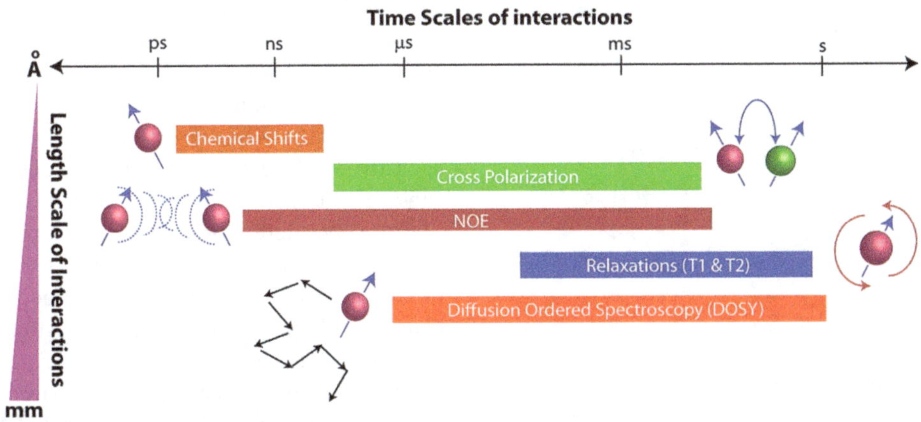

Fig. 1 An illustration of NMR observable parameters/experiments and their associated time/length scale dependence. According to this figure, NMR is able to probe a wide range of interactions over a range of time/length scales. Importantly, in environmental science, a range of interactions exists in soil such that a single NMR experiments will yield only a piece of the puzzle. Therefore, in order to fully understand the dynamics of the system, it is important to perform a range of different NMR experiments, allowing for a better view of the system involved

chemicals on clay surfaces [42–44]. Importantly, SSNMR techniques require that molecules to be bound tightly on the surface and the samples have to be dry. However, less information is available with regards to fully hydrated clay and dynamic interactions. To help bridge this gap in our understanding of clay–organic interactions, water-mediated nuclear overhauser effect (NOE) will be investigated in this study. Owing to the hydration layers on clay surfaces [18, 45], intermolecular NOE from the clay–bound water can be transferred to other molecules if they are in contact with each other, allowing their interactions at the atomistic level to be probed. In addition, ^1H-^1H NOE can be a sensitive technique used to probe weak intermolecular interactions [46]. A diagrammatic representation of a hydrated clay–organic system is shown in Fig. 2. In the present study, we demonstrate the avenue in which water-mediated NOE can be transferred to a clay–bound molecules via an NMR radio frequency (r.f.) sequence called Water Ligand Observed via Gradient Spectroscopy (waterLOGSY) [40, 47, 48]. In order to illustrate the usefulness of this sequence, a series of xenobiotic compounds will be screened for their affinity for clay particles, allowing us to evaluate its effectiveness in organo-clay research.

Methods

Materials

All chemicals, including pesticides, were purchased from Sigma-Aldrich. The mica-montmorillonite was purchased from the Clay society (www.clay.org). The montmorillonite was carefully treated according to the procedure outlined in the subsequent section before use.

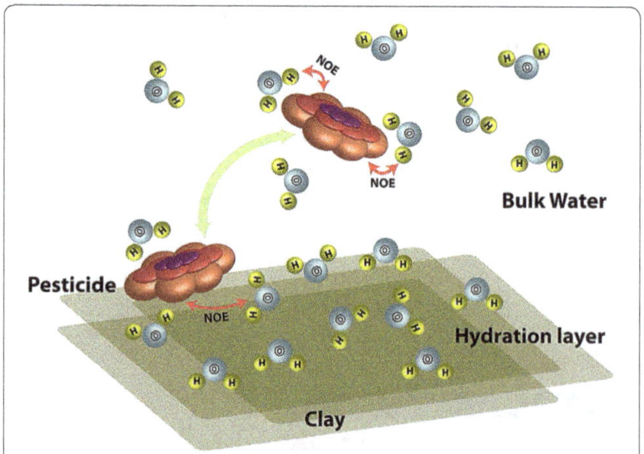

Fig. 2 An illustration of the interaction between pesticides and water on clay surface. In this case, we have two different types of water molecules with different dynamics. Due to the oxygen-rich surface, the clay surface is covered with water due to hydrogen bonding. The clay–bound water exhibits correlation time that is significantly lower compared to that of free water, leading to a different sign in their NOE. If the pesticide is in contact with the water molecules, magnetization from the water will be transferred to the pesticides via NOE

Preparation of colloidal clay particles

Synthetic mica-montmorillonite (<2 μm, NL Industries) was subject to sodium exchange and saturation in order to form a stable suspension. A centrifuge tube was filled to approximately 1.5 cm with clay, and to 8 cm with 1 M sodium chloride solution. The centrifuge tube was continuously inverted until the clay was thoroughly suspended. The mixture was centrifuged for 5 min at 2000 rpm. The supernatant was decanted, and the procedures stated above were repeated two additional times. After the last decantation, the centrifuge tube was filled to 8 cm with distilled water and the clay was re-suspended. This mixture was centrifuged for 5 min at 2000 rpm. The supernatant was decanted, and the distilled water wash was repeated two more times in order to obtain a supernatant which appeared turbid. The Na-saturated sediment was then re-suspended in distilled water and two drops of 5 M sodium chloride solution. The mixture was centrifuged for 6 min at 1000 rpm. The turbid supernatant was pipetted into a 500-mL beaker with caution to prevent disturbing the sediment. This process was repeated seven more times until the supernatant was clear. The bulk supernatant containing Na-saturated clay was freeze dried.

Preparation of NMR samples

5 mg of pesticide and 10 mg of colloidal clay were dissolved in 500 μL of H_2O/D_2O (90:10). The mixture was stirred on a vortexer for 30 min and was subsequently transferred to a 5-mm NMR tube. The pH of the solution was tested via litmus paper.

NMR spectroscopy

The NMR experiments were acquired on a Bruker Avance III NMR spectrometer operating at 11.7 T, observing ^1H at 499.98 MHz. The ^1H field strength used was 25 kHz, with an acquisition length of 4 k data points and 15 ppm spectral width. The ^1H waterLOGSY spectra were averaged over 128 scans with a recycle delay of 5 s. All NMR spectra were recorded at 297 K using a 4-channel 5-mm QXI inverse detection probe tuned to ^1H, ^{13}C, ^{15}N, and ^{19}F. Typical parameters used for waterLOGSY experiment were as follows: 4-ms 180° selective pulse (Φ_2) with Gaussian shape along with a 5-ms sinc-shaped 90° pulse (Φ_5) (water flipback) were used for selecting the water resonance; 1 ms squared gradient pairs were used at 40% of the maximum gradient strength (~54 G/cm) to select the water signal; and 2-ms square pulses (Φ_6 and Φ_8) along with gradients at 31 and 11% to dephase the water during water suppression. A gradient at 0.2% was applied throughout the mixing time to dephase water magnetization. A gradient recovery of 200 μs was used. The typical mixing times used in the waterLOGSY experiments were between 10 ms to 1.5 s. The mixing times used should be less than or equal to that of the longitudinal (T_1) of water.

Data processing

The NMR spectra were processed by applying an exponential multiplication of the FIDs by a factor of 5 Hz prior to Fourier transform and zero filled to 8 K. The waterLOGSY NOE build-up curve was fitted according to the following equation and assumed to be a two-spin model for which the NOE intensity is given by [49]

$$\eta = \frac{\sigma}{\rho}[1 - \exp(-\rho t)] \quad (1)$$

where σ is the cross-relaxation rate, ρ is the spin lattice relaxation rate, and t is the mixing time. The epitope maps based on NOE build-up (Fig. 9) were generated through measuring the changes in the cross-relaxation rate in either the presence or the absence of suspended clay particles for each resolvable ^1H resonances in the NMR spectrum. The epitope maps calculated from a single mixing time are simply derived from the relative change in signal intensity on addition of clay. In both cases, the position showing the strongest interaction (highest peak) is set to 100%, whereas the other hydrogens are progressively scaled accordingly.

Discussion

Numerical simulation of NOE based on clay–bound interactions

WaterLOGSY is a commonly used radio frequency (r.f.) NMR sequence for screening ligand–protein interactions [40, 41, 47, 48]. The pulse sequence is shown in Fig. 3. This sequence was conceived to exploit the differences in the dynamics between bound water on protein surfaces and free water in solution. When water molecules are bound on a protein, their correlation times will increase temporarily to match that of the protein, thus leading to a change in the sign of their NOE [40, 47, 48]. Similarly, for the case of clay–bound water molecules, the sign of their NOE will be opposite of that in free water due to the differences in their correlation times. Unlike proteins, the water molecules on clay surface are tightly bound such that a constant hydrogen reservoir can be assumed to be available for NOE transfer [45, 50]. Therefore, in this case, we can assume that for NOE transfer to occur, the residence time of the xenobiotics on clay surface will be comparable to that of clay–bound water. In this case, the relationship between the rate of NOE transfer between xenobiotics and clay–bound water can be described via Eq. (2) [40, 47, 48]

$$\sigma_{wc} = \frac{\gamma^4 \hbar \left(\frac{\mu_o}{4\pi}\right)}{10r^6}\left(\frac{\tau_r \tau_p}{\tau_r + \tau_p}\right)\left(\frac{6}{1 + \frac{4\omega_0^2(\tau_r\tau_p)^2}{(\tau_r+\tau_p)^2}} - 1\right), \quad (2)$$

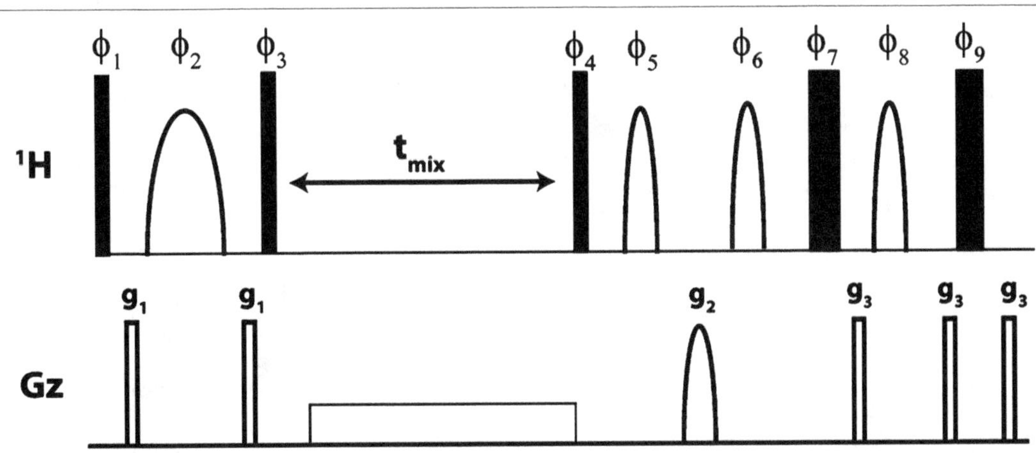

Fig. 3 A schematic illustrating the radio frequency (r.f.) pulse sequence used for a homonuclear one-dimensional WaterLOGSY experiment. A 180° Gaussian shape pulse (Φ_2), in combination with a gradient (g_1), is used to select for the water signal. During the mixing time (t_{mix}), the NOE from the water is transferred to a molecule only if they are in contact with each other. A small continuous gradient is applied during t_{mix} to reduce the effect of radiation damping. Lastly, both water flipback [5-ms sinc-shaped 90° pulse (Φ_5)] and excitation sculpting (Φ_6 and Φ_8) are used for water suppression before signal acquisition. (*Note* The pulse length for excitation sculpting was calibrated to achieve optimal water suppression) The phase cycling for each pulse is as followed: $\Phi_1 = 0$, $\Phi_2 = 0123$, $\Phi_3 = (0)_4\,(2)_4$, $\Phi_4 = 0$, $\Phi_5 = 02$, $\Phi_6 = (2)_8(3)_8(2)_8(3)_8(0)_8(1)_8(0)_8(1)_8$, $\Phi_7 = (0)_8(1)_8(0)_8(1)_8(2)_8(3)_8(2)_8(3)_8$, $\Phi_8 = (2)_4(3)_4(2)_4(3)_4(0)_4(1)_4(0)_4(1)_4$, $\Phi_9 = (0)_4(1)_4(0)_4(1)_4(2)_4(3)_4(2)_4(3)_4$, and $\Phi_{rec} = 0123012323012301$. The gradient amplitudes (%) were set to 40, 31, and 11 for g_1, g_2, and g_3, respectively. The phase cycle requires at least two transients

where γ is the gyromagnetic ratio of hydrogen nuclei, $\hbar$ is Planck's constant divided by 2π, μ_0 is the permeability of free space, r is the distance between the observed ^1H on the molecules and water, τ_r is the residence time of the molecules, τ_p is the rotation correlation times of the clay particles, and ω_0 is the Larmor frequency.

In order to understand the rate of NOE build-up in our system, a numerical simulation is performed. In this case, we assume the average clay particle to be 0.2 μm in diameter and are able to tumble freely in solution. Also, the water molecules are assumed to be tightly bound on the clay surface where they form a rich hydrogen reservoir, facilitating magnetization transfer via NOE. The result of the simulation was shown in Fig. 4. The simulation demonstrates the dependence of σ_{wc} on both r and τ_r. Our simulation illustrates that the effective distance on NOE build-up rate is limited to 2 Å. This is short compared to the typical 5 Å range for NOE, indicating that physical contact appears to be the primary mechanism for magnetization transfer. Interestingly, the sign of the NOE build-up is critically dependent on the residence time of the molecules on the clay surface. For short residence times, we observed a positive build-up rate which became negative as the residence time increased. The crossover point is ~2 ns residence time. Therefore, this demonstrates that a molecule's affinity for clay can be revealed via the sign of its NMR signals. From our simulation, we can conclude the following: (1) the magnetization transfer from clay–bound water can only occur through contact, (2) the sign of NOE build-up of the molecules that are in contact with clay–bound water will be opposite compared to those in contact with free water, (3) only if the residence time of the molecules on clay surface is greater than 2 ns, it is possible to observe a change in the sign of their NMR signals.

^{1}H waterLOGSY NMR spectra of four different xenobiotics in the absence and presence of clay particles

In order to demonstrate the applicability of waterLOGSY, three different pesticides were chosen, namely Diflufenzopyr, Imazapyr, and Nicotine. While Nicotine is most famous for being a key component of tobacco, it is now employed as a pesticide [51]. In addition, DMSO is employed as a control/reference as it is known to form hydrogen bonds with water and break/compete with hydrogen bonds occurring on clay surfaces [52]. Figure 5 shows the structural assignments for the xenobiotics used in this study. The observed ^1H chemical shifts values, and multiplet patterns are consistent with the literature values [12]. The xenobiotics used in this study contain different functional moieties that span most of

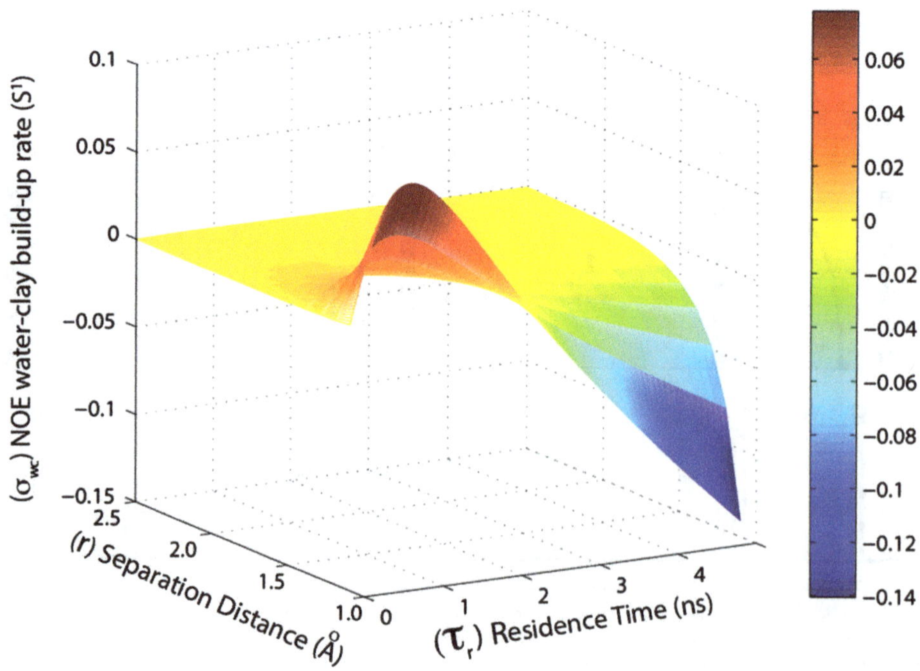

Fig. 4 A numerical simulation of the NOE water–clay build-up rate (σ_{wc}) as a function of both separation distance (r) and residence time (τ_r). This simulation illustrates that the sign of the NOE build-up rate depends on how long the molecules stay on the clay surface. In addition, NOE build-up occurs only within a separation distance of ≈2 Å, which is far less than the typical 5 Å range for NOE. The crossover point in which the build-up rate becomes negative is ~2 ns

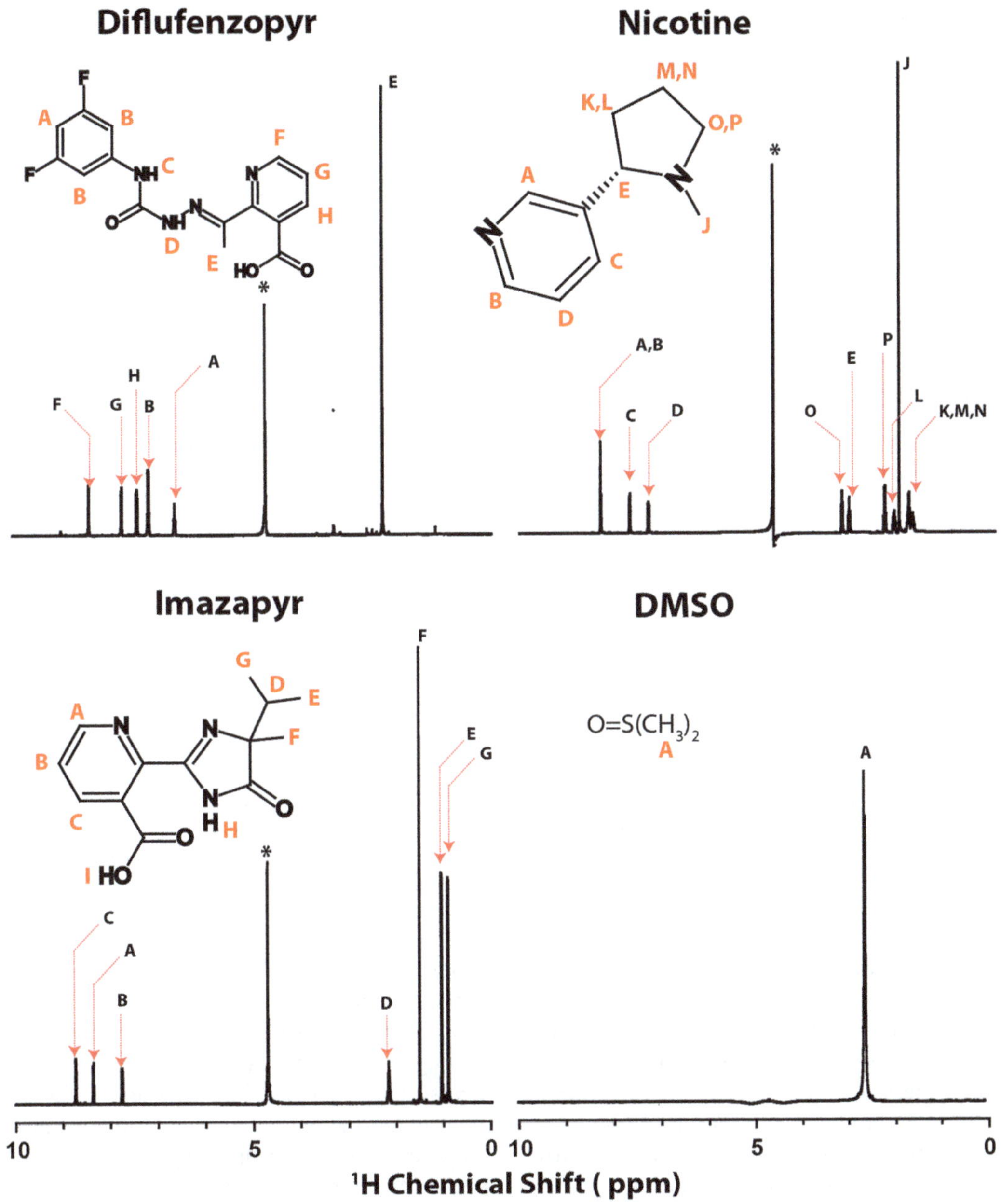

Fig. 5 ^{1}H NMR spectra of four different xenobiotics used in our current study. (*Note*: *Asterisk* denotes residual water)

the ^{1}H chemical shift range. These moieties can influence the manner in which molecules interact with clay surfaces. Importantly, the water-mediated NOE transfer may shed new insights into the parameters that dictate these interactions.

A series of waterLOGSY spectra was acquired for the four different xenobiotics, and it is shown in Fig. 6. These spectra were acquired in the presence and in the absence of clay in order to highlight changes in molecular interactions, which can be detected via waterLOGSY. The presence of clay particles indeed changes some of the ^{1}H spectral features of the xenobiotics in the current studies. For instance, a noticeable increase in ^{1}H resonance line width of Nicotine is observed in the presence of clays as shown in the loss of resolution in many of the multiplet patterns. The changes in the linewidth in some of the xenobiotics are an indication of interactions. Since linewidth can be a measure of mobility, this points to a situation in which the compound temporarily binds to the surface of the clay particles, resulting in an

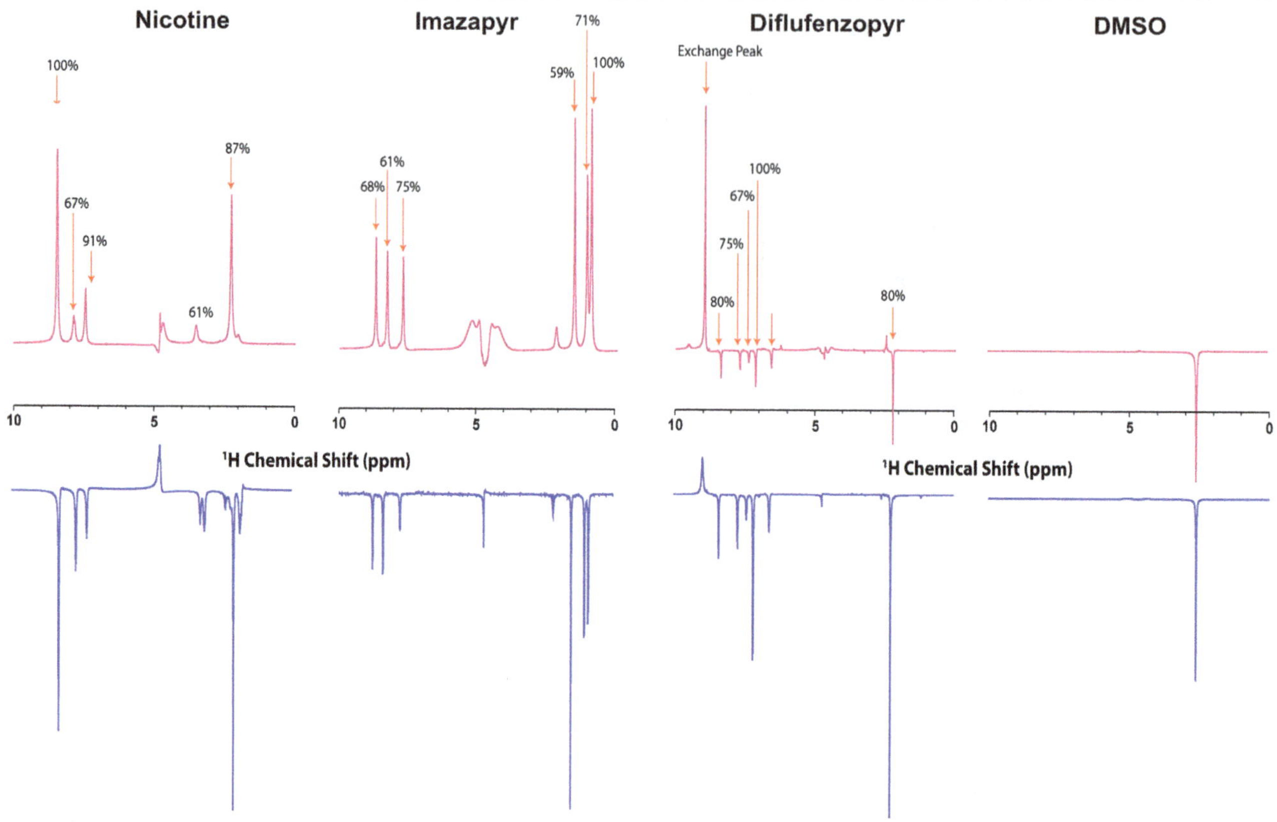

Fig. 6 ¹H NMR WaterLOGSY spectra of four different xenobiotics. The four different compounds used in this current study are Nicotine, Imazapyr, Diflufenzopyr, and dimethyl sulfoxide (DMSO). *Top* WaterLOGSY spectra in the presence of clay. *Bottom* WaterLOGSY spectra in the water. The mixing time used to acquire waterLOGSY spectra was 800 ms. The differences in the intensity between the spectra were attributed to the different NOE build-up rates

increase in their correlation times. Notwithstanding the changes in the linewidth, the resonances in these spectra remain resolved, except for peaks K, L, M, and N as well as O and E of Nicotine which overlapped with each other (Fig. 6). Based on these spectral features, the interactions between the molecules and clay particles are likely in the fast exchange regime where waterLOGSY would prove to be applicable. Another indication that the system is in a fast exchange region is the fact that we only observed a small shifts in all hydrogen resonances, indicating that the interactions were fast in the chemical shift timescale. (Note: if the system is undergoing a slow exchange, two resonances, bound and free, are observed.)

The changes in the sign of the NOE depending on clay's presence confirm that interactions occur for the case of Nicotine and Imazapyr. On the other hand, there is no change in the sign of the NOE for the case of Diflufenzopyr and DMSO; this indicates that their interactions with the clay particle were transient at best. Furthermore, the aromatic ^{1}H peaks of Diflufenzopyr remain inverted despite the absence of clay, while the amide proton (~9 ppm) has an opposite phase due to fast exchange. The different intensities for hydrogen in the NMR spectrum can be attributed to the clay interactions leading to differences in the NOE build-up rate and can be confirmed through experiments with varying mixing times. Epitope maps, which provide information on the binding orientation, can be generated from the data in Fig. 6 as well as from NOE build-up curves. Both these approaches are discussed later in the paper. In conclusion, the results of these experiments confirm our numerical simulations that indeed, clay–bound water-mediated NOE can be used to discriminate binding interactions on clay surfaces.

Strength of clay–bound interaction revealed through NOE build-up

The rate of NOE build-up can provide a qualitative measure on the strength of the interactions between clay particles and various xenobiotics. Since most of the clay particle sizes are in the range of microns, the key factor in determining the NOE build-up rate will be the residence time of a molecule on clay surfaces. In essence, the length of time a molecule spent on clay surfaces reflects the strength of their interactions. A series of NOE build-up curves was collected (Fig. 7), and the build-up rates were extracted for each of the resolvable resonances. The build-up rates for each compound were summarized as in a bar graph in Fig. 8. An overall increase in the build-up rates was observed in the presence of clay irrespective of their affinity for clay surfaces. Importantly, this increase in the build-up rate can be attributed to their weak transient molecular interactions with clay particles. This is consistent with our simulation results that even at short residence time, an increase in the NOE build-up rate will be observed. However, only when the residence time exceeds 2 ns does a change in NOE build-up rate along with a change in the sign occur. For molecules showing an affinity for clay particles such as Nicotine and Imazapyr, we observed a significant increase in the NOE build-up rate. On the other hand, for DMSO and Diflufenzopyr where their interactions with clay surfaces are weak, we observed only a slight change in their respective NOE build-up rate. Interestingly, DMSO indiscriminately formed hydrogen bonds with water and most likely interacted with bulk water molecules. Thus, these experiments provide a qualitative explanation into which the differences of ^{1}H resonance intensity were observed between the presence and absence of suspended clay particles.

Water NOE-driven epitope map of clay–xenobiotic interactions

An epitope map of clay–xenobiotic interactions (Fig. 9) was generated via changes in the NOE build-up rate due to contacts with clay-bound water. The value for the

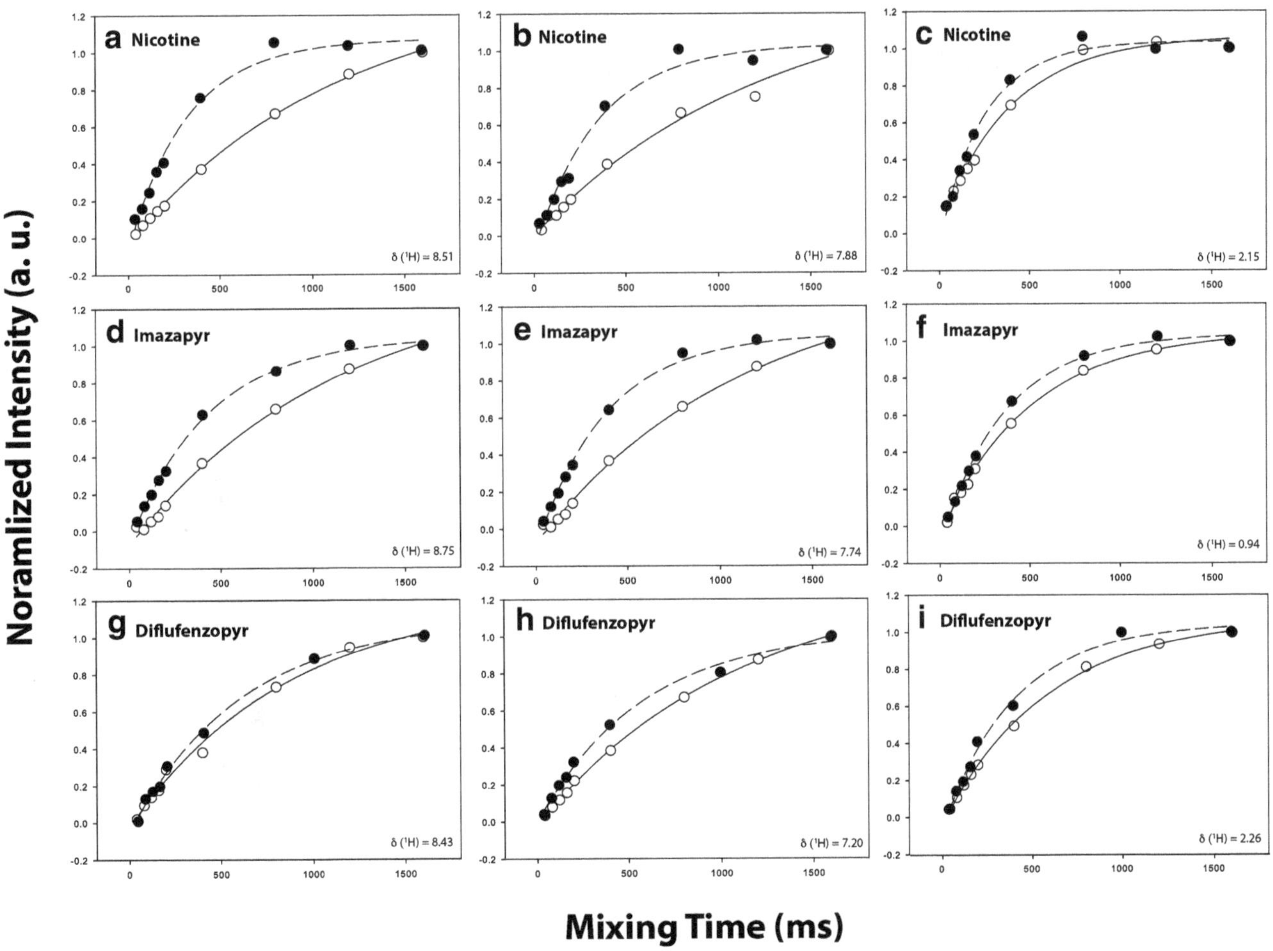

Fig. 7 A series of NOE buildup curves for the various hydrogen resonances in different pesticides. The *filled circle* denotes the NOE buildup in the presence of clay particles, and *open circle* denotes the NOE buildup in the absence of clay. The NOE buildup in the hydrogen of nicotine is shown in Graph (**a–c**). The NOE buildup in the hydrogen of imazapyr is shown in Graph (**d–f**). The NOE buildup in the hydrogen of Diflufenzopyr is shown in Graph (**g–i**). **a–c** Nicotine: hydrogen chemical shifts 8.51, 7.88, and 2.15. **d–f** Imazapyr: hydrogen chemical shifts 8.75, 7.74, and 0.94. **g–i** Diflufenzopyr: hydrogen chemical shifts 8.43, 7.20, and 2.26

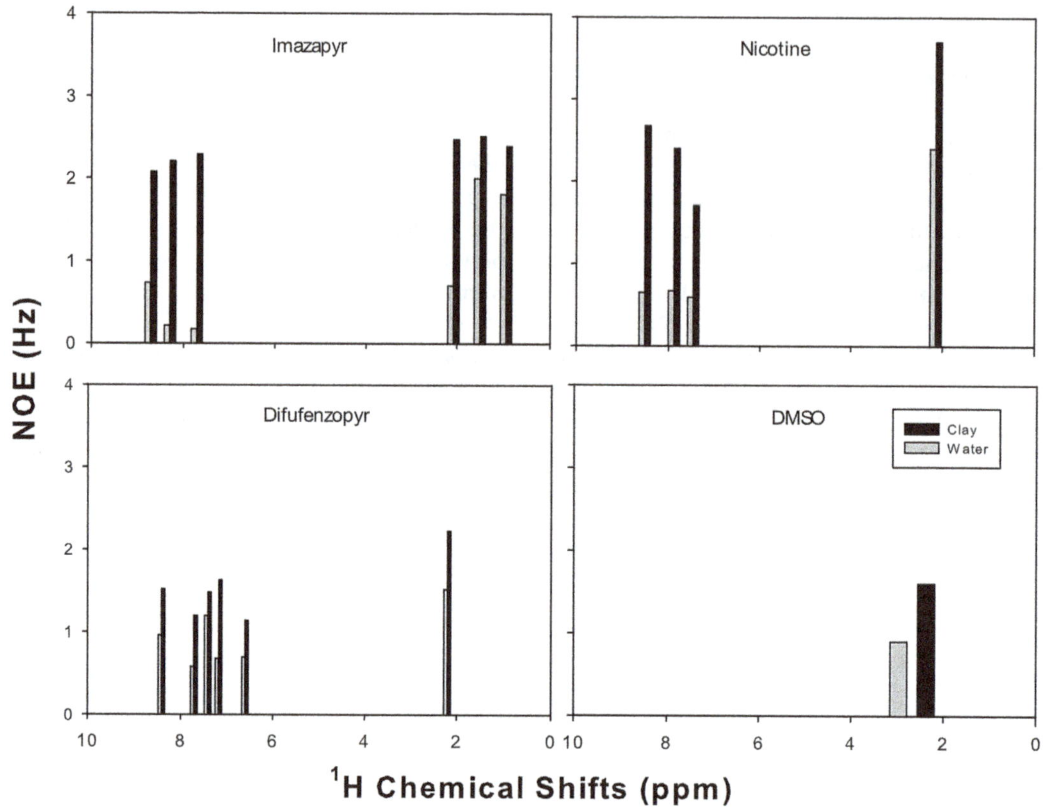

Fig. 8 Bar graphs illustrating the differences in NOE at different ^{1}H chemical shifts in the presence of clay suspension (*black*) and in water (*gray*). In all cases, a general increase in NOE build-up was observed in the presence of clay

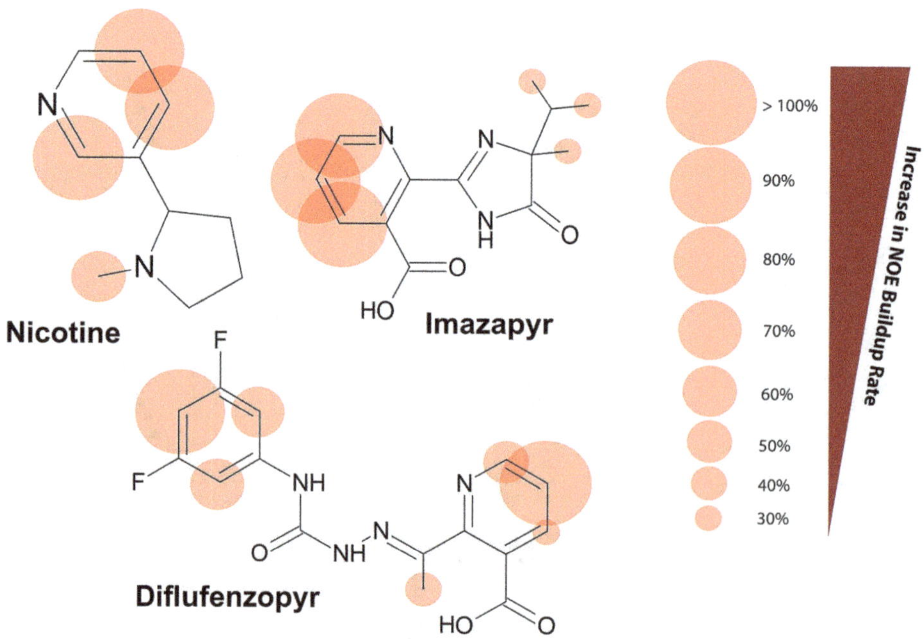

Fig. 9 An epitope map of changes in the water-mediated NOE build-up rate in the presence of suspended clay particles for each of the xenobiotics used in this study Nicotine, Imazapyr, and Diflufenzopyr. Most of the changes in the NOE build-up rate reside in the aromatic region of the molecules

hydrogen peak showing the largest interaction (i.e., largest relative change in presence of clay) is set to 100%, whereas the other hydrogens are progressively scaled accordingly. Such epitope mapping provides an overview of the average molecular orientation of the molecule with respect to the clay. For the case of Nicotine and Imazapyr, most significant changes in the NOE build-up rate were observed in the aromatic moiety. Interestingly, the changes in the NOE build-up rate for aliphatic moieties seem to be moderate as compared to the aromatic counterparts. This can be attributed to a hydrophobic effect, which led to only a moderate change in the water-mediated NOE build-up rate. This was consistent with the literature according to which polycyclic aromatic carbons may bind to clay surfaces [16, 22, 23, 53–55]. In addition, the pyridine moiety, which contains a lone pair of electrons on the nitrogen, is readily available for hydrogen bonding with the water and the hydroxyls groups present on clay surfaces. Therefore, the main contributor to the NOE build-up rate continues to be the close proximity of the hydrogens to the interfacial water.

Furthermore, nitrogen groups, aromatic rings, halogens, and carboxyl group may all represent important contact points with clay, whereas unsubstituted aliphatics appear to have less influence. In addition to calculating the epitope maps from the full NOE build-up curves, it is also possible to calculate maps from a single mixing time, as shown in Fig. 6. Using a single mixing time is advantageous in that only two experiments need to be measured, one in the presence and one in the absence of clay. In turn, this permits molecular binding to be screening in more challenging natural samples or at very low concentration where running multiple experiments across a range of mixing times (i.e. full NOE build-up curves) would be time prohibitive. Figure 10 shows the epitope maps calculated from the data in Fig. 6 collected at a single mixing time. These epitope maps are generally similar to those in Fig. 9 with the main exception of the methyl groups which tend to be over-emphasized in Fig. 10. In Fig. 9, the data are derived from NOE only, which is itself influenced by both exchange and dipolar interactions. Rotational exchange of methyl groups leads to dipolar fluctuations which in turn contribute to NOE. Conversely, the method of calculation used in Fig. 9 is based on cross-relaxation rate (σ) and therefore the interaction itself. In simple terms, a method based on

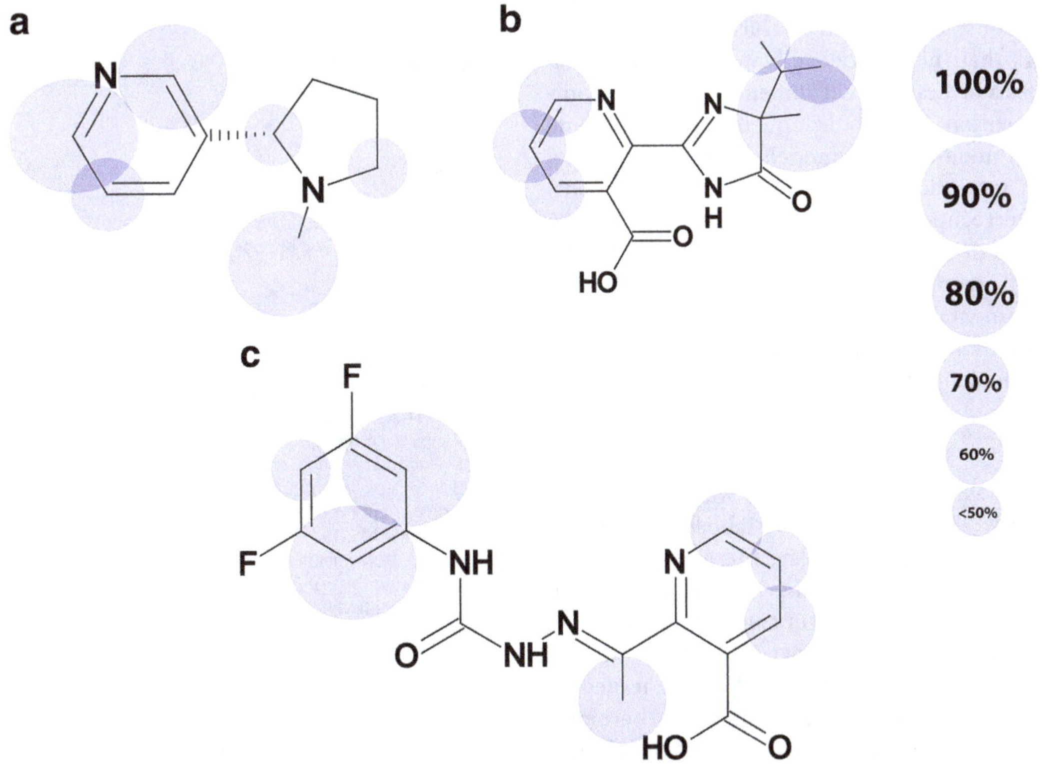

Fig. 10 An epitope map derived from the changes in the intensity of nicotine (**a**), imazapyr (**b**) and diflufenzopyr (**c**) via waterLOGSY in the presence and absence of clay. The diameter of the circles represents the relative strength of interaction at that position in the structure. The largest circle (i.e. strongest interaction) is normalized to 100%, and all other positions set relative to this (see scale on right). Most of the changes take place in the aromatic regions or in the vicinity of the basic moiety of the molecules. The interactions at the methyl positions are over-emphasized as discussed in the main text

cross relaxation derived from a full NOE build-up (i.e. Fig. 9) will give the most accurate mechanism of interactions. Conversely, epitope maps from a single mixing time (Fig. 10) will also provide the dominant mechanism signals, albeit groups with fast rotational exchange may be over-emphasized. However, the fact that the latter method requires only two experiments enables interaction-based studies at low concentration and permits the studies of relatively rapid changes in interactions during processes such as swelling, drying, and remediation.

In summary, NOE-based approaches outlined here should provide a key tool to study the dynamic interactions of organic molecules with mineral surfaces. Further studies using a wider array of structures, different clay, cations, and pH will be needed to fully elucidate the relative influence of functional groups and structural motifs on clay binding. Many of the NMR applications currently present in the literature mainly focus on the irreversible binding of anthropogenic compounds with clay particles. However, the water-mediated NOE approach introduced in this study should represent a further complimentary approach to enable interrogation of systems involving reversible binding in the fast exchange regime. In addition, it allows a better understanding of the role played by interfacial water in the thermodynamics of xenobiotic interactions. Importantly, it has been suggested over the years that this interfacial water, namely vicinal water, provides a favorable environment for organic compounds to partition from bulk aqueous environments [53]. Although none of these models have been verified through high-resolution spectroscopic means, a comprehensive interrogation into these models will provide a better understanding of how clay particles facilitate the partitioning of organics from bulk water environments, which is fundamental to the question of transport, bioavailability, and bioaccessiblity of anthropogenic contaminants and agrochemicals [53].

Conclusion

The water-mediated NOE approach was successfully applied to study the interactions established between several xenobiotics and a model clay. This approach takes advantage of the differences in the NOE build-up rate developing when a compound is in contact with the clay vicinal water, which leads to a change in the sign of NOE. The latter effect manifests itself only when the molecule resides on clay surfaces for longer than 2 ns. Therefore, this is a tool to screen compounds as a function of their affinity with clay. Importantly, the sorption of anthropogenic compounds on clay minerals is a complex multivariable problem with intricate interacting parameters that warrant an in-depth investigation. The water-mediated NOE approach can provide insights into the role that vicinal water played in the binding process whether it is kinetically or thermodynamically driven. Therefore, this further validates the incorporation of this approach in the current repertoire of NMR experiments used in organo-mineral research.

Abbreviations

NMR: nuclear magnetic resonance; NOE: nuclear overhauser effect; SOM: soluble organic matters; SANS: small-angle neutrons scattering; XRD: x-ray powder diffraction; CMP: comprehensive multiphase; STD: saturation transfer difference; SSNMR: solid-state NMR; CP-MAS: cross polarization magic angle spinning; WaterLOGSY: Water Ligand Observed via Gradient Spectroscopy; r.f.: radio frequency; DMSO: dimethyl sulfoxide.

Authors' contributions

RS carried out the experiments and drafted the manuscript. AB assisted in carrying out the experiments and JW prepared the clay particles. HF and DCM assisted in drafting the manuscript. AJS conceived this study and drafted the manuscript. All authors read and approved the final manuscript.

Acknowledgements

AJS thanks NSERC, (Strategic and Discovery Programs), the Canada Foundation for Innovation (CFI), the Ministry of Research and Innovation (MRI), and Krembil Foundation for providing funding. AJS also thanks the Government of Ontario for an Early Researcher Award. AJS would like to thank Dr. Rudraksha Dutta Majumdar for reading the manuscripts and his helpful suggestions.

Competing interests

The authors declare that they have no competing interests.

Funding

This study is supported by Grants from NSERC (Strategic and Discovery Programs), the Canadian Foundation for innovation (CFI), the Ministry of Research and Innovation (MRI), and Krembil Foundation.

References

1. Staples CA, Dorn PB, et al. A review of the environmental fate, effects, and exposures of bisphenol A. Chemosphere. 1998;36(10):2149–73.
2. Erbe MC, Ramsdorf WA, et al. Toxicity evaluation of water samples collected near a hospital waste landfill through bioassays of genotoxicity piscine micronucleus test and comet assay in fish astyanax and ecotoxicity *Vibrio fischeri* and *Daphnia magna*. Ecotoxicology. 2011;20(2):320–8.
3. Milla S, Depiereux S, et al. The effects of estrogenic and androgenic endocrine disruptors on the immune system of fish: a review. Ecotoxicology. 2011;20(2):305–19.
4. Kidd KA, Blanchfield PJ, et al. Collapse of a fish population after exposure to a synthetic estrogen. Proc Natl Acad Sci USA. 2007;104(21):8897–901.
5. Weinberger B, Vetrano AM, et al. Effects of maternal exposure to phthalates and bisphenol a during pregnancy on gestational age. J Matern Fetal Neonatal Med. 2014;27(4):323–7.
6. Zeman FA, Boudet C, et al. Exposure assessment of phthalates in French pregnant women: results of the ELFE pilot study. Int J Hyg Environ Health. 2013;216(3):271–9.
7. Yuk J, McKelvie JR, et al. Comparison of 1-d and 2-d nmr techniques for screening earthworm responses to sub-lethal endosulfan exposure. Environ Chem. 2010;7(6):524–36.
8. Yuk J, Simpson MJ, et al. 1-d and 2-d NMR metabolomics of earthworm responses to sub-lethal trifluralin and endosulfan exposure. Environ Chem. 2011;8(3):281–94.

9. Yuk J, Simpson MJ, et al. 1-d and 2-d NMR-based metabolomics of earthworms exposed to endosulfan and endosulfan sulfate in soil. Environ Pollut. 2013;175:35–44.
10. Lankadurai BP, Simpson AJ, et al. H-1 NMR metabolomics of *Eisenia fetida* responses after sub-lethal exposure to perfluorooctanoic acid and perfluorooctane sulfonate. Environ Chem. 2012;9(6):502–11.
11. Lankadurai BP, Wolfe DM, et al. H-1 NMR-based metabolomic observation of a two-phased toxic mode of action in *Eisenia fetida* after sub-lethal phenanthrene exposure. Environ Chem. 2011;8(2):105–14.
12. Shirzadi A, Simpson MJ, et al. Molecular interactions of pesticides at the soil–water interface. Environ Sci Technol. 2008;42(15):5514–20.
13. Shirzadi A, Simpson MJ, et al. Application of saturation transfer double difference nmr to elucidate the mechanistic interactions of pesticides with humic acid. Environ Sci Technol. 2008;42(4):1084–90.
14. Longstaffe JG, Courtier-Murias D, et al. The ph-dependence of organofluorine binding domain preference in dissolved humic acid. Chemosphere. 2013;90(2):270–5.
15. Simpson MJ, Simpson AJ. The chemical ecology of soil organic matter molecular constituents. J Chem Ecol. 2012;38(6):768–84.
16. Meleshyn A, Tunega D. Adsorption of phenanthrene on Na-montmorillonite: a model study. Geoderma. 2011;169:41–6.
17. De Stefanis A, Tomlinson AAG, et al. Nanostructures of the montmorillonite-derived restructured clays K10 (r), HMO and the Mg^{2+} exchanged analogue Mg-HMO. ASANS, N_2 sorption and XRPD study. J Mater Chem. 2003;13(5):1145–8.
18. Delville A. Modeling the clay water interface. Langmuir. 1991;7(3):547–55.
19. Takeda M, Matsunaga T, et al. Rheo-SANS studies on shear thickening in clay-poly(ethylene oxide) mixed solutions. Macromolecules. 2010;43(18):7793–9.
20. Kelleher BP, Simpson AJ. Humic substances in soils: are they really chemically distinct? Environ Sci Technol. 2006;40(15):4605–11.
21. Simpson AJ, Simpson MJ, et al. The application of ^1H high-resolution magic-angle spinning NMR for the study of clay–organic associations in natural and synthetic complexes. Langmuir. 2006;22(10):4498–503.
22. Schlautman MA, Morgan JJ. Sorption of perylene on a nonporous inorganic silica surface-effects of aqueous chemistry on sorption rates. Environ Sci Technol. 1994;28(12):2184–90.
23. Schwarzenbach RP, Westall J. Transport of non-polar organic-compounds from surface-water to groundwater—laboratory sorption studies. Environ Sci Technol. 1981;15(11):1360–7.
24. Malwitz MM, Butler PD, et al. Orientation and relaxation of polymer-clay solutions studied by rheology and small-angle neutron scattering. J Polym Sci B Polym Phys. 2004;42(17):3102–12.
25. Simpson AJ, Simpson MJ, et al. Nuclear magnetic resonance spectroscopy and its key role in environmental research. Environ Sci Technol. 2012;46(21):11488–96.
26. Farooq H, Courtier-Murias D, et al. HR-MAS NMR spectroscopy: a practical guide for natural samples. Curr Org Chem. 2013;17(24):3013–31.
27. Clemente JS, Gregorich EG, et al. Comparison of nuclear magnetic resonance methods for the analysis of organic matter composition from soil density and particle fractions. Environ Chem. 2012;9(1):97–107.
28. Mazzei P, Piccolo A. Interactions between natural organic matter and organic pollutants as revealed by NMR spectroscopy. Magn Reson Chem. 2015;53(9):667–78.
29. Tadini AM, Constantino IC, et al. Characterization of typical aquatic humic substances in areas of sugarcane cultivation in brazil using tetramethylammonium hydroxide thermochemolysis. Sci Total Environ. 2015;518–519:201–8.
30. Tadini AM, Pantano G, et al. Off-line TMAH-GC/MS and NMR characterization of humic substances extracted from river sediments of northwestern sao paulo under different soil uses. Sci Total Environ. 2015;506–507:234–40.
31. Shirzadi A, Simpson MJ, et al. Application of saturation transfer double difference NMR to elucidate the mechanistic interactions of pesticides with humic acid. Environ Sci Technol. 2008;42(4):1084–90.
32. Simpson AJ, Simpson MJ, et al. Nuclear magnetic resonance spectroscopy and its key role in environmental research. Environ Sci Technol. 2012;46(21):11488–96.
33. Nebbioso A, Piccolo A. Advances in humeomics: enhanced structural identification of humic molecules after size fractionation of a soil humic acid. Anal Chim Acta. 2012;720:77–90.
34. Nebbioso A, Piccolo A. Molecular characterization of dissolved organic matter (DOM): a critical review. Anal Bioanal Chem. 2013;405(1):109–24.
35. Mazzei P, Piccolo A. Quantitative evaluation of noncovalent interactions between glyphosate and dissolved humic substances by NMR spectroscopy. Environ Sci Technol. 2012;46(11):5939–46.
36. Masoom H, Courtier-Murias D, et al. From spill to sequestration: the molecular journey of contamination via comprehensive multiphase NMR. Environ Sci Technol. 2015;49(24):13983–91.
37. Wheeler HL, Soong R, et al. Comprehensive multiphase NMR: a promising technology to study plants in their native state. Magn Reson Chem. 2015;53(9):735–44.
38. Courtier-Murias D, Farooq H, et al. Comprehensive multiphase nmr spectroscopy: basic experimental approaches to differentiate phases in heterogeneous samples. J Magn Reson. 2012;217:61–76.
39. Longstaffe JG, Simpson MJ, et al. Identifying components in dissolved humic acid that bind organofluorine contaminants using $^1H\{^{19}F\}$ reverse heteronuclear saturation transfer difference NMR spectroscopy. Environ Sci Technol. 2010;44(14):5476–82.
40. Dalvit C, Fogliatto G, et al. WaterLOGSY as a method for primary NMR screening: practical aspects and range of applicability. J Biomol NMR. 2001;21(4):349–59.
41. Dalvit C, Pevarello P, et al. Identification of compounds with binding affinity to proteins via magnetization transfer from bulk water. J Biomol NMR. 2000;18(1):65–8.
42. Sanders RL, Mueller KT. Investigating clay mineral surface reactivity across scales using solid-state nuclear magnetic resonance. Geochim Cosmochim Acta. 2009;73(13):A1153-A.
43. Sanders RL, Washton NM, et al. Measurement of the reactive surface area of clay minerals using solid-state NMR studies of a probe molecule. J Phys Chem C. 2010;114(12):5491–8.
44. Sanders RL, Washton NM, et al. Atomic-level studies of the depletion in reactive sites during clay mineral dissolution. Geochim Cosmochim Acta. 2012;92:100–16.
45. Delville A, Grandjean J, et al. Order acquisition by clay platelets in a magnetic-field-NMR-study of the structure and microdynamics of the adsorbed water layer. J Phys Chem. 1991;95(3):1383–92.
46. Antanasijevic A, Ramirez B, et al. Comparison of the sensitivities of water-LOGSY and saturation transfer difference NMR experiments. J Biomol NMR. 2014;60(1):37–44.
47. Gossert AD, Henry C, et al. Time efficient detection of protein–ligand interactions with the polarization optimized PO-waterlogsy NMR experiment. J Biomol NMR. 2009;43(4):211–7.
48. Sun P, Jiang X, et al. Biomolecular ligands screening using radiation damping difference waterLOGSY spectroscopy. J Biomol NMR. 2013;56(3):285–90.
49. Balbach J, Forge V, et al. Detection of residue contacts in a protein folding intermediate. Proc Natl Acad Sci USA. 1997;94(14):7182–5.
50. Sposito G, Prost R. Structure of water adsorbed on smectites. Chem Rev. 1982;82(6):553–73.
51. Booker CJ, Bedmutha R, et al. Experimental investigations into the insecticidal, fungicidal, and bactericidal properties of pyrolysis bio-oil from tobacco leaves using a fluidized bed pilot plant. Ind Eng Chem Res. 2010;49(20):10074–9.
52. Olejnik S, Aylmore AG, et al. Infrared spectra of kaolin mineral-dimethyl sulfoxide complexes. J Phys Chem. 1968;72(1):241–9.
53. Zhu DQ, Herbert BE, et al. Cation-pi bonding: a new perspective on the sorption of polycyclic aromatic hydrocarbons to mineral surfaces. J Environ Qual. 2004;33(4):1322–30.
54. Solc R, Gerzabek MH, et al. Wettability of kaolinite (001) surfaces—molecular dynamic study. Geoderma. 2011;169:47–54.
55. Hur J, Schlautman MA. Influence of humic substance adsorptive fractionation on pyrene partitioning to dissolved and mineral-associated humic substances. Environ Sci Technol. 2004;38(22):5871–7.

21

Extraction and determination of total polyphenols and antioxidant capacity of red coffee (*Coffea arabica* L.) pulp of wet processing plants

Melkayo Geremu[1], Yetenayet Bekele Tola[1*] and Abrar Sualeh[2]

Abstract

Background: Coffee waste disposal from wet processing units is increasing with an increase in coffee consumption throughout the world. This not only has a limitation in terms of environmental pollution, but also high waste volume is an opportunity to extract essential bioactive compounds and antioxidants for food and pharmaceutical industries. The aim of this paper was to investigate the effects of extraction solvents and coffee cherry pulps of different varieties in terms of their yield potential for total polyphenols and antioxidant capacity.

Results: Results show that coffee cherry pulps from different varieties extracted by different solvents showed significant effects. Among the studied varieties, coffee cherry pulp from Ababuna variety was found to be superior in total polyphenols (1809.9 mgGAE/gm) content and antioxidant capacity (70 % inhibition). This value was followed by pulp from the Dessu variety. But the pulp from variety 741, extracted using aqueous ethanol solvent, showed inferior results for both total polyphenol content and antioxidant capacity. Among the solvents studied in this work, aqueous methanol responded better for pulps of all varieties except Dessu which responded better for aqueous acetone solvent.

Conclusions: Red coffee cherry pulp discarded as a waste from wet coffee-processing plants can be used as a raw material to extract important value-added products like polyphenols and other antioxidants. These compounds can be used as biological additives in food and pharmaceutical industries to fight chronic diseases and aging. Furthermore, the study shows the possibility of extracting value-added bioactive compounds from biological wastes.

Background

Bioactive compounds are value-added products, justifying their isolation from industrial wastes [1]. These residues, in fact, could be an alternative source for obtaining natural antioxidants, which are considered completely safe in comparison with synthetic antioxidants [2, 3]. Coffee is one of the copious beverages, and the worldwide production is over 105 million tons annually [4]. Industrial processing of coffee cherry is done to isolate coffee beans by removing pulp and mucilaginous parts from cherries. To get beans for roasting, coffee is subjected to wet (washed) and dry (unwashed) processing. In the former case, red coffee cherry pulp is removed mechanically within a short period of harvesting time to produce better quality coffee beans. For every 2 tons of coffee cherries processed, nearly 1 ton pulp is generated [5]. Most of the coffee pulp generated during wet processing goes without any treatment directly to huge waste disposal sites or river streams. Toxic compounds from disposed fermenting cherries leach to the environment and pollute sources of water in the coffee-producing regions. Such ways of waste disposal have been the major environmental health challenges in coffee-growing and -processing

*Correspondence: yetenayet@gmail.com; yetenayet.bekele@ju.edu.et
[1] Jimma University College of Agriculture and Veterinary Medicine, P.O. BOX 307, Jimma, Ethiopia

areas. Moreover, such unsafe ways of waste disposal have greatly affected the terrestrial and aquatic biota [6].

So far, different attempts have been made to use the pulp from wet processing plants. However, a study showed that coffee pulp cannot be used as an animal feed, mainly due to its toxic components such as caffeine, mineral salts, amino acids, tannins, phenols and other polyphenols [7]. On the contrary, it is used to make compost to enrich soil fertility or as a mulch material to prevent evaporation of moisture and to control weed growth. New aspects concerning the use of this waste as a raw material to extract important bioactive components is the other aspect to minimize environmental pollution through the value addition process. Because of their high value nature and less cost of extraction methods, extraction and use of bioactive compounds from waste pulp as a food additive are gaining interest [8, 9]. Coffee pulp as a by-product contains associated bioactive compounds like polyphenols with good antioxidant properties, which endow additional health benefits. Studies [10, 11] also indicated the possibility of extraction of total polyphenol compounds and antioxidants from coffee pulp to manage coffee waste. However, the effect of varieties and extraction solvents on the total polyphenols and antioxidant capacity has not yet been studied. Therefore, the study aimed to investigate the effect of variety and extraction solvents on total polyphenols and antioxidants capacity of coffee cherry pulp of wet processing stations.

Methods

Experimental materials and sample preparation

Coffee pulp samples from Ababuna (hybrid of 741× Dessu), pure line varieties of 741, Dessu and 74110 were obtained at full maturity and ripening stages from Jimma Agricultural Research Center, Ethiopia. Ababuna was included purposefully for two reasons. First, due to its high yielding capacity, coffee cherry pulp from this variety will be high in the future during massive production. The second reason was to investigate the combined effect of the two lines on the total polyphenol and antioxidant capacity of cherry pulp from the hybrid. Before the pulping process, coffee cherries were sorted to remove over-ripe, deteriorated and bruised ones. Pulping was done on the same date of harvesting using a manually operated coffee pulper. The pulp was drained of excess water, blotted on tissue paper and stored at −18 °C for further investigation.

Extraction of the sample

The coffee cherry pulp sample was extracted according to Saewan et al. [12] and Jagtap and Bapat, [13]. For extraction purpose, three solvents (aqueous 80 % methanol, 80 % acetone and 80 % ethanol) were used. Ten g coffee pulp was mixed with 100 ml of the respective solvent and ground for 3 min in a homogenizer and transferred to conical flask. The ground samples were extracted using the maceration technique by soaking the samples in the solvents for 24 h at room temperature, followed by filtration using Whatman No. 1 filter paper. The filtered extract was used to determine the total polyphenol content and antioxidant capacity.

Determination of total polyphenol content

Total polyphenol content (TPC) of coffee pulp extract was determined using the Folin–Ciocalteu method [14]. Folin–Ciocalteu reagent, a mixture of phosphotungstic ($H_3PW_{12}O_{40}$) and phosphomolybdic ($H_3PMo_{12}O_{40}$) acids, is reduced to blue oxides of tungsten (W_8O_{23}) and molybdenum (Mo_8O_{23}) during phenol oxidation. This reaction occurs under alkaline condition provided by sodium carbonate. The intensity of blue color reflects the quantity of polyphenol compounds, which can be measured using a spectrophotometer [15]. Gallic acid was used as a standard and the total polyphenols were expressed as mg/g gallic acid equivalents (GAE) from the calibration curve ($R^2 = 0.996$) using gallic acid. Gallic acid (0.5 g) was accurately weighed into a 10 ml volumetric flask, dissolved in 10 ml absolute methanol and the solution was made up to 100 ml with 80 % of the same solvent. For standard curve 0, 1, 2, 3, 4 and 5 ml of standard solutions were added into a 100 ml flask and diluted to give 0, 50, 100, 150, 400 and 500 mg/l of gallic acid. Then, 0.5 ml of each sample was introduced into test tubes and mixed with 2.5 ml of a tenfold dilute Folin–Ciocalteu reagent and 2 ml of 7.5 % sodium carbonate. The tubes were covered with aluminum foil and allowed to stand for 30 min at room temperature before the absorbance was read at 765 nm using UV/Vis spectrophotometer (T80, China). The samples were prepared in triplicate for each analysis and the mean value of absorbance was obtained.

Determination of total antioxidant capacity

Free radical scavenging assay or DPPH (2,2-diphenyl-1-picrylhydrazyl) assay was done as per the method of Hemalatha et al. [16] with minor modification. One ml solvent extract of the samples was taken in a test tube and 2 ml of 0.1 mM DPPH was added. The mixture was shaken well and incubated at room temperature in the dark for 30 min. The decrease in absorbance of the resulting solution was then measured spectrophotometrically at 517 nm (UV–Vis spectrophotometer, T80, China). All the experiments were performed in triplicate and the mean ± SE values were reported. The scavenging activity was calculated from the control sample using the following equation:

$$\text{Radical scavenging activity (\%)} = \left(\frac{A_c - A_t/A_s}{A_c}\right) \times 100,$$

where A_c = the absorbance of the control, A_t = the absorbance of the test solution, A_s = the absorbance of the standard solution, and the IC_{50} value was also calculated from the graph of the percentage DPPH free radical scavenging activity versus concentrations of the samples.

Experimental design and data analysis

The experiment was laid out in two-factors factorial arrangement in complete randomized design in triplicate. Factors were coffee cherry pulp of four varieties and three types of extraction solvents as indicated above. ANOVA of the collected data were analyzed using Minitab computer software program (Version 16.0, Inc. Pennsylvania, USA). Means separation was carried out using Tukey's test for means showing significant difference at $p < 0.05$.

Results and discussion

Total polyphenol content of coffee cherry pulp

Polyphenols are secondary metabolites that are produced by plants to protect themselves against plant diseases and insects, but they play important roles in human health in protecting against a number of diseases related to oxidative stress and free radical-induced damage [17]. In this study, considerable variation ($p \leq 0.001$) was observed among the type of solvents used and pulps tested. The amount of total polyphenols recovered from different coffee pulp extracts varied from 1809.9 to 489.5 mg GAE/g. Coffee pulp from Ababuna exhibited the highest polyphenol (1809.9 GAE/g) content when its cherry pulp was extracted with 80 % methanol. However, the lowest (489.5 mg GAE/g) value was obtained from a cherry pulp of 741 extracted with 80 % ethanol. When pulp sources were compared, Ababuna responded better to all solvent types as compared to other varieties except for the considerable effect of acetone on Dessu (Table 1). Since Ababuna is a hybrid of 741 and Dessu, the heterosis effect resulted in creating not only high-yielding hybrid vigor variety, but also superior variety in terms of total polyphenol content. However, the two pure parent lines as well as genotype 74110 and 741 exhibited lower polyphenol values, except Dessu (Table 1). This result is also in agreement with Gorinstein et al. [18] who reported that the total phenol content and antioxidant potential were significantly higher in grapefruit's hybrid than parent lines. As it is known in plant pathology, plants mainly produce phenolic compounds as a means to defend themselves against diseases and insect pests [19]. Therefore, disease or stress resistance of the hybrid may be associated with such type of metabolic boost which has a great significance in improving the bioactive content of plants for the benefit of human health.

When extraction solvents are considered, they show a significant effect on the yield of total polyphenols extracted. Methanol in all cases except acetone for Dessu showed a significant effect in extracting better total polyphenol yield. The total polyphenol content of the methanol extracts varied from 900.6 mg GAE/g (741) (wb) to 1809.9 mg GAE/g (Ababuna) (Table 1). This is in agreement with Siddiq et al. [20]; they reported that 80 % methanol is known to be an efficient and widely used solvent system to extract natural antioxidative components, especially phenolics, from plant materials. This might be because the methanol–water mixture has high polarity and thus has greater efficacy toward extraction of polar phytochemicals such as phenolics and flavonoids. Furthermore, methanol is the most suitable solvent in the extraction of polyphenol compounds due to its ability to inhibit the reaction of polyphenol oxidase that causes the oxidation of phenolics and its ease of evaporation compared to water [21]. Turkmen et al. [22] also reported that solvents with different polarity had significant effect on polyphenol content and antioxidant activity. Similar to Ababuna, 74110 and 741 showed better extraction of total polyphenols when treated with the same solvent. However, Dessu responded better when it was treated with 80 % acetone. This shows that methanol might not serve as a good extraction solvent for all varieties, and hence varietal preference for suitable type of solvent should be checked for better extraction efficiency. But from the current study, considering the concentration of total polyphenols extracted, the extraction efficiency of methanol > acetone > ethanol.

For the purpose of this study, polyphenol yield potential (in percent) of pulps and extraction potential of solvents were compared after taking Ababuna and methanol as a standard pulp variety and solvent type, respectively. As indicated in Table 1 (column 3), 741 showed the lowest polyphenol yield potential, even though it has a significant boosting effect when combined with Dessu. Variety 74110 exhibited almost the same result with 741, but Dessu performed better than these two varieties. Therefore, a hybridization program of coffee not only resulted in better hybrid vigor varieties for quantitative traits like yield, but also offers an opportunity to extract bioactive components from wastes to use in the food and pharmaceutical industries.

Antioxidant activity of coffee cherry pulp

An antioxidant is a molecule stable enough to donate an electron to a rampaging free radical and neutralize it, thus reducing its capacity for cell damage. These antioxidants delay or inhibit cellular damage mainly through their free radical scavenging property [23]. The DPPH scavenging assay was done for all four coffee cherry pulps and three solvent extracts, taking ascorbic acid as

Table 1 Interaction effect of variety and solvents on polyphenol contents of coffee cherry pulp and percent yield

Types of variety as source of pulp	Types of extraction solvent	Total polyphenol content (mg GAE/g)* ± SE	Total polyphenols (%) of varieties as compared to Ababuna taking methanol as a standard solvent
Ababuna	Acetone	1434.3 ± 3.49^{d}	79.2**
	Ethanol	1603.8 ± 3.04^{c}	88.6
	Methanol	1809.9 ± 1.33^{a}	100.0
Average		1616	89.3
Dessu	Acetone	1709.0 ± 1.02^{b}	94.4
	Ethanol	1448.4 ± 1.01^{d}	80.0
	Methanol	929.1 ± 2.64^{f}	51.3
Average		1362.2	75.3
74110	Acetone	662.1 ± 0.65^{i}	36.6
	Ethanol	664.8 ± 0.99^{i}	36.7
	Methanol	993.5 ± 0.68^{e}	54.9
Average		773.5	42.7
741	Acetone	707.4 ± 1.32^{h}	39.1
	Ethanol	489.5 ± 3.98^{j}	27.0
	Methanol	900.6 ± 3.68^{g}	49.8
Average		699.2	38.6
CV (%)		8.8	

* Mean values followed by the same letter in the column are not significantly different at $p < 0.05$

** Values calculated taking 1809.9 of Ababuna and methanol as a base to compare with others

a standard antioxidant. The interaction effect of solvents and coffee pulps showed highly significant ($p < 0.001$) effects on DPPH scavenging capacity. The percentage of DPPH free radical scavenging activity of different types of coffee pulps as affected by the type of extracting solvent is shown in Fig. 1. In the methanolic extract, the scavenging activity ranged from 17.3 % (741) to 70.2 % (Ababuna). In the acetone extract, the scavenging activity ranged from 9.8 % (74110) to 50 % (Dessu). In the ethanolic extract, the scavenging activity ranged from 7.5 % (741) to 37.3 (Ababuna).

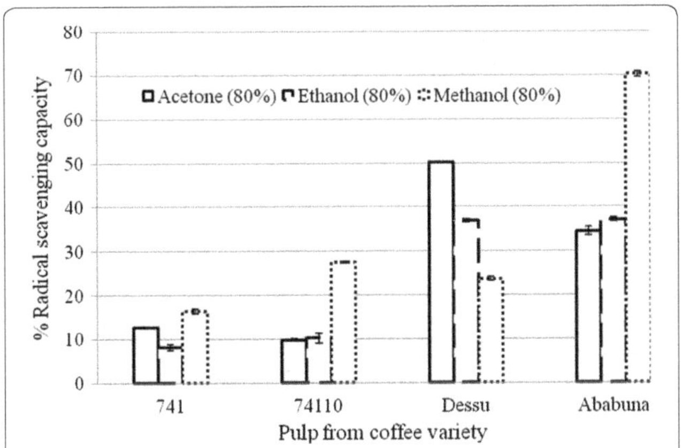

Fig. 1 DPPH radical scavenging activity of aqueous methanol, acetone and ethanol extracts from different varieties of coffee pulps

The methanol extract from the hybrid of Ababuna showed the highest antioxidant activity (70.2 %), which is in line with its high polyphenol content. This shows the potency of the extract from the hybrid for use as a good source of antioxidants. The lowest scavenging activity (7.5 and 10 %) was measured from 80 % ethanol extract of 741 and 80 % acetone extract of 74110 varieties. However, the scavenging activity of the standard (ascorbic acid) (93 %) was higher as compared to the different coffee pulp extracts. The results of this study are also in line with the reports of Murthy and Naidu [4], who reported the percentage free radical scavenging activity of coffee pulp to be 65 %. The discrepancy might be due to the difference in the type of coffee pulp and extraction solvent used, since they used aqueous iso-propanol (60 %) as the extraction solvent.

The extracts that perform the highest antioxidant activity (Fig. 1) have the highest concentration of polyphenols (Table 1). Polyphenols are very important plant constituents, because of their scavenging ability on free radicals due to their hydroxyl groups. This is in good agreement with Sultana and Anwar [24] who reported that, with the increase in concentration of polyphenol compounds, DPPH free radical scavenging capacity increases and thus antioxidant activity increases. Therefore, the polyphenol content of plants may contribute directly to their antioxidant action [25]. Another case extract from the Dessu variety showed higher percentage of radical scavenging

Table 2 Interaction effect of variety and solvents on the DPPH scavenging assay IC_{50} value (mg/ml)

Variety	Solvents	IC_{50} value (mg/ml)* ± SE
Ababuna	Acetone	4.4 ± 0.05^{c}
	Ethanol	2.4 ± 0.05^{b}
	Methanol	0.4 ± 0.01^{a}
Dessu	Acetone	1.1 ± 0.03^{ab}
	Ethanol	4.1 ± 0.17^{c}
	Methanol	8.4 ± 0.72^{e}
74110	Acetone	15.8 ± 0.10^{h}
	Ethanol	15.4 ± 0.16^{h}
	Methanol	6.2 ± 0.39^{d}
741	Acetone	13.6 ± 0.11^{g}
	Ethanol	22.6 ± 0.16^{i}
	Methanol	11.3 ± 0.11^{f}
CV (%)		5

* Mean values followed by the same letter in the column are not significantly different at $p < 0.05$

activity in acetone. This is also in accordance with other studies [26, 27] that reported that 80 % acetone was the most suitable solvent for the extraction of DPPH scavenging compounds from barley and legumes.

The inhibition concentration (IC_{50}) value for the DPPH scavenging assay was calculated for all the four solvent extracts of coffee pulps. IC_{50} value is defined as the concentration of antioxidant required for 50 % scavenging of DPPH radicals. It is a parameter widely used to measure antioxidant activity of biological and non-biological compounds. A smaller IC_{50} value corresponds to a higher antioxidant activity of the plant extract [28], since it requires less concentration to achieve the same value. In the present study, the lowest IC_{50} values of 0. 4 mg/ml was obtained when a hybrid of Ababuna was extracted with methanol, which indicated its powerful free radical scavenging ability followed by the acetone extract of the Dessu pulp (1.1 mg/ml) (Table 2). According to this study, to achieve 50 % equal inhibition of free radicals, the required concentration by 74110 and 741 is 15.5 and 28.3 times more as compared to the extract of Ababuna, respectively. The highest IC_{50} value (22.6 mg/ml) was obtained when 741 was extracted with ethanol, which indicated its least potent antioxidant activity. Due to higher values of total polyphenols and antioxidant activity of methanolic extracts except acetone in the case of Dessu, the lowest IC_{50} value was also estimated when methanol was used as a solvent. Azlim Almey and others [29] showed that the difference in IC_{50} values between methanolic and ethanolic extracts might be due to the intrinsic properties of plant compounds and their interaction with the solvents used. Even though a methanolic extract of Ababuna showed higher antioxidant activity, however, statistically, it was not different ($p > 0.05$) from the acetone extract of Dessu. In this study, the IC_{50} values of cherry pulp from hybrid and other coffee varieties has been determined for the first time, and hence no literature data are available to compare our results with information from other literature.

Conclusions

Coffee is one of the hot beverages consumed copiously every day throughout the world by millions of people. Even though the quality coffee is determined by the type of coffee variety, premium quality coffee comes from washed coffee of wet processing units than unwashed coffee from dry processing methods. However, coffee cherry pulping, fermentation and washing of parchment in the wet processing method discharges million tons of fresh coffee cherry pulp to the environment. But as a waste utilization strategy and value addition process, pulps discarded into the environment can be used to extract valuable ingredients for use as an additive in the food and pharmaceutical industries. The current study showed that total polyphenols, antioxidant capacity and free radical inhibition concentration varied among coffee pulps from different varieties and the type of solvent used for extraction. Waste pulp from Ababuna exhibited better total polyphenol content, antioxidant capacity and IC_{50} value when 80 % methanol was used as an extraction solvent. Among solvent types, the same solvent showed better extraction efficiency for all pulps except Dessu. Therefore, extraction of such type of valuable bioactive compounds from coffee pulp discarded as a waste could add value to improve the economic value of coffee pulp waste. Furthermore, coffee cherry pulp can be used as an alternative source to extract polyphenols and antioxidant compounds for the benefit of human health.

Authors' contributions

MG carried out most of the research works and wrote the paper, YBT was involved in developing the concept and part of the research activities as well as in editing the manuscript, AS was involved in facilitating research work and editing the manuscript. All authors read and approved the final manuscript.

Author details

[1] Jimma University College of Agriculture and Veterinary Medicine, PO. BOX 307, Jimma, Ethiopia. [2] Ethiopian Institute of Agricultural Research, Jimma Agricultural Research Center, POX 192, Jimma, Ethiopia.

Acknowledgements

The authors thank the Jimma Agricultural Research Center for supplying coffee pulps and the Jimma University College of Agriculture and Veterinary Medicine of Ethiopia and PHMIL project of Global Affairs of Canada for financial support to conduct this research.

Competing interests

The authors declare that they have no competing interests.

References

1. Moure A, Cruz JM, Franco D, Dominguez JM, Siniero JMJ, Dominguez H, et al. Natural antioxidants from residual sources. Food Chem. 2001;72:145–8.
2. Chen C, Pearson AM, Gray JI. Effects of synthetic antioxidants (BHA, BHT and PG) on the mutagenicity of IQ-like compounds. Food Chem. 1992;43(3):177–83.
3. Kahl R, Kappus H. Toxicology of the synthetic antioxidants BHA and BHT in comparison with the natural antioxidant vitamin E. Zeitschrift fur Lebensmittel-untersuchung und-forschung. 1993;196(4):329–38.
4. Murthy PS, Naidu MM. Recovery of polyphenols antioxidants and functional compounds from coffee industry by-products. Food Bioprocess Technol. 2012;5(3):897–903.
5. Roussos S, Aquiáhuatl MA, Trejo-Hernández MR, Perraud IG, Favela E, Ramakrishma M, Raimbault M, Viniegragonzález G. Biotechnological management of coffee pulp-isolation, screening, characterization, selection of caffeine-degrading fungi and natural microflora present in coffee pulp and husk. Appl Microbiol Biotech. 1995;42:756–62.
6. Navia PDP, Velasco MRDJ, Hoyos CJL. Production and evaluation of ethanol from coffee processing by-products. Vitae. 2011;18(3):287–94.
7. Braham JF, Bressani S. Coffee pulp: composition, technology and utilization. International Development Research Center, Ottawa, Canada. 1979: IDRC-108e.
8. Oreopoulou V, Tzia C. Utilization of plant by-products for the recovery of proteins, dietary fibers, antioxidants, and colorants. In: Oreopoulou V, Russ W, editors. Utilization of by-products and treatment of waste in the food industry. New York: Springer; 2007. p. 209–32.
9. Arvanitoyannis IS, Tserkezou P. Wheat, barley and oat waste: a comparative and critical presentation of methods and potential uses of treated waste. Int J Food Sci Technol. 2008;43(4):694–725.
10. Torres-Mancera MT, Cordova-López J, Rodríguez-Serrano G, Roussos S, Ramírez-Coronel MA, Favela-Torres E, Saucedo-Castañeda G. Enzymatic extraction of hydroxycinnamic acids from coffee pulp. Food Technol Biotechnol. 2011;49:369–73.
11. Pushpa S, Murthy MR, Manjunatha G, Sulochannama M. Extraction, characterization and bioactivity of coffee anthocyanins. Eur J Biol Sci. 2012;4(1):13–9.
12. Saewan N, Atisakul K, Mana T, Jimtaisong A, Kittigowittana K. Antioxidant capacity of different levels of ripeness and parts of *Coffea arabica* extracts. Pure and applied chemistry international conference. 2012.
13. Jagtap UB, Bapat VA. Antioxidant activities of various solvent extracts of custard apple (*Annona squamosa* L.) fruit pulp. Nutrafoods. 2012;11(4):137–44.
14. Singleton VL, Orthofor R, Raventos RML. Analysis of total phenols and other oxidation substrates and antioxidants by means of Folin–Ciocaltau reagent. Methods Enzymol. 1999;299:152–78.
15. Conforti F, Statti G, Uzunov D, Menichini F. Comparative chemical composition and antioxidant activities of wild and cultivated *Laurus nobilis* L. leaves and *Foeniculum vulgare* subsp. piperitum (Ucria) coutinho seeds. Biol Pharm Bull. 2006;29(10):2056–64.
16. Hemalatha S, Lalitha P, Arulpriya P. Antioxidant activities of the extracts of the aerial roots of *Pothos aurea* (Linden ex Andre). Der Pharma Chemica. 2010;2(6):84–9.
17. Vladimir-Knežević S, Blažeković B, Bival Štefan M, Babac M. Plant polyphenols as antioxidants influencing the human health. Phytochemicals as nutraceuticals–global approaches to their role in nutrition and health; 2012. p. 155–80.
18. Gorinstein S, Zachwieja Z, Katrich E, Pawelzik E, Haruenkit R, Trakhtenberg S, Martin-Belloso O. Comparison of the contents of the main antioxidant compounds and the antioxidant activity of white grapefruit and his new hybrid. LWT-Food Sci Technol. 2004;37(3):337–43.
19. Lattanzio V, Lattanzio VM, Cardinali A. Role of phenolics in the resistance mechanisms of plants against fungal pathogens and insects. Phytochem Adv Res. 2006;661:23–67.
20. Siddiq A, Anwar F, Manzoor M, Fatima A. Antioxidant activity of different solvent extracts of *Moringa oleifera* leaves under accelerated storage of sunflower oil. Asian J Plant Sci. 2005;4(6):630–5.
21. Yao LH, Jiang YM, Datta N, Singanusong R, Liu X, Duan J, Raymont K, Lisle A, Xu Y. HPLC analyses of flavanols and polyphenols acids in the fresh young shoots of tea (*Camellia sinensis*) grown in Australia. Food Chem. 2004;84(2):253–63.
22. Turkmen N, Sari F, Velioglu YS. Effects of extraction solvents on concentration and antioxidant activity of black and black mate tea polyphenols determined by ferrous tartrate and Folin–Ciocalteu methods. Food Chem. 2006;99(4):835–41.
23. Halliwell B, Murcia MA, Chirico S, Aruoma OI. Free radicals and antioxidants in food and in vivo: what they do and how they work. Crit Rev Food Sci Nutr. 1995;35(1–2):7–20.
24. Sultana B, Anwar F. Flavonols (kaempferol, quercetin, myricetin) contents of selected fruits, vegetables and medicinal plants. Food Chem. 2008;108(3):879–84.
25. Tosun M, Ercisli S, Sengul M, Ozer H, Polat T, Ozturk E. Antioxidant properties and total phenolic content of eight Salvia species from Turkey. Biol Res. 2009;42(2):175–81.
26. Zhao H, Dong J, Lu J, Chen J, Li Y, Shan L, Gu G. Effects of extraction solvent mixtures on antioxidant activity evaluation and their extraction capacity and selectivity for free phenolic compounds in barley (*Hordeum vulgare* L.). J Agric Food Chem. 2006;54(19):7277–86.
27. Xu BJ, Chang SKC. A comparative study on phenolic profiles and antioxidant activities of legumes as affected by extraction solvents. J Food Sci. 2007;72(2):S159–66.
28. Maisuthisakul P, Suttajit M, Pongsawatmanit R. Assessment of polyphenols content and free radical scavenging capacity of some Thai indigenous plants. Food Chem. 2007;100(4):1409–18.
29. Azlimalmey AA, Ahmedjalalkhan C, Syedzahir I, Mustaphasuleiman K, Aisyah MR, Kamarulrahim K. Total polyphenols content and primary antioxidant activity of methanolic and ethanolic extracts of aromatic plants' leaves. Int Food Res J. 2010;17(4):1077–83.

Use of potato as carbohydrate source in poultry ration

Fowzia Sultana[1], Hafiza Khatun[2] and Mohammad Amdad Ali[3*]

Abstract

Potato is an extensively cultivated annual crop in Bangladesh. In some years, when the production of potato is surplus, tons of potatoes are damped due to insufficient storage facilities. This study aims at the use of potato in poultry rations that could greatly benefit the potato farmers and poultry industries. Potato meal was incorporated in rations of broiler (Cobb-500) and fayoumi laying hens, respectively. The rations were isonitrogenous and isocaloric, with the only difference being the content of corn and potato. The results showed there were no differences in feed conversion ratio, survivability rate and dressing percentage among the treatment groups of broiler up to 30% replacement of corn with potato. Furthermore, no notable difference was observed in egg production and egg quality of laying hens on potato diets. These results suggest that potato meal could be included up to 30% as substitute of corn meal in both broiler and laying hen diets without having any negative effects. We believe that the results from this study would significantly contribute to the nation where corn production is not as great as that of potato.

Keywords: Potato meal, Poultry feed, Production and quality of meat and egg, Broiler and laying hens

Background

Poultry meat and egg are the cheapest sources of high-quality animal protein and good source of essential polyunsaturated fatty acids (PUFAs), especially rich in omega (n)-3 fatty acids [7]. Furthermore, meat contains lower level of fat compared to those of other animal protein. These factors contribute to the high demand for poultry meal and egg and motivate farmers to extend their present level of poultry farm and establish new one. However, cost of the poultry feed (70–80% of total production cost) increases day by day and significantly affects the poultry production [18]. For the formulation of poultry feed, corn (cereal) is the conventional choice of carbohydrate sources due to its high energy, low fibre, higher palatability, and the presence of pigments and essential fatty acids [21]. It possesses about 50–70% portion in total diet. Inclusion of corn meets up to 70% energy and 30% protein-dietary requirements [21]. However, its applications for use in food industries (such as beverage, juice, soup, and sauce) in respect of human consumption and in biofuel have been expanding rapidly. These competing demands for the corn result in inadequate supply, thus leading to a high price of poultry products [14, 18].

To lower the production cost of poultry products (egg, meat, etc.), an alternative energy source for the poultry feed with lower cost has been investigated. Ideally, the ingredient should be able to substitute corn totally or partially and not have any negative impact on the efficiency and the cost of poultry production [3, 20]. Many alternative ingredients, such as water yam [5], cassava root meal [23, 24], pearl millet [1], sweet potato [9, 12, 18], etc. have been investigated and reported. It was also demonstrated that these cereals are deficient in certain essential amino acids, especially lysine and tryptophan, which are very important for the hens to grow healthily. [18, 21]. Moreover, most of these available alternatives either suffer from processing complexity, such as mouldiness, and dustiness, or contain anti-nutrition factors, such as cyanogenic glycosides in cassava [23].

Our aim of this study is to find the usefulness of potato (*Solanum tuberosum* L.) as an alternative of corn. As far as nutritious value is concerned, it contains substantially greater amount of all essential amino acids, especially lysine [13]. Potato (fresh basis) contains approximately

*Correspondence: amdad53@yahoo.com
[3] Department of Agricultural Chemistry, Faculty of Agriculture, Bangladesh Agricultural University, Mymensingh 2202, Bangladesh
Full list of author information is available at the end of the article

20–25% dry matter with 12% crude protein (50% pure) [18]. It contains reasonably high amounts of most amino acids but is deficient in tryptophan and sulphur-containing amino acids [17]. It is to be noted that synthetic amino acids as supplement in dietary ration could be added to meet the feeding requirement. It is also an important source of both macro and micro minerals. Of the minerals, potassium, phosphorus, magnesium, sulphur, chlorine, iron, and zinc are found in larger quantities. Although calcium, phosphorus and iron are not present in significantly higher amounts, but even with limited contents, they can still provide important nutritional value to potato [4]. Potato also contains a variety of vitamin groups such as, A, B, C (27–36/100 mg) and I3, while cereal does not [4]. Recently, Mozafari et al. [18] reported that up to 25% maize can be replaced by potato meal in broiler ration, without inducing any negative affects on production performance. Except the above study, to the best of our knowledge, the potato meal as an alternative nutritious dietary source has not been investigated for the purpose of feeding laying hens for egg production. The present study therefore investigated the effects of potato meal on the quality and production of broiler meat as well as egg production of laying hens.

Moreover, potato is an extensively cultivated annual crop in Bangladesh showing higher productivity (16.68 t/ha versus that of maize at 4.97 t/ha) with shorter harvesting period (90 days for potato compared to 140 days for maize). As a result, the cost of potato is significantly lower (four times lower than corn) in Bangladesh [10, 11]. The production of potato is surplus in Bangladesh, and several tons of unused potato are dumped each year due to insufficient storage facilities. It would greatly benefit the potato farmers if potato could be used in poultry ration. It is worthy to note that 90% of corn is being imported into Bangladesh. This also leads to serious adverse effects on nation's gross domestic product (GDP). We believe that the results from this study would significantly contribute to the nation's GDP where corn production is not so great as that of potato.

Methods

The study was carried out at the Bangladesh Agricultural University, Mymensingh, Bangladesh during the period from July 2014 to December 2015. Good-quality, dust-free potatoes were collected from local market. They were then sliced, dried in sunlight and finally ground into meal by passing through a 1-mm screen using a WILEY MILL. Before determining the optimal formulation of ration, chemical analysis of potato meal (5 g) was performed according to methods of Association of Official Analytical Chemists (AOAC) [8] to measure nutrient values which are shown in Table 1. All ingredients were weighed and then thoroughly mixed by hand at various dietary levels of potato according to each experimental treatment. The rations were isonitrogenous and isocaloric conforming to the recommendations by the NRC [19]. There was no restriction on feed and water consumption in the both experiments. This study includes two independent experiments. In the first experiment, effects of potato meal were investigated on broiler (Cobb-500) rearing, which is assigned as "Experiment-1: Potato meal-fed broiler". Second experiment was performed to assess the egg production and egg quality of potato meal-fed laying hens, which is assigned as "Experiment-2: Potato meal-fed laying hens".

Experiment 1: potato meal-fed broiler

Birds and rations

Sixty-day-old white broiler (Cobb-500) chicks were divided into four groups. Each group consisted of five birds with three replications. On arrival, the chicks were wing-banded, weighed, and randomly placed into brooding pens. Good bedding materials, sufficient feeders and waterer's space per bird were provided. The brooding pens were equipped with two 250 W infrared fluorescent brooding lamps to maintain temperature in the brooder house. The starting temperature was 35 °C and then subsequently lowered at the rate of 5 °C per week until reaching finally the ambient temperature. Humidity level in the brooder house was ca. RH 75%. After 2 weeks, the birds were transferred to a grower house (in floor). Ration formulations for starter and grower are presented in Table 2. For the four dietary treatment groups, corn was substituted with 0 (control), 20, 25 and 30% of potato meal, respectively. The starter ration was fed for the first 2 weeks, which was then continued with the grower ration for further 5 weeks. Body weights and feed intakes were recorded weekly for the entire study period.

Slaughtering and processing

Feed was withdrawn on day 35, and the birds were slaughtered on day 36. Body weights of 24-h-fasted birds were recorded. The birds were placed in killing cones, and

Table 1 Approximate composition of potato meal (dry matter basis)

Components (%)	Potato meal
Energy (MJ/Kg)	3220
Water	79
Crude fibre	4.9
Crude protein	2.8
Ash	1.4
Ether extract	0.21

Table 2 Compositions of boiler rations

Ingredients (%)	Starter diet (% potato)				Grower diet (% potato)			
	0	20	25	30	0	20	25	30
Maize	51.16	36.00	33	28.3	61.45	45.00	43.00	38.50
Potato meal	0	20	25	30	0	20	25	30
Soybean meal	41.71	37.5	35.8	35.3	31.63	29	27.5	27
Soybean oil	3.38	3.8	3.6	3.8	3.1	3.2	2.8	2.8
DCP	1.63	1.4	1.4	1.4	1.72	1.00	0.50	0.50
Limestone	0.952	0.5	0.5	0.40	0.94	0.70	0.50	0.50
Vit. Min. pre	0.25	0.20	0.20	0.20	0.25	0.25	0.20	0.20
Salt	0.273	0.20	0.20	0.20	0.27	0.25	0.20	0.20
L-Lysine	0.08	0.08	0.08	0.08	0.13	0.13	0.07	0.07
DL-methionine	0.305	0.22	0.12	0.22	0.27	0.25	0.18	0.18
$NaHCO_3$	0.23	0.10	0.10	0.10	0.23	0.22	0.05	0.05
Total	100	100	100	100	100	100	100	100

DCP digestible crude protein, *Vit.Min. Pre* vitamin mineral premix

their throats were cut to facilitate bleeding out. The birds were dipped in a scald for 30 s in hot water (at 62.8 °C) to loosen the feathers. Then, the feathers were removed using a batch defeatherer, and skins were eviscerated. Chicken carcasses were thoroughly washed using water and weighed to obtain dressing percentages. Non-carcass components, such as lungs, liver, spleen, kidney, testicles, heart, and feet were separately weighed. Legs, thighs, and breasts were separated, vacuum-sealed, and stored in a refrigerator for further analyses.

Experiment 2: potato meal-fed laying hens

Birds and rations

Twenty-four 16-week-old healthy fayoumi (ready-to-lay) birds were divided into four groups. Each group consists of two birds with three replications, and all the birds were randomly assigned to an individual cage in an environmentally controlled room with a 16-h photoperiod. Four dietary treatments of potato meal replacing 0 (control), 20, 25 and 30% of the corn in the diets, respectively, were implemented during the experimental period of 14 weeks (30–44 weeks of age). The rations (shown in Table 3) were formulated according to the NRC [19] recommendation.

Egg production performance and quality

Egg production was recorded daily, and feed consumption was recorded weekly throughout the experiment. Eggs from each group were collected biweekly to measure egg weight, shell-breaking strength, shell thickness, shell ratio, albumen ratio, yolk ratio, yolk colour and Haugh units. Egg weight was measured using an electronic digital balance. Shell-breaking strength was measured using an eggshell strength instrument (accuracy: 0.1 kg/cm^2, Fujihira Industry Co., Ltd.). Individual egg was broken on a flat metal plate to measure the thickness of the albumen. Weights of albumen, egg yolk and eggshell were measured using an electronic digital balance. After removing the shell membrane, the shell thickness was measured using a dial gauge (Peacock, Tokyo, Japan) at three locations on the egg (air cell, equator and sharp end), and an average value of the three measurements was recorded. Yolk colour was measured using a Roche Yolk Colour Fan (Roche Ltd., Basel, Switzerland). The shell ratio, albumen ratio, yolk ratio and Haugh units were calculated according to the following formulas:

$$\text{Shell ratio} = \frac{\text{Shell weight} \times 100}{\text{Egg weight}}$$

$$\text{Albumen ratio} = \frac{\text{Albumen weight} \times 100}{\text{Egg weight}}$$

$$\text{Yolk ratio} = \frac{\text{Yolk weight} \times 100}{\text{Egg weight}}$$

$$\text{Haugh units} = 100 \log(H - 1.7W^{0.37} + 7.6),$$

where H and W are the observed height of albumen (mm) and the weight of each egg (g), respectively.

Statistical analysis

All the recorded and calculated parameters were compared among different dietary combinations in a factorial experiment in a completely randomized design (CRD) performing ANOVA, and LSD was used to compare the significant differences among different diets by analysing data using the computer package GENSTAT.

Table 3 Compositions of dietary rations for laying hens

Variable	Laying hens diet (% potato)			
	0	10	20	30
Maize	58.5	50.50	47	42.5
Potato meal	0.00	10.00	20	30
Rice polish	17	17.00	12	9
Soybean meal	17	15.00	14	12
Lime stone	7	7	6.5	6
Common salt	0.25	0.25	0.25	0.25
Premix	0.25	0.25	0.25	0.25
Total	100	100	100	100
	ME = 2845, CP = 15%	ME = 2818, CP = 14.7%	ME = 2808, CP = 14.62%	ME = 2804, CP = 14.24%

NB: NRC recommendation→ [19], Fayoumi Ration (Laying hens)

ME metabolizable energy, *CP* crude protein

Results

Experiment 1: potato meal-fed broiler

Effects of potato meal on production and growth performance of broiler are presented in Table 4. Significant ($P < 0.01$) difference was observed in body weight gain and feed intake among the treatment groups. Higher body weight of the birds was found in control group and with the increasing potato meal portion in the diets, the body weight of the birds decreased linearly, while food intake showed the opposite trend. This observation questions the nutrient utilization of potato by the broilers. Feed conversion ratio of the birds also linearly decreased with the potato meal in the diets. Importantly, no mortality was occurred in this study. Dietary treatments have no notable effects on dressing percentage and weights of other carcass components, such as breast, thigh, drumstick, wing etc., except abdominal fat ($P < 0.01$) as shown in Table 4. Significantly low abdominal fat (~4 times lower than that of control groups) was found in all the potato meal-fed birds.

Experiment 2: potato meal-fed laying hens

The effects of potato meal on the egg production of laying hens with the ages ranging from 30 to 44 week are presented in Table 5. Physical parameters, such as feed consumption, egg mass, feed conversion ratio, feed cost, egg production, egg weight and survivability, were found to be similar among the treatment groups. The statistical differences were recorded only regarding the feed cost. The feed cost was slightly lower when the portion of potato was increased in the feed. The quality of eggs was also assessed, and it was observed that eggshell thickness, shape index, shell weight, shell percentage, egg-breaking strength, albumen index, yolk index, Haugh unit and yolk colour score were all non-significant for all the treatment groups (Table 5).

Discussion

Experiment 1: potato meal-fed broiler

In our study, the control group gained the highest body weight, and it likely decreases slightly with the increasing potato meal, even though the feed intake increases with the potato meal ratio. It might be due to poor nutrient utilization of potato meal. Some reported results in the literature suggested that the inclusion of sweet potato meal in broiler ration causes poor utilization of nutrient compared with maize-based diets [16, 25], thus leading to a lower body weight gain of bird. This might be responsible for the inversely linear fashion body weight gain with the amount of sweet potato meal [3]. In contrast, some other studies, with feeding 25% potato to starter birds [18], 50% sweet potato to broiler starter [9], 36 and 45% sweet potato to starter and finisher birds, respectively [3, 26], reported that either potato or sweet potato in diets did not affect the body weight gain, food conversion ratio, or mortality percentages. Furthermore, we have found that dietary treatments have no notable effects on the dressing percentage and the weights of other carcass components, such as breast, thigh, drumstick, wing etc., except abdominal fat ($P < 0.01$). These results are in agreement with those in the reported literature [18] on dried-cooked potato meal diets up to 35%.

A point to be noted is that the abdominal fat percentage decreases with the increasing potato meal. Since the fats in other limbs in hens is related to the abdominal fat content, it can be said that the fat contents in other limbs of meat are lower as well, compared to those of the control group. We speculated that high fibre content plays a crucial role in this regard. The significance of this result is twofold. First, the lower weight of dressing is likely due to the low content of fat. Second, the reduced fat in meat is definitely of more worth in terms of fetching higher

Table 4 Effect of dietary potato meal on growth performance and meat yields of broiler

Parameter	Potato content (%)				Level of significance
	0	20	25	30	
Body weight (g/broiler)	1247.88^{a}	1033.33^{c}	1013.96^{d}	1106.25^{b}	**
Body weight gain (g/broiler)	923.47^{a}	719.54^{c}	692.85^{d}	787.58^{b}	**
Feed intake (g/broiler)	2050.46^{d}	2096.06^{c}	2172.56^{b}	2228.92^{a}	**
FCR	2.41^{b}	2.91^{a}	3.15^{a}	2.83^{a}	0.0039**
Survivability (%)	100	100	100	100	NS
Body weight (g/broiler) of slaughtered bird	1293.85^{a}	1057.60^{c}	1021.28^{c}	1142.33^{b}	**
Thigh weight	10.50^{a}	9.09^{b}	9.88^{ab}	9.44^{ab}	0.111
Drumstick weight	8.51	8.13	8.38	8.80	0.745
Wing weight	7.98	7.73	7.87	8.08	0.804
Breast weight	9.40	9.04	9.63	8.96	0.2050
Abdominal fat	0.84^{a}	0.18^{c}	0.30^{b}	0.18^{c}	**
Dressing	61.01^{ab}	56.09^{b}	62.69^{a}	59.16^{ab}	0.0690

^{a,b,c,d} Means within a row with different superscripts are significantly different ($P < 0.01$)

NS non-significant, *g* gram

** P < 0.01, significant at 1% level

Table 5 Egg production performance and egg quality of laying hen for a period of 14 weeks (30–44 weeks of age)

Parameter	Potato content (%)				Level of significance
	0	10	20	30	
Feed consumption (g/b/d)	105.14	105.42	104.95	105.07	0.58^{NS}
Egg mass production (g/h/d)	48.10	46.57	43.31	46.08	0.55^{NS}
Feed conversion ratio	2.19	2.27	2.45	2.28	0.54^{NS}
Feed cost (Tk/kg egg) produced	33.46	33.10	32.84	32.46	<0.0001**
Feed cost (hen/d)	3.52	3.49	3.45	3.41	<0.0001**
Egg production (%)	57.07	53.81	51.99	52.90	0.45^{NS}
Egg weight (g/egg)	40.14	39.29	40.06	39.02	0.81^{NS}
Survivability (%)	100	100	100	100	0.00^{NS}
Shell thickness (mm)	0.38	0.35	0.38	0.37	0.33^{NS}
Shape index	78.32	79.03	78.23	79.09	0.22^{NS}
Shell weight (g/egg)	4.18	4.29	3.96	4.22	0.41^{NS}
Shell (%)	10.35	10.67	10.26	11.22	0.12^{NS}
Egg breaking strength	1500.90	1504.77	1500.95	1506.01	0.11^{NS}
Albumen index	0.10	0.10	0.11	0.09	0.18^{NS}
Yolk index	0.46	0.48	0.45	0.46	0.50^{NS}
Haugh unit	89.73	90.67	88.16	88.70	0.28^{NS}
Yolk colour score	8.17	9.67	9.00	8.50	0.34^{NS}

NS non-significant, *g* gram, *b* bird, *d* day, *h* hen, *mm* millimetre

** P < 0.01, significant at 1% level

market prices since high animal fat causes elevated blood cholesterol leading to atherosclerosis or heart failure [6]. A detailed investigation is required to relate the fat content in meat to the fibre in diet. It is also suggested to find an optimal level of potato meal in broiler diets to maintain the fat content in poultry meat within acceptable limits.

Experiment 2: potato meal-fed laying hens

Inclusion of potato slightly reduced the feed cost, although egg production and quality were statistically insignificant and found to be similar among the treatment groups. A slightly higher egg production and better feed conversion ratio were observed in the control group, which might be due to the high feed intake of hens compared to the potato meal-fed hens. Again, this lower egg production may be due to the high fibre content in potato meal arising from the poor nutrient utilization by birds. Direct comparison of production performance and egg quality cannot be made since, to the best of our knowledge, there is no report in the literature investigating the effects of potato meal on the egg production of laying hens. However, the results are in agreement with some of the studies with sweet potato used as a substitute for maize. Ladokun and Tewe [15] reported that egg production reduced due to the high fibre content of sweet potato and the low protein utilizations by birds fed on sweet potato meal diet. They also suggested that sun-dried sweet potato meal at a substitution rate of 50% can safely replace maize in the diets of laying hens without any negative effect on egg production. In contrast, Sobamiwa [22] did not observe any significant difference in the cumulative egg production when sweet potato was incorporated up to 20% in the diets of laying hens. Afolayan et al. [2] also reported that 20% replacement of maize in the diets of laying hens with sweat potato meal led to no significant differences in the feed costs per kg of egg production and laying performances.

Conclusions

The results in this study suggested that there were no significant differences in meat production or meat quality among the treatment groups of broiler with up to 30% replacement of corn with potato. Furthermore, no significant difference was observed on either egg production or egg quality of laying hens fed on potato meal diets. These results suggested that potato meal could be safely included up to 30% as replacement of corn meal in both diets of broiler and laying-hens, without causing any negative effects on the production performance. However, we were able to reduce the production cost by up to 4%. These results would encourage the farmers of the country where price of potato is lower than that of corn. Importantly, we found the benefit of a low content of fat in potato meal-fed broilers. We believe that the results from this study will play a crucial role in the further development of low-fat poultry meat.

Authors' contributions

FS developed the experiment design and most of the laboratory analysis, interpretation of the results and manuscript writing. HK offered advice during the development of the experiment, interpretation of the results and manuscript writing. MAA provided general conception and interpretation of results and manuscript writing. All the authors read and approved the final manuscript.

Authors' information

FS is a professor at the Department of Poultry, Bangladesh Agricultural University, Bangladesh. She holds a Bachelor degree in Animal Husbandry, and M.Sc. and Ph.D. in poultry science. She spent some of her research time at the University of Hawaii, USA during her postdoctoral research. She has to her credit more than 15 years of research experience and 14 years of teaching experience at the graduate and post-graduate levels. She is specialized in poultry digestive and reproduction system. Her articles have been published in more than 10 international journal publications. She has been a guide to many M.Phil. and Ph.D. students. HK has received her Bachelor degree in Animal Husbandry and Master degree in Animal Science from the Bangladesh Agricultural University, Bangladesh. Subsequently she involved herself as an Assistant Professor at the same Department. She is currently on study leave for pursuing her M.Sc. & Ph.D. degree in Animal Science at the Saga University, Japan under Japanese government (Monbukagakusho) scholarship. Her specialization is on animal reproduction and biotechnology. MAA is an assistant professor at the Department of Agricultural Chemistry at the Bangladesh Agricultural University, Bangladesh. He is currently on leave and working as a Postdoctoral Research Associate at the University of Illinois at Urbana-Champaign,USA. He received his Ph.D. degree in Physical Chemistry from the University of Queensland, Australia, after his postgraduate Master degree in Advanced Materials Engineering from the Kookmin University, South Korea and his Bachelor degree in Chemistry from the University of Dhaka, Bangladesh. He was awarded with an International Postgraduate Research Scholarship and University of Queensland Centennial Scholarship for Ph.D. study and had obtained a fellowship provided by the Ministry of Science and Technology/Korea Science and Engineering Foundation to support his Master degree. His research interest involves the fundamental understanding of the physical properties of materials. He is credited with publications in more than twenty international journals that include Nature Communications (IF 11.329), Chemical Communications (IF 6.56), Carbon (IF 6.19), Sensors and Actuator B: Chemical (IF 4.75), ChemPhysChem (IF 3.13).

Author details

[1] Department of Poultry Science, Faculty of Animal Husbandry, Bangladesh Agricultural University, Mymensingh 2202, Bangladesh. [2] Department of Animal Science, Faculty of Animal Husbandry, Bangladesh Agricultural University, Mymensingh 2202, Bangladesh. [3] Department of Agricultural Chemistry, Faculty of Agriculture, Bangladesh Agricultural University, Mymensingh 2202, Bangladesh.

Acknowledgements

This study was supported by the Bangladesh University Grand Commission (UGC) and the Ministry of Science and Information & communication Technology, Government of Peoples` Republic Bangladesh.

Competing interests

The authors declare that they have no competing interests.

References

1. Abd-Elrazig SM, Elzubeir EA. Effects of feeding pearl millet on laying hen performance and egg quality. Anim Feed Sci Technol. 1998;76:89–94.
2. Afolayan SB, Dafwang II, Tegbe TSB, Sekoni A. Effect of dietary maize substitution with sweat potato meal on performance of growers (10–22 weeks) and subsequent egg production (23-35 weeks). Asian J Poult Sci. 2013;7:55–64.
3. Agwunobi LN. Performance of broiler chickens fed sweet potato meal (*Ipomea batatas* L.) diets. Trop Anim Health Prod. 1999;31:383–9.
4. Ahmed KU. Potatoes for the tropics. Dhaka; 1977.
5. Aina ABJ, Animo AO. Substitution of maize with cassava and sweet potato meal as the energy source in the rations of laying hens. Pertanika J Trop Agric Sci. 1997;20:163–7.
6. American Medical Association. Diet and coronary heart disease: a joint policy statement of the AMA council on feeds and nutrition and food and nutrition board of the national academy of sciences. National Research Counciln. JAMA. 1977;222:1647.
7. Anonymous. Poultry and product annual. EU-28 Poultry sector to grow again 2013–14. Required report: public distribution; 2013.
8. AOAC. Official methods of analysis. 16th ed. Arlington: Association of Official Analytical Chemistry; 1995.

9. Ayuk EA, Essien A. Growth and haematological response of broiler chicks fed graded levels of sweet potato (*Ipomoea batata*) meal as replacement for maize. Int J Poult Sci. 2009;8:485–8.
10. Azimuddin MD, Alam QM, Baset MA. Potato for food security in Bangladesh. Int J Sustain Crop Prod. 2009;4(1):94–9.
11. BBS. Yearbook of Agricultural Statistics of Bangladesh. Bangladesh Bureau of Statistics, Statistics Division, Ministry of Planning, Government of People's Republic of Bangladesh, Dhaka; 2012.
12. Beckford RC, Bartlett IR. Inclusion levels of sweet potato root meal in the diet of broilers I. Effect on performance, organ weights, and carcass quality. Poult Sci. 2015;94(6):1316–22.
13. Danielle Donnelly PHD, Stan Kubow PHD. Role of potato in human health. Montréal: Plant Science Department & School of Dietetics and Human Nutrition. McGill University; 2011.
14. Kreutzer DW. Renewable fuel standard, ethanol use, and corn prices. Background, heritage foundation; 2012; No. 2727.
15. Ladokun OA, Tewe OO. Sweet potato as a feed resource for layer production in Nigeria. J Poult Sci. 2003;83:56–61.
16. Maphosa T, Gunduza KT, Kusina J, Mutungamiri A. Evaluation of sweet potato tuber (Ipomeabatatas l.) as a feed ingredient in broiler chicken diets. Livestock Research Rural Development; 2003.
17. Morrison FB. Feeds and feeding. Iowa: Morrison Publishing Company; 1961.
18. Mozafari O, Ghazi S, Moeini MM. The effects of differen levels of edible potat (Solanumtubresum) replacing maize on performance, serum metabolite and immune system of broiler chicks. Iran J. Anim. Sci. 2013;3(3):583–8.
19. National Research Council. Nutrient requirements for poultry. Washington, DC; 1994.
20. Ojewola GS, Olojede AO, Ehiri CG. Evaluation of Livingston potato/ Rizga (Plectranthusesculentus N.Br) and Hausa potato (Solenostemon rotundifolius Poir) as energy sources for broiler chicken. J Anim Vet Adv. 2006;5:472–7.
21. Panda K, Zaidi PH, Rama rao SV, Raju MV. Efficacy of quality protein maize in meeting energy and essential amino acid requirements in broiler chicken production. J Appl Anim Res. 2014;42(2):133–9.
22. Sobamiwa O. The effect of replacing maize with dried sweet potatoes on the performance of layers. Niger J Anim Prod. 1988;15:99–102.
23. Sultan F, Ali MA, Jahan I. Growth performance meat yield and profitability of broiler chickens fed diets incorporating cassava tuber meal. J Environ Sci Nat Resour. 2012;5(1):47–53.
24. Sultana F, Khatun MF, Ali MA. Effect of dietary cassava tuber meal on egg production and egg quality of laying hen. Int J BioRes. 2011;2:1–4.
25. TEWE OO. Sweet potato utilisation in poultry diets, Tropic root crops. a developing economy. Proceedings of the Ninth Symposium of the International Society for Tropical Roots Crops. Accra Ghana; 1991. p. 426–35.
26. Turner WJ, Malynicz GL, Nad H. Effect of feeding rations based on cooked sweet potato and a protein supplement to broiler and crossbreed poultry. Papua New Guinea Agric. J. 1976;27(3):69–72.

Permissions

All chapters in this book were first published in CBTA, by Springer; hereby published with permission under the Creative Commons Attribution License or equivalent. Every chapter published in this book has been scrutinized by our experts. Their significance has been extensively debated. The topics covered herein carry significant findings which will fuel the growth of the discipline. They may even be implemented as practical applications or may be referred to as a beginning point for another development.

The contributors of this book come from diverse backgrounds, making this book a truly international effort. This book will bring forth new frontiers with its revolutionizing research information and detailed analysis of the nascent developments around the world.

We would like to thank all the contributing authors for lending their expertise to make the book truly unique. They have played a crucial role in the development of this book. Without their invaluable contributions this book wouldn't have been possible. They have made vital efforts to compile up to date information on the varied aspects of this subject to make this book a valuable addition to the collection of many professionals and students.

This book was conceptualized with the vision of imparting up-to-date information and advanced data in this field. To ensure the same, a matchless editorial board was set up. Every individual on the board went through rigorous rounds of assessment to prove their worth. After which they invested a large part of their time researching and compiling the most relevant data for our readers.

The editorial board has been involved in producing this book since its inception. They have spent rigorous hours researching and exploring the diverse topics which have resulted in the successful publishing of this book. They have passed on their knowledge of decades through this book. To expedite this challenging task, the publisher supported the team at every step. A small team of assistant editors was also appointed to further simplify the editing procedure and attain best results for the readers.

Apart from the editorial board, the designing team has also invested a significant amount of their time in understanding the subject and creating the most relevant covers. They scrutinized every image to scout for the most suitable representation of the subject and create an appropriate cover for the book.

The publishing team has been an ardent support to the editorial, designing and production team. Their endless efforts to recruit the best for this project, has resulted in the accomplishment of this book. They are a veteran in the field of academics and their pool of knowledge is as vast as their experience in printing. Their expertise and guidance has proved useful at every step. Their uncompromising quality standards have made this book an exceptional effort. Their encouragement from time to time has been an inspiration for everyone.

The publisher and the editorial board hope that this book will prove to be a valuable piece of knowledge for researchers, students, practitioners and scholars across the globe.

List of Contributors

Fábio Lopes Olivares, Lívia da Silva Lima, Natália Oliveira Aguiar and Luciano Pasqualoto Canellas
Núcleo de Desenvolvimento de Insumos Biológicos para a Agricultura (NUDIBA), Universidade Estadual do Norte Fluminense Darcy Ribeiro (UENF), Av. Alberto Lamego, 2000, Campos dos Goytacazes, Rio de Janeiro 28013-602, Brazil

Jader Galba Busato and Alessandra M. de Paula
Faculdade de Agronomia e Medicina Veterinária, Universidade de Brasília, Campus Universitário Darcy Ribeiro, Caixa Postal 4508, CEP 70910-970 Brasília, DF, Brazil

Youssef Moustakime, Zakaria Hazzoumi and Khalid Amrani Joutei
Laboratory of Bioactive Molecules: Structure and Function, Faculty of Science and Technology Fez, B.P. 2202-Road of Imouzzer, Fez, Morocco

David Drissner
Research Group Microbiology of Plant Foods, Agroscope, Schloss 1, 8820 Wädenswil, Switzerland

Florian M. Freimoser
Research Division Plant Protection, Agroscope, Schloss 1, 8820 Wädenswil, Switzerland

Han Yin Sun
Chair of Soil Ecology, Technical University of Munich, Ingolstädter Landstr. 1, Neuherberg, 85764 Munich, Germany
Institute of Network Biology, Helmholtz-Zentrum München, Munich, Germany

Reiner Schroll
Institute of Network Biology, Helmholtz-Zentrum München, Munich, Germany

Philipp Koal and Georg Gerl
Institute of Biochemical Plant Pathology, Helmholtz-Zentrum München, Ingolstädter Landstr. 1, Neuherberg, 85764 Munich, Germany

Rainer Georg Joergensen
Soil Biology and Plant Nutrition, University of Kassel, Nordbahnhofstr. 1a, 37213 Witzenhausen, Germany

Jean Charles Munch
Chair of Grassland, Technical University of Munich, WZW, Alte Akademie 12, 85354 Freising, Germany

Donatella Capitani, Anatoly P. Sobolev, Valeria Di Tullio and Noemi Proietti
Istituto di Metodologie Chimiche, Laboratorio di Risonanza Magnetica, "Annalaura Segre", CNR, Monterotondo, 00015 Rome, Italy

Luisa Mannina
Istituto di Metodologie Chimiche, Laboratorio di Risonanza Magnetica, "Annalaura Segre", CNR, Monterotondo, 00015 Rome, Italy
Dipartimento di Chimica e Tecnologie del Farmaco, Sapienza Università di Roma, Piazzale Aldo Moro 5, 00185 Rome, Italy

Simone Circi and Cinzia Ingallina
Dipartimento di Chimica e Tecnologie del Farmaco, Sapienza Università di Roma, Rome, Italy

Donatella Capitani and Anatoly P. Sobolev
Istituto di Metodologie Chimiche, CNR, Laboratorio di Risonanza Magnetica "Annalaura Segre", Monterotondo, Rome, Italy

Luisa Mannina
Dipartimento di Chimica e Tecnologie del Farmaco, Sapienza Università di Roma, Rome, Italy
Istituto di Metodologie Chimiche, CNR, Laboratorio di Risonanza Magnetica "Annalaura Segre", Monterotondo, Rome, Italy

Antonio Randazzo
Dipartimento di Farmacia, Università di Napoli "Federico II", Naples, Italy

Xiaolong Wang, Zhen Ma, Xiaoping Li, Liu Liu, Xiuxiu Yin, Kailong Zhang, Yu Liu and Xinzhong Hu
College of Food Engineering and Nutrition Science, Shaanxi Normal University, Xi'an 710119, China

Hyo-Seok Chae, Woo Seung Song and Hyang-Hyun Cho
HANIL BIOMED Inc., Gwangju 62363, Korea

Hye-Ji Noh
MEDINUTROL Co. Ltd., Yeonggwang-gun 57024, Korea

Pierluigi Mazzei and Alessandro Piccolo
Centro Interdipartimentale per la Risonanza Magnetica Nucleare per l'Ambiente, l'Agro-Alimentare ed i Nuovi Materiali (CERMANU), Università di Napoli Federico II, Via Università 100, 80055 Portici, Italy

Taye Victor Familoni, Bamidele Juliet Akinyele and Anthony Kayode Onifade
Department of Microbiology, The Federal University of Technology, PMB 704, Akure, Nigeria

Clement Olusola Ogidi
Department of Microbiology, The Federal University of Technology, PMB 704, Akure, Nigeria
Department of Biotechnology, Faculty of Science, Kings University, PMB 555, Odeomu, Nigeria

Saksit Nobsathian
Nakhonsawan Campus, Mahidol University, Nakhonsawan 60130, Thailand

Vasakorn Bullangpoti
Animal Toxicology and Physiology Speciality Research Unit, Zoology Department, Faculty of Science, Kasetsart University, Bangkok 10900, Thailand

Nutchaya Kumrungsee, Natnicha Wongsa and Dussadee Ruttanakum
Biology Department, Faculty of Science and Technology, Rajamangala University of Technology Thanyaburi, Pathumthani, Thailand

Noushin Ghazijahani
Department of Microbiology, Islamic Azad University, Karaj Branch, Karaj, Iran

Ebrahim Hadavi
Department of Horticultural Sciences, Islamic Azad University, Karaj Branch, Karaj, Iran

Moon S. Son and Byoung R. Jeong
Department of Horticulture, Gyeongsang National University, Jinju 660-701, Republic of Korea

Mengistu Welemariam and Bobe Bedadi
College of Agriculture and Environmental Sciences, Haramaya University, Dire Dawa, Ethiopia

Fassil Kebede
Ethiopian Agricultural Transformation Agency, Addis Ababa, Ethiopia

Emiru Birhane
Department of Land Resources Management and Environmental Protection, Mekelle University, Mekelle, Ethiopia
Faculty of Environmental Sciences and Natural Resource Management, Norwegian University of Life Sciences, 1432 Ås, Norway

Alemayehu Mekonnen, Minaleshewa Atlabachew and Banchalem Kassie
Chemistry Department, Science College, Bahir Dar University, Bahir Dar, Ethiopia

Chukwuma S. Ezeonu, Chinedu Imo, Assumpta Iruka and Abraham Joseph
Department of Biochemistry, Federal University Wukari, Wukari, Taraba State, Nigeria

Dawn I. Agwaranze
Department of Microbiology, Federal University Wukari, Wukari, Taraba State, Nigeria

Sneha Jha and H. A. Modi
Department of Life Sciences, University School of Sciences, Gujarat University, Ahmedabad, Gujarat 380009, India

Dainan Zhang, Dandan Duan, Yu Yang and Yong Ran
State Key Laboratory of Organic Geochemistry, Guangzhou Institute of Geochemistry, Chinese Academy of Sciences, Guangzhou 510640, China

Youda Huang
State Key Laboratory of Organic Geochemistry, Guangzhou Institute of Geochemistry, Chinese Academy of Sciences, Guangzhou 510640, China
University of Chinese Academy of Sciences, Beijing 100049, China

Cécile Thonar, Dominika Kundel, Sarah Symanczik and Paul Mäder
Soil Sciences Department, Research Institute of Organic Agriculture FiBL, Ackerstrasse 113, 5070 Frick, Switzerland

Jonas Duus Stevens Lekfeldt and Andreas de Neergaard
Department of Plant and Environmental Sciences, University of Copenhagen, Copenhagen, Denmark

Vincenza Cozzolino and Alessandro Piccolo
CERMANU-University of Naples Federico II, Naples, Italy

Martin Kulhánek
Department of Agro-environmental Chemistry and Plant Nutrition, Czech University of Life Sciences, Prague, Czech Republic.

Carla Mosimann
Soil Sciences Department, Research Institute of Organic Agriculture FiBL, Ackerstrasse 113, 5070 Frick, Switzerland
Agrofutura AG, Brugg, Switzerland

Günter Neumann and Markus Weinmann
Institute of Crop Science 340h, University of Hohenheim, Stuttgart, Germany

Martin Rex
Institut für Baustoff Forschung, Duisburg, Germany

Florian Walder
Soil Sciences Department, Research Institute of Organic Agriculture FiBL, Ackerstrasse 113, 5070 Frick, Switzerland
Agroscope Reckenholz, Zurich, Switzerland

Pierluigi Mazzei and Alessandro Piccolo
Centro Interdipartimentale per la Risonanza Magnetica Nucleare per l'Ambiente, l'Agro-Alimentare ed i Nuovi Materiali (CERMANU), Università di Napoli Federico II, Via Università 100, 80055 Portici, Italy

Ronald Soong, Adolfo Botana, Jasmine Wang, Hashim Farooq, Denis Courtier-Murias and Andre Simpson
Environmental NMR Center, University of Toronto Scarborough, 1265 Military Trails, Toronto, ON M1C 1A4, Canada

Melkayo Geremu
Jimma University College of Agriculture and Veterinary Medicine, Jimma, Ethiopia

Yetenayet Bekele Tola
Jimma University College of Agriculture and Veterinary Medicine, Jimma, Ethiopia

Abrar Sualeh
Ethiopian Institute of Agricultural Research, Jimma Agricultural Research Center, Jimma, Ethiopia

Fowzia Sultana
Department of Poultry Science, Faculty of Animal Husbandry, Bangladesh Agricultural University, Mymensingh 2202, Bangladesh

Hafiza Khatun
Department of Animal Science, Faculty of Animal Husbandry, Bangladesh Agricultural University, Mymensingh 2202, Bangladesh

Mohammad Amdad Ali
Department of Agricultural Chemistry, Faculty of Agriculture, Bangladesh Agricultural University, Mymensingh 2202, Bangladesh

Nidal Amin Jaradat, Abdel Naser Zaid and Fatima Hussein
Department of Pharmacy, Faculty of Medicine and Health Sciences, An-Najah National University, Nablus, Palestine

Chiara Roberta Girelli
Di.S.Te.B.A., University of Salento, Campus Ecotekne, Via Prov.le Lecce-Monteroni, 73100 Lecce, Italy

Laura Del Coco
Di.S.Te.B.A., University of Salento, Campus Ecotekne, Via Prov.le Lecce-Monteroni, 73100 Lecce, Italy

Udomsak Kongmuang
Faculty of Public Health, Bangkokthonburi University, Khet Taweewatana, Bangkok 10170, Thailand

Chaovayut Phornpimolthape
Faculty of Public Health, Bangkokthonburi University, Khet Taweewatana, Bangkok 10170, Thailand

Index

www.ingramcontent.com/pod-product-compliance
Lightning Source LLC
LaVergne TN
LVHW080918210926
848485LV00004B/677